Sequences and the de Bruijn Graph

Sequences and the de Bruijn Graph

Properties, Constructions, and Applications

Tuvi Etzion
Department of Computer Science
Technion – Israel Institute of Technology
Haifa, Israel

ACADEMIC PRESS
An imprint of Elsevier

Academic Press is an imprint of Elsevier
125 London Wall, London EC2Y 5AS, United Kingdom
525 B Street, Suite 1650, San Diego, CA 92101, United States
50 Hampshire Street, 5th Floor, Cambridge, MA 02139, United States

Notices

Knowledge and best practice in this field are constantly changing. As new research and experience broaden our understanding, changes in research methods, professional practices, or medical treatment may become necessary.

Practitioners and researchers must always rely on their own experience and knowledge in evaluating and using any information, methods, compounds, or experiments described herein. In using such information or methods they should be mindful of their own safety and the safety of others, including parties for whom they have a professional responsibility.

To the fullest extent of the law, neither the Publisher nor the authors, contributors, or editors, assume any liability for any injury and/or damage to persons or property as a matter of products liability, negligence or otherwise, or from any use or operation of any methods, products, instructions, or ideas contained in the material herein.

ISBN: 978-0-443-13517-0

For information on all Academic Press publications
visit our website at https://www.elsevier.com/books-and-journals

Publisher: Mica Haley
Acquisitions Editor: Mara Conner
Editorial Project Manager: Sara Greco
Production Project Manager: Kumar Anbazhagan
Cover Designer: Matthew Limbert

Typeset by VTeX

Working together
to grow libraries in
developing countries

www.elsevier.com • www.bookaid.org

Devoted to the memory of:

Abraham Lempel (February 10, 1936 – February 4, 2023)

My Ph.D. Advisor at the Technion from April 1982 until June 1984.

Solomon W. Golomb (May 30, 1932 – May 1, 2016)

My post-doc advisor at University of Southern California from September 1985 until August 1987.

Alexander Vardy (November 12, 1963 – March 11, 2022)

My number one collaborator from April 1992 until February 2022 when we last met.

Contents

Preface

The de Bruijn graph is credited to the Dutch mathematician Nicolaas Govert de Bruijn who defined this graph in 1946 with the motivation to count the number of binary cyclic sequences of length 2^n in which each binary n-tuple is contained exactly once in a window of length n. These sequences were also called by the name of de Bruijn. Although it was defined for mathematical purposes, the graph has been used throughout the years for many applications. It was first heavily used to develop the theory of shift-register sequences used for many practical applications, particularly for the NASA space program. These sequences were also used in other unrelated areas such as cryptology, VLSI testing, and wireless communication. The graph itself was also used for new applications. When parallel computation began, the graph was an inspiration for networks such as the shuffle-exchange network, the omega network, and other related networks. The de Bruijn sequences and shift-register sequences were generalized for two-dimensional arrays and these arrays were applied for pattern recognition and computer vision, and related arrays were used also as self-locating patterns. In the late 1980s the human genome project attracted much research and the de Bruijn graph was also used in this project for the DNA sequencing of the genome assembly. The method used, called the de Bruijn graph method, is based on paths along the edges of the graph. Finally, at the beginning of the 21st century DNA storage was developed and some of the related theories also made use of the de Bruijn graph and its sequences.

In parallel with the research on the graph, the theory of sequences was developed rapidly, starting with the theory of shift-register sequences. This theory is well documented in a book by Solomon W. Golomb who developed the theory of shift-register sequences. His book on these sequences, "Shift-Register Sequences", is the "bible" of this area. Other types of sequences were developed to supply the demands for sequences with special properties to areas like cryptography, sequence design for radar and sonar, constrained sequences, DNA storage sequences, etc. This book's focus and scope are the de Bruijn graph and its sequences, their generalizations, and their applications. However, it will also cover some of the topics associated with the graph and its sequences, and their generalizations that were suggested. Although the book is quite thick, it will be impossible to cover everything. On the other hand, we tried to be as comprehensive as possible.

The de Bruijn graph and digital sequences were used in many disciplines as noted above and, in many cases, the different disciplines have not interacted. These disciplines include combinatorics, graph theory, communication, data storage, computer science, pattern recognition, computer vision, bioinformatics, biology, and more. One of the main goals of this book is to make these interactions and to bring to each discipline the knowledge, the methods, and the applications that were discovered and used in the other disciplines. We will not be able to cover everything, but we intend to present a few different angles and directions. The book can be categorized as an algebraic and combinatorial work on sequences and the de Bruijn graph, although there are also some algorithmic sections in the book. Moreover, this book is the first attempt to put together digital sequences, the de Bruijn graph, and their combination in the center of the exposition, spreading over the various disciplines. These disciplines made use of these concepts and also various mathematical concepts that arise from the graph and its sequences. We will also try to bring to the attention of all researchers, the diverse amount of literature on the de Bruijn graph, its sequences, properties, applications, and the generalizations of both the graph and its sequences.

The book is a monograph that can be used by researchers in various research fields, but it can be also used as a textbook for a combined course for graduate and undergraduate students. The book contains many open problems that can motivate research and inspire the theses of graduate students. Most of the claims in the book are proved in the text and as such the book is self-contained, but the reader must have certain basic knowledge in algebra and combinatorics to understand the book. Similarly, from time to time throughout the book, we will use also well-known concepts that need no introduction or definition. References for each chapter are presented only in the last section, which is always titled "Notes". In this section, the credit of the results in the section are given to the appropriate references. The section also contains additional results, some of which are proved in the section and some of which are presented with no proof. The number of references on de Bruijn sequences, the de Bruijn graph, sequences, and related topics is enormous and many such references were left out. We apologize to those authors whose important manuscripts were left out. Although most of the book is based on existing research work done over the years, some parts are novel. Some of the results do not appear elsewhere and for some results, new original proofs are provided.

Each chapter of the book can be used in a course as a basis for two to four hours of a lecture or two, respectively. While writing the book, I have used the book for a course that I developed in parallel and some parts of the book are a consequence of my experience from this course. To teach the whole material of each chapter at least four hours are required and some chapters require at least six hours. The course title can be the same as the title of the book "Sequences and the de Bruijn Graph: Properties, Constructions, and Applications". The course can be more oriented towards sequences and it can be titled "Digital

Sequences and the de Bruijn Graph" or more oriented toward graph theory and it can be titled "The de Bruijn Graph and its Applications".

The material in this book can be partitioned in a few different ways and I will describe one of these ways. The first part in Chapter 1 is an introduction to some of the main topics covered in the book including some basic concepts in algebra, number theory, and combinatorics on the one hand, and a brief outline of the various chapters on the other hand. The second part, which starts in Chapter 2 and ends in Chapter 8, is devoted to the one-dimensional theory of sequences, including their properties, enumeration, constructions, complexity measures, classifications, and applications. The third part of the book, given in Chapters 9 and 10, is devoted to the generalization of the one-dimensional sequences into multi-dimensional arrays and in particular two-dimensional arrays. The fourth part of the book, which starts in Chapter 11 and ends in Chapter 12, considers generalizations of de Bruijn graph from the point of view of graph theory and especially interconnection networks.

The tools that will be used in the book are mainly from algebra and combinatorics as mentioned, but there will be also material based on number theory, some basic graph theory, and some algorithms. Also, coding theory, cryptography, and games will be used, applied, and developed in the process. Out of all the topics in the book, there is comprehensive coverage of the following topics:

1. The linear and nonlinear theory of shift registers and their sequences.
2. Constructions and properties of de Bruijn sequences.
3. Linear complexity of sequences whose length and alphabet size are powers of the same prime.
4. Two-dimensional de Bruijn sequences and arrays with distinct differences.
5. Generalizations and applications for the de Bruijn graph and its sequences.
6. de Bruijn graph type of interconnection networks – constructions, routing, and layouts.
7. Graphs with a unique path property and their associated networks.

Some large parts of this book are based on my own research work and research done by my colleagues. My Ph.D. advisor, Abraham Lempel, at the Technion, introduced me to the de Bruijn graph and its sequences. This led to my Ph.D. titled "Sequences with special properties" that was devoted to de Bruijn sequences and the shuffle-exchange network. I am grateful to him for his guidance and support throughout the years. My post-doc advisor, Solomon W. Golomb, at the University of Southern California (USC), broadened my knowledge and my interest in all the related topics. I have learned so much from him on these topics and how to make use of every mathematical "gem" either for a practical use or just to develop an interesting theory. My work on two-dimensional de Bruijn arrays and arrays with distinct differences started during my time at USC. At USC, I was also introduced to graphs with a unique path property and their connections to interconnection networks. Finally, I spent a considerable time from 1994 at Royal Holloway University of London, where I worked on linear complexity of sequences, single-track Gray codes, and applications of two-dimensional arrays with distinct differences.

Some of my colleagues and students worked with me on some of the topics mentioned in this book. I should thank all of them and they are listed alphabetically as follows: Israel Bar-David, Simon R. Blackburn, Alfred M. Bruckstein, Yeow Meng Chee, Johan Christnata, Raja Giryes, David Goldfeld, Han Mao Kiah, Sagi Marcovich, Keith M. Martin, Chris J. Mitchell, Kenneth G. Paterson, Maura B. Paterson, Moshe Schwartz, Herbert Taylor, Alexander Vardy, and Eitan Yaakobi.

I should give special thanks to Moshe Schwartz and Eitan Yaakobi who have given important criticism on some chapters of the first draft that helped me to improve the draft considerably. Last, but not least, I am indebted to my Ph.D. student Daniella Bar-Lev who read almost the whole text and her endless remarks, comments, and suggestions helped me to improve all the chapters of this book and to remove many hidden errors.

Tuvi Etzion

September, 2023

Chapter 1

Introduction

Preliminaries, de Bruijn graph, shift registers

The de Bruijn graph G_n was defined in 1946 by Nicolaas Govert de Bruijn. His purpose in defining the graph was to find the number of binary cyclic sequences of length 2^n in which each binary n-tuple is contained exactly once as a window of length n in the sequence. The graph was defined in parallel by Irving John Good to generate the same sequences and hence it is sometimes called the de Bruijn–Good graph. It was also discovered later by de Bruijn that Flye-Sainte Marie found the number of these sequences 50 years earlier before his discovery. Later, de Bruijn and Aardenne-Ehrenfest generalized these results and defined the de Bruijn graph $G_{\sigma,n}$ over an alphabet Σ whose size is σ. It was also mentioned by Good that his definition of the graph can be generalized for any alphabet.

During the years since their introduction, both the graph and its sequences were subject to extensive research. In the beginning, the graph and its sequences, which include many families of sequences and not only de Bruijn sequences, were mainly used for their combinatorial analysis and applications based on feedback shift-registers theory. These shift registers and their associated sequences were used in the space program of NASA. Later, more applications for the graph and its sequences were found. Starting in the 1970s, the graph was used and also inspired research on parallel computing with interconnection networks. This research also motivated researchers to find the embedding of the graph for different purposes. The human genome project, which was started towards the end of the 1980s, also used paths of the de Bruijn graph for DNA sequencing, which is part of the genome assembly. At the beginning of the 21st century, research on nonvolatile memories and in particular flash memories and DNA storage has inspired some new research associated with the graph and its sequences. Moreover, the sequences of the de Bruijn graph have found applications in coding theory, VLSI testing, cryptography, pattern recognition, and also in other disciplines. As the disciplines that used the de Bruijn graph and its sequences are not always contained in the same research field, the results and the requirements used in one research area were not always known in the other research areas. One of the main goals of this book is to bridge this gap.

The goal of this chapter is to provide a short introduction to the main topics of the book and to present some preliminaries from other mathematical areas

Sequences and the de Bruijn Graph. https://doi.org/10.1016/B978-0-44-313517-0.00007-X

that will be frequently used to obtain the results for the main subject of this book. The rest of this chapter is organized as follows.

Concepts in number theory such as primes, congruences, the Euler function, the Möbius function, quadratic residues, and more, play an important role in the exposition of various chapters. In Section 1.1 we present a short introduction to basic number theory. Although we mainly concentrate on binary sequences, the theory is applied also to non-binary sequences and mainly sequences over finite fields. Moreover, the binary case is not always a special case of the more general case. For this, basic concepts of finite fields must be supplied and these will be also presented at the beginning of Section 1.1.

In Section 1.2 we will give a brief introduction to a few other concepts that appear throughout the book. Graphs will appear in many chapters, e.g., the de Bruijn graph and its generalization, UPP graphs, the shuffle-exchange network, etc. Sequences will appear throughout the book and their basic definitions will be provided in Section 1.2. Finally, although this is not a book on coding theory, there are some connections between codes and sequences, and hence the basics for the theory of error-correcting codes will be given in this section.

In Section 1.3 the major connection between the de Bruijn graph and its sequences will be discussed. This connection is through shift registers and their sequences. These concepts will be presented and in particular, some theory of nonsingular feedback shift registers will be given.

A comprehensive overview of the specific chapters of this book will be given in Section 1.4.

1.1   Some concepts in finite fields and number theory

This section introduces basic concepts used in the definitions and the techniques used throughout the book, namely groups, finite fields, and number theory. Finite fields play a major role in part of the exposition. We will assume the knowledge only of very basic concepts in finite fields, linear algebra, and number theory, such as prime numbers, divisibility, functions, equivalence relations, polynomials, etc., although some will appear in the rather extensive introduction to number theory. Two concepts, groups and rings, will lead to the definition of a finite field.

Definition 1.1. A pair $(\mathcal{G}, \circ)$ is called a *group* if $\mathcal{G}$ is a nonempty set, $\circ$ is a binary operation defined on $\mathcal{G}$, and the following three properties are satisfied:

1. $(a \circ b) \circ c = a \circ (b \circ c)$ for all $a, b, c \in \mathcal{G}$.
2. There is an *identity* element $e \in \mathcal{G}$ such that $a \circ e = e \circ a = a$ for all $a \in \mathcal{G}$.
3. For each $a \in \mathcal{G}$ there exists an element $a^{-1} \in \mathcal{G}$ called the *inverse* such that $a \circ a^{-1} = a^{-1} \circ a = e$.

For an additive group, the identity element will be denoted by 0.

The group $(\mathcal{G}, \circ)$ is called an *Abelian group* (or a *commutative group*) if $a \circ b = b \circ a$ for all $a, b \in \mathcal{G}$.

The group $(\mathcal{G}, \circ)$ is a *cyclic group* if there exists an element $a \in \mathcal{G}$, such that each $b \in \mathcal{G}$ is equal to $a^i \triangleq \overbrace{a \circ a \circ \cdots \circ a}^{i \text{ times}}$ for some integer i. The element a is called a *generator* of the group. For $b \in \mathcal{G}$, the smallest $i > 0$ such that $b^i = e$ is called the *order* of b. It is easy to verify that a is a generator of the group $\mathcal{G}$ if and only if the order of a is the size of the group, i.e., $|\mathcal{G}|$.

A *subgroup* H of a group G is a subset H of G that is also a group. If $x \in G \setminus H$, then $x \circ H \triangleq \{x \circ h : h \in H\}$ is called a *coset* of H in G. This coset is a left coset and similarly, we have a right coset $H \circ x$ for each $x \in G \setminus H$. Also, $e \circ H = H \circ e = H$, where e is the identity element of G is a coset of H in G. For an Abelian group, the left cosets and the right cosets coincide. Henceforth, we assume that all our groups are Abelian. The cosets of H in G define a group called the *quotient group* and denoted by G/H. The *index* of the subgroup H in a finite group G, denoted by $[G : H]$ is the number of cosets of H in G. It is easy to verify that if H is a subgroup of G then the relation R defined on the elements of G by $(a, b) \in R$ if $a \circ b^{-1} \in H$, is an equivalence relation.

The following theorem is called Lagrange's theorem.

Theorem 1.1. *If H is a subgroup of a finite group G, then*

$$|G| = [G : H] \cdot |H|.$$

Proof. The cosets of H in G are the equivalent classes of the equivalence relation. Therefore the cosets form a partition of G. Each coset has the same size and the number of cosets is $[G : H]$. Therefore

$$|G| = [G : H] \cdot |H|. \qquad \square$$

When a finite group G of operators are acting on a finite set U, an equivalence relation is defined on U by these operators, where $x, y \in U$ are related if there exists an operator $g \in G$ such that $y = g(x)$. The next theorem is known as Burnside's lemma.

Theorem 1.2. *The number of equivalence classes to which a set U is partitioned by a finite group G of operators acting on U is*

$$\frac{1}{|G|} \sum_{g \in G} \mathrm{Fix}(g),$$

where $\mathrm{Fix}(g)$ *is the number of points of U that remain fixed by g.*

Proof. Let x be an element in the space U and let G_x be the subgroup (called the *stabilizer group*) of G defined by the elements of the group that fix the element x, i.e.,

$$G_x \triangleq \{g : g \in G, \ g(x) = x\}$$

and let $G(x)$ be the coset of x, i.e.,

$$G(x) \triangleq \{g(x) \ : \ g \in G\}.$$

By Lagrange's theorem, we have that

$$|G(x)| = [G \ : \ G_x] = \frac{|G|}{|G_x|}.$$

Clearly, by these definitions

$$\sum_{g \in G} \text{Fix}(g) = |\{(g,x) \ : \ g \in G, x \in U, \ g(x) = x\}| = \sum_{x \in U} |G_x| = \sum_{x \in U} \frac{|G|}{|G(x)|}.$$

It is also easy to verify that

$$\sum_{x \in U} \frac{1}{|G(x)|} = \sum_{A \in U/G} \sum_{x \in A} \frac{1}{|A|} = \sum_{A \in U/G} 1 = |U/G|,$$

where U/G is the quotient set defined by the relation on U, and hence

$$\sum_{g \in G} \text{Fix}(g) = \sum_{x \in U} \frac{|G|}{|G(x)|} = |G| \sum_{x \in U} \frac{1}{|G(x)|} = |G| \cdot |U/G|.$$

Thus

$$|U/G| = \frac{1}{|G|} \sum_{g \in G} \text{Fix}(g),$$

which completes the proof. □

Definition 1.2. A triple $(\mathcal{R}, +, \cdot)$ is called a *ring* if $\mathcal{R}$ is a nonempty set, $+$ and $\cdot$ are two binary operations defined on $\mathcal{R}$, and the following four properties are satisfied:

1. $(\mathcal{R}, +)$ is an Abelian group.
2. $(a \cdot b) \cdot c = a \cdot (b \cdot c)$ for all $a, b, c \in \mathcal{R}$.
3. There is a unique element $1 \in \mathcal{R}$ such that $a \cdot 1 = 1 \cdot a = a$ for all $a \in \mathcal{R}$.
4. $a \cdot (b + c) = a \cdot b + a \cdot c$ and $(a + b) \cdot c = a \cdot c + b \cdot c$ for all $a, b, c \in \mathcal{R}$.

The identity element of the group $(\mathcal{R}, +)$ is denoted by 0. The ring $(\mathcal{R}, +, \cdot)$ is called a *commutative ring* if $a \cdot b = b \cdot a$ for all $a, b \in \mathcal{R}$.

Note that $(\mathcal{R} \setminus \{0\}, \cdot)$ might not have an inverse for each element of $\mathcal{R} \setminus \{0\}$, and hence it is not necessarily a group.

Definition 1.3. A ring $(\mathbb{F}, +, \cdot)$ is called a *field* if the pair $(\mathbb{F} \setminus \{0\}, \cdot)$ is an Abelian group. The element 0 is the identity element of the Abelian group $(\mathbb{F}, +)$ and 1 is the identity element of the Abelian group $(\mathbb{F} \setminus \{0\}, \cdot)$.

We denote the set $\mathcal{G} \setminus \{0\}$, where $\mathcal{G}$ is a group (also for a ring or a field) and 0 is the additive identity element, by $\mathcal{G}^*$. The group $(\mathbb{F}, +)$ is called the **additive group** of the field and the group $(\mathbb{F}^*, \cdot)$ is called the **multiplicative group** of the field.

Our main interest is in **finite fields**, i.e., fields with a finite number of elements. All such fields with the same number of elements are isomorphic and they are called **Galois fields**. The number of elements in such a field is q, where q is a power of a prime and it is denoted by GF(q) or $\mathbb{F}_q$. The Abelian group $(\mathbb{F}_q^*, \cdot)$ is a cyclic group.

The ring of integers modulo m will be denoted by $\mathbb{Z}_m$. Addition and multiplication in the ring are performed modulo m. This ring, $\mathbb{Z}_p$, is a field if p is a prime integer. It contains the set of integers $\{0, 1, \ldots, p - 1\}$ (or equivalently the set of p distinct residues modulo p) where addition and multiplication are performed modulo p.

The finite field $\mathbb{F}_{q^k}$, where q is a power of a prime, has q^k elements. The multiplicative group of $\mathbb{F}_{q^k}^*$ is a cyclic group with a generator α. The generator α is a root of some irreducible polynomial

$$c(x) = x^k - \sum_{i=1}^{k} c_i x^{k-i}, \quad c_i \in \mathbb{F}_q$$

called a **primitive polynomial** and each one of its roots α is called a **primitive element**. The elements of $\mathbb{F}_{q^k}$ can be represented as the q^k vectors of length k over $\mathbb{F}_q$, i.e., $\mathbb{F}_q^k$. For two elements α^i, α^j, represented by the vectors $x = (x_1, x_2, \ldots, x_k) \in \mathbb{F}_q^k$ and $y = (y_1, y_2, \ldots, y_k) \in \mathbb{F}_q^k$, respectively, we have that $\alpha^i \cdot \alpha^j = \alpha^{i+j}$, where superscripts are taken modulo $q^k - 1$, and

$$\alpha^i + \alpha^j = x + y = (x_1 + y_1, x_2 + y_2, \ldots, x_k + y_k) = \alpha^\ell \,,$$

where the addition $x_i + y_i$ is performed in $\mathbb{F}_q$ and α^ℓ is represented by the vector $(x_1 + y_1, x_2 + y_2, \ldots, x_k + y_k) \in \mathbb{F}_q^k$.

Since α is a root of $c(x)$, it follows that

$$0 = c(\alpha) = \alpha^k - \sum_{i=1}^{k} c_i \alpha^{k-i}$$

and $\alpha^k = \sum_{i=1}^{k} c_i \alpha^{k-i}$. The element $\alpha^0 = 1$ is represented by the vector $(00 \cdots 001)$, the element α by the vector $(00 \cdots 010)$, and so on, where α^{k-1} is represented by the vector $(10 \cdots 000)$.

The element α^k is represented by the vector $(c_1, c_2, \ldots, c_k)$. Similarly, if $\alpha^i = (a_1, a_2, \ldots, a_k)$, then

$$\alpha^{i+1} = (a_2, \ldots, a_k, 0) \text{ if } a_1 = 0$$

and

$$\alpha^{i+1} = (a_2, \ldots, a_k, 0) + a_1 \alpha^k = (a_2, \ldots, a_k, 0) + (a_1 c_1, \ \ldots, a_1 c_k) \ \text{if} \ a_1 \neq 0.$$

Recall that an irreducible polynomial $c(x)$ is a primitive polynomial if each of its roots (which are primitive elements) generates the field, i.e., the $q^k - 1$ powers of any root α, of $c(x)$, are distinct elements as q-ary vectors in this computation. Since usually, there is a large number of primitive polynomials (see Theorem 3.4), it follows that the vector representation of the finite field is not unique. The representation of a finite field and its connection to the main topic of the book will be discussed in Chapter 2. An example of $\mathbb{F}_{16}$ is given in Example 2.3.

The representation of the elements of $\mathbb{F}_{q^k}$ by the q-ary vectors of length k, over $\mathbb{F}_q$, induces a bijection between $\mathbb{F}_{q^k}$ and $\mathbb{F}_q^k$. This bijection is used to simplify the representation of some structures.

Groups, finite fields, and other concepts associated with graphs and sequences make use of many concepts in number theory. In the rest of this section, we will present some of the basic theory of numbers that is used in this book.

A *prime* number is a positive integer greater than 1 that is divisible only by itself and by 1. In other words, p is a prime number if it does not have any divisor d such that $1 < d < p$.

Two integers x and y are said to be *congruent* modulo a positive integer $m > 1$ if $y = x + jm$ for some integer j. This relation between y and x is an equivalence relation and it will be denoted by $y \equiv x \pmod{m}$.

Going back to groups, there is a special interest in the group $\mathbb{Z}_m$, $m \geq 2$ (same notation as the ring $\mathbb{Z}_m$), which contains the set $\{0, 1, \ldots, m-1\}$ of integers, where the binary operation is addition modulo m. The elements of $\mathbb{Z}_m$ can also be considered as the m distinct residues modulo m; $\mathbb{Z}_m$ will be also used to denote an alphabet with m elements.

Prime numbers and divisibility of numbers are two of the most basic concepts in number theory. A positive integer $d > 1$ that divides a positive integer n is called a *factor* of n. One of the most basic questions is to factor an integer n into its prime factors. We start with a basic discussion on prime numbers.

Theorem 1.3. *There are infinitely many primes.*

Proof. Assume, on the contrary, that there are only t primes, $p_1, p_2, \ldots, p_t$. Consider the integer

$$m = \prod_{i=1}^{t} p_i + 1.$$

Clearly, m is greater than 1 and not divisible by any of the primes $p_1, p_2, \ldots, p_t$, and hence by definition m is a prime, a contradiction. Thus there are infinitely many primes. $\qquad\square$

Finding primes is an important problem since primes have many applications in diverse areas, some theoretical and some practical. It is also important for these applications to know how sparse, or dense, is the set of primes. Such applications will appear later in the book.

Divisors of integers and common divisors have an important role in obtaining various results. The **greatest common divisor** of two positive integers a and b is the largest integer k, such that k divides a and k divides b. The greatest common divisor of a and b is denoted by **g.c.d.**(a,b). Two positive integers a and b are said to be **relatively primes** if g.c.d.$(a, b) = 1$. For s positive integers $m_1, m_2, \ldots, m_s$ the **greatest common divisor** denoted by g.c.d.$(m_1, m_2, \ldots, m_s)$ is the largest integer k that divides each m_i, $1 \leq i \leq s$. The **least common multiple** of the positive integers $m_1, m_2, \ldots, m_s$ is the smallest possible integer k such that m_i divides k for each i, $1 \leq i \leq s$. The least common multiple is denoted by $[m_1, m_2, \ldots, m_s]$. The following lemma is a straightforward claim inferred from the definitions.

Lemma 1.1. *If a and b are positive integers, then $[a, b] = \frac{a \cdot b}{\text{g.c.d.}(a,b)}$.*

There is a simple algorithm for computing the greatest common divisor k of two distinct integers a and b.

Euclid's algorithm:

The inputs to the algorithm are two distinct positive integers a and b and w.l.o.g. (without loss of generality) assume that $b \leq a$.

(E1) Set $c_1 := b$, $c_2 := a$, and $i := 1$.
(E2) If $c_2 = mc_1$, for some $m \geq 1$, then $k := c_1$ and stop.
(E3) Let $c_2 = c_1 m_i + r_i$, where $m_i \geq 1$ and $1 \leq r_i < c_1$.
(E4) Set $c_2 := c_1$, $c_1 := r_i$, and $i := i + 1$; go to **(E2)**. ∎

Theorem 1.4. *If a and b are two positive integers, then, when Euclid's algorithm terminates, the value k obtained in the algorithm is the greatest common divisor of a and b.*

Proof. It is readily verified that throughout the algorithm we have that $c_2 \geq c_1$ and the values of c_2 and c_1 are always positive and reduced in **(E4)**. Hence, the algorithm will stop at **(E2)**.

First, note that if in **(E2)** we have that $c_2 = mc_1$, for some $m \geq 1$, then g.c.d.$(c_1, c_2) = c_1$, which is the value for k assigned in **(E2)**. Moreover, since throughout the algorithm $c_1 < c_2$, it follows that step **(E3)** is valid.

Now, it is clear that to prove the claim of the theorem it is sufficient to show that the greatest common divisor of a and b is the same as the greatest common divisor of c_1 and c_2 throughout the algorithm. The claim will be proved by induction on i. The basis is done in **(E1)**, where $i = 1$, c_1 and c_2 are assigned with the values of b and a, respectively. Assume that the claim is true at some value of i before **(E3)** is performed. If an integer κ divides c_1 and c_2, then since $c_2 = c_1 m_i + r_i$, it follows that κ also divides r_i and hence we have that κ divides c_1 and r_i. Assume now that an integer τ divides c_1 and r_i. Again, since $c_2 = c_1 m_i + r_i$, it follows that τ also divides c_2. Therefore the greatest common divisor of c_1 and c_2 is also the greatest common divisor of c_1 and r_i. Since in **(E4)**, c_2 and c_1 are replaced by c_1 and r_i, respectively, it follows that the induction step is proved and hence k in **(E2)** is the greatest common divisor of a and b.

Thus when the algorithm terminates we have that the obtained k is the greatest common divisor of a and b. $\qquad\square$

Theorem 1.5. *If $k = $ g.c.d.(a, b), then there exist two integers x and y such that $k = ax + by$.*

Proof. The proof of this claim is based again on Euclid's algorithm. If $a = b$, then the claim is trivial and hence we will continue to assume w.l.o.g. that $b < a$. Moreover, the claim is also true if b divides a.

We will prove that each r_i computed in **(E3)** can be always written as $r_i = z_1 a + z_2 b$ for some integers z_1 and z_2. The claim will be proved again by induction on i. After the assignments of b to c_1 and a to c_2, if c_1 is not a divisor of c_2, then in **(E3)** we have that $r_1 = c_2 - c_1 m_1 = a - b m_1$ and the claim is proved. At **(E4)** c_1 is assigned to c_2 and hence the value of c_2 is b; r_1 is assigned to c_1 and hence the value of c_1 is $a - b m_1$; i now equals 2 and we are back at **(E2)**.

We continue with two cases depending on whether the algorithm stops at **(E2)** or not.
Case 1: If the algorithm stops at (E2), then $k = c_1$ is the greatest common divisor, and since $c_1 = r_1 = a - b m_1$, it follows that $x = 1$ and $y = -m_1$ are the values such that $k = ax + by$.
Case 2: The algorithm does not stop at **(E2)**. We continue at **(E3)**, where the assignment implies that

$$r_2 = c_2 - c_1 m_2 = b - r_1 m_2 = b - (a - b m_1) m_2 = -a m_2 + (1 + m_1 m_2) b$$

and the claim regarding r_2 is proved.

Assume now for the general step $i + 1$, $i > 1$, we have that $r_i = z_1 a + z_2 b$ and $r_{i-1} = z_3 a + z_4 b$ for some $i \geq 2$. If r_i divides c_1, then after the assignments at **(E4)**, i.e., $c_2 := c_1 = r_{i-1}$ and $c_1 := r_i$ the algorithm will stop at **(E2)**, x will be z_1 and y will be z_2. If the algorithm does not stop at **(E2)** and continues to **(E3)**, then we have

$$
\begin{aligned}
r_{i+1} &= c_2 - c_1 m_{i+1} = r_{i-1} - r_i m_{i+1} \\
&= z_3 a + z_4 b - (z_1 a + z_2 b) m_{i+1} = (z_3 - z_1 m_{i+1}) a + (z_4 - z_2 m_{i+1}) b
\end{aligned}
$$

and the induction step is proved.

Thus the claim regarding r_i, i.e., $r_i = z_1 a + z_2 b$ for some integers z_1 and z_2, is correct and hence when the algorithm stops at **(E2)** we have that x will be z_1 and y will be z_2. $\qquad\square$

Corollary 1.1. *If $k = $ g.c.d.(a, b), then there do not exist two integers x and y such that $\gamma = ax + by$, where $0 < |\gamma| < k$.*

The greatest common divisor is naively generalized for polynomials. Let $g(x)$ and $h(x)$ be two distinct polynomials over a field $\mathbb{F}$. The greatest common divisor of $g(x)$ and $h(x)$ is a polynomial $f(x)$ of the largest degree in $\mathbb{F}$ such that $f(x)$ divides $g(x)$ and $f(x)$ divides $h(x)$. Similarly, we define the least common multiple for polynomials. Euclid's algorithm is derived in a naive way to find the greatest common divisor of two polynomials over the same field $\mathbb{F}$. W.l.o.g. we will assume that all polynomials that will be discussed are **monic**, i.e., the leading coefficient of their highest degree is 1. The reason for taking monic polynomials is to have a unique solution to the greatest common divisor. When Euclid's algorithm is applied to polynomials, we can obtain the following result that generalizes Theorem 1.5.

Theorem 1.6. *Let $g(x)$ and $h(x)$ be two monic polynomials over $\mathbb{F}_q$ and let $f(x) = $ g.c.d.$(g(x), h(x))$. Then, there exist two polynomials $\alpha(x)$ and $\beta(x)$ over $\mathbb{F}_q$, where $\deg \alpha(x) < \deg h(x)$ and $\deg \beta(x) < \deg g(x)$, such that $f(x) = \alpha(x)g(x) + \beta(x)h(x)$.*

Next, the following results are required to prove the Chinese remainder theorem that will be stated later.

Lemma 1.2. *If t is a common multiple of $m_1, m_2, \ldots, m_s$, then $[m_1, m_2, \ldots, m_s]$ divide t.*

Proof. If t is a common multiple of $m_1, m_2, \ldots, m_s$ and $m = [m_1, m_2, \ldots, m_s]$, then $t \geq m$. If m does not divide t, then we can write $t = jm + r$, where j is some positive integer and $0 < r < m$. Since m_i divides both m and t for each $1 \leq i \leq s$ and $t = jm + r$, it follows that m_i divides r, and hence r is a common multiple of $m_1, m_2, \ldots, m_s$, a contradiction since $r < m$ and m is the least common multiple of $m_1, m_2, \ldots, m_s$.

Thus $[m_1, m_2, \ldots, m_s]$ divides t. $\qquad\square$

Lemma 1.3. *If $y \equiv x \pmod{m}$ and the positive integer d divides m, then $y \equiv x \pmod{d}$.*

Proof. The congruence $y \equiv x \pmod{m}$ implies that $y = x + jm$ for some integer j. Furthermore, d divides m implies that $y = x + j\ell d$ for nonzero integer ℓ. However, this also implies that $y \equiv x \pmod{d}$. $\qquad\square$

Theorem 1.7.
(1) g.c.d.$(a, m) = 1$ *implies that* $ay \equiv ax \pmod{m}$ *if and only if* $y \equiv x \pmod{m}$.
(2) *For* s *distinct integers,* $m_1, m_2, \ldots, m_s$, *we have that* $y \equiv x \pmod{m_i}$ *for each* $1 \leq i \leq s$, *if and only if* $y \equiv x \pmod{[m_1, m_2, \ldots, m_s]}$.

Proof.

(1) Let g.c.d.$(a, m) = 1$ and assume that $ay \equiv ax \pmod{m}$.
Assume, on the contrary, that $y \not\equiv x \pmod{m}$. This implies that

$$y = x + jm + r, \text{ where } 1 \leq r \leq m - 1 \text{ and } j \in \mathbb{Z}.$$

Since $ay \equiv ax \pmod{m}$, it follows that

$$ax + ajm + ar = a(x + jm + r) = ay \equiv ax \pmod{m}$$

and hence

$$ar = \ell m,$$

for some $\ell \in \mathbb{Z}$. However, since g.c.d.$(a, m) = 1$ and $1 \leq r \leq m - 1$, it follows that $ar \neq \ell m$, a contradiction. Thus $y \equiv x \pmod{m}$.
Now, let g.c.d.$(a, m) = 1$ and assume that $y \equiv x \pmod{m}$.
By definition, $y \equiv x \pmod{m}$ implies that $y = x + jm$ for some $j \in \mathbb{Z}$ and hence

$$ay = ax + ajm.$$

Therefore $ay \equiv ax \pmod{m}$, as required.
(2) Let $m_1, m_2, \ldots, m_s$ be s distinct integers. If $y \equiv x \pmod{m_i}$ for $1 \leq i \leq s$, then m_i divides $y - x$. This implies that $y - x$ is a common multiple of $m_1, m_2, \ldots, m_s$ and therefore by Lemma 1.2 we have that $[m_1, m_2, \ldots, m_s]$ divides $y - x$. It follows that $y \equiv x \pmod{[m_1, m_2, \ldots, m_s]}$.
If we assume that $y \equiv x \pmod{[m_1, m_2, \ldots, m_s]}$, then by Lemma 1.3 we have that $y \equiv x \pmod{m_i}$ for each i, $1 \leq i \leq s$, since m_i divides $[m_1, m_2, \ldots, m_s]$. $\qquad\square$

Theorem 1.8. *If* g.c.d.$(a, m) = 1$, *then the equation* $ax \equiv b \pmod{m}$ *has a solution* $x = x_1$. *All the solutions for the equation are given by* $x = x_1 + jm$, *where* $j \in \mathbb{Z}$.

Proof. Since g.c.d.$(a, m) = 1$, it follows by Theorem 1.5 that there exist two integers z and y such that

$$az + my = 1$$

and therefore

$$abz + mby = b.$$

This implies that $abz = b - bym$, i.e., the equation $ax \equiv b$ (mod m) has a solution $x = x_1 = bz$.

Assume now that $x = x_2$ is also a solution to the equation $ax \equiv b$ (mod m), where $x_2 \not\equiv x_1$ (mod m). We can write

$$ax_1 = b + j_1 m \text{ and } ax_2 = b + j_2 m,$$

which implies that

$$a(x_2 - x_1) = (j_2 - j_1)m. \tag{1.1}$$

Since g.c.d.$(a, m) = 1$, it follows from Eq. (1.1) that m divides $x_2 - x_1$ and hence $x_2 = x_1 + jm$, where j is an integer.

If $x_2 = x_1 + jm$, where $j \in Z$, then $ax_2 = ax_1 + ajm \equiv b$ (mod m), and hence each x_2 of this form is a solution for the equation. □

The next theorem is the ***Chinese remainder theorem***.

Theorem 1.9. *Let $m_1, m_2, \ldots, m_s$ denote s positive integers greater than 1 that are pairwise relatively prime and let $a_1, a_2, \ldots, a_s$ denote any s integers. Then, the s congruences*

$$x \equiv a_1 \text{ (mod } m_1)$$
$$x \equiv a_2 \text{ (mod } m_2)$$
$$\vdots \tag{1.2}$$
$$x \equiv a_s \text{ (mod } m_s)$$

have common solutions and for any two such solutions x_1 and x_2 we have that $x_2 \equiv x_1$ (mod $m_1 \cdot m_2 \cdot \cdots \cdot m_s$).

Proof. Since $m_1, m_2, \ldots, m_s$ are pairwise relatively prime, it follows that

$$m = \prod_{i=1}^{s} m_i = [m_1, m_2, \ldots, m_s].$$

Therefore $\frac{m}{m_j}$ is an integer and g.c.d.$(\frac{m}{m_j}, m_j) = 1$ for each j, $1 \le j \le s$. Therefore by Theorem 1.8, there exists an integer b_j such that $\frac{m}{m_j} b_j \equiv 1$ (mod m_j). Clearly, $\frac{m}{m_j} b_j \equiv 0$ (mod m_i) for each $1 \le i \le s$, where $i \ne j$. Now, if we define x_0 as

$$x_0 = \sum_{j=1}^{s} \frac{m}{m_j} b_j a_j,$$

then for each $1 \le i \le s$ we have that

$$x_0 \equiv \sum_{j=1}^{s} \frac{m}{m_j} b_j a_j \equiv \frac{m}{m_i} b_i a_i \equiv a_i \text{ (mod } m_i),$$

where the second equality is due to $\frac{m}{m_j} b_j \equiv 0 \pmod{m_i}$, for each $i \neq j$, and the third equality is due to Theorem 1.7(1) since $\frac{m}{m_i} b_i \equiv 1 \pmod{m_i}$. Hence, x_0 is a common solution of the congruences in Eq. (1.2).

If x_1 and x_2 are both solutions for x in $x \equiv a_i \pmod{m_i}$ for all $1 \leq i \leq s$, then $x_2 \equiv x_1 \pmod{m_i}$ for all $1 \leq i \leq s$, and hence by Theorem 1.7(2) we have that $x_2 \equiv x_1 \pmod{m}$ and the proof is completed. □

Corollary 1.2. *Let $m_1, m_2, \ldots, m_s$ be s pairwise relatively prime positive integers greater than 1. Let*

$$m = \prod_{i=1}^{s} m_i.$$

If $0 \leq x < m$, then the system of equations

$$
\begin{aligned}
x &\equiv i_1 \pmod{m_1} \\
x &\equiv i_2 \pmod{m_2} \\
&\vdots \\
x &\equiv i_s \pmod{m_s}
\end{aligned}
\tag{1.3}
$$

has a unique solution, where $0 \leq i_j < m_j$.

Proof. Let $j_1, j_2, \ldots, j_s$ be s integers such that $0 \leq j_k < m_k$ for $1 \leq k \leq s$, and consider the set of equations

$$
\begin{aligned}
y &\equiv j_1 \pmod{m_1} \\
y &\equiv j_2 \pmod{m_2} \\
&\vdots \\
y &\equiv j_s \pmod{m_s}
\end{aligned}
$$

By Theorem 1.9 this set of equations has a unique solution when y is taken modulo $m_1 \cdot m_2 \cdots m_s$. Assume that y is also a solution for

$$
\begin{aligned}
y &\equiv \ell_1 \pmod{m_1} \\
y &\equiv \ell_2 \pmod{m_2} \\
&\vdots \\
y &\equiv \ell_s \pmod{m_s}
\end{aligned}
$$

where $0 \leq \ell_k < m_k$ for $1 \leq k \leq s$. This implies that $y \equiv j_k \pmod{m_k}$ and $y \equiv \ell_k \pmod{m_k}$, where $0 \leq j_k, \ell_k < m_k$, for each $1 \leq k \leq s$ and hence $j_k = \ell_k$ for each $1 \leq k \leq s$.

There are $\prod_{k=1}^{s} m_k$ distinct substitutions for the variables $i_1, i_2, \ldots, i_s$ in Eq. (1.3) and $\prod_{k=1}^{s} m_k$ possible values of y, where $0 \leq y < \prod_{k=1}^{s} m_k$, which implies that the system of equations in Eq. (1.3) has a unique solution for x, where $0 \leq i_k < m_k, 1 \leq k \leq s$. $\qquad\square$

The concepts in number theory are useful in the construction of sequences that satisfy certain properties. For example, Corollary 1.2 will be used in one of our constructions of some two-dimensional arrays.

There are a few interesting functions associated with number theory. The two that will be required for our exposition are the Euler's totient function and the Möbius function, which are presented now.

Euler's function $\phi(n)$, known also as ***Euler's totient function***, where n is a positive integer, is the number of integers between 1 to n that are relatively prime to n. In other words

$$\phi(n) \triangleq |\{i \ : \ 1 \leq i \leq n, \ \text{g.c.d.}(i, n) = 1\}|.$$

We also define the set n_ϕ of $\phi(n)$ residues modulo n that are relatively prime to n as follows:

$$n_\phi \triangleq \{i \ : \ 1 \leq i < n \text{ and g.c.d.}(i, n) = 1\}. \tag{1.4}$$

These definitions imply that $\phi(n) = |n_\phi|$. The following lemma can be easily verified (for the proof of the third claim, the Chinese remainder theorem is applied).

Lemma 1.4.

(1) *If p is a prime number, then $\phi(p) = p - 1$.*
(2) *If p is a prime and $e > 1$ is an integer, then $\phi(p^e) = (p-1)p^{e-1}$.*
(3) *If $n_1 > 1$ and $n_2 > 1$ are two integers such that g.c.d.$(n_1, n_2) = 1$, then $\phi(n_1 n_2) = \phi(n_1)\phi(n_2)$.*

Corollary 1.3. *Let $p_1, p_2, \ldots, p_r$ be r distinct primes and let $e_1, e_2, \ldots, e_r$ be r positive integers. If $n = p_1^{e_1} p_2^{e_2} \ldots p_r^{e_r}$, then*

$$\phi(n) = n \prod_{i=1}^{r} \left(1 - \frac{1}{p_i}\right)$$

and

$$\phi(n) = n - \sum_{i=1}^{r} \frac{n}{p_i} + \sum_{0 < i < j \leq r} \frac{n}{p_i p_j} - \cdots + (-1)^r \frac{n}{p_1 p_2 \cdots p_r}.$$

Proof. By applying Lemma 1.4(3) several times and then applying Lemma 1.4(2) we have that

$$\phi(n) = \phi(p_1^{e_1} p_2^{e_2} \cdots p_r^{e_r}) = \prod_{i=1}^{r} \phi(p_i^{e_i}) = \prod_{i=1}^{r} \left(p_i^{e_i - 1}(p_i - 1)\right)$$

and therefore,

$$\phi(n) = n \frac{\prod_{i=1}^{r} \left(p_i^{e_i-1}(p_i - 1) \right)}{n} = n \prod_{i=1}^{r} \frac{p_i - 1}{p_i} = n \prod_{i=1}^{r} \left(1 - \frac{1}{p_i} \right).$$

Now, by developing the right side of the equation we have that

$$\phi(n) = n \prod_{i=1}^{r} \left(1 - \frac{1}{p_i} \right) = n - \sum_{i=1}^{r} \frac{n}{p_i} + \sum_{0 < i < j \leq r} \frac{n}{p_i p_j} - \cdots + (-1)^r \frac{n}{p_1 p_2 \cdots p_r}.$$

$\square$

We continue with a second function of number theory. The **Möbius function** $\mu(n)$ is defined by

$$\mu(n) = \begin{cases} 1 & \text{if } n = 1 \\ 0 & \text{if } a^2 | n \text{ for some } a > 1 \\ (-1)^r & \text{if } n = p_1 p_2 \cdots p_r, \text{ for } r \text{ distinct primes} \end{cases}.$$

Lemma 1.5. *If n is a positive integer, then*

$$\sum_{d|n} \mu(d) = \begin{cases} 1 & \text{if } n = 1 \\ 0 & \text{if } n > 1 \end{cases},$$

where $d|n$ stands for d divides n.

Proof. If $n = 1$, then the only divisor of n is $d = 1$ and since $\mu(1) = 1$ the claim of the lemma for $n = 1$ follows.

If $n > 1$, then n can be written as $n = p_1^{e_1} p_2^{e_2} \dots p_r^{e_r}$, where the r p_is are distinct primes and $e_i \geq 1$ for $1 \leq i \leq r$. A divisor d of n has the form

$$d = p_1^{\varepsilon_1} p_2^{\varepsilon_2} \dots p_r^{\varepsilon_r},$$

where $0 \leq \varepsilon_i \leq e_i$ for each $1 \leq i \leq r$. If $\varepsilon_i > 1$ for some i, then by definition we have that $\mu(d) = 0$. Hence, we have to consider in the sum $\sum_{d|n} \mu(d)$ only the divisors for which $\varepsilon_i \leq 1$ for each $1 \leq i \leq r$. For each k, $0 \leq k \leq r$, there are $\binom{r}{k}$ such divisors, each one is a product of k distinct primes. By the definition of the Möbius function we have that each such divisor contributes $(-1)^k$ to the sum $\sum_{d|n} \mu(d)$. This implies that for $n > 1$

$$\sum_{d|n} \mu(d) = 1 - \binom{r}{1} + \binom{r}{2} - \cdots + (-1)^r \binom{r}{r} = \sum_{k=0}^{r} (-1)^k \binom{r}{k}.$$

By Newton's binomial theorem we have that $(1-1)^r = \sum_{i=0}^{r}(-1)^i \binom{r}{i}$ and hence we have that

$$\sum_{d|n}\mu(d) = \sum_{k=0}^{r}(-1)^k \binom{r}{k} = (1-1)^r = 0,$$

which completes the proof of the lemma. □

The next theorem is well known as the **Möbius inversion formula** or the **Möbius inversion theorem**.

Theorem 1.10. *If for each positive integer n and two arithmetic functions f and g we have that*

$$g(n) = \sum_{d|n} f(d),$$

then

$$f(n) = \sum_{d|n} \mu(d) \cdot g\left(\frac{n}{d}\right).$$

Proof. If $g(n) = \sum_{d|n} f(d)$ for every positive integer n, then for each positive integer d that divides n, we have

$$g\left(\frac{n}{d}\right) = \sum_{d'|\frac{n}{d}} f(d')$$

and hence

$$\sum_{d|n}\mu(d) \cdot g\left(\frac{n}{d}\right) = \sum_{d|n}\mu(d)\sum_{d'|\frac{n}{d}} f(d').$$

This double summation ranges over all positive integers d and d' such that $d \cdot d'$ divides n. If we choose d' first, then d ranges over all divisors of n/d'. Thus

$$\sum_{d|n}\mu(d) \cdot g\left(\frac{n}{d}\right) = \sum_{d'|n} f(d')\sum_{d|\frac{n}{d'}}\mu(d).$$

By Lemma 1.5 we have that $\sum_{d|\frac{n}{d'}}\mu(d) = 0$ unless $d = 1$, i.e., $n = d'$, in which case

$$\sum_{d|\frac{n}{d'}}\mu(d) = 1.$$

This implies the claim of the theorem. □

Theorem 1.10 will be used several times in our enumerations that will be given mainly in Chapter 3. However, first, we will state a simple lemma that will present the strength of this formula.

Lemma 1.6. *If n is a positive integer, then*

$$\phi(n) = n \sum_{d|n} \frac{\mu(d)}{d}.$$

Proof. We partition the set of integers in $\{1, 2, \ldots, n\}$ into ℓ subsets, a subset for each divisor of n. For a divisor d of n, the subset S_d of this partition is defined by $S_d \triangleq \{i : \text{g.c.d.}(i, n) = d, \ 1 \le i \le n\}$. The integer i is contained in S_d if and only if i is of the form jd, where $1 \le j \le \frac{n}{d}$ and g.c.d.$(j, \frac{n}{d}) = 1$. Hence, there are exactly $\phi\left(\frac{n}{d}\right)$ elements in S_d. Since there are n elements in $\{1, 2, \ldots, n\}$, it follows that

$$n = \sum_{d|n} \phi\left(\frac{n}{d}\right) = \sum_{d|n} \phi(d).$$

Now, we apply the Möbius inversion formula (Theorem 1.10), where $g(n) = n$ and $f(n) = \phi(n)$. The outcome is

$$\phi(n) = \sum_{d|n} \mu(d) \cdot \frac{n}{d} = n \sum_{d|n} \frac{\mu(d)}{d}. \qquad \square$$

Corollary 1.4. *For each positive integer n*

$$\frac{\phi(n)}{n} = \sum_{d|n} \frac{\mu(d)}{d}.$$

Theorem 1.10 for the Möbius inversion formula is one direction of a more general result. Although there is no use for the general result in our exposition, it is given for completeness.

Theorem 1.11. *For each positive integer n and two arithmetic functions f and g we have that*

$$g(n) = \sum_{d|n} f(d),$$

if and only if

$$f(n) = \sum_{d|n} \mu(d) \cdot g\left(\frac{n}{d}\right).$$

Proof. One direction of the proof was proved in Theorem 1.10. In the other direction, assume that

$$f(n) = \sum_{d|n} \mu(d) \cdot g\left(\frac{n}{d}\right).$$

This implies that

$$\sum_{d|n} f(d) = \sum_{d|n} \sum_{d'|d} \mu(d') \cdot g\left(\frac{d}{d'}\right) = \sum_{d'd''|n} \mu(d') \cdot g(d'') = \sum_{d''|n} \sum_{d'|\frac{n}{d''}} \mu(d') \cdot g(d''),$$

where $d'' = \frac{d}{d'}$. Changing some order in the equation implies that

$$\sum_{d|n} f(d) = \sum_{d''|n} \sum_{d'|\frac{n}{d''}} \mu(d') \cdot g(d'') = \sum_{d''|n} g(d'') \sum_{d'|\frac{n}{d''}} \mu(d').$$

By Lemma 1.5 the only value for which $\sum_{d'|\frac{n}{d''}} \mu(d') \neq 0$ is when $\frac{n}{d''} = 1$, i.e., $d'' = n$. Thus

$$\sum_{d|n} f(d) = \sum_{n|n} g(n) \sum_{d'|1} \mu(d') = g(n),$$

which completes the proof. □

The next result is known as Euler's generalization for Fermat's theorem (which will be given as a consequence of this generalization).

Theorem 1.12. *If a and m are positive integers such that* g.c.d.$(a, m) = 1$, *then*

$$a^{\phi(m)} \equiv 1 \ (\mathrm{mod} \ m).$$

Proof. Let $r_1, r_2, \ldots, r_{\phi(m)}$ be the $\phi(m)$ distinct integers of the set m_ϕ. If g.c.d.$(r_i, m) = 1$ and g.c.d.$(a, m) = 1$, then also g.c.d$(ar_i, m) = 1$. Moreover, by Theorem 1.7(1) $ar_i \equiv ar_j \ (\mathrm{mod} \ m)$ if and only if $r_i \equiv r_j \ (\mathrm{mod} \ m)$ and hence the set $\{ar_1, ar_2, \ldots, ar_{\phi(m)}\}$ contains the same $\phi(m)$ distinct residues modulo m, namely, $r_1, r_2, \ldots, r_{\phi(m)}$. Therefore we have

$$a^{\phi(m)} \prod_{i=1}^{\phi(m)} r_i = \prod_{i=1}^{\phi(m)} (ar_i) \equiv \prod_{j=1}^{\phi(m)} r_j \ (\mathrm{mod} \ m).$$

Now, since g.c.d.$(r_i, m) = 1$ for each i, $1 \leq i \leq \phi(m)$, it follows again by Theorem 1.7(1) that $a^{\phi(m)} \equiv 1 \ (\mathrm{mod} \ m)$. □

As an immediate corollary from Theorem 1.12, we have what is known as Fermat's theorem.

Corollary 1.5. *Let p by a prime and a be a positive integer such that p does not divide a, then* $a^{p-1} \equiv 1 \ (\mathrm{mod} \ p)$ *and* $a^p \equiv a \ (\mathrm{mod} \ p)$.

The next consequence is known as Euler's criterion.

Lemma 1.7. *If p is an odd prime and $a \not\equiv 0 \pmod{p}$, then $x^2 \equiv a \pmod{p}$ has two solutions or no solutions modulo p according to whether $a^{(p-1)/2} \equiv 1$ or $-1 \pmod{p}$. In particular, $x^2 \equiv -1 \pmod{p}$ has two solutions if $p = 4k+1$, but no solutions if $p = 4k+3$.*

Proof. Assume that for $a \not\equiv 0 \pmod{p}$ we have two distinct integers modulo p, x and y, such that

$$y^2 \equiv x^2 \equiv a \pmod{p}.$$

This implies that $(y - x)(y + x) \equiv 0 \pmod{p}$. Therefore $y + x \equiv 0 \pmod{p}$, i.e., $y \equiv -x \pmod{p}$ and hence the equation $x^2 \equiv a \pmod{p}$ has two solutions or no solutions modulo p.

By Corollary 1.5 we have that

$$(a^{(p-1)/2} - 1)(a^{(p-1)/2} + 1) = a^{p-1} - 1 \equiv 0 \pmod{p} \tag{1.5}$$

and hence $a^{(p-1)/2} \equiv \pm 1 \pmod{p}$.

Now, if $x^2 \equiv a \pmod{p}$, then

$$a^{(p-1)/2} \equiv (x^2)^{(p-1)/2} \equiv x^{p-1} \equiv 1 \pmod{p}. \tag{1.6}$$

Hence, for each such a we have that $a^{(p-1)/2} - 1$ equals 0 modulo p in Eq. (1.5).

Since $x^2 \equiv (-x)^2 \pmod{p}$, the set $\{x^2 \pmod{p} : 1 \le x \le p - 1\}$ contains $\frac{p-1}{2}$ residues modulo p and hence by the first part of the proof we have that if $a^{(p-1)/2} \equiv 1 \pmod{p}$, then $x^2 \equiv a \pmod{p}$ has two solutions.

Thus $x^2 \equiv a \pmod{p}$ has two solutions or no solutions modulo p according to whether $a^{(p-1)/2} \equiv 1$ or $-1 \pmod{p}$. Since $(-1)^{(p-1)/2} = 1$ when $p = 4k + 1$ and $(-1)^{(p-1)/2} = -1$ when $p = 4k + 3$, it follows by Eq. (1.6) that $x^2 \equiv -1 \pmod{p}$ has two solutions if $p = 4k + 1$, but no solutions if $p = 4k + 3$. $\qquad \square$

The last two important concepts in number theory that will be introduced are two of the most interesting ones, the quadratic residues and the Legendre symbol. They will be used later to form some interesting sequences and also to prove that some types of two-dimensional sequences (arrays) do not exist.

Let p be an odd prime. An integer r, $r \not\equiv 0 \pmod{p}$, is a **quadratic residue** modulo p if there exists an integer x such that $x^2 \equiv r \pmod{p}$. An integer n, $n \not\equiv 0 \pmod{p}$, which is not a quadratic residue modulo p is a **quadratic non-residue** modulo p. For the set of residues modulo p, we denote by $\mathcal{R}_p$ the set of quadratic residue modulo p and $\mathcal{N}_p$ the set of quadratic non-residues modulo p.

The **Legendre symbol** $\left(\frac{m}{p}\right)$ is defined as follows. $\left(\frac{m}{p}\right) = 1$ if m is a quadratic residue modulo p and $\left(\frac{m}{p}\right) = -1$ if m is a quadratic non-residue modulo p. If $m \equiv 0 \pmod{p}$, then $\left(\frac{m}{p}\right) = 0$.

There are many interesting properties of the Legendre symbol. The most basic ones will be introduced now in two lemmas. The claims in the first lemma are easy to verify by the definition of the Legendre symbol.

Lemma 1.8. *Let p be an odd prime and a, b two integers. Then,*

(1) $\left(\frac{a}{p}\right)\left(\frac{b}{p}\right) = \left(\frac{ab}{p}\right)$.

(2) *If $a \equiv b \pmod p$, then $\left(\frac{a}{p}\right) = \left(\frac{b}{p}\right)$.*

The second lemma follows from Lemma 1.7.

Lemma 1.9. *If p is an odd prime and a is relatively prime to p, then*

$$\left(\frac{a}{p}\right) \equiv a^{(p-1)/2} \pmod p.$$

Corollary 1.6.

(1) *If p is a prime of the form $4k - 1$, then -1 is a quadratic non-residue residue modulo p.*
(2) *If p is a prime of the form $4k + 1$, then -1 is a quadratic residue modulo p.*

Proof. By Lemma 1.9 we have that

$$\left(\frac{-1}{p}\right) = (-1)^{(p-1)/2}.$$

(1) If $p = 4k - 1$, then $\frac{p-1}{2} = 2k - 1$ and hence $(-1)^{(p-1)/2} = -1$, i.e., -1 is a quadratic non-residue modulo p.
(2) If $p = 4k + 1$, then $\frac{p-1}{2} = 2k$ and hence $(-1)^{(p-1)/2} = 1$, i.e., 1 is a quadratic residue modulo p. □

Corollary 1.7.

- *If p is a prime of the form $4k - 1$, then a is a quadratic residue modulo p if and only if $p - a$ is a quadratic non-residue modulo p.*
- *If p is a prime of the form $4k + 1$, then a is a quadratic residue modulo p if and only if $p - a$ is a quadratic residue modulo p.*

Corollary 1.8. *If $p = 2k + 1$ is a prime, then there are k quadratic residues modulo p and k quadratic non-residues modulo p.*

The following interesting properties of the set of quadratic residues and the set of quadratic non-residues will be very useful.

Theorem 1.13. *Let p be a prime of the form $4k - 1$, r an arbitrary quadratic residue modulo p, n an arbitrary quadratic non-residue modulo p. Each one of the sets $r + \mathcal{N}_p$ and $n + \mathcal{R}_p$ consists of 0, $k - 1$ quadratic residues, and $k - 1$ quadratic non-residues.*

Proof. Let $p = 4k - 1$ and consider the set of expressions $\mathcal{H}_p$ (not the result of the expression) of the form $r_i + n_j$, $1 \leq i, j \leq \frac{p-1}{2}$, where r_i is a quadratic residue and n_j is a quadratic non-residue. By Corollary 1.7(1), the value 0 is represented $\frac{p-1}{2}$ times in $\mathcal{H}_p$ since $r \in \mathcal{R}_p$ if and only if $p - r \in \mathcal{N}_p$ when $p \equiv 3 \pmod 4$.

We show that all nonzero residues modulo p are represented equally often in $\mathcal{H}_p$. Every representation of 1, $1 = r + n$, $r \in \mathcal{R}_p$, $n \in \mathcal{N}_p$, corresponds to a unique representation of g, $g = r' + n'$, where $r' = gr$, $n' = gn$, when g is a quadratic residue and $r' = gn$, $n' = gr$, when g is a quadratic non-residue. Conversely, every representation of g, $g = r + n$, corresponds to a unique representation of 1, $1 = r' + n'$, where $r' = g^{-1}r$, $n' = g^{-1}n$, when g is a quadratic residue, and $r' = g^{-1}n$, $n' = g^{-1}r$, when g is a quadratic non-residue. Thus in $\mathcal{H}_p$ there exists a one-to-one correspondence between the representation of 1 and the representations of any other nonzero residue modulo p. Hence, $\mathcal{H}_p$ contains as many representations of quadratic residues and quadratic non-residues.

Suppose now that the set $1 + \mathcal{N}_p$ contains more (fewer) quadratic residues than quadratic non-residues. Let $r \in \mathcal{R}_p$ be any quadratic residue modulo p. Then, the set $r + \mathcal{N}_p = r(1 + r^{-1}\mathcal{N}_p) = r(1 + \mathcal{N}_p)$ would also contain more (fewer, respectively) quadratic residues than quadratic non-residues. Consequently,

$$\mathcal{H}_p = \bigcup_{r \in \mathcal{R}_p} (r + \mathcal{N}_p)$$

would contain more (fewer, respectively) quadratic residues than quadratic non-residues, a contradiction. It follows that the set $1 + \mathcal{N}_p$ contains as many quadratic residues as quadratic non-residues; the sets $r + \mathcal{N}_p = r(1 + \mathcal{N}_p)$ and $n + \mathcal{R}_p = n(1 + \mathcal{N}_p)$ also have this property, where $r \in \mathcal{R}_p$ and $n \in \mathcal{N}_p$. $\qquad\square$

The proof of the next theorem is very similar to the proof of Theorem 1.13.

Theorem 1.14. *Let p be a prime of the form $4k + 1$, r an arbitrary quadratic residue, n an arbitrary quadratic non-residue. The sets $r + \mathcal{N}_p$ and $n + \mathcal{R}_p$ consist of k quadratic residues and k quadratic non-residues.*

1.2 Codes, graphs, and sequences

This section is devoted to the two concepts that are the main goals of this book, sequences and graphs, as suggested by the title of the book. We start with another concept, codes that are associated with sequences in information theory. Some concepts on codes will be used in the exposition of the book and others are given for completeness. The same will apply also to concepts on graphs.

An $[n, k]_q$ *(linear) code* is a linear subspace of dimension k over $\mathbb{F}_q^n$, i.e., a linear subspace, whose dimension is k, from the set of all words (vectors) of *length* n over $\mathbb{F}_q$.

An $[n, k]_q$ code C can be represented by some matrices. The first one is a
generator matrix G, which is a $k \times n$ matrix over $\mathbb{F}_q$, whose rows form a basis
for the code, i.e., the linear span of the rows of G is C. The second matrix is
a ***parity-check matrix*** H, which is an $(n - k) \times n$ matrix over $\mathbb{F}_q$, whose rows
form a basis for the ***dual subspace*** $C^{\perp}$ of the code C. The dimension $r = n - k$
of this dual subspace is called the ***redundancy*** of the code.

A generator matrix of an $[n, k]_q$ code is in ***standard form*** if its first k
columns form an identity matrix of order k, i.e.,

$$G = [\, I_k \mid A \,],$$

where I_k is the $k \times k$ identity matrix. The related parity-check matrix is given by

$$H = [\, -A^{\mathrm{tr}} \mid I_{n-k} \,].$$

It is readily verified that for these two matrices, we have

$$G \cdot H^{\mathrm{tr}} = 0$$

and

$$H \cdot G^{\mathrm{tr}} = 0,$$

where 0 is an all-zeros matrix of the appropriate size and A^{tr} is the ***transpose*** of
the matrix A (and the same notation when 0 or A are vectors).

The following proposition is a simple observation.

Proposition 1. *The parity-check matrix H of an $[n, k]_q$ code C is a generator
matrix of an $[n, n - k]_q$ code.*

If G is the generator matrix of an $[n, k]_q$ code C, then the $[n, n - k]_q$ code $C^{\perp}$
whose generator matrix is the parity-check matrix H of C is called the ***dual
code*** of C. A code C is called ***self-dual*** if $C = C^{\perp}$. These definitions imply the
following.

Lemma 1.10. *For an $[n, k]_q$ self-dual code we have that $n = 2k$.*

There is another representation of the parity-check matrix. Let α be a primi-
tive element in $\mathbb{F}_{q^r}$ and let $H = [h_1, h_2, \ldots, h_n]$ be an $r \times n$ parity check-matrix
for the code C, where all the h_is are nonzero column vectors. Assume that h_j is
the q-ary representation of the element α^{i_j}, $1 \leq j \leq n$, in $\mathbb{F}_{q^r}$. The parity-check
matrix can be written as $H = [\alpha^{i_1}, \alpha^{i_2}, \ldots, \alpha^{i_n}]$. Finally, note that the word
$x = (x_1, x_2, \ldots, x_n) \in \mathbb{F}_q^n$ is a codeword in C if and only if $H \cdot x^{\mathrm{tr}} = 0$.

We call k coordinates in a code C, over $\mathbb{F}_q$, ***systematic*** if, in the projection
on these k coordinates of C, each of the q^k vectors of length k over $\mathbb{F}_q$ appears
exactly once. Clearly, in an $[n, k]_q$ code whose generator matrix is in standard
form, the first k coordinates are systematic. By definition, one can easily verify
the following lemma.

Lemma 1.11. *k coordinates in the generator matrix G of an* $[n,k]_q$ *code* C *are systematic coordinates if and only if the related* k *vector columns of* G *are linearly independent.*

Definition 1.4. Let C be a linear code over $\mathbb{F}_q$ with an $r \times n$ parity-check matrix H. For any word $x = (x_1, x_2, \ldots, x_n) \in \mathbb{F}_q^n$, the **syndrome** of x, $S(x)$, is defined by

$$S(x) = H \cdot x^{\mathrm{tr}}.$$

The syndromes are column vectors of length r, the redundancy of the code. Hence, there are q^r possible distinct syndromes. The first important property related to the syndromes is associated with the syndromes of the codewords. The value of these syndromes can be verified from the definition of the parity-check matrix of a code C.

Lemma 1.12. *The syndrome of a codeword in a linear code is equal to the all-zeros vector.*

Linear codes are used to transmit information on a noisy channel. The syndromes are very useful in correcting errors that occurred during this transmission.

The **Hamming distance** (or distance in short) between two given words $x = (x_1, x_2, \ldots, x_n)$ and $y = (y_1, y_2, \ldots, y_n)$, over $\mathbb{F}_q$, $d(x, y)$, is the number of coordinates in which x and y differ. In other words

$$d(x, y) \triangleq |\{i \ : \ x_i \neq y_i\}|.$$

The **minimum distance** of a code C, is the smallest integer δ, such that there exist two distinct codewords $x, y \in C$ for which $d(x, y) = \delta$.

Definition 1.5. An $[n, k, d]_q$ **code** is an $[n, k]_q$ code whose minimum Hamming distance is at least d. When $q = 2$, we can write an $[n, k, d]$ code.

The **weight** of a word x, denoted by $\mathrm{wt}(x)$, is the number of nonzero entries in x. Since the codewords of an $[n, k, d]_q$ code form a linear subspace we have the following result.

Lemma 1.13. *The minimum distance of an* $[n, k, d]_q$ *code* C *is the minimum weight of its nonzero codewords.*

Corollary 1.9. *The minimum distance d of an* $[n, k, d]_q$ *code* C *is the minimum number of linearly dependent columns of its parity-check matrix* H.

Proof. The claim follows immediately from the fact that $c \in C$ if and only if $H \cdot c^{\mathrm{tr}} = 0$ and hence the minimum number of linearly dependent columns of H is the minimum weight of a nonzero codeword in C. $\square$

The $[n, k, d]_q$ code C has a generator matrix G and a parity-check matrix H. The code C is used to transmit information words of length k over $\mathbb{F}_q$, via a channel that accepts words of length n. An information word $z = (z_1, z_2, \ldots, z_k)$ is transformed into a codeword $c = (c_1, c_2, \ldots, c_n)$ of length n, where $c = z \cdot G$. Since c is generated as a linear combination of rows from G and the rows of H span a subspace orthogonal to the linear span of the rows of G, it follows, as also implied by Lemma 1.12, that $\mathcal{S}(c) = H \cdot c^{\text{tr}} = \mathbf{0}$. Assume that in the channel, an error $\varepsilon \in \mathbb{F}_q^n$ has occurred in the codeword c and instead of the codeword c, the word $c + \varepsilon$ was received. The syndrome of $c + \varepsilon$ is

$$\mathcal{S}(c + \varepsilon) = H \cdot (c + \varepsilon)^{\text{tr}} = H \cdot c^{\text{tr}} + H \cdot \varepsilon^{\text{tr}} = H \cdot \varepsilon^{\text{tr}} \, .$$

This implies that if it is assumed that only an error from a set $\mathcal{E}$ can occur and each of the elements in the set $\mathcal{E}$ has a different syndrome, then using the value of the syndrome of the received word we have the syndrome of the error. This syndrome should be unique to this error and hence we can find the exact error and recover the codeword that was transmitted over the channel. This implies the following observation.

Corollary 1.10. *A linear code can correct e errors if and only if all the syndromes of the distinct words with weight at most e are distinct.*

The next basic concepts in the book are associated with graphs. Graph theory is a major area in mathematics, computer science, and electrical engineering, as well as other disciplines. Although most concepts should be well known to the reader we define most of them, but we will assume throughout the book that some other basic facts are well known to the reader. For this reason, we rarely present theorems on graphs.

A graph $G = (V, E)$ has a set of vertices V and a set of edges E. Our first assumption is that all our graphs are finite, i.e., the number of vertices is finite and the number of edges is finite. There are two types of graphs, undirected and directed.

For an **undirected graph** $G = (V, E)$, the set V contains the **vertices** of the graph and the set E contains the **edges** of the graph, where an edge $e = \{u, v\}$ is an unordered pair of two vertices $u, v \in V$. Such two vertices are called **adjacent** unless $u = v$. We also say that u is **incident** to e and v is incident to e. An edge $e = \{u, u\}$, i.e., an edge from a vertex to itself is called a **self-loop**. It is a self-loop edge e and also a self-loop vertex u. The **degree** of a vertex v is the number of edges in which v participates, i.e., the number of edges to which v is incident. A self-loop is considered twice for the degree of its vertex. Two edges between the same two vertices are called **parallel edges**. A **path** in G is a sequence of edges $e_1 e_2 \cdots e_\ell$, where $e_i = \{v_{i-1}, v_i\}$, $1 \le i \le \ell$, is an edge in E. If the graph has no parallel edges, then such a path can be described by the sequence of vertices $v_0 v_1 \cdots v_\ell$. The **length** ℓ of the path is the number of edges in the path. If $v_i \ne v_j$ for $0 \le i < j \le \ell$, we say that the path is **simple**.

The path can be described also by its set of consecutive vertices $v_0v_1 \cdots v_\ell$. If $v_\ell = v_0$, then the path is called a **cycle**. If the cycle contains k edges (not necessarily distinct), then it will be also called a **k-cycle**. It should be noted that a cycle can be described by starting with any vertex or edge on it. This implies that there is no starting position and no ending position for the cycle. A simple cycle is a cycle with no repeated vertices (when described as a path, the first and the last vertex, which are the same, will not be considered as repeated vertices). A self-loop is considered to be a simple cycle. A **factor** in an undirected graph is a set of vertex-disjoint (simple) cycles that contain all the vertices of the graph. A **simple graph** is a graph with no parallel edges. An undirected graph is called a **connected graph** if there exists a path between any two distinct vertices of the graph. The **distance**, $d_G(u, v)$ ($d(u, v)$ if the graph is understood from the context) between two vertices $u, v \in V$ is the length of the shortest path between u and v.

A **directed graph** (**digraph**) $G = (V, E)$ consists of a set V of vertices and a set E of **directed edges**, i.e., $e = (u, v) \in E$ implies that there is a directed edge from the vertex $u \in V$ to the vertex $v \in V$. This edge is an **out-edge** (or **outgoing edge**) for u and an **in-edge** (or **incoming edge**) for v. The edge $e = (u, v)$ is also denoted by $u \rightarrow v$. The vertex u is called the **start-point** of the edge e and the vertex v is the **end-point** of e. The two vertices u and v that form the edge e are called **adjacent**, unless $u = v$. The **in-degree** of a vertex $v \in V$, is the number of edges whose end-point is v. The **out-degree** of a vertex $v \in V$, is the number of edges whose starting-point is v. A **directed path** in G is a sequence of edges $e_1e_2 \cdots e_\ell$, where $e_i = (v_{i-1}, v_i)$ is an edge in E. If the graph has no parallel edges, then such a path can be described by the sequence of vertices $v_0v_1 \cdots v_\ell$. The **length** ℓ of the path is the number of edges in the path. If $v_i \neq v_j$ for $0 \leq i < j \leq \ell$ we say that the path is **simple**. The path can be described also by its set of consecutive vertices. If $v_\ell = v_0$, then the path is called a **directed cycle**. The first vertex in a cycle is not defined as the cycle can start in any vertex of the cycle. The **distance**, $d_G(u, v)$ ($d(u, v)$ if the graph is understood from the context) from a vertex u to a vertex v, where $u, v \in V$, in a directed graph, is the length of the shortest directed path from u to v. A cycle with no repeated vertices is called **simple** and a cycle with no repeated edges is called a **tour**.

A **closed path** is a cycle in which the first vertex is determined. If the cycle is not simple and it is not several repetitions of the same simple cycle, then there exists a vertex $v \in V$ (not necessarily unique) that can be chosen as the first vertex in the associated closed path and defines a few different closed paths from the same cycle.

Example 1.1. The cycle

$$u_1 \rightarrow u_2 \rightarrow u_3 \rightarrow u_4 \rightarrow u_1 \rightarrow u_2 \rightarrow u_3 \rightarrow u_4 \rightarrow u_1,$$

where the u_is are distinct, is a double repetition of the simple cycle

$$u_1 \rightarrow u_2 \rightarrow u_3 \rightarrow u_4 \rightarrow u_1.$$

The cycle

$$v_1 \to v_2 \to v_3 \to v_4 \to v_5 \to v_2 \to v_6 \to v_7 \to v_1,$$

where the v_is are distinct, is a closed path that can start with the vertex v_2 in two different ways as follows:

$$v_2 \to v_3 \to v_4 \to v_5 \to v_2 \to v_6 \to v_7 \to v_1 \to v_2$$

and

$$v_2 \to v_6 \to v_7 \to v_1 \to v_2 \to v_3 \to v_4 \to v_5 \to v_2.$$

∎

The edge $e = (u, u)$ is a ***self-loop*** edge e and a self-loop vertex u. Two edges e_1 and e_2 are called ***parallel edges*** if their start-points are the same and their end-points, are the same. Two edges (u, v) and (v, u) are called ***anti-parallel edges***. The ***underline graph*** $G' = (V, E')$ of a directed graph $G = (V, E)$ is an undirected graph with the same vertices as in G, and the same edges, but with no direction to the edges, i.e., if $(u, v) \in E$, then $\{u, v\} \in E'$. Two anti-parallel edges (or parallel edges) in the directed graph will become two parallel edges in the underline graph. A directed graph is connected if its underline graph is connected. A directed graph is called ***strongly connected*** if there exists a directed path from u to v for every two distinct vertices u and v of G. An ***independent set*** in a graph is a set of vertices for which no two are adjacent (in the underline graph if the graph is directed).

An ***undirected tree*** is an undirected connected graph with no simple cycles. A ***directed tree*** is a directed graph with one vertex, called ***the root*** whose in-degree is zero and there is a unique directed path from the root to each vertex in the graph. A ***leaf*** in an undirected tree is a vertex whose degree is one. A ***leaf*** in a directed tree is a vertex whose out-degree is zero. A ***binary directed tree*** is a directed tree in which each vertex that is not a leaf has out-degree two. A ***balanced binary directed tree*** is a binary directed tree in which all the paths from the root to the leaves have the same length.

Given a graph $G = (V, E)$, a ***subgraph*** $G' = (V', E')$ of G, is a graph for which $V' \subseteq V$ and $E' \subseteq E$. A ***spanning tree*** $T = (V', E')$ of a graph $G = (V, E)$ is a subgraph of G, for which $V' = V$ and T is a tree.

An ***undirected bipartite graph*** $G = (V, E)$ is a graph whose vertices can be partitioned into two parts A and B, where $V = A \cup B$, $A \cap B = \varnothing$, and for each edge $\{u, v\} \in E$ we have that either $u \in A$ and $v \in B$ or $v \in A$ and $u \in B$. A ***directed bipartite graph*** $G = (V, E)$ is a graph whose vertices can be partitioned into two parts A and B, where $V = A \cup B$, $A \cap B = \varnothing$, and for each edge $(u, v) \in E$ we have that $u \in A$ and $v \in B$.

A ***matching*** in a graph $G = (V, E)$ is a set of disjoint edges (with no common vertices and no self-loops). A ***perfect matching*** in a graph $G = (V, E)$ is a matching for which each vertex of V is contained in one of the edges.

An **undirected complete graph** on n vertices, denoted by K_n, is a graph $G = (V, E)$ with n vertices and $\binom{n}{2}$ edges, where $E = \{\{u, v\} \; : \; u, v \in V, \; u \neq v\}$. A **directed complete graph** on n vertices, K_n^* is a directed graph $G = (V, E)$ with n vertices and $n(n-1)$ edges, where $E = \{(u, v) \; : \; u, v \in V, \; u \neq v\}$.

A **connected component** in an undirected graph $G = (V, E)$ is a connected subgraph $G' = (V', E')$ of G such that if $v \in V$, $v \notin V'$, then there is no path between v to some vertex of V'.

Two graphs $G_1 = (V_1, E_1)$ and $G_2 = (V_2, E_2)$ are **isomorphic** if there exists a bijective function $f : V_1 \to V_2$, such that for an undirected graph, $\{u, v\} \in E_1$ if and only if $\{f(u), f(v)\} \in E_2$ and for a directed graphs, $(u, v) \in E_1$ if and only if $(f(u), f(v)) \in E_2$.

For a digraph $G = (V, E)$, the **line graph** $L(G) = (V', E')$, is a digraph defined as follows. The set of vertices V' is equal to the set of edges E of G, i.e., $V' = E$. The set of edges E' of $L(G)$ is defined by

$$E' \triangleq \{(e_1, e_2) \; : \; e_1 = (v_1, v_2) \in E, \; e_2 = (v_2, v_3) \in E\},$$

i.e., an edge in E' is a directed path of length two in G. The concept of the line graph is very important and it can be defined similarly to undirected graphs, but it will be used only for directed graphs in the book.

There are several connections between graphs and matrices, some of which will be discussed in our exposition. The **adjacency matrix** A of an undirected graph $G = (V, E)$ in a $|V| \times |V|$ symmetric matrix, whose rows and columns are indexed by the vertices of V, and $A(u, v) = k$ if and only there are exactly k edges between u and v. For a directed graph $G = (V, E)$ the **adjacency matrix** A is a $|V| \times |V|$ binary matrix, whose rows and columns are indexed by the vertices of V, and $A(u, v) = k$ if and only there are exactly k directed edges from u to v. The following important property can be easily verified by this definition.

Lemma 1.14. *If A is the adjacency matrix of a directed graph $G = (V, E)$, then $A^\ell(u, v) = k$ if and only if there are exactly k distinct directed paths of length ℓ from u to v.*

An example of an adjacency matrix will be given in Example 1.6. While the adjacency matrix describes which vertices are adjacent, the second matrix, the **incidence matrix** indicates which vertices are incident to each edge. In other words, for an undirected graph $G = (V, E)$, without self-loops the incidence matrix is a binary $|V| \times |E|$ matrix whose rows are indexed by the vertices of V and its columns are indexed by the edges of E. $A(u, e) = 1$, where $u \in V$ and $e \in E$ if and only if $e = \{u, v\}$ for some $v \in V$. In this case also $A(v, e) = 1$. For a directed graph, we omit the definition as it will not be used in the book.

A **factor** in a directed graph G is a subgraph of G that contains a set of vertex-disjoint simple cycles that contain all the vertices of the graph. The factor can be described by its set of edges since each vertex in a cycle has in-degree *one* and out-degree *one*. Factors, Hamiltonian cycles, and Eulerian cycles, in a graph will play an important role throughout the book.

A **Hamiltonian path** in a graph G is a path that visits each vertex of G exactly once. A **Hamiltonian cycle** in a graph G is a cycle that visits each vertex of G exactly once. An **Eulerian path** in a graph G is a path that visits each edge of G exactly once. An **Eulerian cycle** in a graph G is a cycle that visits each edge of G exactly once.

Theorem 1.15. *A strongly connected directed graph $G = (V, E)$ has an Eulerian cycle if and only if for each $v \in V$, the in-degree of v equals the out-degree of v.*

Proof. Let $G = (V, E)$ be a strongly connected directed graph.

Assume that at one vertex $v \in V$, the in-degree is not equal to the out-degree. In any tour C going through the vertices of V, the number of in-edges of v on the tour equals the number of out-edges of v in the tour C. Thus C cannot contain all the edges of G and hence it is not an Eulerian cycle.

Assume now that for each vertex of V, the in-degree equals the out-degree. Let C be any tour of the largest length in G. If C is not an Eulerian cycle, then, since G is strongly connected, there exists a vertex $v \in V$ that is also on the tour C for which not all the out-edges of v are on C. Let C be written as

$$v \to u_1 \to u_2 \to \cdots \to u_\ell \to v.$$

Since not all the out-edges of v are on C, it follows that we can enlarge the path C from the last written appearance of v until it ends in v again (this is because for each in-edge to a vertex u in V, which is not on C, there exists also an out-edge from u, not on C). The outcome is a new tour C' larger than C. This contradicts our assumption that C is a tour of the largest length.

Therefore G has an Eulerian cycle if and only if for each $v \in V$, the in-degree of v equals the out-degree of v. $\qquad\square$

Note that the proof of Theorem 1.15 can be used for an efficient algorithm to find an Eulerian cycle in any strongly connected directed graph in which for each vertex the in-degree equals the out-degree.

Finally, we are going to present some basic concepts for sequences that are the second main topic of this book. Usually, the definitions are given in a general way to fit into binary sequences and non-binary sequences, but some of them are suitable only for binary sequences.

Definition 1.6. A **sequence** S over an alphabet Σ of size σ is an ordered list of symbols from Σ.

There are many types of sequences and hence there are also various notations for sequences. There are finite sequences and there are infinite sequences. There are cyclic sequences and there are acyclic sequences. An infinite sequence can be denoted by $a_0a_1a_2\cdots$, or $\{a_0a_1a_2\cdots\}$ or just $\{a_k\}$. A sequence with k consecutive αs, where α is an alphabet letter, will be denoted by α^k. The starting point (index of the first element) of the sequence does not have to be *zero*. We will not consider sequences that do not have a starting point. In a **cyclic sequence** $S = [s_0s_1\cdots s_{n-1}]$ the first symbol s_0 follows the last symbol s_{n-1} and hence each symbol can be taken as the starting point. Nevertheless, also in cyclic sequences, there are cases when the starting point is important. This is very similar to the distinction between cycles and closed paths. Two cyclic sequences S_1 and S_2 for which S_2 is a cyclic shift of S_1, but do not have the same starting point, are said to be **equivalent** and will be denoted by $S_2 \simeq S_1$. The **length** of the sequence is the number of symbols written in the sequence. Hence, the length of the sequence $S = [s_0s_1\cdots s_{n-1}]$ is n. **The period** $\pi(S)$ of the cyclic sequence $S = [s_0s_1\cdots s_{n-1}]$ or an infinite sequence $S = s_0s_1s_2\cdots$, is the least positive integer π such that $s_i = s_{\pi+i}$, for each $i \geq 0$ if the sequence is infinite, and for each $0 \leq i \leq n - 1$, where indices are taken modulo n, when the sequence is cyclic. For a cyclic sequence $S = [s_0s_1\cdots s_{n-1}]$ any integer π, $1 \leq \pi \leq n$, such that π divides n and $s_i = s_{\pi+i}$ for $0 \leq i \leq n - 1 - \pi$ is **a period** of S, but the period is the smallest such integer. The same applies to an infinite sequence. To avoid confusion, sometimes we refer to the period as **the least period**. If the least period of a cyclic sequence is equal to its length, then we say that the sequence is not periodic (**aperiodic**). The **weight** of a sequence (cyclic or acyclic, periodic or aperiodic) is the number of nonzero entries in the sequence.

For two sequences (finite or infinite) of the same length $A = a_0, a_1, a_2, a_3, \ldots$ and $B = b_0, b_1, b_2, b_3, \ldots$, over a group, a ring, or a field, we define the addition $A + B$ by

$$A + B \triangleq a_0 + b_0, a_1 + b_1, a_2 + b_2, a_3 + b_3, \ldots \quad ,$$

where $a_i + b_i$ is the addition in the group, the ring, or the field, respectively.

Usually, when addition or subtraction is done, it would be understood from the context if it is done in the finite group $\mathbb{Z}_m$, the finite field $\mathbb{F}_q$, or in the ring of integers $\mathbb{Z}$. When binary addition and addition over the integers are mixed, $\oplus$ will be used for the binary addition.

Example 1.2. The infinite sequence $S = s_0s_1s_2\cdots$ defined by $s_k = s_{k-6}$ with initial conditions $s_0 = s_2 = s_3 = s_5 = 0$ and $s_1 = s_4 = 1$ can be written as

$$010010010010\cdots .$$

The period of S is 3, but it also has periods, 6, 9, and similarly each positive multiple of 3.

The cyclic sequence

$$[010010010]$$

is of length 9, but its period is 3. It also has periods 6 and 9.

The two sequences

$$[0011101111001] \quad \text{and} \quad [1011110010011]$$

are equivalent sequences as one is a cyclic shift of the other, i.e.,

$$[0011101111001] \simeq [1011110010011].$$

These two sequences can be added to obtain a new sequence, i.e.,

$$[0011101111001] + [1011110010011] = [1000011101010].$$

■

For an *acyclic sequence* $A = (a_1, a_2, \ldots, a_n)$, there is always a starting symbol and a last symbol. The period $\pi(A)$ (or just π) of the sequence A (will be rarely used in our context) is the smallest positive integer π such that $a_i = a_{i+\pi}$ for each $1 \leq i \leq n - \pi$. If no such π exists, then the period is defined to be n.

A cyclic sequence is said to be of *full-order* if the period of the sequence is equal to the length of the sequence. If it is not of a full-order, then it is a *degenerated* sequence.

Example 1.3. For the acyclic sequence

$$S = (10010010010011)$$

of length 14, the period is 13.

The acyclic sequence

$$S = (1100101000100)$$

of length 13 has period 13.

For the acyclic sequence

$$S = (1001001001001)$$

of length 13, the period is 3. It also has periods 6, 9, and 12.

For the cyclic sequence

$$S = [1001001001001]$$

of length 13, the period is 13, i.e., it is aperiodic.

The acyclic sequence

$$S = (100100100100)$$

of length 12 has period 3. It also has periods 6 and 9.

■

Usually, words are considered to be acyclic sequences. Sequences that represent cycles in a graph, which will be considered in the next section, are represented by cyclic sequences. However, while sequences are represented by their consecutive digits, the cycles can be also represented by their consecutive vertices. Note the difference in brackets between cyclic sequences and acyclic sequences. Without brackets, the type of the sequence (cyclic or acyclic) should be understood from the context. Note that also for acyclic sequences we can define cyclic shifts. This will be used, for example, in enumerations performed in Chapter 3.

The **characteristic vector** of a subset $\{s_1, s_2, \ldots, s_k\}$ taken from an n-set, say $\mathbb{Z}_n$ is a binary vector (word) of length n, where its jth entry is *one* if and only if $j \in \{s_1, s_2, \ldots, s_k\}$.

Example 1.4. Let $S = \{0, 3, 5, 6, 7, 9, 12\}$ be a set taken from $\mathbb{Z}_{13}$ that has the characteristic vector (1001011101001). ■

For a binary sequence $S = s_1, s_2, \ldots, s_n$ the **complement sequence** $\bar{S}$ is defined by

$$\bar{S} \triangleq \bar{s}_1, \bar{s}_2, \ldots, \bar{s}_n,$$

where $\bar{x}$ is the binary complement of x, i.e., $\bar{x} = 1 - x$. The **reverse sequence** S^R of a sequence $S = s_1, s_2, \ldots, s_n$ is defined by

$$S^R \triangleq s_n, \ldots, s_2, s_1.$$

A cyclic binary sequence S is called a **complement-reverse** (or a CR sequence) if $S = \bar{S}^R$, which is equivalent to $S^R = \bar{S}$.

Finally, a cyclic sequence S is called a **self-dual** sequence if $\bar{S} \simeq S$. It will be proved later that a self-dual sequence S can be written as $S = [X \ \bar{X}]$. Self-dual sequences will have an important role throughout the book.

Now, we will define a few operators for sequences that will be used throughout the book. The first important operator that will be used throughout our discussion is the **shift operator** $\mathbf{E}$. When applied on a cyclic sequence $[s_0 s_1 \cdots s_{n-1}]$ it is defined by $\mathbf{E} s_i = s_{i+1}$, where indices are taken modulo n, i.e.,

$$\mathbf{E}[s_0, s_1, \ldots, s_{n-1}] = [s_1, s_2, \cdots, s_{n-1}, s_0].$$

The sequence remains the same but in another cyclic shift, i.e., an equivalent sequence is obtained. If the operator is applied on an acyclic sequence (word) $(s_1, s_2, \ldots, s_n)$, then

$$\mathbf{E}(s_1, s_2, \ldots, s_n) = (s_2, \ldots, s_n, s_1).$$

A second operator is the **derivative operator** $\mathbf{D}$. We distinguish now between cyclic sequences and acyclic sequences. For a cyclic sequence $[s_0 s_1 \cdots s_{n-1}]$ it is defined by

$$\mathbf{D}[s_0, s_1, \ldots, s_{n-1}] = [s_1 - s_0, s_2 - s_1, \cdots, s_{n-1} - s_{n-2}, s_0 - s_{n-1}],$$

where indices are taken modulo n, i.e., $\mathbf{D} = \mathbf{E} - \mathbf{1}$, and $\mathbf{1}$ is the identity operator. If the sequence is binary, then also $\mathbf{D} = \mathbf{E} - \mathbf{1} = \mathbf{E} + \mathbf{1}$. If the operator is applied on an acyclic sequence (word) $(s_1, s_2, \ldots, s_n)$, then

$$\mathbf{D}(s_1, s_2, \ldots, s_n) = (s_2 - s_1, s_3 - s_2, \ldots, s_n - s_{n-1}).$$

Two more related operators are the prefix and the suffix of a sequence. For a sequence $S = (s_1, s_2, \ldots, s_{n-1}, s_n)$, the **prefix** of S, $\mathbf{L}S$, is defined by

$$\mathbf{L}(s_1, s_2, \ldots, s_{n-1}, s_n) = (s_1, s_2, \ldots, s_{n-1})$$

and the **suffix** of S, $\mathbf{R}S$, is defined by

$$\mathbf{R}(s_1, s_2, \ldots, s_{n-1}, s_n) = (s_2, \ldots, s_{n-1}, s_n).$$

Clearly, by the definitions, for an acyclic sequence S, we have that

$$\mathbf{D}S = \mathbf{R}S - \mathbf{L}S = (\mathbf{R} - \mathbf{L})S.$$

It is also important to note that the operators $\mathbf{D}$, $\mathbf{E}$, $\mathbf{L}$, and $\mathbf{R}$, are linear operators, i.e., $\mathbf{D}(S_1 + S_2) = \mathbf{D}S_1 + \mathbf{D}S_2$ and the same applies to the other three operators. Each operator can be applied several times, e.g., $\mathbf{E}^i S = \mathbf{E}(\mathbf{E}^{i-1}S)$, for $i > 1$. Finally, the operator $\mathbf{E}$ can be used as a parameter in a polynomial as follows.

Example 1.5. If $f(x) = x^3 + x + 1$ is a polynomial and $S = [0110110001]$, then

$$\begin{aligned}
f(\mathbf{E})S &= (\mathbf{E}^3 + \mathbf{E} + 1)S = \mathbf{E}^3 S + \mathbf{E}S + S \\
&= [0110001011] + [1101100010] + [0110110001] = [1101011000].
\end{aligned}$$

∎

1.3 The de Bruijn graph and feedback shift registers

The graph $G_{\sigma,n} = (V_{\sigma,n}, E_{\sigma,n})$, is called **the de Bruijn graph of order** n over an alphabet Σ of size σ, usually taken as $\{0, 1, \ldots, \sigma - 1\}$. It is a directed graph with σ^n vertices in $V_{\sigma,n}$, associated by the σ^n words of length n over Σ. The graph has σ^{n+1} edges in $E_{\sigma,n}$, associated by the σ^{n+1} words of length $n + 1$ over Σ. From the vertex $x = (x_0, x_1, \ldots, x_{n-1})$, where $x_i \in \Sigma$, $0 \le i \le n - 1$, there is a directed edge to each one of the σ vertices of the form $y = (x_1, \ldots, x_{n-1}, x_n)$, where $x_n \in \Sigma$. The associated edge is represented by the $(n + 1)$-tuple $(x_0, x_1, \ldots, x_{n-1}, x_n)$.

A path of length ℓ in $G_{\sigma,n}$ can be represented by its consecutive vertices or consecutive edges as in a directed graph, but it can be also represented by a sequence of $\ell + n - 1$ consecutive symbols from Σ,

$$x_1 x_2 x_3 \cdots x_\ell x_{\ell+1} \cdots x_{\ell+n-1},$$

where $(x_i, x_{i+1}, \ldots, x_{i+n-1})$, $1 \le i \le \ell$, is the ith vertex in the path. A cycle of length ℓ, in $G_{\sigma,n}$, can be represented similarly by ℓ consecutive symbols of the cyclic sequence

$$[x_0 x_1 \cdots x_{\ell-1}],$$

where $(x_i, x_{i+1}, \ldots, x_{i+n-1})$, $0 \le i \le \ell - 1$, is the ith vertex in the cycle and indices are taken modulo ℓ.

Lemma 1.15. *Given two vertices $u, v \in G_{\sigma,n}$, there is a unique path of length n from u to v.*

Proof. The representation of the paths in $G_{\sigma,n}$ by the symbols from Σ implies that from the vertex $u = (u_1, u_2, \ldots, u_n)$ to the vertex $v = (v_1, v_2, \ldots, v_n)$ there exists a unique path that contains $2n$ consecutive symbols from Σ. This path consists of $n + 1$ vertices (each one is represented by n consecutive symbols) and n edges (each one is represented by $n + 1$ consecutive symbols), is given by

$$u_1 u_2 \cdots u_n v_1 v_2 \cdots v_n. \qquad \qquad \square$$

The following theorem can be easily verified by the definitions of the de Bruijn graph and the line graph.

Theorem 1.16. *The Line graph of $G_{\sigma,n}$ is $G_{\sigma,n+1}$, i.e., $G_{\sigma,n+1} = L(G_{\sigma,n})$.*

The graph $G_{2,n}$ will be denoted by $G_n = (V_n, E_n)$, and its vertices are represented by binary words of length n. The vertices can be also represented by the integers from 0 to $2^n - 1$. We will also denote $N = 2^n$ and hence the two self-loops of G_n are at vertices 0 and $N - 1$. From the vertex whose integer value is k, there are two edges, one to vertex $2k$ and a second to vertex $2k + 1$, where the computation is performed modulo N if $2k \ge N$. The graph G_3 is depicted in Fig. 1.1 in both its binary and integer representations.

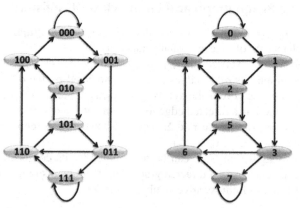

FIGURE 1.1 The de Bruijn graph G_3 in binary representation on the left and integer representation on the right.

We will demonstrate the adjacency matrix on G_3 in the following example.

Example 1.6. Consider the de Bruijn graph G_3, which is a directed graph, depicted in two different ways in Fig. 1.1. The adjacency matrix of G_3, where vertices are ordered by their number (binary or integer), is

$$\begin{bmatrix} 1 & 1 & 0 & 0 & 0 & 0 & 0 & 0 \\ 0 & 0 & 1 & 1 & 0 & 0 & 0 & 0 \\ 0 & 0 & 0 & 0 & 1 & 1 & 0 & 0 \\ 0 & 0 & 0 & 0 & 0 & 0 & 1 & 1 \\ 1 & 1 & 0 & 0 & 0 & 0 & 0 & 0 \\ 0 & 0 & 1 & 1 & 0 & 0 & 0 & 0 \\ 0 & 0 & 0 & 0 & 1 & 1 & 0 & 0 \\ 0 & 0 & 0 & 0 & 0 & 0 & 1 & 1 \end{bmatrix}.$$

■

Given the de Bruijn graph G_n defined by binary words, it is quite natural to define the **complement of G_n**, $\bar{G}_n$, to be the graph defined by complementing the labeling on the vertices and edges of G_n. In other words, $\bar{G}_n = (\bar{V}_n, \bar{E}_n)$, where

$$\bar{V}_n \triangleq \{\bar{v} = (\bar{v}_1, \bar{v}_2, \ldots, \bar{v}_n) : v = (v_1, v_2, \ldots, v_n) \in V_n\}$$

and

$$\bar{E}_n \triangleq \{(u, v) : (\bar{u}, \bar{v}) \in E_n\}.$$

Lemma 1.16. *For each $n \geq 1$, $\bar{G}_n = G_n$.*

Proof. Clearly, by the definition, we have that $\bar{V}_n = V_n$, i.e., each of these sets contains all the 2^n binary words of length n. By the definition of $\bar{E}_n$ we have that

$$\begin{aligned} \bar{E}_n &= \{((\bar{u}_1, \bar{u}_2, \ldots, \bar{u}_n), (\bar{u}_2, \ldots, \bar{u}_n, \bar{u}_{n+1})) : u_i \in \{0, 1\}, \ 1 \leq i \leq n+1\} \\ &= \{((u_1, u_2, \ldots, u_n), (u_2, \ldots, u_n, u_{n+1})) : u_i \in \{0, 1\}, \ 1 \leq i \leq n+1\} = E_n \end{aligned}$$

and hence $\bar{G}_n = G_n$.   □

The **reverse graph** G^R of the digraph $G = (V, E)$, is a digraph defined as $G^R = (V, E^R)$, where

$$E^R \triangleq \{(u, v) : (v, u) \in E\}.$$

Lemma 1.17. *For each $n \geq 1$, G_n^R is isomorphic to G_n.*

Proof. By the definition of G_n^R, we have that

$$E_n^R = \{((x_1, \ldots, x_{n-1}, x_n), (x_0, x_1, \ldots, x_{n-1})) : x_i \in \{0, 1\}, \ 0 \leq i \leq n\}.$$

Now, we apply the mapping $g_R : V_n \to V_n$ defined by

$$g_R(x_1, x_2, \ldots, x_n) = (x_n, \ldots, x_2, x_1),$$

which yields a mapping $g'_R : E_n^R \to E_n$ defined by

$$g'_R(e) = (u^R, v^R), \quad \text{where } e = (u, v) \in E_n^R,$$
$$u = (x_1, \ldots, x_{n-1}, x_n), \text{ and } v = (x_0, x_1, \ldots, x_{n-1}).$$

This definition implies that

$$g'_R(E_n^R) = \{((x_n, x_{n-1}, \ldots, x_1), (x_{n-1}, \ldots, x_1, x_0)) \; : \; x_i \in \{0, 1\}, \; 0 \le i \le n\}.$$

Clearly, $g'_R(E_n^R) = E_n$ and hence the definition of g'_R is consistent and G_n^R is isomorphic to G_n. □

A *span* n de Bruijn sequence (or a de Bruijn sequence of *order* n, also called a *de Bruijn cycle* of order n) over Σ is a cyclic sequence $S = [s_0, s_1, \ldots, s_{\sigma^n - 1}]$ in which each n-tuple over Σ appears exactly once as a window in the sequence (where a *window* consists of consecutive elements of the sequence and it can also start at the end of the sequence and end at its beginning). The cyclic sequence $S = [s_0, s_1, \ldots, s_{\sigma^n - 1}]$ can be described as an acyclic sequence of length $\sigma^n + n - 1$,

$$S' = (s_0, s_1, \ldots, s_{\sigma^n - 1}, s_0, \ldots, s_{n-1}),$$

where each n-tuple over Σ appears exactly once as a window in the sequence and also a first n-tuple is defined. We also define a span n **shortened de Bruijn sequence** of length $\sigma^n - 1$ to be a cyclic sequence in which each nonzero n-tuple over Σ appears exactly once as a window in the sequence. The *span* of a sequence S is the least n for which all the consecutive n-tuples are distinct.

A de Bruijn sequence of order n is associated with two types of cycles in the graph. On the one hand, this sequence forms an Eulerian cycle in the graph $G_{\sigma, n-1}$. This sequence also forms a Hamiltonian cycle in the graph $G_{\sigma, n}$. Hence, such cycles in the de Bruijn graph are also called de Bruijn cycles. As a Hamiltonian cycle, a de Bruijn cycle will be also called a *full cycle*. The cycles (either Eulerian or Hamiltonian) are represented by σ^n consecutive symbols, where each n consecutive symbols either represents an edge in $G_{\sigma, n-1}$ or a vertex in $G_{\sigma, n}$. Given an Eulerian cycle in $G_{\sigma, n-1}$ or a Hamiltonian cycle in $G_{\sigma, n}$, the de Bruijn sequence is generated, for example, by the first symbol of the consecutive edges in the Eulerian cycle in $G_{\sigma, n-1}$ or the first symbol of the consecutive vertices in the Hamiltonian cycle in $G_{\sigma, n}$. In other words, given a de Bruijn sequence $S = [s_0 s_1 s_2 \cdots s_{\sigma^n - 1}]$, a Hamiltonian cycle in $G_{\sigma, n}$ is constructed where the ith vertex in this cycle is $(s_i, s_{i+1}, \ldots, s_{i+n-1})$, and subscripts are taken modulo σ^n. Clearly, $(s_i, s_{i+1}, \ldots, s_{i+n-1}) \to (s_{i+1}, \ldots, s_{i+n-1}, s_{i+n})$ is an edge in $G_{\sigma,n}$ and hence S represents a Hamiltonian cycle in $G_{\sigma,n}$. On the other hand, $(s_i, s_{i+1}, \ldots, s_{i+n-1})$ can be viewed as the edge

$$(s_i, s_{i+1}, \ldots, s_{i+n-2}) \to (s_{i+1}, \ldots, s_{i+n-2}, s_{i+n-1})$$

in $G_{\sigma,n-1}$ and since each n-tuple appears exactly once as a window in S, it follows that each edge of $G_{\sigma,n-1}$ will appear exactly once in the cycle. Therefore we have the following two theorems.

Theorem 1.17. *The number of span n de Bruijn sequences equals the number of Eulerian cycles in $G_{\sigma,n-1}$.*

Theorem 1.18. *The number of span n de Bruijn sequences equals the number of Hamiltonian cycles in $G_{\sigma,n}$.*

Corollary 1.11. *The number of Eulerian cycles in $G_{\sigma,n-1}$ equals the number of Hamiltonian cycles in $G_{\sigma,n}$.*

Similarly to a de Bruijn sequence, each cycle in $G_{\sigma,n}$ can be represented by a cyclic sequence. A cycle of length k in $G_{\sigma,n}$ is represented by a cyclic sequence of length k, $S = [s_0, s_1, \ldots, s_{k-1}]$, where each n consecutive symbols of S represents a vertex in $G_{\sigma,n}$.

Example 1.7. Consider the graph G_6 and the four sequences, $S_1 = [01]$, $S_2 = [0101]$, $S_3 = [00101]$, and $S_4 = [0001101]$.

The sequence S_1 is the following simple cycle of length 2 and weight 1

$$(010101) \to (101010) \to (010101).$$

The sequence S_2 is the following cycle

$$(010101) \to (101010) \to (010101) \to (101010) \to (010101)$$

of length 4. This cycle is not simple and its weight is 2. It is periodic with period 2, where in one period we have the cycle S_1.

S_3 is the cycle of length 5, period 5, and weight 2

$$(001010) \to (010100) \to (101001) \to (010010) \to (100101) \to (001010) .$$

S_4 is the following cycle of length 7, period 7, and weight 3

$$(000110) \to (001101) \to (011010) \to (110100)$$
$$\to (101000) \to (010001) \to (100011) \to (000110) .$$

The introduction of the de Bruijn graph was motivated by the interesting combinatorial problem of enumerating the number of binary cyclic sequences of length 2^n in which each binary n-tuple appears exactly once as a window of n consecutive symbols. The introduction of nonsingular feedback shift registers that are associated with factors in the de Bruijn graph was motivated by many practical applications. For example, it was first used in NASA missions for space exploration.

The theory of shift-register sequences was developed in parallel with the theory of cycles, with different lengths and properties, in the de Bruijn graph. In this section, we consider only the binary case although many of the results can be generalized to the non-binary case. However, in some cases, the generalization for the non-binary case is slightly more complicated. Nevertheless, in Chapter 2 the linear theory will be discussed for any finite field, so some of the definitions will be given for any finite field.

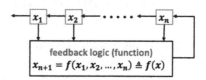

FIGURE 1.2 Feedback shift register of order n.

A *feedback shift register* of order n (an FSR$_n$ in short) has 2^n *states*, represented by the set of 2^n binary words of length n. The register has n cells (which are binary storage elements, e.g., flip-flops that are positioned on a delay line), where each cell stores at each stage one of the bits of the current state. Such an FSR$_n$ is depicted in Fig. 1.2. Given the word $(x_1, x_2, \ldots, x_n)$ that is a state in the FSR$_n$, x_i is stored in the ith cell of the FSR$_n$. The n cells are connected to another logic element that computes a Boolean feedback function $f(x_1, x_2, \ldots, x_n)$. At periodic intervals that are controlled by a master clock, x_2 is transferred to x_1, x_3 to x_2, and so on until x_n is transferred to x_{n-1}. The value of the feedback function is transferred to x_n and hence it is common to denote $x_{n+1} = f(x_1, x_2, \ldots, x_n)$. The register starts to work with an *initial state* $(a_1, a_2, \ldots, a_n)$, where a_i, $1 \leq i \leq n$, is the initial value stored in the ith cell. The feedback function f is a Boolean function and hence it also has a truth table. There are 2^n possible distinct values for $x_1, x_2, \ldots, x_n$, i.e., there are 2^n distinct states and each one can have a value of either 0 or 1. Hence, there are 2^{2^n} different FSR$_n$, but not all of these functions are of interest.

A *linear feedback shift register* of order n (an LFSR$_n$ in short) over $\mathbb{F}_q$ is an FSR$_n$ whose feedback function f is linear, i.e.,

$$x_{n+1} = f(x_1, x_2, \ldots, x_n) = \sum_{i=1}^{n} c_i x_i, \quad c_i \in \mathbb{F}_2.$$

It is possible to define feedback shift registers also over $\mathbb{F}_q$, where $c_i \in \mathbb{F}_q$. In this section, our exposition on shift registers will be only over $\mathbb{F}_2$. However, when we consider the representation of linear shift registers and their sequences as polynomials, also sequences and polynomials over $\mathbb{F}_q$ will be discussed. All these will be done without an associated feedback shift register over $\mathbb{F}_q$. However, it should be clear that once a factor in $G_{\sigma,n}$ is given, then there exists a bijective function from $V_{\sigma,n}$ to $V_{\sigma,n}$ associated with this factor.

Each FSR$_n$ has a *state diagram* that is a graph with 2^n vertices (the states of the FSR$_n$). Given an FSR$_n$ with a feedback function $f(x_1, x_2, \ldots, x_n)$, the vertex $(x_1, x_2, \ldots, x_n)$ in the associated state diagram has an edge to the vertex $(x_2, \ldots, x_n, x_{n+1})$ if $x_{n+1} = f(x_1, x_2, \ldots, x_n)$. This implies that the state diagram with 2^n states (vertices) has exactly 2^n edges. A *nonsingular feedback shift register* has a feedback function whose state diagram contains only cycles. Such a state diagram is associated with a factor in G_n. These are the only FSR$_n$s that are of interest and hence from the next chapter, all FSR$_n$s will always be nonsingular. The length of the cycles (sequences) in the state diagram of the FSR$_n$ will be always considered equal to their period, i.e., all these cycles are simple. The same convention will be made in a factor of a graph, i.e., all the cycles will be simple.

The feedback function of an FSR$_n$ is a Boolean (binary) function and hence, as said before, it can be represented by a truth table. A *truth table* of a Boolean function with n variables contains 2^n rows. Each row contains a distinct binary n-tuple $(x_1, x_2, \ldots, x_n)$ and also the binary value $f(x_1, x_2, \ldots, x_n)$ of the feedback function. When we refer to the truth table we usually refer to the values of the function f.

Lemma 1.18. *A (binary)* FSR$_n$ *is nonsingular if and only if for each* $x_i \in \{0, 1\}$, $2 \le i \le n$, *we have*

$$f(0, x_2, \ldots, x_n) \ne f(1, x_2, \ldots, x_n).$$

Proof. A state diagram of an FSR$_n$ contains only cycles if and only if the in-degree and the out-degree of each vertex is one. By the definition of an FSR$_n$, we have that each vertex of the state diagram has an out-degree of one. Given a vertex $(x_2, \ldots, x_n, x_{n+1})$, its in-degree is one if either $x_{n+1} = f(0, x_2, \ldots, x_n)$ or $x_{n+1} = f(1, x_2, \ldots, x_n)$. Hence, a binary FSR$_n$ is nonsingular if and only if for each $x_i \in \{0, 1\}$, $2 \le i \le n$, we have that $f(0, x_2, \ldots, x_n) \ne f(1, x_2, \ldots, x_n)$. $\square$

Corollary 1.12. *The last* 2^{n-1} *terms in the truth table of the feedback function of a nonsingular* FSR$_n$ *are the complements of the first* 2^{n-1} *terms.*

The *weight of the feedback function* $f(x_1, \ldots, x_n)$ of an FSR$_n$ is the number of *ones* of the first 2^{n-1} terms (of the function f) in the truth table. This will be also called the weight of the truth table (both halves of the table are complements and together have weight 2^{n-1} and hence only one half is important in describing the weight of the table). The truth table represents the function and some properties of the function and its associated state diagram can be obtained from the truth table. The function f of the FSR_n can be also described by the binary sequence whose entries are the consecutive values of f. Therefore the truth table is important in our discussion.

Theorem 1.19. *A (binary)* FSR_n *is nonsingular if and only if its feedback function* $f(x_1, x_2, \ldots, x_n)$ *satisfies*

$$f(x_1, x_2, \ldots, x_n) = x_1 + g(x_2, \ldots, x_n),$$

where $g(x_2, \ldots, x_n)$ *is any Boolean function.*

Proof. Clearly for $x = (x_1, x_2, \ldots, x_n)$ we have,

$$f(x) = x_1 f(1, x_2, \ldots, x_n) + (x_1 + 1) f(0, x_2, \ldots, x_n). \qquad (1.7)$$

Assume first that $f(x_1, x_2, \ldots, x_n) = x_1 + g(x_2, \ldots, x_n)$. Hence, for each $x_i \in \{0, 1\}$, $2 \leq i \leq n$, we have that

$$f(0, x_2, \ldots, x_n) \neq f(1, x_2, \ldots, x_n)$$

and thus, by Lemma 1.18 f is nonsingular.

Assume now that f is nonsingular, i.e., by Lemma 1.18 for each $x_i \in \{0, 1\}$, $2 \leq i \leq n$, we have that

$$f(1, x_2, \ldots, x_n) \neq f(0, x_2, \ldots, x_n),$$

which is equivalent to

$$f(1, x_2, \ldots, x_n) = 1 + f(0, x_2, \ldots, x_n).$$

Therefore by Eq. (1.7) we have that

$$\begin{aligned} f(x) &= x_1(f(0, x_2, \ldots, x_n) + 1) + (x_1 + 1) f(0, x_2, \ldots, x_n) \\ &= x_1 + f(0, x_2, \ldots, x_n) = x_1 + g(x_2, \ldots, x_n). \end{aligned} \qquad \square$$

Corollary 1.13. *The number of distinct nonsingular* FSR_ns *is* $2^{2^{n-1}}$.

Corollary 1.14. *The number of distinct factors in* G_n *is* $2^{2^{n-1}}$.

Definition 1.7. The *companion*, x' of a state $x = (x_1, x_2, \ldots, x_{n-1}, x_n)$ is the state $x' = (x_1, x_2, \ldots, x_{n-1}, \bar{x}_n)$, i.e., x and x' differ exactly on their last bit.

Definition 1.8. The *conjugate*, $\hat{x}$ of a state $x = (x_1, x_2, \ldots, x_{n-1}, x_n)$ is the state $\hat{x} = (\bar{x}_1, x_2, \ldots, x_{n-1}, x_n)$, i.e., x and $\hat{x}$ differ exactly on their first bit.

Given a state diagram of a nonsingular FSR_n, one might want to reduce the number of cycles in the state diagram and obtain a new state diagram for another FSR_n with another feedback function. For this purpose, a simple merging method called ***the merge-or-split method*** can be applied. Assume that we are given a nonsingular FSR_n with a feedback function

$$f(x) = f(x_1, x_2, \ldots, x_n) = x_1 + g(x_2, \ldots, x_n)$$

and a state diagram (factor) $\mathcal{F}$ of G_n that is associated with the function $f(x_1, x_2, \ldots, x_n)$. Consider now the two conjugate states $(0, z_1, \ldots, z_{n-1})$ and $(1, z_1, \ldots, z_{n-1})$, and also consider the two companion states $(z_1, \ldots, z_{n-1}, 0)$ and $(z_1, \ldots, z_{n-1}, 1)$, for any set of values for $z_1, \cdots, z_{n-1} \in \{0, 1\}$. These four states are associated with four edges in G_n as follows:

$$(0, z_1, \ldots, z_{n-1}) \rightarrow (z_1, \ldots, z_{n-1}, 0),$$
$$(1, z_1, \ldots, z_{n-1}) \rightarrow (z_1, \ldots, z_{n-1}, 1),$$
$$(0, z_1, \ldots, z_{n-1}) \rightarrow (z_1, \ldots, z_{n-1}, 1),$$
$$(1, z_1, \ldots, z_{n-1}) \rightarrow (z_1, \ldots, z_{n-1}, 0).$$

The factor $\mathcal{F}$ of G_n contains exactly two of these edges, either the first two or the last two. Removing these two edges from $\mathcal{F}$ and adding the other two yields a set $\mathcal{F}_1$. We claim that $\mathcal{F}_1$ is also a factor in G_n. This is a simple observation as the exchange of these pairs of edges preserves the vertices of the factor and also preserves the in-degree *one* and out-degree *one* of all the vertices. The new function of the nonsingular FSR_n associated with the factor $\mathcal{F}_1$ is

$$f_1(x) = f_1(x_1, x_2, \ldots, x_n) = x_1 + g(x_2, \ldots, x_n) + x_2^{z_2} x_3^{z_3} \cdots x_n^{z_n}, \quad (1.8)$$

where $x_i^1 = x_i$ and $x_i^0 = \bar{x}_i$. The function f_1 differs from the function f only in the rows of the conjugate states $(0, z_2, \ldots, z_n)$ and $(1, z_2, \ldots, z_n)$. The weights of the truth table of the functions f and f_1 differ by 1. If $\mathcal{F}$ contains the first two edges, then its weight is smaller by 1 from the weight of the truth table of the function f_1 and vice versa. Finally, this exchange of edges either splits one cycle in the state diagram of $\mathcal{F}$ into two cycles or merges two cycles in the state diagram of $\mathcal{F}$ into one cycle. This merge-or-split method is depicted in Fig. 1.3. When the predecessors of the companion states x and y (which are the conjugate states u and v) are interchanged, either one cycle is split into two cycles or two cycles are merged into one cycle. The two states x and y are called the ***bridging states*** of the join (or the split). Bridging states can be taken either as companion states (which will be usually the case in our discussion) or as conjugate states. If in the state diagram, the state y follows the state v and the state x follows the state u (one cycle), and the merge-or-split method is applied, then one cycle is split into two. If in the state diagram the state y follows the state u and the state x follows the state v (two cycles) and the merge-or-split method is applied, then two cycles are merged into one. This implies the following consequences.

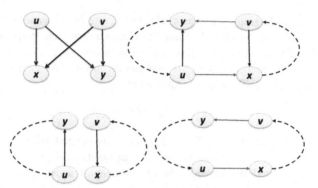

FIGURE 1.3 The merge-or-split method. The conjugate states are u and v; the companion states are x and y; the split and join of the cycles are depicted in the lower figures.

Lemma 1.19. *For each state diagram of an* FSR_n*, we can find a pair of bridging states x and x', such that interchanging the predecessors of x and x' either*
(1) merges two cycles C_1 and C_2 into one cycle and reduces the number of cycles in the state diagram by one; or
(2) splits one cycle C into two cycles C_1 and C_2 and increases the number of cycles in the state diagram by one.

If two companion states x and x' are on two different cycles, then by applying the merge-or-split method with x and x' as bridging states, the two cycles will be merged into one cycle. If two companion states x and x' are on the same cycle, then by applying the merge-or-split method with x and x' as bridging states, the cycle will be split into two cycles.

Lemma 1.20. *Assume $\mathcal{F}$ is a factor in G_n and assume we are changing a* zero *to a* one *or a* one *to a* zero *in the top half of the associated truth table and the opposite change in the bottom half of the truth table (in the same related row position modulo 2^{n-1}). Then, either the number of cycles is increased by 1 compared to $\mathcal{F}$ (one cycle is split into two cycles) or the number of cycles is decreased by 1 (two cycles are merged into one cycle).*

Proof. The claim of the lemma follows immediately from the fact that such a change in the truth table is equivalent to one application of the merge-or-split method. □

Corollary 1.15. *Each application of the merge-or-split method changes the weight of the truth table by one.*

To summarize the merge-or-split method as described, we have the following theorem.

Theorem 1.20. *If a factor $\mathcal{F}$ contains k cycles, then the minimum number of applications of the merge-or-split method to form a full cycle is $k-1$. Moreover, there exist $k-1$ applications of the merge-or-split method that form a full cycle.*

Proof. By Lemma 1.20 each application of the merge-or-split method can join two cycles of a factor into one. Therefore at least $k-1$ applications of the merge-or-split method are required to form a full cycle. To complete the proof it is sufficient to prove that we can join two of the cycles and obtain a factor with $k-1$ cycles. Since by Lemma 1.15 we have that G_n is a strongly connected graph, it follows that there exists an edge $e=(u,v)$ of G_n, which is not an edge of $\mathcal{F}$, with u on one cycle C_1 of $\mathcal{F}$ and v on another cycle C_2 of $\mathcal{F}$. This implies that (u,v') is an edge in $\mathcal{F}$ on C_1, $(\hat{u},v)$ is an edge in $\mathcal{F}$ on C_2, and $(\hat{u},v')$ is an edge of G_n that is not in $\mathcal{F}$. Applying the merge-or-split method with these four edges, (u,v), (u,v'), $(\hat{u},v)$, and $(\hat{u},v')$, will join two cycles of $\mathcal{F}$ and form a new factor with $k-1$ cycles. By induction, a total of $k-1$ applications of this process with the merge-or-split method will yield a full cycle. $\square$

1.4 Overview of the chapters

The rest of the material in the book is organized into eleven chapters and is built in a way that it can be used as a textbook for a course on this topic. Many chapters contain material that can be split into two lectures or some of the material can be skipped. Therefore one can prepare a course that can mention multidimensional arrays only briefly or not mention them at all. Similarly, one can reduce the enumeration methods considerably. This depends on the flavor that the teachers want to give to their course. Each chapter will contain a short introduction to the sections of the chapter, as was done at the beginning of the current chapter.

Chapter 2 is devoted to the linear theory of feedback shift-register sequences. In particular, the chapter discusses the length of the cycles obtained by a given $LFSR_n$. The traditional way to compute the length is by using a polynomial representation of the feedback function and also a polynomial representation of the obtained sequences. We will present a second method to compute the length of the cycles by factoring the associated polynomial and either adding sequences associated with the factored polynomials or interleaving them. The scenario can be sometimes very clear if we add into the analysis the shift operator $\mathbf{E}$ and consider the polynomial that generates the sequence as a polynomial with the variable $\mathbf{E}$. In particular, we consider the sequences of maximum length obtained by an $LFSR_n$, called M-sequences. These sequences have many interesting properties, some of which are exhibited in this chapter. For example, these sequences are associated with patterns with distinct differences that will be also discussed in the chapter. M-sequences are generated by primitive polynomials. These polynomials and irreducible polynomials play an important role in our exposition and they will be discussed in this chapter.

In Chapter 3 the nonlinear theory of feedback shift-register sequences is considered. In particular, we introduce a few FSR_ns and consider the number of cycles that are produced by these feedback shift registers (some of these FSR_ns are linear, but the number of cycles in these FSR_n is not computed by linear methods). Counting the number of irreducible polynomials and primitive polynomials of a given degree will be done in this chapter. We continue and present some enumeration methods for the number of cycles with some given properties and given length in de Bruijn graphs. In particular, we consider the number of self-dual cycles. Finally, in this chapter, we show what is the largest number of cycles induced by a state diagram of an FSR_n.

Constructions of de Bruijn sequences are the topic of Chapter 4. First, constructions based on Eulerian cycles in the de Bruijn graph are presented. Based on these constructions we will present an enumeration method to compute the total number of de Bruijn sequences in G_n. Next, a recursive construction based on a homomorphism of the de Bruijn graph, called the **D**-morphism (associated with the operator **D**) will be presented. The properties of this operator in the binary case and the non-binary case will be discussed. The feedback functions of the sequences obtained by the recursive construction will be given. Finally, constructions based on Hamiltonian cycles in the graph are presented. The basic idea of the constructions is to start with a factor in the graph and merge its cycles via the merge-or-split method. This method is also the basis for the first naive constructions of de Bruijn sequences. Some other constructions that compete with the first naive construction will be discussed. In these methods, the next symbol of the sequence is determined by some rule associated with the last constructed n symbols of the sequence. These methods include the prefer one (for the next bit) construction, the prefer same construction, and the prefer opposite construction.

The linear complexity of sequences whose length is a power of a prime is the topic of Chapter 5. The linear complexity is the length of the shortest LFSR that generates the sequence. The linear complexity is an important parameter to determine the predictability of the sequence (low linear complexity makes the sequence of high predictability, but high linear complexity does not guarantee low predictability). In particular, the linear complexity of de Bruijn sequences is discussed. The basic concepts of linear complexity will be presented and an efficient algorithm to compute the linear complexity of sequences whose length is a power of a prime, and their alphabet is over a power of the same prime, will be discussed. Constructions for de Bruijn sequences with specified complexity, e.g., minimal and maximal, will be presented.

The linear complexity of sequences is one criterion from which sequences are classified. Chapter 6 is devoted to several other methods to classify sequences. In particular, we classify balanced sequences, i.e., binary sequences that have either the same number of *ones* and *zeroes* or the difference in their number is one. M-sequences and de Bruijn sequences form two important families of these sequences. The classifications will be based on some properties associated with these sequences. Sequences that satisfy some of the basic properties of M-sequences, which will be discussed in Chapter 2, will be classified. Classification of de Bruijn sequences by the weight of the truth table will be discussed too. Furthermore, the linear complexity and the complexity distribution of shortened de Bruijn sequences will be examined too. Generalizations of de Bruijn sequences, to sequences in which each ℓ-tuple appears the same number of times in the sequence, will be given. In this context, the derivative operator $\mathbf{D}$, will be used to classify balanced sequences based on the number of their balanced derivatives. The basic property that each n-tuple appears as a window is a constraint implied by the properties of a de Bruijn sequence. Finally, we continue and further generalize the concept of linear complexity discussed earlier to acyclic sequences of any given length. With this generalization, we will also classify linear codes and for this purpose, another measure called the depth of the sequence will be defined.

Chapter 7 completes most of the discussion on one-dimensional sequences by presenting several diverse applications of these sequences and the techniques used for these sequences. The applications will come from different areas of science. Stream ciphers form the cryptographic application of long sequences in the de Bruijn graph (such as de Bruijn sequences). Testing and verification of many Boolean functions with a small number of inputs for each function, but a large number of inputs for all the functions is the next application. It will be shown how to use sequences of irreducible polynomials for this testing. Gray codes are the subject of much research as they have many applications, on the one hand, and are of interest from a mathematical point of view, on the other hand. One family of Gray codes, namely single-track Gray codes, has some nice applications. The properties of such codes will be discussed. Two types of constructions of such codes, which resemble constructions of de Bruijn sequences, will be presented, and their properties will be discussed too. The constructions are based on either full-order sequences or full-order self-dual sequences. Proof of the nonexistence of some codes based on the linear complexity of sequences will be also given. The concepts of linear complexity and depth can be applied also in combinatorial games and in particular in one game called the rotating-table game. An analysis of a winning strategy for the game will be presented.

Chapter 8 is dedicated to two applications associated with DNA. The first one is associated with the human genome project. The human genome project started in the late 1980s and had as its target to find the human genome from subsequences of the genome. Many methods were proposed for this task. In particular, one of the most efficient methods is based on paths in the de Bruijn graph, and much research was done in this direction. An introduction and a brief illustration of the methods and in particular those using the de Bruijn graph will be presented. The second application is associated with storage. The amount of storage required at the beginning of the 21st century is increasing day to day and new reliable and space-saving storage is required. It will not be too long before we will suffer from a shortage of storage and especially reliable storage. Non-volatile memories and in particular DNA storage was suggested as a solution to the high demand for storage. This is the second topic of this chapter. Some solutions for these applications require certain generalizations of the de Bruijn graph. The first one is to consider a de Bruijn sequence based only on the words of the same weight w (or words with a few consecutive weights) in the graph. As the subgraph of G_n that contains these words of length n and weight w is not a connected graph, the words of length $n - 1$ and weights $w - 1$ and w, which can represent these words of length n and weight w, will be considered in G_{n-1}. The existence of an Eulerian cycle in this subgraph will be shown. Another type of problem that will be discussed is the reconstruction of sequences from their subsequences (either some complete windows in the sequence or some projections of the sequence). Finally, a few types of codes in which suffixes do not overlap prefixes will be considered too. These types of codes include comma-free codes and non-overlapping codes and they have some direct connection to our overall exposition.

Chapter 9 presents generalizations of the one-dimensional sequences into two-dimensional arrays. In particular, there is a generalization for de Bruijn sequences, called de Bruijn arrays or perfect maps, an analog generalization for M-sequences, called pseudo-random arrays, and an analog for shortened de Bruijn sequences that will be called shortened de Bruijn arrays (or shortened perfect maps). There are also associated generalizations for the de Bruijn graph. Direct constructions and recursive constructions for such arrays will be presented. There are also several generalizations for patterns with distinct differences into two dimensions. These generalizations are also motivated by several applications that are presented in Chapter 10.

The first application presented in Chapter 10 is to use two-dimensional patterns with window property in a global positioning system (GPS), i.e., to design an instrument for self-location in some areas. The constructed arrays have a two-dimensional window property. Another application that will be discussed is in assigning cryptographic keys for secure communication in a sensor network. Patterns with distinct differences play an important role in the construction of such schemes and systems. Finally, in this chapter, we present a method of folding a one-dimensional sequence with either the span n property or with distinct differences into a two-dimensional array (or a multi-dimensional array) with similar properties as in the one-dimensional sequence. These directions will force us to consider some results associated with integer lattices and tiling of the multi-dimensional grid.

Chapter 11 is devoted to one generalization of the de Bruijn graph into a family of digraphs in which there is a unique directed path of length n between any pair of vertices. These graphs can be described also in terms of the properties of their adjacency matrices. Several properties of these graphs are discussed and based on some of these properties an algorithm to decide whether two such graphs are isomorphic or not, will be given. Constructions for large sets of these graphs will be presented and an asymptotic enumeration of the number of non-isomorphic graphs obtained by the constructions is provided. In this chapter we will call the line graph also the integral graph, since it can be described as an operation on the graph and the inverse operation can be regarded as its derivative. The properties of these two operations will be discussed. It will be also shown how factors in these graphs can be associated with state diagrams of nonsingular FSR_ns. These graphs can be also used to form a class of networks in one of the important families of interconnection networks.

Interconnection networks are the topic of Chapter 12. The chapter takes a completely different direction, defining networks that are used for parallel computations. The discussion will be on networks that are either defined from the de Bruijn graph or defined by a modification and a generalization of the de Bruijn graph. This chapter is concentrated on the definitions of these networks and in particular the shuffle-exchange network, which is derived directly from the de Bruijn graph. Multistage interconnection networks, in which the basic element is a 2×2 switching box, such as the omega network, the flip network, the modified data manipulator network, and the baseline network, will be defined. It will be shown that these networks are isomorphic. The connection between these networks and the graphs with a unique path property will be discussed. Interconnection networks will be considered also for the realization of permutations and the minimum number of stages as well as the minimum number of switching elements in a permutation network, will be considered too. Finally, in this chapter, we consider problems associated with the implementations of some of these networks with hardware. Implementation of interconnection networks is on a board with appropriate wires between vertices that are connected in the graph. Such an implementation forms an embedding of the network into another graph and it is a layout of the network into electronic chips. Several layouts for the shuffle-exchange network will be presented.

1.5   Notes

This chapter is very general and since it presents various concepts from different areas, some material from each section appears in other books. Sometimes, the definitions might be slightly different in the different books, but they all lead to the same results. The chapter is an introduction to various concepts, but many other important concepts are used in the book. For example, the well-known Stirling's formula, e.g., see MacWilliams and Sloane [47, p. 319], is required for the approximation of formulas.

Theorem 1.21. *For large enough n we have*

$$n! \sim \sqrt{2\pi n} \left(\frac{n}{e}\right)^n$$

and hence

$$\log_2(n!) = n \cdot \log_2 n - n \cdot \log_2 e + \Theta(\log_2 n),$$

where e is Euler's number, i.e.,

$$e = \lim_{n \to \infty} \left(1 + \frac{1}{n}\right)^n \approx 2.71828\ldots .$$

Section 1.1. This section presents the first preliminaries in groups, rings, and finite fields. These are very important algebraic structures for some parts of the book. An excellent reference book for algebra was written by MacLane and Birkhoff [46] and also such material can be found in the excellent book of Lidl and Niederreiter [35].

An important result that is frequently used in the discussion on polynomials is the fundamental theorem of algebra. In our exposition, it is required in the following setup.

Theorem 1.22. *An irreducible polynomial $f(x)$ of degree n over $\mathbb{F}_q$ has n distinct roots in the field $\mathbb{F}_{q^n}$.*

Basic properties of finite fields will be used throughout the book. An excellent reference on finite fields is the book by Lidl and Niederreiter [35]. Such material also exists in most books on coding theory, e.g., see MacWilliams and Sloane [47]. As an example, we have the following theorem whose proof can be found in MacWilliams and Sloane [47, p. 107].

Theorem 1.23. *If p is a prime number and m a positive integer, then $x^{p^m} - x$ is equal to the product of all monic polynomials irreducible over $\mathbb{F}_p$, whose degree divides m.*

Similarly to Theorem 1.23, the following theorem can be also proved, as was mentioned in MacWilliams and Sloane [47, p. 107].

Theorem 1.24. *If q is a prime power and m a positive integer, then $x^{q^m} - x$ is equal to the product of all monic polynomials irreducible over $\mathbb{F}_q$, whose degree divides m.*

Corollary 1.16. *If q is a prime power, m a positive integer, and α a primitive element in $\mathbb{F}_{q^m}$, then*

$$x^{q^m-1} - 1 = \prod_{i=0}^{q^m-1} (x - \alpha^i).$$

Burnside's lemma was credited to the work of Burnside [13,14], but Burnside himself wrote that the credit for the result should be given to Frobenius [22].

The results we have presented on number theory are classic ones and appear in any basic book on number theory such as Andrews [5], Niven and Zuckerman [50], and Rosen [56]. A large number of open problems in basic number theory and references to the work that was done on these problems can be found in the book by Guy [27].

One step in finding primes is the well-known Dirichlet's theorem.

Theorem 1.25. *If a and b are two positive relatively prime integers, then the arithmetic progression of terms $ai + b$, for $i = 1, 2, ...$, contains an infinite number of primes.*

The proof of Theorem 1.25 is not simple and it is also beyond what is generally proved in the basic books on number theory. A proof of the theorem can be found in Ireland and Rosen [31], Rosen [56], and Serre [57]. Even a simple case of the theorem such as $b = 1$ requires a slightly complicated proof in Niven and Zuckerman [50, pp. 226–228]. What is the gap between two consecutive primes? This question occupied the thoughts of mathematicians for more than a hundred years, e.g., see Tchudakoff [59]. A very important result in this direction was given by Ingham [30] from which many consequences were obtained. Such a consequence is the following result proved by Cheng [15].

Theorem 1.26. *For any given integer n sufficiently large, there is always a prime p such that $n < p < n + n^{5/8}$.*

It is worthwhile noting that Corollary 1.3 can be also proved by using the well-known combinatorial principle of inclusion–exclusion, see for example van Lint and Wilson [38, p. 89].

The Euler function $\phi(\cdot)$ and the Möbius function $\mu(\cdot)$ are basic functions in number theory and their analysis and applications can be found in any basic book on number theory and also in some combinatorics books.

Theorems 1.13 and 1.14 are well-known theorems in number theory. The converse to these theorems was proved later by Kelly [32] in two associated theorems as follows.

Theorem 1.27. *Let p be an integer of the form $4k + 1$. Let the 4k least positive residues modulo p be partitioned into two mutually exclusive classes of 2k elements each. Call these two classes A and B. Suppose that A and B may be chosen so that:*

(a) $1 \in A$.
(b) *For every choice of $a^* \in A$, the set $a^* + B$ contains k elements of A and k elements of B.*
(c) *For every choice of $b^* \in B$, the set $b^* + A$ contains k elements of A and k elements of B.*

Then,

(1) *p is a prime.*
(2) *A consists of the quadratic residues modulo p and B consists of the quadratic non-residues modulo p.*

Theorem 1.28. *Let p be an integer of the form $4k - 1$. Let the $4k - 2$ least positive residues modulo p be partitioned into two mutually exclusive classes of $2k - 1$ elements each. Call these two classes A and B. Suppose that A and B may be chosen so that:*

(a) $1 \in A$.
(b) *For every choice of $a^* \in A$, the set $a^* + B$ contains 0, $k - 1$ elements of A and $k - 1$ elements of B.*

(c) *For every choice of $b^* \in B$, the set $b^* + A$ contains 0, $k - 1$ elements of A and $k - 1$ elements of B.*

Then,

(1) *p is a prime.*
(2) *A consists of the quadratic residues modulo p and B consists of the quadratic non-residues modulo p.*

There are many interesting families of primes of which some will be mentioned and used in the book. One such family is the set of Mersenne primes. A prime number of the form $2^n - 1$ is called a ***Mersenne prime***. Mersenne primes are called after the French mathematician and monk Martin Mersenne who studied these primes in the 17th century. If $2^n - 1$ is a prime number, then n must be a prime. The main interest in these primes arose since the test for their primality is easier than for other possible primes. Such tests have been known for more than a hundred years, see for example Lehmer [34] and Lucas [45], and there is a volunteers project "great internet Mersenne prime search" (GIMPS) to find Mersenne primes. Other primality tests for Mersenne primes were considered during the years, e.g., see Gross [26]. Hence, it is no surprise that the largest known primes are Mersenne primes. As of 2022, there are 51 known Mersenne primes. The last interesting fact about Mersenne primes is their connection with perfect numbers. A number n is called a ***perfect number*** if n equals the sum of its divisor. It is well known that n is an even perfect number if and only if n is of the form $(2^p - 1)2^{p-1}$, where $2^p - 1$ is a Mersenne prime. It is not known whether there exists an odd perfect number greater than 1. This is a topic of extensive research, see Ochem and Rao [51], Pollack and Shevelev [54] and references in these papers and papers that cite them.

Section 1.2. Coding theory and the theory of sequences are parts of two related areas in information theory. We already mentioned the excellent book on error-correcting codes by MacWilliams and Sloane [47]. There are other books on coding theory, such as those of Berlkamp [7], Blahut [10], Blake and Mullin [11], Etzion [19], Lin and Costello [36], McEliece [49], Pless [53], and Roth [55].

Graph theory is also a very rich discipline and a sample of the theory can be found in many books such as those of van Lint and Wilson [38], West [61], and Wilson [62]. The research area of sequences is rich as sequences are also used in many disciplines that are not connected at all to our exposition. For example, one can consult a book on the combinatorics of words, e.g., see the volumes under the pseudonym Lothaire [41–44]. The history with the basic results of combinatorics on words was given by Berstel and Perrin [9].

Although this book is not about algorithms, it will be impossible to avoid considering some of the associate algorithms, e.g., Euclid's algorithm or algorithms for constructing de Bruijn cycles. Algorithms to find Hamiltonian paths and Eulerian paths are two types of algorithms that will be mentioned. In a general graph, the complexity of finding an Eulerian path, if such a path exists, depends on the data structure that is used. The process can be done in time complexity $O(|E|)$. On the other hand, finding a Hamiltonian path in a general graph is considered to be a difficult problem and it is in the class of problems called NP-complete, see Garey and Johnson [23]. Nevertheless, in specific graphs like G_n, both problems are equivalent and are relatively easy to solve.

Section 1.3. The de Bruijn graph G_n was defined by de Bruijn [12] and in parallel by Good [25]. The number of span n de Bruijn sequences that will be proved in Corollary 4.3 was computed by de Bruijn [12].

Theorem 1.29. *The number of span n de Bruijn sequences is* $2^{2^{n-1}-n}$.

The generalization to $G_{\sigma,n}$ was done by van Aardenne-Ehrenfest and de Bruijn [1]. The de Bruijn graph and its sequences are the concepts that tie together all the topics in our book. No book is devoted to the graph, although some books have a chapter or two associated with the graph. Examples of such books are those by Hall [29] and van Lint and Wilson [38]. There is no book whose title and content indicate that the de Bruijn graph is at its center. There are books on the genome assembly that have chapters on the de Bruijn graph or mention the concept, e.g., see Compeau and Pevzner [16] and Mäkinen, Belazzougui, Cunial, and Tomescu [48]. These chapters are noted since one of the applications of the de Bruijn graph is for the genome assembly; subgraphs of the de Bruijn graph are considered for this application and more specifically, paths and cycles in the graph are considered.

As for feedback shift registers and their sequences, there is no other excellent book like that of Golomb [24]. The book also contains some material associated with the de Bruijn graph. However, the material in this book is limited in scope, and since more than fifty years have passed from the time that it was written and only some limited material was added to the book in later editions, an updated and more comprehensive book was required. Only the binary case is covered in the book of Golomb [24] and the current book also expands in this direction. As was mentioned when the definition of a shift register was given, the shift register is made from some flip-flops. These are electronic elements in logic design that were considered in many books, such as that by Patterson and Hennessy [52].

Section 1.4. The book contains diverse material from different areas of mathematics, computer science, and electrical engineering. Some of the material in this book is in the front line of the research and some material is slightly outdated. However, the slightly outdated material also has its mathematical interest and it includes techniques that might be used in future areas of research, as the general topic of the book has a habit to reinvent itself over and over again in new research areas. There is also new material in some of the chapters. For example, the material in Chapters 2 and 3 is classic shift-register theory that is covered in Golomb [24], but the first edition of that book was published in 1967 and only a "selective update" was added for the third edition. We have added new material that is not covered in that book. Some of the enumerations did not appear in Golomb [24], patterns with distinct differences were not covered there, and the material on sequences from powers of irreducible polynomials is presented from a different point of view. Chapters 4, 5, and 6 continue with some classic theory on sequences and in particular de Bruijn sequences and some more recent material. Chapter 7 presents various applications in which the graph, its sequences, and their generalizations, are involved. These applications show that the theory that will be developed in the book can be used in practice. Chapter 8 is an example of more recent applications of the graph and its sequences. These applications have a direct line to DNA, either as DNA sequencing used in the genome assembly project or for new storage technology, namely DNA storage. Chapters 9 and 10 present the two-dimensional sequences, material that currently does not appear in any book. Chapters 11 and 12 are motivated by research that was done on the slightly outdated interconnection networks, but they contain material of theoretical value as well as some material that is still of use in communication for out-of-the-universe space. Also, the material of Chapter 11 has its own algebraic and combinatorial interest.

Many topics on the de Bruijn graph and its sequences were not covered by this book. For example, a nice connection between the de Bruijn graph and the well-investigated fix-free codes was considered in Ahlswede, Balkenhol, and Khachatrian [2] and Deppe and Schnetteler [18]. A partial list of some other references for such papers includes Alhakim [3], Alhakim and Akinwande [4], Au [6], Bermond and Fraigniaud [8], Deng and Wu [17], Etzion and Bar-David [20], Fraigniaud and Gauron [21], Hales and Hartsfield [28], Klasing, Monien, Peine, and Stöhr [33], Lin and Zhang [37], Liu, Hildebrandt, and Cavin [39], Liu, Lee, and Jordan [40], Tan, Xu, and Qi [58], and Wang, Zheng, Wang, and Qi [60].

References

[1] T. van Aardenne-Ehrenfest, N.G. de Bruijn, Circuits and trees in ordered linear graphs, Simon Steven 28 (1951) 203–217.

[2] R. Ahlswede, B. Balkenhol, L. Khachatrian, Some properties of fixed-free codes, in: Proc. 1st Int. Seminar on Coding Theory and Combinatorics, Thahkadzor, Armenia, 1996, pp. 20–33.

[3] A. Alhakim, Spans of preference functions for de Bruijn sequences, Discrete Appl. Math. 160 (2012) 992–998.

[4] A. Alhakim, M. Akinwande, A multiple stream generator based on de Bruijn digraph homomorphisms, J. Stat. Comput. Simul. 79 (2009) 1371–1380.

[5] G.E. Andrews, Number Theory, Saunders, Philadelphia, 1971.

[6] Y.H. Au, Generalized de Bruijn words for primitive words and powers, Discrete Math. 338 (2015) 2320–2331.

[7] E.R. Berlkamp, Algebraic Coding Theory, McGraw-Hill, New York, 1968, World Scientific, Singapore, 2015.

[8] J.-C. Bermond, P. Fraigniaud, Broadcasting and gossiping in de Bruijn networks, SIAM J. Comput. 23 (1994) 212–225.

[9] J. Berstel, D. Perrin, The origins of combinatorics on words, Eur. J. Comb. 28 (2007) 996–1022.

[10] R.E. Blahut, Theory and Practice of Error Control Codes, Addison-Wesley, Reading, MA, 1983.

[11] I.F. Blake, R.C. Mullin, The Mathematical Theory of Coding, Academic Press, New York, 1975.

[12] N.G. de Bruijn, A combinatorial problem, Ned. Akad. Wet. 49 (1946) 758–764.

[13] W. Burnside, Theory of Groups of Finite Order, Cambridge Univ. Press, London, 1897, Cambridge Univ. Press, United Kingdom, 2012.

[14] W. Burnside, Theory of Groups of Finite Order, Cambridge Univ. Press, London, 1911, Dover, New York, 1955.

[15] Y.-Y.F.-R. Cheng, Explicit estimate on primes between consecutive cubes, Rocky Mt. J. Math. 40 (2010) 117–153.

[16] P. Compeau, P. Pevzner, Bioinformatics Algorithms: An Active Learning Approach, Active Learning Publishers, La Jolla, CA, 2018.

[17] A. Deng, Y. Wu, de Bruijn digraphs and affine transformations, Eur. J. Comb. 26 (2005) 1191–1206.

[18] C. Deppe, H. Schnetteler, On q-ary fix-free codes and directed de Bruijn graphs, in: IEEE Int. Symp. Infor. Theory (ISIT), 2016, pp. 1482–1485.

[19] T. Etzion, Perfect Codes and Related Structures, World Scientific, Singapore, 2022.

[20] T. Etzion, I. Bar-David, An explicit construction of Euler circuits in shuffle nets and related networks, Networks 22 (1992) 523–529.

[21] P. Fraigniaud, P. Gauron, D2B: a de Bruijn based content-addressable network, Theor. Comput. Sci. 355 (2006) 65–79.

[22] F.G. Frobenius, Ueber die Congruenz nach einem aus zwei endlichen Gruppen gebildeten Doppelmodul, J. Reine Angew. Math. 101 (1887) 273–299.

[23] M.R. Garey, D.S. Johnson, Computers and Intractability: A Guide to the Theory of NP-Completeness, W. H. Freeman and Company, New York, 1979.

[24] S.W. Golomb, Shift Register Sequences, Holden Day, San Francisco, 1967, Aegean Park, Laguna Hills, CA, 1980, World Scientific, Singapore, 2017.

[25] I.J. Good, Normally recurring decimals, J. Lond. Math. Soc. 21 (1946) 167–169.

[26] B.H. Gross, An elliptic curve test for Mersenne primes, J. Number Theory 110 (2005) 114–119.

[27] R.K. Guy, Unsolved Problems in Number Theory, Springer-Verlag, New York, 1981.

[28] A.W. Hales, N. Hartsfield, The directed genus of the de Bruijn graph, Discrete Math. 309 (2009) 5259–5263.

[29] M. Hall, Combinatorial Theory, Blaisdell, Waltham, MA, 1967, John Wiley and Sons, Inc., New York, 1998.

[30] A.E. Ingham, On the difference between consecutive primes, Q. J. Math., Oxford Ser. 8 (1937) 255–266.

[31] K. Ireland, M. Rosen, A Classical Introduction to Modern Number Theory, Springer-Verlag, New York, 1982.

[32] J.B. Kelly, A characteristic property of quadratic residues, Proc. Am. Math. Soc. 5 (1954) 38–46.

[33] R. Klasing, B. Monien, R. Peine, E.A. Stöhr, Broadcasting in butterfly and de Bruijn networks, Discrete Appl. Math. 53 (1994) 183–197.

[34] D.H. Lehmer, On Lucas's test for the primality of Mersenne's numbers, J. Lond. Math. Soc. 10 (1935) 162–165.

[35] R. Lidl, H. Niederreiter, Introduction to Finite Fields and Their Applications, Cambridge Univ. Press, Cambridge, United Kingdom, 1994.

[36] S. Lin, D.J. Costello Jr., Error Correcting Coding: Fundamentals and Applications, Prentice-Hall, Upper Saddle River, NJ, 2004.

[37] R. Lin, H. Zhang, Matching preclusion and conditional edge-fault Hamiltonicity in binary de Bruijn graphs, Discrete Appl. Math. 233 (2017) 104–117.

[38] J.H. van Lint, R.M. Wilson, A Course in Combinatorics, Cambridge University Press, New York, 2001.

[39] W. Liu, T.H. Hildebrandt, R. Cavin III, Hamiltonian cycles in the shuffle-exchange network, IEEE Trans. Comput. 38 (1989) 745–750.

[40] G. Liu, K.Y. Lee, H.F. Jordan, TDM and TWDM de Bruijn networks and shuffle nets for optical communications, IEEE Trans. Comput. 46 (1997) 695–701.

[41] M. Lothaire, Combinatorics on Words, Encyclopedia of Mathematics and Its Applications, vol. 17, Addison-Wesley, Reading, MA, 1983.

[42] M. Lothaire, Combinatorics on Words, corrected reprint, Cambridge University Press, MA, 1997.

[43] M. Lothaire, Algebraic Combinatorics on Words, Encyclopedia of Mathematics and Its Applications, vol. 90, Cambridge University Press, MA, 2002.

[44] M. Lothaire, Applied Combinatorics on Words, Cambridge University Press, MA, 2005.

[45] E. Lucas, Nouveaux théorèmes d'Arithmétique supérieure, C. R. Acad. Sci. Paris 83 (1876) 1286–1288.

[46] S. MacLane, G. Birkhoff, Algebra, Chelsea Publishing Company, New York, 1988.

[47] F.J. MacWilliams, N.J.A. Sloane, The Theory of Error-Correcting Codes, North-Holland, Amsterdam, 1977.

[48] V. Mäkinen, D. Belazzougui, F. Cunial, A.I. Tomescu, Genome-Scale Algorithm Design, Cambridge Univ. Press, Cambridge, United Kingdom, 2015.

[49] R.J. McEliece, The Theory of Information and Coding, Addison-Wesley, Reading, MA, 1977, Cambridge Univ. Press, Cambridge, United Kingdom, 2002.

[50] I. Niven, H.S. Zuckerman, An Introduction to the Theory of Numbers, Cambridge Univ. Press, United Kingdom, 1972.

[51] P. Ochem, M. Rao, Odd perfect numbers are greater than 10^{1500}, Math. Comput. 81 (2012) 1869–1877.

[52] D.A. Patterson, J.L. Hennessy, Computer Organization and Design: the Hardware/Software Interface, Morgan Kaufmann Publishers Inc., San Francisco, CA, 2017.

[53] V. Pless, Introduction to the Theory of Error-Correcting Codes, John Wiley and Sons Inc., New York, 1989.

[54] P. Pollack, V. Shevelev, On perfect and near-perfect numbers, J. Number Theory 132 (2012) 3037–3046.

[55] R.M. Roth, Introduction to Coding Theory, Cambridge Univ. Press, Cambridge, United Kingdom, 2005.

[56] K.H. Rosen, Elementary Number Theory and Its Applications, Addison-Wesley, Reading, MA, 1984.

[57] J.-P. Serre, Linear Representations of Finite Groups, Springer-Verlag, New York, 1977.

[58] L. Tan, H. Xu, W.-F. Qi, Preliminary results on the minimal polynomial of modified de Bruijn sequences, Finite Fields Appl. 50 (2018) 356–365.

[59] N.G. Tchudakoff, On the difference between two neighbouring primes numbers, Rec. Math. (Mat. Sb.) NS 1 (1936) 799–814.

[60] H.-Y. Wang, Q.-X. Zheng, Z.-X. Wang, W.-F. Qi, The minimal polynomials of modified de Bruijn sequences revisited, Finite Fields Appl. 68 (2020) 101735.

[61] D.B. West, Introduction to Graph Theory, Prentice-Hall, Upper Saddle River, NJ, 1996.

[62] R.J. Wilson, Introduction to Graph Theory, Pearson Education Limited, Harlow, United Kingdom, 2010.

Chapter 2

LFSR sequences

Polynomials, M-sequences, cycles, difference sets

This chapter is devoted to the most important shift registers, the linear shift registers, and to the linear theory of shift registers. There are several interesting questions about the state diagrams of these shift registers. The first is about the length of cycles in the state diagrams of the related shift registers. This question is partially answered in Section 2.1 by using a polynomial representation for both the shift-register functions that are linear and also for the sequences generated by the associated functions of the shift registers.

The most interesting shift-register sequences are those for which their function is derived from a primitive polynomial. Primitive polynomials form a subset of the irreducible polynomials. Sequences associated with primitive polynomials are called M-sequences (for maximal length linear shift-register sequences) or PN sequences (for pseudo-noise sequences). These sequences, which have the maximum possible length for LFSR_ns, contain all the states of the state diagram, except for the all-zeros state that is always isolated, with a self-loop, in the state diagram, since the function is linear. These sequences have many interesting properties, some of which are discussed in Section 2.2.

Section 2.3 considers sequences that are generated by powers of irreducible polynomials. We mainly concentrate on counting the number of sequences of each length that are generated from a power of an irreducible polynomial. With the initial discussion presented in Section 2.1, this completes the exposition on this topic for any given polynomial.

One of the properties that M-sequences have is a good autocorrelation function. As such they can be used to form difference sets. The type of difference set that they form is called a Hadamard difference set. These combinatorial structures are discussed in Section 2.4.

2.1 Sequence length and polynomial representation

Recall that an LFSR_n has a feedback function of the form

$$f(x) = f(x_1, x_2, \ldots, x_n) = \sum_{i=1}^{n} c_i x_{n+1-i}, \quad c_i \in \mathbb{F}_q.$$

Sequences and the de Bruijn Graph. https://doi.org/10.1016/B978-0-44-313517-0.00008-1

As only nonsingular FSRs are considered we have that $c_n \neq 0$. Moreover, w.l.o.g. we will assume later that $c_n = 1$ and it will imply that our polynomials in the following are monic. In other words, similar to Theorem 1.19 we have the following theorem.

Theorem 2.1. *An* LFSR$_n$ *over* $\mathbb{F}_q$ *is nonsingular if and only if its feedback function* $f(x_1, x_2, \ldots, x_n)$ *satisfies*

$$f(x_1, x_2, \ldots, x_n) = cx_1 + g(x_2, \ldots, x_n), \quad c \in \mathbb{F}_q \setminus \{0\},$$

where $g(x_2, \ldots, x_n)$ *is any linear function with* $n - 1$ *variables over* $\mathbb{F}_q$.

Corollary 2.1. *The number of nonsingular* LFSR$_n$ *over* $\mathbb{F}_q$ *is* $(q - 1)q^{n-1}$. *The number of nonsingular binary* LFSR$_n$ *is* 2^{n-1}.

Note that sometimes we will use FSR$_n$ for plural. When used as plural it will be readily understood from the context.

An LFSR$_n$ sequence $\{a_k\}_{k=-n}^{\infty}$, where the initial state is $(a_{-n}, a_{-n+1}, \ldots, a_{-1})$, satisfies a linear recursion

$$a_k = \sum_{i=1}^{n} c_i a_{k-i}, \quad k = 0, 1, \ldots. \tag{2.1}$$

The **generating function** of $\{a_k\}$ is defined by

$$G(x) = \sum_{k=0}^{\infty} a_k x^k.$$

The **characteristic polynomial** of the sequence $\{a_k\}$ is defined by

$$c(x) = 1 - \sum_{i=1}^{n} c_i x^i. \tag{2.2}$$

We say that the characteristic polynomial $c(x)$ **generates** the sequence $\{a_k\}$.

Theorem 2.2. *Let* $\{a_0, a_1, a_2, \ldots\}$ *be a shift-register sequence with initial state* $(a_{-n}, a_{-n+1}, \ldots, a_{-1})$. *If*

$$\gamma(x) \triangleq \sum_{i=1}^{n} c_i x^i (a_{-i} x^{-i} + a_{-i+1} x^{-i+1} + \cdots + a_{-1} x^{-1})$$

$$= \sum_{i=1}^{n} c_i x^i \sum_{k=-i}^{-1} a_k x^k,$$

then $\deg \gamma(x) \leq n - 1$ *and* $G(x) = \frac{\gamma(x)}{c(x)}$.

Proof. The fact that deg $\gamma(x) \leq n - 1$ is easily verified by the structure of $\gamma(x)$. Now,

$$G(x) = \sum_{k=0}^{\infty} a_k x^k = \sum_{k=0}^{\infty} \left(\sum_{i=1}^{n} c_i a_{k-i} \right) x^k = \sum_{i=1}^{n} c_i x^i \sum_{k=0}^{\infty} a_{k-i} x^{k-i}$$

$$= \sum_{i=1}^{n} c_i x^i \sum_{k=-i}^{\infty} a_k x^k = \sum_{i=1}^{n} c_i x^i \left(G(x) + \sum_{k=-i}^{-1} a_k x^k \right).$$

Therefore

$$G(x) = G(x) \sum_{i=1}^{n} c_i x^i + \sum_{i=1}^{n} c_i x^i \sum_{k=-i}^{-1} a_k x^k = G(x) \sum_{i=1}^{n} c_i x^i + \gamma(x).$$

Hence,

$$G(x) \left(1 - \sum_{i=1}^{n} c_i x^i \right) = \gamma(x),$$

which implies that

$$G(x) = \frac{\gamma(x)}{1 - \sum_{i=1}^{n} c_i x^i} = \frac{\gamma(x)}{c(x)},$$

and the proof of the theorem is completed. $\qquad\qquad\square$

By the definition of $\gamma(x)$ in Theorem 2.2 and since $\deg \gamma(x) \leq n - 1$, it follows that

$$\gamma(x) = \sum_{i=1}^{n} c_i x^i (a_{-i} x^{-i} + a_{-i+1} x^{-i+1} + \cdots + a_{-1} x^{-1}) = \sum_{i=0}^{n-1} \gamma_i x^i. \quad (2.3)$$

The coefficients of the polynomial $\gamma(x)$ in Eq. (2.3) can be written in a matrix form as follows:

$$\begin{pmatrix} \gamma_0 \\ \gamma_1 \\ \vdots \\ \gamma_{n-2} \\ \gamma_{n-1} \end{pmatrix} = \begin{pmatrix} c_n & c_{n-1} & \cdots & c_2 & c_1 \\ 0 & c_n & \cdots & c_3 & c_2 \\ \vdots & \vdots & \ddots & \vdots & \vdots \\ 0 & 0 & \cdots & c_n & c_{n-1} \\ 0 & 0 & \cdots & 0 & c_n \end{pmatrix} \cdot \begin{pmatrix} a_{-n} \\ a_{-n+1} \\ \vdots \\ a_{-2} \\ a_{-1} \end{pmatrix}. \quad (2.4)$$

If the LFSR$_n$ is nonsingular, then $c_n \neq 0$ (and w.l.o.g. $c_n = 1$), and hence the $n \times n$ matrix is invertible. Therefore for every polynomial $h(x)$ whose degree is less than n, there exists an initial state for which $\gamma(x) = h(x)$. This polynomial is obtained by solving the associated equations system obtained from Eq. (2.4).

Each sequence of a nonsingular FSR_n over $\mathbb{F}_q$ is periodic since there are only q^n possible states. Hence, the period of the sequence is $\pi \leq q^n$ and for $LFSR_n$ we have that $\pi \leq q^n - 1$ since the all-zeros state is in the separate sequence $[0]$.

Definition 2.1. The *exponent* $e(h(x))$ of a polynomial $h(x)$ over $\mathbb{F}_q$ is the least integer e such that $h(x)$ divides $x^e - 1$.

The next result is a key theorem for computing the length of the sequences generated by an $LFSR_n$. It presents the connection between the exponent of a polynomial and the period of the sequences that it generates.

Theorem 2.3. *Let $\{a_k\}$ be a nonzero sequence over $\mathbb{F}_q$ whose characteristic polynomial is $c(x)$. If g.c.d.$(\gamma(x), c(x)) = 1$, then $\pi(\{a_k\}) = e(c(x))$.*

Proof. Denote $\pi(\{a_k\})$ by π. By Theorem 2.2, the definition of $G(x)$, and since $\frac{1}{1-x} = \sum_{i=0}^{\infty} x^i$, it follows that

$$\frac{\gamma(x)}{c(x)} = G(x) = \sum_{k=0}^{\infty} a_k x^k = \left(\sum_{k=0}^{\pi-1} a_k x^k\right)\left(\sum_{i=0}^{\infty} x^{i\pi}\right) = \frac{\sum_{k=0}^{\pi-1} a_k x^k}{1 - x^\pi} = \frac{A(x)}{1 - x^\pi}.$$

Hence,

$$\gamma(x)(1 - x^\pi) = A(x)c(x).$$

Since g.c.d.$(\gamma(x), c(x)) = 1$, it follows that $c(x)$ divides $1 - x^\pi$, which implies that $e(c(x)) \leq \pi(\{a_k\})$.

Let e be the exponent of $c(x)$. Since $c(x)$ divides $1 - x^e$, the degree of $\gamma(x)$ is at most $n - 1$, and the degree of $c(x)$ is n, it follows that we can define

$$\beta(x) \triangleq \gamma(x)\frac{1 - x^e}{c(x)} = \sum_{k=0}^{e-1} \beta_k x^k.$$

Now, by Theorem 2.2, we have that

$$\sum_{k=0}^{\infty} a_k x^k = G(x) = \frac{\gamma(x)}{c(x)} = \frac{\beta(x)}{1 - x^e} = \beta(x)\sum_{i=0}^{\infty} x^{ie}.$$

Hence, $\pi(\{a_k\}) \leq e(c(x))$.

Thus we have that $\pi(\{a_k\}) = e(c(x))$. $\qquad\qquad\square$

Corollary 2.2. *If $c(x)$ is an irreducible polynomial, then the period of its associated sequence $\{a_k\}$ is the same for each initial state, except for the all-zeros state.*

Corollary 2.3. *If $c(x)$ is an irreducible polynomial of degree n, over $\mathbb{F}_q$, then all the nonzero sequences that it generates have the same period, which is a factor of $q^n - 1$.*

Corollary 2.4. *If $2^n - 1$ is a prime, then every irreducible polynomial of degree n, over $\mathbb{F}_2$, corresponds to a binary linear shift-register sequence of maximum length, i.e., length $2^n - 1$.*

Assume that a sequence contains all the $q^n - 1$ nonzero states before it returns to the initial state. This implies that each nonzero n-tuple appears as a window of length n in the sequence. Hence, we can choose the n-tuple (10^{n-1}) as the initial state. This implies that $\gamma(x) = c_n = 1$, g.c.d.$(\gamma(x), c(x)) = 1$ and the period of the sequence $q^n - 1$ is the exponent of $c(x)$.

Theorem 2.4. *If the period of an LFSR_n sequence is $q^n - 1$, then its characteristic polynomial $c(x)$ is irreducible.*

Proof. Assume that $c(x)$ is a reducible polynomial and $c(x) = s(x)t(x)$, where g.c.d.$(s(x), t(x)) = 1$. Hence, by Theorem 1.6 there exist two polynomials $\alpha(x)$ and $\beta(x)$, where $\deg \alpha(x) < \deg s(x)$ and $\deg \beta(x) < \deg t(x)$, such that

$$1 = \alpha(x)t(x) + \beta(x)s(x).$$

Therefore

$$\frac{1}{c(x)} = \frac{\alpha(x)}{s(x)} + \frac{\beta(x)}{t(x)}.$$

Assume that $s(x)$ and $t(x)$ have degrees $n_1 > 0$ and $n_2 > 0$, respectively, where $n = n_1 + n_2$, $\frac{\alpha(x)}{s(x)}$ is a power series with a period π_1, which is at most $q^{n_1} - 1$, $\frac{\beta(x)}{t(x)}$ is a power series with period π_2, which is at most $q^{n_2} - 1$. Hence, $\frac{1}{c(x)} = \frac{\alpha(x)}{s(x)} + \frac{\beta(x)}{t(x)}$ is a power series with a period equal to $[\pi_1, \pi_2]$. Clearly, $[\pi_1, \pi_2]$ is at most $(q^{n_1} - 1)(q^{n_2} - 1) < q^n - 1$, a contradiction. If $c(x) = (s(x))^\ell$, then the claims will follow from Corollary 2.25 (proved in Section 2.3), which enumerates all the sequences obtained from this characteristic polynomial. $\square$

Definition 2.2. An irreducible polynomial of degree n over $\mathbb{F}_q$, which is a characteristic polynomial of an LFSR_n that generates a sequence of period $q^n - 1$, is called a ***primitive polynomial***. The sequence of period $q^n - 1$ that it generates is called an **M-*sequence***.

Another definition for a primitive polynomial was given in Section 1.1 and from the analysis of M-sequences it will be understood that the two definitions are equivalent. In the rest of this section and also in Section 2.3 we will complete the examination of the question regarding the number of cycles generated by a given LFSR_n whose characteristic polynomial is $c(x)$, where $c(x)$ is not an irreducible polynomial. We examine first the sequences obtained from the multiplication of some irreducible polynomials, not necessarily of the same degree.

The following simple lemma is very significant in understanding the behavior of sequences generated by an LFSR_n.

Lemma 2.1. *If $f(x_1, x_2, \ldots, x_n)$ is a function of an $LFSR_n$, $y = (y_1, y_2, \ldots, y_n)$ and $z = (z_1, z_2, \ldots, z_n)$ are two binary n-tuples, then $f(y+z) = f(y) + f(z)$, i.e.,*

$$f(y_1 + z_1, y_2 + z_2, \ldots, y_n + z_n) = f(y_1, y_2, \ldots, y_n) + f(z_1, z_2, \ldots, z_n).$$

Proof. Assume that

$$f(x_1, x_2, \ldots, x_n) = \sum_{i=1}^{n} a_i x_i, \quad a_i \in \mathbb{F}_q.$$

This implies that

$$f(y_1 + z_1, y_2 + z_2, \ldots, y_n + z_n) = \sum_{i=1}^{n} a_i (y_i + z_i)$$

$$= \sum_{i=1}^{n} a_i y_i + \sum_{i=1}^{n} a_i z_i = f(y_1, y_2, \ldots, y_n) + f(z_1, z_2, \ldots, z_n). \qquad \square$$

Corollary 2.5. *If S_1 and S_2 are two sequences of the same period that are generated by an $LFSR_n$ with a feedback function f, then $S_1 + S_2$ is also generated by the same $LFSR_n$.*

Corollary 2.6. *If $f_i(x)$, $1 \le i \le r$, are r different primitive polynomials of degree n over $\mathbb{F}_q$, then the exponent of the polynomial $\prod_{i=1}^{r} f_i(x)$ is $q^n - 1$.*

We note that Corollary 2.6 can be also proved as a consequence of Theorem 1.24 and Definition 2.1.

We consider again the linear recursion of a sequence as given in Eq. (2.1) that is associated with the characteristic polynomial $c(x)$ in Eq. (2.2). The following lemma is a direct consequence of the definition of the shift operator $\mathbf{E}$.

Lemma 2.2. *If a sequence $S = [s_0, s_1, \ldots, s_{\pi-1}]$ satisfies a linear recurrence of the degree m,*

$$s_{i+m} = a_{m-1} s_{i+m-1} + a_{m-2} s_{i+m-2} + \cdots + a_1 s_{i+1} + a_0 s_i, \quad a_i \in \mathbb{F}_q$$

for each $i \ge 0$, then

$$\mathbf{E}^m s_i = \left(\sum_{j=0}^{m-1} a_j \mathbf{E}^j \right) s_i$$

for each $i \ge 0$.

We rephrase now the definition for a polynomial that generates a sequence.

Definition 2.3. The polynomial $c(x)$ **generates the sequence** S if $c(\mathbf{E})S = 0$ or equivalently if $c(x)S(x) \equiv 0 \pmod{x^\pi - 1}$, where $S(x)$ is the generating function of S, i.e., the representation of S by a polynomial.

Definition 2.4. If $c(x)$ is the polynomial of the least degree that generates the sequence S, then $c(x)$ will be also called the **minimal polynomial** of S.

Definition 2.5. The set of nonzero sequences that are generated by a polynomial $f(x)$ will be denoted by $S(f(x))$.

Example 2.1. For the irreducible polynomial $f(x) = x^4 + x^3 + x^2 + x + 1$ over $\mathbb{F}_2$, we have that $S(f(x)) = \{[00011], [01010], [11011]\}$. ∎

By the definitions, we have that the characteristic polynomial $c(x)$ of a sequence S generates the sequence S. Since the sequence S is a cyclic sequence and $c(x)$ generates the sequence S, it follows that $c(x)$ generates any cyclic shift $\mathbf{E}^i S$ for each $i \geq 0$.

Corollary 2.7. *For any nonzero polynomial $f(x)$ and any cyclic nonzero sequence S, $f(\mathbf{E})S = 0$, if and only if the sequence S is generated by $f(x)$.*

Corollary 2.8. *If S is a nonzero cyclic sequence generated by the nonzero polynomial $f^m(x)$ and $f^{m-1}(\mathbf{E})S = R$, then S is not generated by $f^{m-1}(x)$ if and only if the sequence R is a nonzero sequence generated by $f(x)$.*

Each sequence S can be generated by several distinct polynomials, but it appears that the structure of the polynomials that generate S is determined by the minimal polynomial of S.

Lemma 2.3. *If the two polynomials $f(x)$ and $g(x)$ generate the same sequence S, then $h(x) = \text{g.c.d.}(f(x), g(x))$ also generates S.*

Proof. By Corollary 2.7, $f(x)$ generates S if and only if $f(\mathbf{E})S = 0$. Similarly, $g(x)$ generates S if and only if $g(\mathbf{E})S = 0$. By the Euclidean algorithm, there exist two polynomials $\alpha(x)$ and $\beta(x)$ such that $h(x) = \alpha(x)f(x) + \beta(x)g(x)$. Hence, $h(\mathbf{E})S = \alpha(\mathbf{E})f(\mathbf{E})S + \beta(\mathbf{E})g(\mathbf{E})S$, and therefore since $f(\mathbf{E})S = 0$ and $g(\mathbf{E})S = 0$, it follows that $h(\mathbf{E})S = 0$. Therefore by Corollary 2.7, the polynomial $h(x)$ generates S. □

Corollary 2.9. *If S is a nonzero sequence, then it has a unique minimal polynomial.*

Corollary 2.10. *If S is a nonzero sequence that is generated by the polynomial $f(x)$, and $g(x)$ is any nonzero polynomial, then S is also generated by $f(x)g(x)$.*

Corollary 2.11. *If $f(x)$ is a polynomial that generates the nonzero sequence S, then the minimal polynomial of S is a factor of $f(x)$.*

Corollary 2.12. *If $f(x)$ is an irreducible polynomial and $f^m(x)$ generates the nonzero sequence S, then the minimal polynomial that generates S is $f^k(x)$ for some $1 \leq k \leq m$.*

Lemma 2.4. *If $g(x)$ is an irreducible polynomial and R is a sequence in $\mathcal{S}(g(x))$, then for any polynomial $f(x)$, such that g.c.d.$(g(x), f(x)) = 1$, we have that $f(\mathbf{E})R \in \mathcal{S}(g(x))$.*

Proof. Suppose that $S = f(\mathbf{E})R$, for some sequence $R \in \mathcal{S}(g(x))$.

Assume first, on the contrary, that $S = \mathbf{0}$, where $\mathbf{0}$ denotes a sequence of zeros. By Corollary 2.7, this implies that $f(x)$ generates R. Since $g(x)$ also generates R, it follows by Lemma 2.3 that g.c.d.$(g(x), f(x))$ generates R. However, g.c.d.$(g(x), f(x)) = 1$ and hence g.c.d.$(g(x), f(x))$ cannot generate a nonzero sequence R, a contradiction. Thus S is a nonzero sequence.

Since $R \in \mathcal{S}(g(x))$ it follows that

$$g(\mathbf{E})S = g(\mathbf{E})(f(\mathbf{E})R) = f(\mathbf{E})(g(\mathbf{E})R) = f(\mathbf{E})\mathbf{0} = \mathbf{0}.$$

Since S is a nonzero sequence, it follows that $S = f(\mathbf{E})R \in \mathcal{S}(g(x))$. $\square$

The second part of Lemma 2.4 can be proved also by using Corollary 2.5. Lemma 2.4 leads to three interesting consequences.

Corollary 2.13. *If $g(x)$ is an irreducible polynomial and R is a nonzero sequence generated by $g(x)$, then for any polynomial $f(x)$, $f(\mathbf{E})R = \mathbf{0}$ if and only if $f(x)$ is divisible by $g(x)$.*

Proof. If $f(x)$ is divisible by $g(x)$, i.e., $f(x) = h(x)g(x)$, then

$$f(\mathbf{E})R = h(\mathbf{E})g(\mathbf{E})R = h(\mathbf{E})\mathbf{0} = \mathbf{0}.$$

Now, suppose that $f(\mathbf{E})R = \mathbf{0}$ and assume that $f(x)$ is not divisible by $g(x)$. Since $g(x)$ is an irreducible polynomial, it implies that g.c.d.$(g(x), f(x)) = 1$. Therefore by Lemma 2.4, $f(\mathbf{E})R \in \mathcal{S}(g(x))$ and hence $f(\mathbf{E})R \neq \mathbf{0}$, a contradiction. Thus $f(x)$ is divisible by $g(x)$. $\square$

Corollary 2.14. *Assume that $g(x)$ and $f(x)$ are two polynomials over $\mathbb{F}_q$. If g.c.d.$(g(x), f(x)) = 1$, then the only sequence that is generated by both $g(x)$ and $f(x)$ is the all-zeros sequence.*

The third consequence from Lemma 2.4 is a generalization of the shift-and-add property that will be discussed in the next section. This version, which is a consequence of Lemma 2.4, generalizes the one that is usually used.

Corollary 2.15. *Let S be an M-sequence generated by a polynomial $g(x)$. If $f(x)$ is any other polynomial for which g.c.d.$(g(x), f(x)) = 1$, then $f(\mathbf{E})S \simeq S$.*

Proof. Since S is an M-sequence, it follows that $g(x)$ is a primitive polynomial and S is the only sequence in $\mathcal{S}(g(x))$. The claim is now an immediate consequence from Lemma 2.4. $\qquad\square$

We continue to discuss the length of the sequences that are generated by an LFSR and the number of sequences for each length. Sequences will be also referred to as cycles since they represent cycles in the associated states diagrams.

Lemma 2.5. *Assume an* LFSR$_{n_1}$ *with a characteristic polynomial* $c_1(x)$ *generates a cycle* C_1 *whose period is* r, *i.e.,* $C_1 = [a_0 a_1 \cdots a_{r-1}]$ *and an* LFSR$_{n_2}$ *with a characteristic polynomial* $c_2(x)$ *generates a cycle* C_2 *whose period is* s, $s \geq r$, *i.e.,* $C_2 = [b_0 b_1 \cdots b_{s-1}]$. *If* g.c.d.$(c_1(x), c_2(x)) = 1$, *then the* LFSR$_{n_1+n_2}$ *with characteristic polynomial* $c_1(x)c_2(x)$ *has* g.c.d.$(r, s) = \frac{r \cdot s}{[r,s]}$ *distinct cycles of period* $[r, s]$ *of the form*

$$C_1 + \mathbf{E}^j C_2 \triangleq [a_0 + b_j, a_1 + b_{j+1}, a_2 + b_{j+2} \cdots a_{r-1} + b_{j+r-1}, a_0 + b_{j+r}, \cdots],$$

where $0 \leq j <$ g.c.d.(r, s) *and the indices in* ς_ℓ *are taken modulo* s.

Proof. For a given j, where $0 \leq j < \frac{r \cdot s}{[r,s]}$ consider the cycle $C_1 + \mathbf{E}^j C_2$. Since $c_1(x)c_2(x) = c_2(x)c_1(x)$, and by definition $c_1(\mathbf{E})C_1 = \mathbf{0}$ and $c_2(\mathbf{E})\mathbf{E}^j C_2 = \mathbf{0}$, it follows that

$$c_1(\mathbf{E})c_2(\mathbf{E})(C_1 + \mathbf{E}^j C_2) = c_2(\mathbf{E})c_1(\mathbf{E})C_1 + c_1(\mathbf{E})c_2(\mathbf{E})\mathbf{E}^j C_2 = \mathbf{0}$$

and hence $C_1 + \mathbf{E}^j C_2$ is a cycle generated by the characteristic polynomial $c_1(x)c_2(x)$.

It is straightforward to see that $[r, s]$ is a period of $C \triangleq C_1 + \mathbf{E}^j C_2$. We will show now that the least period of the cycle C is $[r, s]$. Since the defined cycle has a period that is the least common multiple of r and s, it follows that if it has some smaller period, then this period should be a divisor of $[r, s]$. Assume, on the contrary, that there exists such a period $r's't$, where $t =$ g.c.d.(r, s), r' divides r/t, and s' divides s/t. This implies that the cycle C also has a period $p' = \delta r < [r, s]$ or a period $p' = \delta s < [r, s]$, where δ divides s or δ divides r, respectively. We distinguish now between these two cases.

Case 1: $p' = \delta r < [r, s]$.

Assume first that $n_1 \geq n_2$. Consider the first n_1 bits, $(x_1, \ldots, x_{n_1})$, of C and these n_1 bits, $(x_1, \ldots, x_{n_1})$, after the period p'. These n_1 bits are obtained from the same n_1 bits of C_1 (since the period r of C_1 divides p') added to n_1 bits of C_2 that are located in two different positions of C_2 (the positions are different since p' is not a multiple of s) and since $n_2 \leq n_1$ these n_1 bits on the two different locations of C_2 cannot be the same. Therefore the addition in these two locations will yield two different strings of length n_1, contradicting the periodicity p'.

Assume now that $n_2 \geq n_1$. Consider the first n_2 bits of the cycle C and these n_2 bits after the period p'. These bits are combined from the same n_2 bits of C_1 added to n_2 bits of C_2 that are from two different positions of C_2. Therefore the addition will yield two different strings of length n_2, contradicting the periodicity p'.

Case 2: $p' = \delta s < [r, s]$.

As in Case 1, consider the first n_1 bits (or n_2 bits, respectively) of the cycle and the n_1 bits (or n_2 bits, respectively) after period p'. Now, the bits coming from C_2 are the same, while those coming from C_1 are different, yielding the same contradiction.

To complete the proof we have to show that for each $0 \leq j_2 < j_1 <$ g.c.d.(r, s) the cycles $C_1 + \mathbf{E}^{j_1} C_2$ and $C_1 + \mathbf{E}^{j_2} C_2$ are distinct cycles (not equivalent cycles, i.e., $C_1 + \mathbf{E}^{j_1} C_2 \not\simeq C_1 + \mathbf{E}^{j_2} C_2$).

Assume, on the contrary, that $C_1 + \mathbf{E}^{j_1} C_2 \simeq C_1 + \mathbf{E}^{j_2} C_2$, i.e., there exists an integer j, where $0 \leq j < [r, s]$, such that

$$\mathbf{E}^j \left(C_1 + \mathbf{E}^{j_1} C_2 \right) = C_1 + \mathbf{E}^{j_2} C_2,$$

which implies that

$$\mathbf{E}^j C_1 - C_1 = \mathbf{E}^{j_2} C_2 - \mathbf{E}^{j+j_1} C_2.$$

By Corollary 2.5, the sequence $\mathbf{E}^j C_1 - C_1$ is also a sequence generated by the polynomial $c_1(x)$ and the sequence $\mathbf{E}^{j_2} C_2 - \mathbf{E}^{j+j_1} C_2$ is also a sequence generated by the polynomial $c_2(x)$. Since g.c.d.$(c_1(x), c_2(x)) = 1$, it follows by Corollary 2.14 that the only sequence that is generated by both $c_1(x)$ and $c_2(x)$ is the all-zeros sequence. If $\mathbf{E}^j C_1 - C_1$ is the all-zeros sequence, then we have that r divides j. If $\mathbf{E}^{j_2} C_2 - \mathbf{E}^{j+j_1} C_2$ is the all-zeros sequence, then we have that s divides $j + j_1 - j_2$. Since $0 \leq j < $ g.c.d.$(r, s) \leq [r, s]$ and r divides j, it follows that $j = i \cdot r$, where $0 \leq i < \frac{[r,s]}{r}$. Similarly, since s divides $j + j_1 - j_2$, it follows that $j + j_1 - j_2 = \ell \cdot s$, i.e., $j_1 - j_2 = \ell \cdot s - i \cdot r$, where $0 \leq \ell < \frac{[r,s]}{s}$.

However, by Lemma 1.1 we have that $rs = $ g.c.d.$(r, s) \cdot [r, s]$. Moreover, by Corollary 1.1 we have that

$$|\ell \cdot s - i \cdot r| \geq \text{g.c.d.}(r, s) = \frac{rs}{[r, s]}$$

and hence since $0 \leq j_2 < j_1 < $ g.c.d.(r, s), it follows that $j_1 - j_2 \neq \ell \cdot s - i \cdot r$, a contradiction.

Thus the claims of the lemma follow. $\qquad\square$

The set of g.c.d.(r, s) cycles obtained from the two cycles C_1 and C_2 generated from the two polynomials $c_1(x)$ and $c_2(x)$, respectively, as described in Lemma 2.5, will be denoted by $\Pi(C_1, C_2)$ and will be called the **product** of the cycles C_1 and C_2. For a simple cycle C, let $|C|$ denote the number of states in C, i.e., the period of C.

Lemma 2.6. *With the same conditions as in Lemma 2.5, the number of distinct states, of length $n_1 + n_2$, in the cycles of the product $\Pi(C_1, C_2)$ is $|C_1| \cdot |C_2|$.*

Proof. This follows immediately from the definition of $\Pi(\mathcal{C}_1, \mathcal{C}_2)$, Lemma 2.5, and the observation that if two cycles have a state in common, then by the recursion implied by the feedback function we have that the two cycles are identical. $\qquad\square$

Theorem 2.5. *Assume an* $LFSR_{n_1}$ *with a characteristic polynomial* $c_1(x)$ *has cycles* $\mathcal{C}_1$ *and* $\mathcal{C}_2$ *(not necessarily distinct) and an* $LFSR_{n_2}$ *with a characteristic polynomial* $c_2(x)$ *has cycles* $\mathcal{S}_1$ *and* $\mathcal{S}_2$ *(not necessarily distinct). Assume further that* g.c.d.$(c_1(x), c_2(x)) = 1$. *If* $\mathcal{C}_1 \not\equiv \mathcal{C}_2$ *or* $\mathcal{S}_1 \not\equiv \mathcal{S}_2$, *then the sets of cycles* $\Pi(\mathcal{C}_1, \mathcal{S}_1)$ *and* $\Pi(\mathcal{C}_2, \mathcal{S}_2)$ *contain distinct states. As a consequence, all the cycles, of the* $LFSR_{n_1+n_2}$ *whose characteristic polynomial is* $c_1(x)c_2(x)$, *are generated in this way.*

Proof. By Lemmas 2.5 and 2.6, we have that $\Pi(\mathcal{C}_i, \mathcal{S}_j)$ have $\frac{rs}{[r,s]}$ distinct cycles with a total of rs distinct states, where r is the period of $\mathcal{C}_1$ and s is the period of $\mathcal{C}_2$. All these cycles are generated by $LFSR_{n_1+n_2}$ and hence by Lemma 2.5 each of the two cycles are either equivalent or disjoint (in states). We will show now that the cycles contained in $\Pi(\mathcal{C}_{i_1}, \mathcal{S}_{j_1})$ and the cycles contained in $\Pi(\mathcal{C}_{i_2}, \mathcal{S}_{j_2})$ are not equivalent when $i_1 \neq i_2$ or $j_1 \neq j_2$. Assume, on the contrary, that there exists $i_1, j_1, i_2, j_2 \in \{1, 2\}$, where $i_1 \neq i_2$ or $j_1 \neq j_2$, such that

$$\mathcal{C}_{i_1} + \mathcal{S}_{j_1} = \mathcal{C}_{i_2} + \mathcal{S}_{j_2},$$

when we assume that the four sequences are given in the appropriate shift to have this equality. However, this implies that $\mathcal{C}_{i_3} = \mathcal{C}_{i_1} - \mathcal{C}_{i_2} = \mathcal{S}_{j_2} - \mathcal{S}_{j_1} = \mathcal{S}_{j_3}$, where $\mathcal{C}_{i_3}$ is generated by $LFSR_{n_1}$ and $\mathcal{S}_{j_3}$ is generated by $LFSR_{n_2}$, which implies that this is the all-zeros sequence, i.e., $\mathcal{C}_{i_1} = \mathcal{C}_{i_2}$ and $\mathcal{S}_{j_1} = \mathcal{S}_{j_2}$, a contradiction.

The number of states in G_{q,n_1} is q^{n_1} and the number of states in G_{q,n_2} is q^{n_2}. Assume that $c_1(x)$ generates k distinct cycles $\mathcal{C}_1, \mathcal{C}_2, \ldots, \mathcal{C}_k$ and $c_2(x)$ generates m distinct cycles $\mathcal{S}_1, \mathcal{S}_2, \ldots, \mathcal{S}_m$. Since by Lemma 2.6 each product of cycles $\Pi(\mathcal{C}_i, \mathcal{S}_j)$ forms sequences with a total of $|\mathcal{C}_i| \cdot |\mathcal{S}_j|$ distinct states in G_{q,n_1+n_2} and no two products have the same states, it follows that the total number of states in

$$\bigcup_{i=1}^{k}\bigcup_{j=1}^{m} \Pi(\mathcal{C}_i, \mathcal{S}_j)$$

is

$$\sum_{i=1}^{k}\sum_{j=1}^{m} |\mathcal{C}_i| \cdot |\mathcal{S}_j| = \left(\sum_{i=1}^{k}|\mathcal{C}_i|\right) \cdot \left(\sum_{j=1}^{m}|\mathcal{S}_j|\right) = q^{n_1}q^{n_2} = q^{n_1+n_2},$$

which implies that all states of G_{q,n_1+n_2} are formed in this way.

Thus the set of cycles $\Pi(\mathcal{C}_1, \mathcal{S}_1)$, $\Pi(\mathcal{C}_2, \mathcal{S}_2)$ contains distinct states and all the cycles of the $LFSR_{n_1+n_2}$ with characteristic polynomial $c_1(x)c_2(x)$ are generated in this way. $\qquad\square$

Theorem 2.5 suggests a simple recursive method to compute the number of cycles for each length generated by a polynomial $f(x)$ that is not irreducible. We just have to factorize $f(x)$ into two polynomials $c_1(x)$ and $c_2(x)$ such that g.c.d.$(c_1(x), c_2(x)) = 1$ and the number of cycles of each length generated by each polynomial is known. The basis of the recursion are polynomials of the form $f(x)^m$, where $f(x)$ is an irreducible polynomial and $m \geq 1$. When $m = 1$ these polynomials are associated with irreducible polynomials, as discussed in this section. Some of them will be also analyzed in the next section. When $m \geq 2$ these polynomials are discussed in Section 2.3.

To end this section we will consider two binary linear shift registers that have reversed feedback functions. For a feedback function

$$f(x_1, x_2, \ldots, x_n) = x_1 + \sum_{i=2}^{n} a_i x_i,$$

the *reversed function* $f^R(x_1, x_2, \ldots, x_n)$ is defined by

$$f^R(x_1, x_2, \ldots, x_n) = x_1 + \sum_{i=2}^{n} a_{n-i+2} x_i.$$

Theorem 2.6. *The sequences generated by the reversed function, f^R, of a feedback function $f(x_1, x_2, \ldots, x_n)$ are the reverse sequences of those generated by the feedback function f.*

Proof. By the function f we have that for each state $(x_1, x_2, \ldots, x_n)$,

$$x_{n+1} = f(x_1, x_2, \ldots, x_n) = x_1 + \sum_{i=2}^{n} a_i x_i .$$

Hence, we have that for the state $(x_{n+1}, x_n, \ldots, x_2)$,

$$f^R(x_{n+1}, x_n, \ldots, x_2) = x_{n+1} + \sum_{i=0}^{n-2} a_{n-i} x_{n-i} = x_{n+1} + \sum_{i=2}^{n} a_i x_i = x_1 .$$

This implies that the sequences generated by f^R are the reverse sequences generated by f. □

2.2 Maximum length linear shift-register sequences

Recall that a nonzero sequence whose characteristic polynomial is primitive is called an **M-*sequence*** (for maximal length sequence) or **PN *sequence*** (for pseudo-noise sequence). For an M-sequence $\{a_k\}$ we define

$$A_0 = (a_0, a_1, \ldots, a_{\pi-1}), \quad \alpha_0 = (a_0, a_1, \ldots, a_{n-1})$$

and

$$A_k = (a_k, a_{k+1}, \ldots, a_{\pi-1}, a_0, \ldots, a_{k-1}), \quad \alpha_k = (a_k, a_{k+1}, \ldots, a_{k+n-1})$$

for each $0 \le k < \pi$.

Clearly, by Corollary 2.5 and Corollary 2.15 we have that $A_i + A_j = A_\ell$ for $0 \le i < j < q^n - 1$ since the A_i and A_j satisfy the same recursion and there is only one nonzero (cyclic) sequence that satisfies this recursion. This is the *shift-and-add* property of M-sequences, a property that was generalized in Corollary 2.15. If the polynomial $f(x) = x^{j-i} + 1$ used for this purpose in Lemma 2.4 and in Corollary 2.15, then we have as in Corollary 2.15, $f(E)S = (E^{j-i} + 1)S = E^{j-i}S + S \simeq S$, which is exactly the shift-and-add property.

Example 2.2. Consider the characteristic polynomial $c(x) = x^4 + x + 1$ over $\mathbb{F}_2$ and the associated recursion $a_4 = a_3 + a_0$. The M-sequence generated by this recursion, when $a_0 = a_1 = a_2 = 0$, and $a_3 = 1$ is 000111101011001. Now,

$$A_0 + A_5 = 000111101011001 + 110101100100011 = 110010001111010 = A_{10}.$$

∎

Random sequences, such as those formed by coin flipping are very important in many applications. The search for associated pseudo-random sequences and the desire to define their properties were always targeted by mathematicians and computer scientists. The following properties defined by Solomon W. Golomb were perhaps the first such attempt.

Golomb's randomness postulates:

For a periodic sequence $\{a_m\}_{m=1}^{\pi}$ over $\{-1, +1\}$

R-1 $\left| \sum_{m=1}^{\pi} a_m \right| \le 1$.

R-2 In every period, half of the runs have length one, a quarter have length two, one-eighth have length three, etc., as long as there are at least two runs. For each of these lengths, the number of runs of -1s and $+1$s are equal.

R-3 Two values for the autocorrelation function

$$C(\tau) = \sum_{m=1}^{\pi} a_m a_{m+\tau} = \begin{cases} \pi & \tau = 0 \\ K & 0 < \tau < \pi \end{cases},$$

where the computation is over the reals.

Theorem 2.7. *A binary M-sequence A satisfies* **R-1**, **R-2**, *and* **R-3**, *when* 0 *is replaced by* $+1$ *and* 1 *is replaced by* -1.

Proof. The first property of a binary M-sequence S of length $2^n - 1$ is that each nonzero binary n-tuple appears exactly once in each period of the M-sequence S.

R-1 is implied since S has 2^{n-1} values of -1s and $2^{n-1} - 1$ values of $+1$s.

R-2   let r_k denote the number of runs of length k in S. Consider the n-tuples starting with $10^k 1$. There are 2^{n-k-2} such n-tuples. Hence, $r_k = 2^{n-k-1}$, where $1 \le k \le n-2$; $r_{n-1} = r_n = 1$. This is for the total of 2^{n-1} runs.

R-3   is implied by the shift-and-add property, which yields $K = -1$. $\square$

Consider the characteristic polynomial $c(x)$ over $\mathbb{F}_q$ given by

$$c(x) = 1 - \sum_{i=1}^{n} c_i x^i$$

and assume that the polynomial

$$\hat{c}(x) \triangleq x^n c(x^{-1}) = x^n - \sum_{i=1}^{n} c_i x^{n-i}, \tag{2.5}$$

called the **companion polynomial**, is a primitive polynomial. The **companion matrix** C of $c(x)$ is defined by

$$C = \begin{bmatrix} 0 & 0 & \cdots & 0 & c_n \\ 1 & 0 & \cdots & 0 & c_{n-1} \\ 0 & 1 & \cdots & 0 & c_{k-2} \\ \vdots & \vdots & \ddots & \vdots & \vdots \\ 0 & 0 & \cdots & 1 & c_1 \end{bmatrix}.$$

If β is a root of $\hat{c}(x)$, then from Eq. (2.5) we have that

$$\beta^n = \sum_{i=1}^{n} c_i \beta^{n-i} = \sum_{i=0}^{n-1} c_{n-i} \beta^i$$

and for each $k \ge 0$ we have by definition that

$$\beta^k = \sum_{i=0}^{n-1} b_i \beta^i.$$

In vector notation, we have

$$\left(\beta^k \right) = \begin{pmatrix} b_0 \\ b_1 \\ \vdots \\ b_{n-1} \end{pmatrix}.$$

Therefore we have that

$$\beta^{k+1} = \sum_{i=0}^{n-1} b_i \beta^{i+1} = \sum_{i=1}^{n-1} b_{i-1} \beta^i + b_{n-1} \beta^n$$

$$= \sum_{i=1}^{n-1} b_{i-1} \beta^i + b_{n-1} \sum_{i=0}^{n-1} c_{n-i} \beta^i = b_{n-1} c_n + \sum_{i=1}^{n-1} (b_{i-1} + b_{n-1} c_{n-i}) \beta^i .$$

This implies that

$$\left(\beta^{k+1} \right) = C \left(\beta^k \right).$$

Therefore we have that $\left(\beta^k \right) = C^k \left(\beta^0 \right)$, and it follows that

$$C^k I = C^k \left[\left(\beta^0 \right) \left(\beta^1 \right) \cdots \left(\beta^{n-1} \right) \right] = \left[\left(\beta^k \right) \left(\beta^{k+1} \right) \cdots \left(\beta^{k+n-1} \right) \right]$$

and hence

$$\beta^i + \beta^j = \beta^\ell \Rightarrow \beta^{i+k} + \beta^{j+k} = \beta^{\ell+k} \Rightarrow C^i + C^j = C^\ell.$$

Theorem 2.8. *The mapping φ defined by $\varphi(\beta^i) = C^i$ for each $0 \le i < \pi$ and $\varphi(0) = 0$, is an isomorphism.*

Proof. The claim follows from the following equalities:

$$\varphi(\beta^i + \beta^j) = \varphi(\beta^\ell) = C^\ell = C^i + C^j = \varphi(\beta^i) + \varphi(\beta^j)$$
$$\varphi(\beta^i \cdot \beta^j) = \varphi(\beta^{i+j}) = C^{i+j} = C^i \cdot C^j = \varphi(\beta^i) \cdot \varphi(\beta^j). \qquad \square$$

Corollary 2.16. *The set $\{0^n\} \cup \{\alpha_k\}_{k=0}^{\pi-1}$ with vector addition and multiplication defined by $\alpha_i \cdot \alpha_j = \alpha_{i+j} \pmod{\pi}$ for $0 \le i, j < \pi$ and $0^n \cdot \gamma = \gamma \cdot 0^n = 0^n$, is isomorphic to $\mathbb{F}_{\pi+1}$. The same holds when α_k is replaced by A_k and 0^n by 0^π.*

Theorem 2.9. *A set with n shifts of an M-sequence (as a row vector) of length $2^n - 1$, which are linearly independent vectors, written by an $n \times (2^n - 1)$ matrix (each shift as one of the rows), contains each nonzero column vector defined by the $2^n - 1$ nonzero n-tuple as one of the columns.*

Proof. It is easy to verify that the first n shifts have this property. If some other n shifts have this property, then by replacing one row with its addition to another row, the new n shifts have this property. Any n linearly independent such shifts can be obtained in this way and the claim is proved. $\qquad \square$

Theorem 2.10. *A set with n shifts of an M-sequence (as a row vector) of length $q^n - 1$, which are linearly independent vectors, written as an $n \times (q^n - 1)$ matrix, contains each nonzero column vector defined by the $q^n - 1$ nonzero n-tuple as one of the columns.*

Theorem 2.11. *For all prime powers q, $n \geq 1$, and $1 \leq k \leq q^n$, there exists an FSR$_n$ over $\mathbb{F}_q$ that generates a cycle of length k.*

Proof. If $k = q^n$, then let S be an M-sequence of length $q^n - 1$ over $\mathbb{F}_q$. By definition, we have the edge $\alpha_0 \rightarrow \alpha_1$, where $\alpha_0 = (1, 0, \ldots, 0)$ and $\alpha_1 = (0, \ldots, 0, 1)$. Clearly, the only state not in S is $(0, 0, \ldots, 0)$ and we can insert it between α_0 and α_1, i.e., $\alpha_0 \rightarrow (0, 0, \ldots, 0) \rightarrow \alpha_1$ to obtain a cycle of length q^n.

Now, assume that $1 \leq k < q^n - 1$ and consider a primitive element $\beta \in \mathbb{F}_{q^n}$. We have that $\beta^k - 1 = \beta^t$ for some t, which implies that $\beta^{k-t} = \beta^0 + \beta^{-t}$, and hence by Corollary 2.16 $\alpha_{k-t} = \alpha_0 + \alpha_{-t}$. Therefore we have that in $G_{q,n}$ the vertices α_{k-t} and α_{-t} have the same successors, i.e., they are conjugate states. Thus we can apply the merge-or-split method to obtain two cycles from the M-sequence S, as depicted in Fig. 2.1.

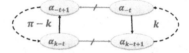

FIGURE 2.1 The merge-or-split of Theorem 2.11.

The first cycle has the following sequence of edges

$$\cdots \rightarrow \alpha_{k-t-1} \rightarrow \alpha_{k-t} \rightarrow \alpha_{-t+1} \rightarrow \alpha_{-t+2} \rightarrow \cdots$$

and hence its length is $\pi - k = q^n - 1 - k$ (see Fig. 2.1). The second cycle has the following sequence of edges

$$\cdots \rightarrow \alpha_{-t-1} \rightarrow \alpha_{-t} \rightarrow \alpha_{k-t+1} \rightarrow \alpha_{k-t+2} \rightarrow \cdots$$

and hence its length is k (see Fig. 2.1). $\square$

Example 2.3. For $q = 2$ and $n = 4$, let $c(x) = x^4 + x^3 + 1$ be the characteristic polynomial and hence $\hat{c}(x) = x^4 + x + 1$ is its companion polynomial and

$$\begin{pmatrix} 0 & 0 & 0 & 1 \\ 1 & 0 & 0 & 1 \\ 0 & 1 & 0 & 0 \\ 0 & 0 & 1 & 0 \end{pmatrix}$$

is its associated companion matrix. This implies that $\{a_k\}$ is obtained from the recurrence $a_k = a_{k-3} + a_{k-4}$ and also $\alpha_k = \alpha_{k-3} + \alpha_{k-4}$. Let β be a root of $\hat{c}(x)$, i.e., $\beta^4 + \beta + 1 = 0$ or $\beta^4 = \beta + 1$ and therefore $\mathbb{F}_{16}$ formed from $\hat{c}(x)$ has the following structure

k	β^3	β^2	β	β^0	a_k	a_{k+1}	a_{k+2}	a_{k+3}
0	0	0	0	1	1	0	0	0
1	0	0	1	0	0	0	0	1
2	0	1	0	0	0	0	1	0
3	1	0	0	0	0	1	0	0
4	0	0	1	1	1	0	0	1
5	0	1	1	0	0	0	1	1
6	1	1	0	0	0	1	1	0
7	1	0	1	1	1	1	0	1
8	0	1	0	1	1	0	1	0
9	1	0	1	0	0	1	0	1
10	0	1	1	1	1	0	1	1
11	1	1	1	0	0	1	1	1
12	1	1	1	1	1	1	1	1
13	1	1	0	1	1	1	1	0
14	1	0	0	1	1	1	0	0

where $\alpha_k = a_k a_{k+1} a_{k+2} a_{k+3}$.

The associated binary M-sequence of length $\pi = 15$ is

$$\{a_k\} = 10001001101011$$

and as a cycle in the de Bruijn graph, it is depicted in Fig. 2.2. This cycle of length 15 is split into two cycles of length 6 and 9, as explained in the proof of Theorem 2.11. ∎

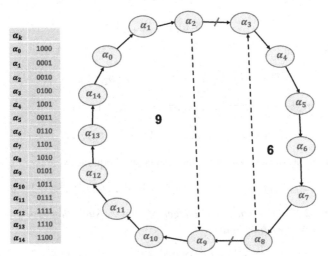

FIGURE 2.2 The merge-or-split of Theorem 2.11 to form cycles of length 6 and 9 from an M-sequence of length 15.

Next, we will try to find some operations that will partition a set of sequences into equivalence classes and will be able to have some classification of the sequences. A more detailed discussion on possible classification will be given in Chapter 6. For this classification, two operations – the shift and the decimation, will be defined. Specifically, these operations will work on M-sequences and difference sets that will be presented in Section 2.4.

Definition 2.6. The t-*shift* of a sequence $\{a_k\}$, with period π, is defined by

$$\mathcal{S}^t\{a_k\} \triangleq \mathbf{E}^t\{a_k\} = \{a_{t+k}\} = \{a_t, a_{t+1}, a_{t+2}, \ldots\}.$$

For an integer r, such that $0 < r < \pi$ and g.c.d.$(r, \pi) = 1$, the r-*decimation* of a sequence $\{a_k\}$ is defined by

$$\mathcal{D}^r\{a_k\} \triangleq \{a_{kr}\} = \{a_0, a_r, a_{2r}, \ldots\}.$$

The following two theorems can be easily verified.

Theorem 2.12. *The π shift operators $\mathcal{S}^0, \mathcal{S}^1, \mathcal{S}^2, \ldots, \mathcal{S}^{\pi-1}$ form an Abelian group under the decomposition*

$$\mathcal{S}^{t_1}\mathcal{S}^{t_2} = \mathcal{S}^{t_3}, \quad t_3 \equiv t_1 + t_2 \ (\text{mod } \pi).$$

Theorem 2.13. *The $\phi(\pi)$ decimation operators $\mathcal{D}^r$, such that $1 \le r < \pi$ and g.c.d.$(r, \pi) = 1$, form an Abelian group under the composition*

$$\mathcal{D}^{r_1}\mathcal{D}^{r_2} = \mathcal{D}^{r_3}, \quad r_3 \equiv r_1 r_2 \ (\text{mod } \pi).$$

Lemma 2.7. *For each integer t and an integer r, such that $0 < r < \pi$ and g.c.d.$(r, \pi) = 1$,*

$$\mathcal{D}^r\mathcal{S}^t\{a_k\} = \{a_{t+kr}\}.$$

Proof. Note that

$$\mathcal{D}^r\{b_0, b_1, b_2, b_3, \ldots\} = \{b_0, b_r, b_{2r}, b_{3r}, \cdots\}$$

and if $\{b_0, b_1, b_2, b_3, \ldots\} = \{a_t, a_{t+1}, a_{t+2}, a_{t+3}, \ldots\}$, then

$$\mathcal{D}^r\{a_t, a_{t+1}, a_{t+2}, \ldots\} = \{a_t, a_{t+r}, a_{t+2r}, \ldots\}.$$

The lemma follows now from the sequence of equalities

$$\mathcal{D}^r\mathcal{S}^t\{a_k\} = \mathcal{D}^r\{a_{t+k}\} = \mathcal{D}^r\{a_t, a_{t+1}, a_{t+2}, \ldots\} = \{a_t, a_{t+r}, a_{t+2r}, \ldots\} = \{a_{t+kr}\}. \ \square$$

Corollary 2.17. *For each two integers t and r, such that $0 < r < \pi$ and g.c.d.$(r, \pi) = 1$, we have that*

$$\mathcal{D}^r\mathcal{S}^{tr}\{a_k\} = \mathcal{D}^r\{a_{tr}, a_{tr+1}, a_{tr+2}, \ldots\}$$
$$= \{a_{tr+r}, a_{tr+2r}, a_{tr+3r}, \ldots\} = \{a_{tr+kr}\} = \{a_{(t+k)r}\}.$$

Lemma 2.8. *For each nonnegative integer t and an integer r, such that $0 < r < \pi$ and g.c.d.$(r, \pi) = 1$,*

$$\mathcal{S}^t \mathcal{D}^r \{a_k\} = \mathcal{D}^r \mathcal{S}^{tr} \{a_k\}.$$

Proof. The claim follows from the sequence of equalities using Corollary 2.17:

$$\mathcal{S}^t \mathcal{D}^r \{a_k\} = \mathcal{S}^t \{a_{kr}\} = \mathcal{S}^t \{a_0 a_r a_{2r} \cdots\}$$
$$= \{a_{tr} a_{(t+1)r} a_{(t+2)r} \cdots\} = \{a_{(t+k)r}\} = \mathcal{D}^r \mathcal{S}^{tr} \{a_k\}. \qquad \square$$

Corollary 2.18. *For each integer t and an integer r, such that $0 < r < \pi$ and g.c.d.$(r, \pi) = 1$, then*

$$\mathcal{D}^r \mathcal{S}^t = \mathcal{S}^{r^{-1} t} \mathcal{D}^r.$$

Proof. By Lemma 2.8 we have that

$$\mathcal{S}^{r^{-1} t} \mathcal{D}^r = \mathcal{D}^r \mathcal{S}^{r^{-1} tr} = \mathcal{D}^r \mathcal{S}^t. \qquad \square$$

Lemma 2.9. *The set π_ϕ (see Eq. (1.4)) is a multiplicative Abelian group of the nonzero residues modulo π.*

Proof. Clearly $(a \cdot b) \cdot c = a \cdot (b \cdot c)$ for each $a, b, c \in \pi_\phi$. It is also clear that $1 \in \pi_\phi$ is an identity element. Finally, if g.c.d.$(a, \pi) = 1$, then by Theorem 1.5 there exist two integers r and s such that

$$a \cdot r + \pi \cdot s = 1,$$

which implies that

$$a \cdot r \equiv 1 \pmod{\pi}.$$

Therefore the residue modulo π between 0 and $\pi - 1$ that is congruent to r modulo π is relatively prime to π and is the inverse of a. Thus π_ϕ is a group and it is readily verified that the group is Abelian. $\qquad \square$

Lemma 2.10. *Let A be a subset of π_ϕ such that for every $r \in A$ there exists an integer t, $0 \le t < \pi$, such that $\mathcal{D}^r \{a_k\} = \mathcal{S}^t \{a_k\}$. Then, A is a subgroup of π_ϕ.*

Proof. Let $r_1, r_2 \in A$, i.e., by the definition of A, there exist t_1, t_2, such that

$$\mathcal{D}^{r_i} \{a_k\} = \mathcal{S}^{t_i} \{a_k\} \text{ for } i \in \{1, 2\}.$$

Now, we have the following sequence of equalities (using Theorems 2.12 and 2.13 and Corollary 2.18)

$$\mathcal{D}^{r_1 r_2} \{a_k\} = \mathcal{D}^{r_1} \mathcal{D}^{r_2} \{a_k\} = \mathcal{D}^{r_1} \mathcal{S}^{t_2} \{a_k\}$$
$$= \mathcal{S}^{r_1^{-1} t_2} \mathcal{D}^{r_1} \{a_k\} = \mathcal{S}^{r_1^{-1} t_2} \mathcal{S}^{t_1} \{a_k\} = \mathcal{S}^{t_1 + r_1^{-1} t_2} \{a_k\}.$$

This implies by the definition of the subset A, that A is a subgroup of π_ϕ. $\qquad \square$

Theorem 2.14. *If $r_1, r_2 \in \pi_\phi$, then $\mathcal{D}^{r_1}\{a_k\}$ and $\mathcal{D}^{r_2}\{a_k\}$ are shifts of one another if and only if r_1 and r_2 are in the same coset of A in π_ϕ.*

Proof. If r_1 and r_2 are in the same coset of A, then $r_2 = m r_1$, where $m \in A$. Hence, using the definition of the subgroup A we have that

$$\mathcal{D}^{r_2}\{a_k\} = \mathcal{D}^{m r_1}\{a_k\} = \mathcal{D}^m \mathcal{D}^{r_1}\{a_k\} = \mathcal{S}^t \mathcal{D}^{r_1}\{a_k\}.$$

If $\mathcal{D}^{r_2}\{a_k\} = \mathcal{S}^t \mathcal{D}^{r_1}\{a_k\}$, then since g.c.d.$(r_1, \pi) = 1$, it follows by Lemma 2.8 that $\mathcal{D}^{r_2}\{a_k\} = \mathcal{D}^{r_1} \mathcal{S}^{t r_1}\{a_k\}$. Therefore we have that

$$\mathcal{D}^{r_1^{-1} r_2}\{a_k\} = \mathcal{D}^{r_1^{-1}} \mathcal{D}^{r_2}\{a_k\} = \mathcal{D}^{r_1^{-1}} \mathcal{D}^{r_1} \mathcal{S}^{t r_1}\{a_k\} = \mathcal{S}^{t r_1}\{a_k\},$$

which implies that $r_1^{-1} r_2$ is an element of A and therefore $r_2 \in r_1 A$, i.e., r_1 and r_2 are in the same coset of A in π_ϕ. $\qquad\square$

Now, we can use Corollary 2.16 to obtain the following result.

Theorem 2.15. *Let $\{a_k\}$ be an M-sequence of length $q^n - 1$ and let r be an integer, $2 \le r \le q^n - 2$, such that g.c.d.$(r, q^n - 1) = 1$. Then, $\{a_{rk}\}$ is also an M-sequence of length $q^n - 1$.*

Definition 2.7. If $\{a_k\}$ is an M-sequence and $\{a_{rk}\}$ is also an M-sequence, then r is called a **multiplier** of the M-sequence.

Example 2.4. Consider $n = 4$ and the span 4 M-sequence of Example 2.3

$$\{a_k\} = 100010011010111 = a_0 a_1 \cdots a_{14}.$$

The period of the sequence is $\pi = 15$ and $\phi(15) = 8$, where

$$\pi_\phi = \{1, 2, 4, 7, 8, 11, 13, 14\}.$$

For this M-sequence, we have the following decimations

$$\mathcal{D}^2\{a_k\} = \{a_{2k}\} = 101011110001001 = a_0 a_2 \cdots a_{14} a_1 a_3 \cdots a_{13} = \mathcal{S}^8\{a_k\},$$
$$\mathcal{D}^4\{a_k\} = \{a_{4k}\} = 111100010011010 = a_0 a_4 a_8 a_{12} a_1 a_5 \cdots a_{11} = \mathcal{S}^{12}\{a_k\}$$

and

$$\mathcal{D}^8\{a_k\} = \{a_{8k}\} = 110001001101011 = a_0 a_8 a_1 a_9 a_2 \cdots a_{14} a_7 = \mathcal{S}^{14}\{a_k\}.$$

Clearly, $\{1, 2, 4, 8\}$ is a subgroup of π_ϕ. ∎

2.3   Powers of irreducible polynomials

In Section 2.1 we found the length of the cycles generated by a given polynomial. We also gave a method to compute the number of cycles of each length. To complete these computations it is required to find the length of the cycles generated by the polynomial $g^n(x)$, where $g(x)$ is an irreducible polynomial and $n > 1$. It is also required to find the number of cycles for each length. The results in this section will be proved only over $\mathbb{F}_2$, but they can be generalized along the same lines to sequences over any field $\mathbb{F}_q$.

Consider any irreducible polynomial $g(x) = \sum_{i=0}^{k} a_i x^i$ over $\mathbb{F}_2$, where $a_0 = a_k = 1$ and $k > 1$. The period of sequences generated by $g(x)$ was given by Theorem 2.3. All the nonzero sequences generated by $g(x)$ have the same period e, where e is the smallest integer such that $g(x)$ divides $x^e - 1$. The number of sequences of such length is $\frac{2^k - 1}{e}$. The sequences generated by the polynomial $g^n(x)$ yield an interesting hierarchy that will be explored in this section. The polynomial $g(x) = x + 1$ yields further properties and they will be discussed separately in Section 5.1.

For any polynomial $g(x)$ of degree k and any binary sequence S we define $\mathbf{D}_g S \triangleq g(\mathbf{E})S$.

Let $\Omega_g(n)$ be the set of sequences that are generated by the polynomial $g^n(x)$ and are not generated by the polynomial $g^{n-1}(x)$, where $n \geq 1$ and $\Omega_g(0)$ is the set that contains the all-zeros sequence. Clearly, by the definitions we have the following observation.

Lemma 2.11. *The set $\Omega_g(n)$ can be defined recursively by the operators $\mathbf{D}_g$ and $g(\mathbf{E})$ as follows:*

$$\Omega_g(n) = \{S \ : \ \exists S' \in \Omega_g(n-1), \ S' = \mathbf{D}_g S = g(\mathbf{E})S\},$$

where $\Omega_g(0) = \{[0]\}$.

We will present two examples for the sets $\Omega_g(\cdot)$, one for a primitive polynomial and a second one for an irreducible polynomial that is not primitive.

Example 2.5. For the primitive polynomial $g(x) = x^2 + x + 1$, the first few sets of $\Omega_g(n)$ are as follows:

$\Omega_g(0) = \{[0]\}$.

$\Omega_g(1) = \{[011]\}$,

$\Omega_g(2) = \{[000101], [011110]\}$,

$\Omega_g(3) = \{[000001110111], [011010101100], [000100110010], [011111101001]\}$

$\Omega_g(4) = \{[000000010001], [011011001010], [101101111100], [110110100111],$
$\qquad\quad [000101010100], [011110001111], [101000111001], [110011100010],$
$\qquad\quad [000001100110], [011010111101], [101100001011], [110111010000],$
$\qquad\quad [000100100011], [011111111000], [101001001110], [110010010101]\}$.

■

Example 2.6. For the irreducible polynomial $g(x) = x^4 + x^3 + x^2 + x + 1$ that is not a primitive polynomial, the first few sets of $\Omega_g(n)$ are as follows:

$\Omega_g(0) = \{[0]\}$,

$\Omega_g(1) = \{[00011], [00101], [01111]\}$,

$\Omega_g(2) = \{[0010101111], [0010111110], [0011111010], [0111101010], [0110101011],$

$\qquad [0010111011], [0011101110], [0110111010], [0011101011], [0110101110],$

$\qquad [0000011011], [0000111001], [0010110001], [1010010001], [1000010011],$

$\qquad [0000000101], [0000010001], [0001010101], [0000001111], [0000101101],$

$\qquad [0000110011], [0010011001], [0011111111], [0110111111]\}$.

■

We will derive now some relations between the various $\Omega_g(i)$s. The proofs of the following claims are left as an exercise. A detailed analysis for another polynomial $g(x) = x + 1$, will be presented in Section 5.1. Recall that the degree of $g(x)$ is k.

Lemma 2.12. *The sequences in* $\bigcup_{i=0}^{n} \Omega_g(i)$ *are generated by the LFSR$_{kn}$ whose characteristic polynomial is* $g^n(x)$.

Lemma 2.13. *The sequences in* $\bigcup_{i=0}^{n} \Omega_g(i)$ *contain each (kn)-tuple exactly once as a window in exactly one of the sequences.*

Lemma 2.14. *A sequence $S = [s_0, s_1, s_2, \ldots, s_{\ell-1}]$ is contained in $\Omega_g(n)$ if and only if* $g^{n-1}(E)S \in S(g(x))$.

Corollary 2.19. *The period of the sequences that are contained in $\Omega_g(2^m + 1)$ is* $2^{m+1} \cdot e(g(x))$.

Definition 2.8. A set of sequences S are *t-interleaved* if they form all the sequences of the form

$$[s_1, s_2, s_3, \ldots, s_{rt}],$$

where $[s_i, s_{i+t}, s_{i+2t}, \ldots, s_{i+(r-1)t}]$ is a sequence in S for each $1 \leq i \leq t$. Any such constructed sequence is said to be *t-interleaved* from S, and the operation done on the sequences is called *t-interleaving*.

Lemma 2.15. *The sequences that are generated by the polynomial $g^{2^m}(x)$ are the sequences in the set of all 2^m-interleaved sequences generated from* $S(g(x)) \cup \{[0]\}$.

Proof. Let S be the set of sequences generated by $g(x)$ (including the all-zeros sequence).

The equality

$$g^{2^m}(x) = \left(\sum_{i=0}^{k} a_i x^i\right)^{2^m} = \sum_{i=0}^{k} a_i x^{i \cdot 2^m} = g(x^{2^m})$$

implies that any 2^m-interleaved sequence generated from S is a sequence generated by $g^{2^m}(x)$.

On the other hand, if $S = [s_1, s_2, s_3, \ldots, s_{r2^m}]$ is a 2^m-interleaved sequence obtained from the sequences of S, then since

$$g^{2^m}(x) = g(x^{2^m}),$$

it follows that for each i, $1 \le i \le 2^m$, the sequence

$$[s_i, s_{i+2^m}, s_{i+2 \cdot 2^m}, \ldots, s_{i+(r-1)2^m}]$$

is a sequence generated by $g(x)$. $\qquad\qquad\square$

Lemma 2.16. *Let S_1 and S_2 be two sequences whose periods are n_1 and n_2, respectively, where n_2 divides n_1. The 2-interleaved sequence of S_1 and S_2 has period $2n_1$ if $S_1 \not\simeq S_2$. If $S_1 \simeq S_2$, then usually the period is $2n_1$, except for one case in which the period is n_1.*

Proof. If $S_1 = [a_1, a_2, \ldots, a_{n_1}]$ and $S_2 = [b_1, b_2, \ldots, b_{n_2}]$, then the 2-interleaved sequences of S_1 and S_2 are of the form

$$T_i = [a_1, b_i, a_2, b_{i+1}, a_3, b_{i+2}, \ldots, a_{n_1}, b_{i-1}], \quad 1 \le i \le n_2.$$

It is easy to verify that since the period of S_1 is n_1 and n_2 divides n_1, it follows that the period of T_i is either n_1 or $2n_1$. Moreover, if n_1 is even, then the only possible period of T_i is $2n_1$ since period n_1 of T_i implies that S_1 has period $\frac{n_1}{2}$. If n_1 is odd, then the possible periods of the T_is are n_1 and $2n_1$, where the period n_1 occurs only when $S_1 = S_2$ and $i = \frac{n_1+3}{2}$. $\qquad\square$

Example 2.7. Let S_1 be the M-sequence [000111101011001] and S_2 be the same M-sequence. The sequences T_1 and T_9 are given by

$$T_1 = [000000111111110011001111000011],$$

whose period is 30, and

$$T_9 = [010001111010110010001110101110],$$

whose period is 15. $\qquad\qquad\blacksquare$

Lemma 2.17. *The period of the sequences that are contained in $\Omega_g(2n)$ is twice the period of the sequences that are contained in $\Omega_g(n)$.*

Proof. The sequences that are contained in $\Omega_g(n)$ are those sequences generated by the polynomial $g^n(x)$ and are not generated by the polynomial $g^{n-1}(x)$, while the sequences that are contained in $\Omega_g(2n)$ are those generated by $g^{2n}(x)$ and are not generated by $g^{2n-1}(x)$. By Lemma 2.15, the sequences generated by the polynomial $g^{2n}(x)$ are the 2-interleaving of the sequences generated by the polynomial $g^n(x)$. As a consequence, the claim of the lemma can be proved by induction using Lemma 2.16. $\qquad\square$

Corollary 2.20. *The period of the sequences that are contained in $\Omega_g(2^m)$ is* $e(g(x)) \cdot 2^m$.

Corollary 2.21. *The period of the sequences that are contained in $\Omega_g(2^m + 1)$ is twice the period of the sequences that are contained in $\Omega_g(2^m)$.*

Corollary 2.22. *If $g(x)$ is an irreducible polynomial of degree k, then the number of sequences that are contained in $\Omega_g(n + 1)$ is $2^k \cdot |\Omega_g(n)|$, for $2^m + 1 \leq n \leq 2^{m+1} - 1$.*

Corollary 2.23. *If $g(x)$ is a primitive polynomial of degree k, then the number of sequences that are contained in $\Omega_g(2^m + 1)$ is $2^{k-1} \cdot |\Omega_g(2^m)|$.*

Corollary 2.24. *All the sequences contained in $\bigcup_{i=2^m+1}^{2^{m+1}} \Omega_g(i)$ have the same period.*

Proof. By Corollary 2.19, we have that the period of the sequences in $\Omega_g(2^m + 1)$ is $e(g(x)) \cdot 2^{m+1}$. By Corollary 2.20, we have that the period of the sequences in $\Omega_g(2^{m+1})$ is $e(g(x)) \cdot 2^{m+1}$. By Lemma 2.11, we have that a sequence S is contained in $\Omega_g(n)$ if there exists a sequence S' in $\Omega_g(n+1)$ such that $S = D_g S'$. Therefore since all the sequences in $\Omega_g(2^m + 1) \cup \Omega_g(2^{m+1})$ have the same period $e(g(x)) \cdot 2^{m+1}$, it follows that the sequences contained in $\bigcup_{i=2^m+2}^{2^{m+1}-1} \Omega_g(i)$ also have the same period $e(g(x)) \cdot 2^{m+1}$. $\square$

Corollary 2.25. *Let $g(x)$ be an irreducible polynomial of degree k that generates $t = \frac{2^k-1}{\pi}$ sequences of period π. The number of sequences in $\Omega_g(n)$, $n \geq 1$, is $\frac{2^k-1}{\pi} 2^{(n-1)k-\lceil \log n \rceil}$ and the period of a sequence in $\Omega_g(n)$ is $\pi \cdot 2^{\lceil \log n \rceil}$.*

A detailed proof of Corollary 2.25 for $g(x) = x + 1$ will be presented in Section 5.1 (see Lemma 5.12).

2.4 Patterns with distinct differences

The autocorrelation property, **R-3**, of the randomness postulates, is associated with another combinatorial structure called a difference set. This combinatorial structure yields sequences with the randomness postulate **R-3**.

A (v, k, λ)-*difference set* $\mathbb{D} = \{d_1, d_2, \ldots, d_k\}$ is a collection of k residues modulo v of the set $\mathbb{Z}_v = \{0, 1, \ldots, v - 1\}$, such that for any nonzero residue α modulo v the equation

$$\alpha \equiv d_i - d_j \pmod{v}$$

has exactly λ solution pairs $(d_i, d_j) \in \mathbb{D} \times \mathbb{D}$.

Lemma 2.18. *For any (v, k, λ)-difference set we have $k(k - 1) = \lambda(v - 1)$.*

Proof. The number of ordered pairs $k(k - 1)$ can be counted by the number of times, λ, that each of the $v - 1$ nonzero residues modulo v appears as a difference. $\square$

The **complementary set** $\mathbb{D}^*$ of a (v, k, λ)-difference set is defined by the set of residues $\mathbb{Z}_v \setminus \mathbb{D}$.

Example 2.8. The following four examples are trivial difference sets:

1. $\mathbb{D}_0 = \varnothing$ is a $(v, 0, 0)$-difference set.
2. $\mathbb{D}_1 = \{i\}$, $0 \leq i \leq v - 1$, is a $(v, 1, 0)$-difference set.
3. $\mathbb{D}_0^* = \{0, 1, \cdots, v - 1\} = \mathbb{Z}_v$ is a (v, v, v)-difference set.
4. $\mathbb{D}_1^* = \mathbb{Z}_v \setminus \{i\}$, $0 \leq i \leq v - 1$, is a $(v, v - 1, v - 2)$-difference set. ∎

Theorem 2.16. *If $\mathbb{D}$ is a (v, k, λ)-difference set, then $\mathbb{D}^*$ is a (v^*, k^*, λ^*)-difference set, where $v^* = v$, $k^* = v - k$, and $\lambda^* = v - 2k + \lambda$.*

Proof. Since the size of $\mathbb{D}$ is k, it follows that each nonzero element of $\mathbb{Z}_v$ appears as a difference in $\mathbb{D} \times \mathbb{Z}_v$ (or $\mathbb{Z}_v \times \mathbb{D}$) exactly k times. For the pairs in $\mathbb{D} \times \mathbb{D}$ each nonzero element of $\mathbb{Z}_v$ appears as a difference exactly λ times. Since $\mathbb{Z}_v = \mathbb{D} \cup \mathbb{D}^*$ and $\mathbb{D} \cap \mathbb{D}^* = \varnothing$, it follows that each nonzero element of $\mathbb{Z}_v$ appears as a difference in $\mathbb{D} \times \mathbb{D}^*$ (or $\mathbb{D}^* \times \mathbb{D}$) exactly $k - \lambda$ times. Hence, each nonzero element of $\mathbb{Z}_v$ appears as a difference in $\mathbb{D}^* \times \mathbb{D}^*$ exactly $\lambda^* = (v - k) - (k - \lambda) = v - 2k + \lambda$ times and $\mathbb{D}^*$ is a (v^*, k^*, λ^*)-difference set, where $v^* = v$, $k^* = v - k$, and $\lambda^* = v - 2k + \lambda$. □

Example 2.9. For $v = 7$ we have the following two difference sets

1. $\mathbb{D} = \{1, 2, 4\}$ is a $(v, k, \lambda) = (7, 3, 1)$-difference set.
2. $\mathbb{D}^* = \{0, 3, 5, 6\}$ is a $(v^*, k^*, \lambda^*) = (7, 4, 2)$-difference set.

Note that the characteristic vector of $\mathbb{D}^*$ is the M-sequence 1001011. ∎

Consider the correspondence between a subset of a set and its associated characteristic vector. Taking this correspondence in mind, the two operators, i.e., the **shift** and the **decimation**, which were defined on sequences, are defined similarly on difference sets. This leads to the following theorem.

Theorem 2.17. *If $\mathbb{D} = \{d_1, d_2, \cdots, d_k\}$ is a (v, k, λ)-difference set, then so are its t-shift*

$$t + \mathbb{D} \triangleq \{t + d_1, t + d_2, \ldots, t + d_k\}$$

and its r-decimation

$$r \cdot \mathbb{D} \triangleq \{rd_1, rd_2, \ldots, rd_k\},$$

where g.c.d.$(r, v) = 1$ and the computations are performed modulo v.

Proof. The claim for the t-shift is trivial since $a - b \equiv (t + a) - (t + b) \pmod{v}$. Similarly, if g.c.d.$(r, v) = 1$, then $r \cdot (\mathbb{Z}_v \setminus \{0\}) = \mathbb{Z}_v \setminus \{0\}$, and the claim is satisfied for the r-decimation. □

Two difference sets $\mathbb{D}$ and $\hat{\mathbb{D}}$ with the same parameters (v, k, λ) are said to be equivalent if there exist t and r, where g.c.d.$(r, v) = 1$ and $\hat{\mathbb{D}} = t + r\mathbb{D}$. Each such r is called a ***multiplier*** of the difference set (compare with Definition 2.7).

Lemma 2.19. *If $n \triangleq k - \lambda$, then v and n are invariant under complement.*

Proof. The claim follows directly from the equation

$$k^* - \lambda^* = (v - k) - (v - 2k + \lambda) = k - \lambda. \qquad \square$$

The relation between the size of the difference set v and $n = k - \lambda$ is very important, as we will see in the following.

Theorem 2.18. *If $\lambda \geq 1$, then $4n - 1 \leq v \leq n^2 + n + 1$.*

Proof. By Lemma 2.18 we have that $k(k - 1) = \lambda(v - 1)$, which implies that

$$n = k - \lambda = k - \frac{k(k - 1)}{v - 1} = \frac{k(v - k)}{v - 1}.$$

Hence, by also using Theorem 2.16 we have that

$$k \cdot k^* = k(v - k) = n(v - 1).$$

Now, by using Theorem 2.16 again, we have that

$$\lambda \cdot \lambda^* = \lambda(v - 2k + \lambda) \geq v - 2n - 1 \quad (\text{since } \lambda \geq 1) \qquad (2.6)$$

and also since $k = n + \lambda$ we have that

$$\lambda \cdot \lambda^* = (k - n)(v - k - n) = k(v - k) - nv + n^2 = n(n - 1) \qquad (2.7)$$

and

$$\lambda + \lambda^* = (k - n) + (v - k - n) = v - 2n. \qquad (2.8)$$

This implies by Eqs. (2.7) and (2.8) that

$$\left(\frac{v - 2n}{2}\right)^2 = \left(\frac{\lambda + \lambda^*}{2}\right)^2 = \left(\frac{\lambda - \lambda^*}{2}\right)^2 + \lambda \cdot \lambda^* \geq \lambda \cdot \lambda^* = n(n - 1). \quad (2.9)$$

Since by Eq. (2.6) we have $\lambda \cdot \lambda^* \geq v - 2n - 1$, it follows now from Eq. (2.9) that $n(n - 1) \geq v - 2n - 1$ and hence

$$v \leq n^2 + n + 1$$

and the upper bound on v is proved.

Moreover, by Eq. (2.9), we have that $\left(\frac{v-2n}{2}\right)^2 \geq n(n-1)$, which implies that $(v-2n)^2 \geq 4n(n-1)$, i.e.,

$$v - 2n \geq \left\lceil \sqrt{4n(n-1)} \right\rceil = 2n - 1$$

and therefore $4n - 1 \leq v$. $\quad\square$

The two extremes of the bound in Theorem 2.18 can be attained.

- If for a (v, k, λ)-difference set we have that $v = 4n - 1$, then the difference set is a Hadamard-type difference set.
- If for a (v, k, λ)-difference set we have that $v = n^2 + n + 1$, then the difference set is a finite projective plane-type difference set.

In our context, the interesting case is $v = 4n - 1$. A $(4n - 1, 2n - 1, n - 1)$ difference set is called **Hadamard difference set**.

Theorem 2.19. *The set* $\mathbb{D}$ *whose characteristic vector is the complement of an M-sequence of length* $2^n - 1$ *forms a* $(2^n - 1, 2^{n-1} - 1, 2^{n-2} - 1)$ *Hadamard difference set.*

Proof. Let $S = s_0 s_1 \cdots s_{v-1}$ be an M-sequence, where $v = 2^n - 1$ and $\bar{S}$ be the characteristic vector of a set $\mathbb{D}$. The number of pairs $(d_i, d_j) \in \mathbb{D} \times \mathbb{D}$ such that $t \equiv d_i - d_j \pmod{v}$, where t is nonzero residue modulo $2^n - 1$, equals the number of pairs (s_i, s_{i+t}) such that $s_i = s_{i+t} = 0$.

By the shift-and-add property, the number of pairs in S, (s_i, s_{i+t}) such that $s_i = s_{i+t} = 1$ or $s_i = s_{i+t} = 0$ is $2^{n-1} - 1$ since $0+0 = 1+1 = 0$ and the number of *zeros* in S is $2^{n-1} - 1$. This implies that the total number of pairs (s_i, s_{i+t}) such that $s_i = 0$ and $s_{i+t} = 1$ or $s_i = 1$ and $s_{i+t} = 0$ is 2^{n-1}. By symmetry arguments the number of pairs (s_i, s_{i+t}) such that $s_i = 0$ and $s_{i+t} = 1$ is 2^{n-2}. Similarly, the number of pairs (s_i, s_{i+t}) such that $s_i = 1$ and $s_{i+t} = 0$ is 2^{n-2}. Since the number of *zeros* in S is $2^{n-1} - 1$, it follows that the number of pairs (s_i, s_{i+t}) such that $s_i = s_{i+t} = 0$ is $(2^{n-1} - 1) - 2^{n-2} = 2^{n-2} - 1$. $\quad\square$

Let p be an odd prime of the form $4m - 1$ and let $\{a_i\}_{i=0}^{p-1}$ the sequence defined by $a_i = \left(\frac{i}{p}\right)$, for $1 \leq i < p - 1$ and a_0 is either $+1$ or -1, where $\left(\frac{i}{p}\right)$ is the Legendre symbol. This sequence is called the **Legendre sequence** or the **quadratic residues sequence**. Its characteristic vector is obtained when $+1$ is replaced by 0 and -1 is replaced by 1.

The following theorem is obtained as a consequence of Theorem 1.13.

Theorem 2.20. *The set* $\mathbb{D}$ *whose characteristic vector is the Legendre sequence of length* $p = 4n - 1$, *where* p *is a prime (the set of quadratic residues modulo* p) *forms a* $(4n - 1, 2n - 1, n - 1)$ *Hadamard difference set.*

We continue to consider other Hadamard difference sets that are associated with sequences satisfying **R-3**. There are three known parameters for families of Hadamard difference sets.

(H.1) $v = 2^n - 1$ and this family is associated mainly with M-sequences, but there are other constructions for this set of parameters.

(H.2) $v = 4n - 1$ is a prime and this type is associated with quadratic residues sequences.

(H.3) $v = p(p+2)$, where p and $p+2$ are primes; these primes are called twin primes, and the associated sequences are called twin primes sequences.

Can one of the parameters be obtained via two families? The only v that can be covered by both **(H.1)** and **(H.3)** is $v = 15$. Clearly, v cannot be covered by both **(H.2)** and **(H.3)**. Finally, the values of v that are covered by both **(H.1)** and **(H.2)** are primes of the form $v = 2^n - 1$, i.e., Mersenne primes.

2.5 Notes

The linear theory of shift register is well covered by Golomb [36]. Some items appear there and are not covered in our exposition. These items include analysis of irreducible polynomials, factorization of polynomials, cyclotomic cosets, and Fourier analysis of M-sequences. A chapter on the linear theory of shift registers also appears in the book of Lidl and Niederreiter [60].

Section 2.1. The approach in this chapter to compute the length of the cycles generated by an LFSR$_n$ is based on a combination of their algebraic and combinatorial structure, while in most papers and books the approach is purely algebraic. Some of the material covered in this section appears in the book of Golomb [36]. Some of the proofs that we gave cover cases that were omitted by Golomb [36]. Cycles generated from a reducible polynomial were considered by Lidl and Niederreiter [60]. The structure of cycles from LFSR$_n$ was also discussed by Elspas [26]. Further analysis of sequences generated by the corresponding polynomials was given by Chee, Chrisnata, Etzion, and Kiah [16].

Section 2.2. M-sequences are probably the most important set of sequences and there is an enormous amount of research that was carried out on these sequences, their properties, and their applications. The book of Golomb [36] should be the first source on these sequences. Applications of M-sequences to coding theory can be found in Etzion [28] and Weng [82], to pseudorandom number generators in Arvillias and Maritsas [3], to orthogonal sequences in Kirimoto and Oh-Hashi [55], to ophthalmic electrophysiology in Müller and Meigen [68], to linear time-invariant systems for signal processing in Xiang [86]. They are used to construct other sequences with good correlation properties, see for example Canteaut, Charpin, and Dobbertin [11], Games [31], Helleseth [44], Helleseth, Lahtonen, and Rosendahl [45], Helleseth and Rosendahl [46], Hollmann and Xiang [47], Lindholm [61], Ness and Helleseth [69], and Sutter [76]. For more information on M-sequences, the reader is referred to the extensive literature, see for example Briggs and Godfrey [8], Chang [12], Chang and Ho [14], Godfrey [33], Gold [34,35], Harvey [42], Mayo [65], Pursley and Sarwate [73], Tretter [80], Tsao [81], Willett [83,84], and Zierler [88].

Theorem 2.11 was proved by Bryant, Heath, and Killick [9] and Yoeli [87] for $\mathbb{F}_2$ and G_n. The theorem was extended for any de Bruijn graph $G_{\sigma,n}$ for any positive integer $\sigma > 1$ by Lempel [59]. He proved that in $G_{\sigma,n}$ there exists a cycle of length k for each k such that $1 \le k \le \sigma^n$. Another proof for the same claim and an algorithm to generate the associated sequences was presented by Etzion [27]. The algorithm was presented only for the binary case with a straightforward generalization for any alphabet.

Section 2.3. The analysis of the length of the cycles obtained by powers of primitive polynomials over $\mathbb{F}_2$ can be generalized for $\mathbb{F}_q$, where $q > 2$ with a similar analysis as was done in this section. This is left as an exercise for the reader.

Section 2.4. Difference sets form an important branch of block design. They are considered in several books, such as Baumert [4], Ding [24], and Jungnickel [53]. There are more constructions for the family **(H.1)**, where $v = 2^n - 1$, of Hadamard difference sets. For some values of composite n we have such sets associated with GMW sequences. These sequences were considered first by Gordon, Mills, and Welch [39] and later by Scholtz and Welch [75]. Baumert and Fredricksen [5] found three Hadamard difference sets for $v = 127$. Cheng [19] found two Hadamard difference sets for $v = 255$. An exhaustive search for $v = 511$ was carried out by Dreier and Smith [25].

As for the family **(H.2)** the only known difference sets except for the quadratic residues difference sets, are the difference sets constructed by Hall [41] for primes of the form $p = x^2 + 27$. These difference sets are called Hall sextic residue difference sets.

The twin primes Hadamard difference sets were constructed by Stanton and Sprott [78]. The generated sequences are also called Jacobi sequences and they are generalizations of the quadratic residues (Legendre) sequences. These are the only known sequences of the family **(H.3)**.

More information on the existence of Hadamard difference sets can be found in the papers by Golomb and Song [38,77] and the most updated information in the work of Charpáin [15]. Constructions and applications of difference sets can be found in Golomb [37]. The most important question associated with Hadamard difference sets is the following.

Problem 2.1. Do there exist more parameters for which there exists a Hadamard difference set?

Hadamard difference sets obtained their name as they can be used to form Hadamard matrices. An $n \times n$ matrix H of $\{-1, +1\}$ is a **Hadamard matrix** of order n if $H \cdot H^{\mathrm{tr}} = H^{\mathrm{tr}} \cdot H = nI_n$.

From a Hadamard difference set, it is straightforward to generate a related Hadamard matrix. Let S be the characteristic vector of a $(4m - 1, 2m - 1, m - 1)$ Hadamard difference set and form a $(4m - 1) \times (4m - 1)$ matrix A from all the cyclic shifts of S. Let H be the $(4m) \times (4m)$ matrix obtained from A by replacing each "0" by "+1", each "1" by "−1", adding a row of "−1"s at the end, and a column of "−1"s at the right. The matrix obtained in this way is a Hadamard matrix of order $4m$.

Theorem 2.21. *If there exists a $(4m − 1, 2m − 1, m − 1)$ Hadamard difference set, then there exists a Hadamard matrix of order $4m$.*

Theorem 2.22. *The $2^n − 1$ cyclic shifts of an M-sequence of length $2^n − 1$ with the all-zeros word form a $[2^n − 1, n, 2^{n-1}]$ code.*

Proof. Let C be a code obtained from the $2^n − 1$ cyclic shift of an M-sequence and the all-zeros sequence. The length of the code $2^n − 1$ is an immediate consequence of the length of the M-sequence. From the shift-and-add property, we have that C is closed under addition and since it has 2^n codewords, it follows that its dimension is n. All the nonzero codewords have weight 2^{n-1} and hence the minimum distance of C is 2^{n-1}.   □

Definition 2.9. A $[2^n − 1, n, 2^{n-1}]$ code is a well-known code called the *simplex code*.

Each nonzero n-tuple appears exactly once as a column in any generator matrix of the $[2^n − 1, n, 2^{n-1}]$ simplex code and hence any n consecutive shifts of a span n M-sequences can form a generator matrix of the code.

A *normalized Hadamard matrix* is a Hadamard matrix in which the first row and the first column have only +1s. One can readily verify that each Hadamard matrix can be made a normalized Hadamard matrix by multiplying each row and each column starting with a −1 by −1.

Theorem 2.23. *If a Hadamard matrix of order n exists, then n is 1, 2, or a multiple of 4.*

Proof. It is easy to verify that Hadamard matrices of order 1 and 2 exist and a Hadamard matrix of order 3 does not exist. If $n > 3$ and a Hadamard matrix H of order n exists, w.l.o.g. assume that H is a normalized Hadamard matrix and consider the first three rows of A. The triples, generated by the columns, in these three rows, can be $(+1, +1, +1)^{tr}$, $(+1, +1, −1)^{tr}$, $(+1, −1, +1)^{tr}$ or $(+1, −1, −1)^{tr}$. Assume that there are i_1 triples of the form $(+1, +1, +1)^{tr}$, i_2 triples of the form $(+1, +1, −1)^{tr}$, i_3 triples of the form $(+1, −1, +1)^{tr}$, and i_4 triples of the form $(+1, −1, −1)^{tr}$. Since $H^{tr}H = nI_n$ implies that every two distinct rows are orthogonal, i.e., their inner product is equal to *zero*, it follows that

$$i_1 + i_2 − i_3 − i_4 = 0, \quad \text{from the first row and second row,}$$
$$i_1 − i_2 + i_3 − i_4 = 0, \quad \text{from the first row and third row,}$$
$$i_1 − i_2 − i_3 + i_4 = 0, \quad \text{from the second row and third row.}$$

The solution for this set of three equations implies that $i_1 = i_2 = i_3 = i_4$, i.e., n is divisible by 4.   □

There are many constructions for Hadamard matrices and it is conjectured that for each n divisible by 4, there exists such a matrix.

The most celebrated construction of Hadamard matrices is Sylvester's construction [79], which is a simple doubling construction. Let H be any $n \times n$ Hadamard matrix. The following matrix

$$\begin{bmatrix} H & H \\ H & -H \end{bmatrix}$$

is a $(2n) \times (2n)$ Hadamard matrix.

This construction immediately yields a Hadamard matrix for each order that is a power of 2, but it is also applied to other orders for which Hadamard matrices are known to exist.

Sylvester's construction can be generalized as follows. Let A be an $n \times n$ Hadamard matrix and B be an $m \times m$ Hadamard matrix. The Kronecker product of A and B defined by

$$\begin{bmatrix} a_{11}B & a_{12}B & \cdots & a_{1n}B \\ a_{21}B & a_{22}B & \cdots & a_{2n}B \\ \vdots & \vdots & \ddots & \vdots \\ a_{n1}B & a_{n2}B & \cdots & a_{nn}B \end{bmatrix}$$

is an $(nm) \times (nm)$ Hadamard matrix. It generalizes Sylvester's construction in which the matrix A is the following 2×2 Hadamard matrix

$$A = \begin{bmatrix} 1 & 1 \\ 1 & -1 \end{bmatrix}$$

in the Kronecker product.

Hadamard matrices were defined by Hadamard in [40], where it was proved that any real $n \times n$ matrix A, with real entries between -1 and $+1$, satisfies $|\det A| \leq n^{n/2}$. Hadamard matrices meet this bound. These matrices have many applications in coding theory and various areas of communication, information theory, and computer science, as well as in other areas, see Hedayat and Wallis [43], Horadam [48,49], and Seberry, Wysocki, and Wysocki [74]. They are part of a family of matrices called weighing matrices. Coding with Hadamard matrices and related weighing matrices was done, for example, by Etzion, Vardy, and Yaakobi [29]. The first order divisible by 4, for which no Hadamard matrix is known to exist (as of 2023) is 668, after the previous order of 428 was constructed by Kharaghani and Tayfeh-Rezaie [54]. An excellent book for information on Hadamard matrices is that written by Geramita and Seberry [32] and on the matrices and their applications by Horadam [48].

Difference sets with parameters $(v, k, 1)$ are of special interest as each nonzero residue modulo v is obtained exactly once as a difference. A related concept is a Golomb ruler. A **Golomb ruler** of order k (or a k-**mark Golomb ruler**) is a sequence with k nonnegative integers $(d_1 = 0, d_2, \ldots, d_k)$ such that $d_i < d_j$ when $i < j$, and all the $\binom{k}{2}$ differences $\{d_j - d_i : 0 \le i < j \le k\}$, which are the distances measured by the ruler, are distinct. Each difference set yields a ruler, but usually a ruler does not yield a difference set. The main goal is to find such a ruler with the smallest possible largest number d_k. If also $d_2 < d_k - d_{k-1}$, then such a Golomb ruler is called optimal. For some work on Golomb rulers see Meyer and Jaumard [66]. Another interesting concept in this context is a **difference triangle**. For a sequence of k integers, $b_1 < b_2 < \cdots < b_k$, the difference triangle has $k - 1$ rows, where in the rth row, $1 \le r \le k - 1$, we have the differences $b_{j+r} - b_j$, for $j = 1, 2, \ldots, k - r$, ordered by their appearance in the sequence. Each distance (difference) measured by a ruler appears in a unique row of the triangle. The difference triangle computes the differences in the finite sequence and not just in a ruler. The definition of a difference triangle is modified to a definition of a **difference cylinder** when the sequence is cyclic and we have to consider the difference $b_{j+r} - b_j$ of the rth row for $j = 1, 2, \ldots, k$, where subscripts are taken modulo k. In the difference cylinder, each row has k differences.

The difference triangle (and the difference cylinder) is an important tool to analyze the properties of structures with distinct differences like a ruler. It provides all the differences of the combinatorial structure and also the difference in the positions of the structure where the associated difference appears. There are many other related concepts and some are associated with graphs, as explained in an early survey by Bloom and Golomb [7].

Example 2.10. The unique optimal 8-mark Golomb ruler is

$$(0, 1, 4, 9, 15, 22, 32, 34).$$

The difference triangle of the ruler is

0	1	4	9	15	22	32	34
	1	3	5	6	7	10	2
		4	8	11	13	17	12
			9	14	18	23	19
				15	21	28	25
					22	31	30
						32	33
							34

∎

The concept of difference sets was generalized to difference families, where there are several sets $\mathbb{D}_1, \mathbb{D}_2, \ldots$, each of size k, where each nonzero residue modulo n occurs as a difference in exactly λ of these sets. Constructions and bounds for such a family of designs can be found, for example, in Bitan and Etzion [6], Buratti [10], and Wilson [85]. This family of designs, with $\lambda = 1$, is also related to a family of codes called optical orthogonal codes or constant-weight cyclically permutable codes, see A, Gyorfi, and Massey [1], Bitan and Etzion [6], Chung and Kumar [22], Chung, Salehi, and Wei [23], and Moreno, Zhang, Kumar, and Zinoviev [67]. There are also generalizations for the concept of difference family. One of the most interesting generalizations is the concept of external difference sets that have several variants. There are several sets whose elements are taken from an Abelian group G and the differences that are considered are between the different sets. Without going into the different interesting variants, some work on this topic can be found, for example, in Chang and Ding [13], Fujiwara and Tonchev [30], Huang and Wu [50], Huczynska and Paterson [51], Jedwab and Li [52], Lu, Niu, and Cao [64], Ogata, Kurosawa, Stinson, and Saido [71], and Paterson and Stinson [72]. The motivations for all these variants of external difference sets are from different concepts in cryptography and security, as was presented by Ogata, Kurosawa, Stinson, and Saido [71] and later by Paterson and Stinson [72].

A survey on all types of difference families was given by Ng and Paterson [70].

Sets of sequences with distinct differences can be analyzed using sets of disjoint difference triangles. A comprehensive work on such sets was carried out by Kløve [56–58] and for other work on this topic see Chee and Colbourn [17] Chen [18], Chu, Colbourn, and Golomb [20], Chu and Golomb [21], Ling [62], and Lorentzen and Nilsen [63]. An application of difference triangle sets to constructions of LDPC codes was suggested by Alfarano, Lieb, and Rosenthal [2].

References

[1] N.Q. A, L. Gyorfi, J.L. Massey, Constructions of binary constant-weight cyclic codes and cyclically permutable codes, IEEE Trans. Inf. Theory 38 (1992) 940–949.

[2] G.N. Alfarano, J. Lieb, J. Rosenthal, Construction of LDPC convolutional codes via difference triangle sets, Des. Codes Cryptogr. 89 (2021) 2235–2254.

[3] A.C. Arvillias, D.G. Maritsas, Partitioning the period of a class of m-sequences and application to pseudorandom number generation, J. ACM 25 (1978) 675–686.

[4] L.D. Baumert, Cyclic Difference Sets, Springer, Berlin, 1971.

[5] L.D. Baumert, H. Fredricksen, The cyclotomic numbers of order eighteen with applications to difference sets, Math. Comput. 21 (1967) 204–219.

[6] S. Bitan, T. Etzion, Constructions for optimal constant weight cyclically permutable codes and difference families, IEEE Trans. Inf. Theory 41 (1995) 77–87.

[7] G.S. Bloom, S.W. Golomb, Applications of numbered undirected graphs, Proc. IEEE 65 (1977) 562–570.

[8] P.A.N. Briggs, K.R. Godfrey, Autocorrelation function of a 4-level m sequence, Electron. Lett. 4 (1968) 232–233.

[9] P.R. Bryant, F.G. Heath, R.D. Killick, Counting with feedback shift registers by means of a jump technique, IRE Trans. Electron. Comput. 11 (1962) 285–286.

[10] M. Buratti, Improving two theorems of Bose on difference families, J. Comb. Des. 3 (1995) 15–24.

[11] A. Canteaut, P. Charpin, H. Dobbertin, Binary m-sequences with three-valued crosscorrelation: a proof of Welch's conjecture, IEEE Trans. Inf. Theory 46 (2000) 4–8.

[12] J.A. Chang, Generation of 5-level maximal-length sequences, Electron. Lett. 2 (1966) 258.

[13] Y. Chang, C. Ding, Constructions of external difference families and disjoint difference families, Des. Codes Cryptogr. 40 (2006) 167–185.

[14] R.W. Chang, E.Y. Ho, On fast start-up data communication systems using pseudo-random training sequences, Bell Syst. Tech. J. 51 (1972) 2013–2027.

[15] P.Ó. Charpáin, Inequivalent of difference sets: on a remark of Baumert, Electron. J. Comb. 20 (2013) #P38.

[16] Y.M. Chee, J. Chrisnata, T. Etzion, H.M. Kiah, Efficient algorithm for the linear complexity of sequences and some related consequences, in: Proc. IEEE Int. Symp. Infor. Theory (ISIT), 2020, pp. 2897–2902.

[17] Y.M. Chee, C.J. Colbourn, Constructions for difference triangle sets, IEEE Trans. Inf. Theory 43 (1997) 1346–1349.

[18] Z. Chen, Further results on difference triangle sets, IEEE Trans. Inf. Theory 40 (1994) 1268–1270.

[19] U. Cheng, Exhaustive construction of $(255, 127, 63)$-cyclic difference sets, J. Comb. Theory, Ser. A 35 (1983) 115–125.

[20] W. Chu, C.J. Colbourn, S.W. Golomb, A recursive construction for regular difference triangle sets, SIAM J. Discrete Math. 18 (2005) 741–748.

[21] W. Chu, S.W. Golomb, A note on the equivalence between strict optical orthogonal code and difference triangle sets, IEEE Trans. Inf. Theory 49 (2003) 759–761.

[22] H. Chung, P.V. Kumar, Optical orthogonal codes – new bounds and an optimal construction, IEEE Trans. Inf. Theory 36 (1990) 866–873.

[23] F.R.K. Chung, J.A. Salehi, V.K. Wei, Optical orthogonal codes: design, analysis, and applications, IEEE Trans. Inf. Theory 35 (1989) 595–604.

[24] C. Ding, Codes from Difference Sets, World Scientific, Singapore, 2014.

[25] R.B. Dreier, K.W. Smith, Exhaustive determination of $(511, 255, 127)$-cyclic difference sets, unpublished manuscript.

[26] B. Elspas, Theory of autonomous linear sequential networks, IRE Trans. Circuit Theory 6 (1959) 45–60.

[27] T. Etzion, An algorithm for generating shift-register cycles, Theor. Comput. Sci. 44 (1986) 209–224.

[28] T. Etzion, Perfect byte-correcting codes, IEEE Trans. Inf. Theory 44 (1998) 3140–3146.

[29] T. Etzion, A. Vardy, E. Yaakobi, Coding for the Lee and Manhattan metrics with weighing matrices, IEEE Trans. Inf. Theory 59 (2013) 6712–6723.

[30] Y. Fujiwara, V.D. Tonchev, High-rate self-synchronizing codes, IEEE Trans. Inf. Theory 59 (2013) 2328–2335.

[31] R.A. Games, Crosscorrelation of M-sequences and GMW-sequences with the same primitive polynomial, Discrete Appl. Math. 12 (1985) 139–146.

[32] A.V. Geramita, J. Seberry, Orthogonal Designs: Quadratic Forms and Hadamard Matrices, Mercel Dekker, New York, 1979.

[33] K.R. Godfrey, Three-level m-sequences, Electron. Lett. 2 (1966) 241–243.

[34] R. Gold, Characteristic linear sequences and their coset functions, J. SIAM Appl. Math. 14 (1966) 980–985.

[35] R. Gold, Maximal recursive sequences with 3-valued recursive cross-correlation functions, IEEE Trans. Inf. Theory 14 (1968) 154–156.

[36] S.W. Golomb, Shift Register Sequences, Holden Day, San Francisco, 1967, Aegean Park, Laguna Hills, CA, 1980, World Scientific, Singapore, 2017.

[37] S.W. Golomb, Cyclic Hadamard difference sets - constructions and applications, in: Proc. Sequences Their Appl. (SETA), 1998, pp. 39–48.

[38] S.W. Golomb, H.-Y. Song, A conjecture on the existence of cyclic Hadamard difference sets, J. Stat. Plan. Inference 62 (1997) 39–41.

[39] B. Gordon, W.H. Mills, L.R. Welch, Some new difference sets, Can. J. Math. 14 (1962) 614–625.

[40] J. Hadamard, Résolution d'une question relative aux déterminants, Bull. Sci. Math. 17 (1893) 240–248.

[41] M. Hall Jr., A survey of difference sets, Proc. Am. Math. Soc. 7 (1956) 975–986.

[42] J.T. Harvey, High-speed m-sequences generation, Electron. Lett. 10 (1974) 480–481.

[43] A. Hedayat, W.D. Wallis, Hadamard matrices and their applications, Ann. Stat. 6 (1978) 1184–1238.

[44] T. Helleseth, Correlation of m-sequences and related topics, in: Proc. Sequences Their Appl. (SETA), 1998, pp. 49–66.

[45] T. Helleseth, J. Lahtonen, P. Rosendahl, On Niho type cross-correlation functions of m-sequences, Finite Fields Appl. 13 (2007) 305–317.

[46] T. Helleseth, P. Rosendahl, New pairs of m-sequences with 4-level cross-correlation, Finite Fields Appl. 11 (2005) 674–683.

[47] H.D.L. Hollmann, Q. Xiang, A proof of the Welch and Niho conjectures on cross-correlations of binary m-sequences, Finite Fields Appl. 7 (2001) 253–286.

[48] K.J. Horadam, Hadamard Matrices and Their Applications, Princeton University Press, Princeton, NJ, 2007.

[49] K.J. Horadam, Hadamard matrices and their applications: progress 2007 - 2010, Cryptogr. Commun. 2 (2010) 129–154.

[50] B. Huang, D. Wu, Cyclotomic constructions of external difference families and disjoint difference families, J. Comb. Des. 17 (2009) 333–341.

[51] S. Huczynska, M.B. Paterson, Existence and non-existence results for strong external difference families, Discrete Math. 341 (2018) 85–97.

[52] J. Jedwab, S. Li, Construction and nonexistence of strong external difference families, J. Algebraic Comb. 49 (2019) 21–48.

[53] D. Jungnickel, Difference sets, in: J.H. Dinitz, D.R. Stinson (Eds.), Contemporary Design Theory: A Collection of Surveys, Wiley, New York, 1992, pp. 241–324.

[54] H. Kharaghani, B. Tayfeh-Rezaie, A Hadamard matrix of order 428, J. Comb. Des. 13 (2004) 435–448.

[55] T. Kirimoto, Y. Oh-Hashi, Orthogonal periodic sequences derived from M-sequences on GF(q), IEEE Trans. Inf. Theory 40 (1994) 526–532.

[56] T. Kløve, Bounds on the size of optimal difference triangle sets, IEEE Trans. Inf. Theory 34 (1988) 355–361.

[57] T. Kløve, Bounds and constructions for difference triangle sets, IEEE Trans. Inf. Theory 35 (1989) 879–886.

[58] T. Kløve, Bounds and constructions of disjoint sets of distinct difference sets, IEEE Trans. Inf. Theory 36 (1990) 184–190.

[59] A. Lempel, m-ary closed sequences, J. Comb. Theory 10 (1971) 253–258.

[60] R. Lidl, H. Niederreiter, Introduction to Finite Fields and Their Applications, Cambridge Univ. Press, Cambridge, United Kingdom, 1994.

[61] J.H. Lindholm, An analysis of the pseudo-randomness properties of subsequences of long m-sequences, IEEE Trans. Inf. Theory 14 (1968) 569–576.

[62] A.C.H. Ling, Difference triangle sets from affine planes, IEEE Trans. Inf. Theory 48 (2002) 2399–2401.

[63] R. Lorentzen, R. Nilsen, Application of linear programming to the optimal difference triangle set problem, IEEE Trans. Inf. Theory 37 (1991) 1486–1488.

[64] X. Lu, X. Niu, H. Cao, Some results on generalized strong external difference families, Des. Codes Cryptogr. 86 (2018) 2857–2868.

[65] E.A. Mayo, Efficient computer decoding of pseudorandom radar signal codes, IEEE Trans. Inf. Theory 18 (1972) 680–681.

[66] C. Meyer, B. Jaumard, Equivalence of some LP-based lower bounds for the Golomb ruler problem, Discrete Appl. Math. 154 (2006) 120–144.

[67] O. Moreno, Z. Zhang, P.V. Kumar, V.A. Zinoviev, New constructions of optimal cyclically permutable constant weight codes, IEEE Trans. Inf. Theory 41 (1995) 448–455.

[68] P.L. Müller, T. Meigen, M-sequences in ophthalmic electrophysiology, J. Vis. 16 (2016) 1–19.

[69] G.J. Ness, T. Helleseth, Cross correlation of m-sequences of different lengths, IEEE Trans. Inf. Theory 52 (2006) 1637–1648.

[70] S.-L. Ng, M.B. Paterson, Disjoint difference families and their applications, Des. Codes Cryptogr. 78 (2016) 103–127.

[71] W. Ogata, K. Kurosawa, D.R. Stinson, H. Saido, New combinatorial designs and their applications to authentication codes and secret sharing schemes, Discrete Math. 279 (2004) 383–405.

[72] M.B. Paterson, D.R. Stinson, Combinatorial characterizations of algebraic manipulation detection codes involving generalized difference families, Discrete Math. 339 (2016) 2891–2906.

[73] M.B. Pursley, D.V. Sarwate, Bounds on aperiodic crosscorrelation for binary sequences, Electron. Lett. 12 (1976) 304–305.

[74] J. Seberry, B.J. Wysocki, T.A. Wysocki, On some applications of Hadamard matrices, Metrika 62 (2005) 221–239.

[75] R.A. Scholtz, L.R. Welch, GMW sequences, IEEE Trans. Inf. Theory 30 (1984) 548–553.

[76] E.E. Sutter, The fast m-transform: a fast computation of crosscorrelations with binary m-sequences, SIAM J. Comput. 20 (1991) 686–694.

[77] H.Y. Song, S.W. Golomb, On the existence of cyclic Hadamard difference sets, IEEE Trans. Inf. Theory 40 (1994) 1266–1268.

[78] R.G. Stanton, D.A. Sprott, A family of difference sets, Can. J. Math. 10 (1958) 73–77.

[79] J.J. Sylvester, Thoughts on inverse orthogonal matrices, simultaneous signsuccessions, and tessellated pavements in two or more colours, with applications to Newton's rule, ornamental tile-work, and the theory of numbers, Philos. Mag. 34 (1867) 461–475.

[80] S.A. Tretter, Properties of PN^2 sequences, IEEE Trans. Inf. Theory 20 (1974) 295–297.

[81] S.H. Tsao, Generation of delayed replicas of maximal-length linear binary sequences, Proc. IEE 111 (1964) 1803–1806.

[82] L.-J. Weng, Decomposition of M-sequences and its applications, IEEE Trans. Inf. Theory 17 (1971) 457–463.

[83] M. Willett, The minimum polynomial for a given solution of a linear recursion, Duke Math. J. 39 (1972) 101–104.

[84] M. Willett, The index of an m-sequence, SIAM J. Appl. Math. 25 (1973) 24–27.

[85] R.M. Wilson, An existence theory for pairwise balanced designs II. The structure of PBD-closed sets and the existence conjectures, J. Comb. Theory, Ser. A 13 (1972) 246–273.

[86] N. Xiang, Using M-sequences for determining the impulse responses of LTI-systems, Signal Process. 28 (1992) 139–152.

[87] M. Yoeli, Binary ring sequences, Am. Math. Mon. 69 (1962) 852–855.

[88] N. Zierler, Linear recurring sequences, J. Soc. Ind. Appl. Math. 7 (1959) 31–48.

Chapter 3

Cycles and the nonlinear theory

Necklaces, self-dual sequences, enumeration

After discussing in Chapter 2 the theory of linear feedback shift registers, the length of the cycles in the state diagram of such registers, and the class of M-sequences, we turn our attention to specific shift-register feedback functions and the nonlinear theory of shift registers.

In Section 3.1 we define a few feedback functions for some of the most interesting shift registers. We consider the structure of their cycles and the relations between the cycles of the different state diagrams.

Enumerations for the number of cycles in the state diagrams of these shift registers are presented in Section 3.2. These enumerations involve the Euler function and the Möbius function discussed in Section 1.1. We discuss the applications of these functions in counting the number of irreducible and primitive polynomials. Special attention in the first two sections will be given to self-dual sequences, i.e., sequences that are invariant under complements of their bits.

Section 3.3 is devoted to the following intriguing question: what is the maximum number of cycles in a state diagram of an FSR_n? The answer to this question is given by a coloring of the vertices in the de Bruijn graph. This coloring is defined based on the position of the center of mass of a vertex. The center of mass is based on the bits of its representation that are located on the two-dimensional plane. It appears that one of the specific shift registers defined in Section 3.1 attains the maximum number of cycles, but quite surprisingly, by using the merge-or-split method many other such shift registers can be found.

3.1 Cycles from feedback shift registers

There are a few families of feedback shift registers that are of special interest. Each such family is associated with a different feedback function. Some of these functions are linear functions and some of them are nonlinear functions. The first such family, which was discussed in Section 2.2, is the family of M-sequences that are generated from linear feedback functions associated with primitive polynomials.

The next four families of FSR_n are associated with four simple feedback functions: two linear functions and two nonlinear ones. The first one is called the

PCR$_n$ for the **pure cycling register** of order n, and its linear feedback function is

$$f(x_1, x_2, \ldots, x_n) = x_1 .$$

This definition implies the following result.

Lemma 3.1. *The weight function of the* PCR$_n$ *is* zero.

Let $Z(n)$ be the number of cycles in the state diagram of the PCR$_n$. The computation for the exact value of $Z(n)$ will be presented in Section 3.2.

The **extended representation** $E(\mathcal{C})$ of a cycle $\mathcal{C}$ from the PCR$_n$ is given by an n-tuple $E(\mathcal{C}) \triangleq [x_1 \cdots x_{n-1} x_n]$, where $x = (x_1, x_2, \ldots, x_n)$ is any state on $\mathcal{C}$ and also each cyclic shift of x is a state of $\mathcal{C}$. If the cycle $\mathcal{C}$ is of length smaller than n, i.e., it is not of a full-order, then some of these states formed by cyclic shifts are equal. Each state in the PCR$_n$ is represented by a cyclic shift of the extended representation of its cycle. Hence, the weight of a state is the same as the weight of the extended representation of its cycle. The **extended weight** of a cycle $\mathcal{C}$, wt$_E(\mathcal{C})$ is the weight of any extended representation of $\mathcal{C}$ (for other FSR$_n$ too). Using the extended weight, we have the following simple lemma.

Lemma 3.2. *A state and its companion cannot be states on the same* PCR$_n$ *cycle.*

Lemma 3.3. *The length of a cycle in the state diagram of the* PCR$_n$ *is a divisor of* n.

Proof. By the definition of the feedback function $f(x_1, x_2, \ldots, x_n)$ we have that the edges in the state diagram of the PCR$_n$ are of the form

$$(x_1, x_2, \ldots, x_n) \rightarrow (x_2, \ldots, x_n, x_1),$$

i.e., the states in the cycle that contain the state $x = (x_1, x_2, \ldots, x_n)$ are cyclic shifts of $(x_1, x_2, \ldots, x_n)$. If there are no repeated states in the cycle that contains all the n cyclic shifts of x, then the length of the cycle in the state diagram is n. If there is a repeated state, then w.l.o.g. we assume that the state x is repeated. Assume further that it is repeated μ times on the cycles and let P_i, $1 \le i \le \mu$, the path that starts with the ith appearance of x and ends at the vertex before its $(i+1)$th appearance (where the $(\mu+1)$th appearance is the first appearance). Clearly, each path $P_i P_{i+1} \cdots P_\mu P_1 \cdots P_{i-1}$, $i \ge 2$ (as the path $P_1 P_2 \cdots P_\mu$), is a path of length n from x to x and hence by Lemma 1.15 all these paths are equal. This implies that $P_i = P_j$ for $1 \le i < j \le \mu$ and P_1 is the associated cycle in the state diagram. If ℓ is the length of P_1, then $n = \mu \cdot \ell$ and hence ℓ divides n. $\qquad\square$

Any cyclic sequence of period π can be viewed as a sequence in PCR$_n$ for any n that is divisible by π and hence we have the following consequence.

Corollary 3.1. *The period of a cyclic sequence of length m is a divisor of m, i.e., a sequence of length m is associated with a cycle of length d in G_m, for some d that is a divisor of m.*

Corollary 3.2. *If a state on a* PCR_n *cycle of length ℓ has weight w, then the weight of the cycle (not its extended weight) is* $\frac{w \cdot \ell}{n}$.

It should be noted that the length of a cycle in the state diagram of an FSR_n is the period of the associated sequence. If the sequence S is of length n and its period d is smaller than n, then as a cycle S is not a simple cycle. It is a cycle with $\frac{n}{d}$ repetitions of the cycle of length d that is contained in the state diagram generated by an FSR_n that generates the sequence S. By Lemma 3.3 the length of a cycle from the PCR_n divides n, but we also have that for each d that divides n, there is at least one cycle of length d in the PCR_n. If $d = 1$, then there are two cycles of length d in the PCR_n, the cycles $[0]$ and $[1]$. If $d > 1$ is a divisor of n then the cycle $[0^{d-1}1]$ is one of the cycles of length d in the PCR_n.

Theorem 3.1. *The weight of the feedback function of a span n de Bruijn sequence with the minimum weight function is* $Z(n) - 1$.

Proof. First, recall that by Lemma 3.1, we have that the weight function of the PCR_n is 0 and the state diagram of the PCR_n has $Z(n)$ cycles.

Assume that S is a de Bruijn sequence of order n with weight function $\omega(n)$. By Lemma 1.20 we have that changing a *one* to a *zero* in the truth table of the state diagram either splits one cycle into two cycles or merges two cycles into one cycle. Therefore changing all the $\omega(n)$ *ones* of the truth table of S to *zeros* yields a state diagram with at most $\omega(n) + 1$ cycles. However, this truth table is that of the PCR_n since it has weight function *zero*. Therefore this truth table has $Z(n)$ cycles and $\omega(n) + 1 \geq Z(n)$, i.e., $\omega(n) \geq Z(n) - 1$.

To complete the proof we have to show that there exists a span n de Bruijn sequence S whose weight function is at most $Z(n) - 1$. Applying appropriately the merge-or-split method $Z(n) - 1$ times on the cycles of the state diagram of the PCR_n by using Theorem 1.20 yields a span n de Bruijn sequence whose weight function is $Z(n) - 1$.

The two parts of the proof imply that the weight of the feedback function of a span n de Bruijn sequence with the minimum weight function is $Z(n) - 1$. $\square$

The state diagram of the PCR_n is also called the ***necklaces factor*** as each cycle can be viewed as a necklace with beads of two colors (a ***necklace*** is a cycle in the state diagram of the PCR_n). The number of beads is the length of the cycle. The same feedback function can be defined over any alphabet with σ letters. The associated state diagram for the larger alphabet is also called the necklaces factor of order n, over σ beads, and each cycle has also a length that is a divisor of n with identical proof to the one of Lemma 3.3. The necklaces factor will be used in other parts of this book for various applications. The number of cycles for each length in the state diagram will be computed in the next section.

The second FSR_n is called the CCR_n for the ***complemented cycling register*** of order n, and its nonlinear feedback function is

$$f(x_1, x_2, \ldots, x_n) = x_1 + 1.$$

By definition and also by Lemma 3.1 the weight function of the PCR_n is the minimum possible for a weight function. For the CCR_n, the weight function is the maximum possible one. The sequence associated with the truth table of the CCR_n is the complement of the sequence associated with the truth table of the PCR_n.

Lemma 3.4. *The weight function of the CCR_n is 2^{n-1}.*

Clearly, by the definition of the CCR_n, for each sequence $x_1 x_2 \cdots x_n$ of length n, the cyclic sequence $[x_1, x_2, \ldots, x_n, \bar{x}_1, \bar{x}_2, \ldots, \bar{x}_n]$ is generated by the CCR_n. This sequence is associated with a cycle of length $2n$ in G_n and by Corollary 3.1 the period of the sequence is a divisor of $2n$. Therefore the length of the cycles contained in the state diagram of the CCR_n is a divisor of $2n$. Let $Z^*(n)$ be the number of cycles in the state diagram of the CCR_n. The computation for the exact value of $Z^*(n)$ will be presented in the next section.

The **extended representation** $E(\mathcal{C})$ of a cycle $\mathcal{C}$ from the CCR_n is given by a $(2n)$-tuple $E(\mathcal{C}) \triangleq [x_1 \cdots x_{n-1} x_n \bar{x}_1 \cdots \bar{x}_{n-1} \bar{x}_n]$, where $(x_1, x_2, \ldots, x_n)$ is a state on $\mathcal{C}$. By the definition of the feedback function, it follows that any n consecutive bits (also cyclically) of $E(\mathcal{C})$ form a state on $\mathcal{C}$.

The path of length n from the state $x = (x_1, x_2, \ldots, x_n)$ to its complement $\bar{x} = (\bar{x}_1, \bar{x}_2, \ldots, \bar{x}_n)$ can be represented by the sequence $x_1 x_2 \cdots x_n \bar{x}_1 \bar{x}_2 \cdots \bar{x}_n$, where each n consecutive bits represent a state on this path. Since we have that $f(\bar{x}_1, \bar{x}_2, \ldots, \bar{x}_n) = x_1$, it follows that the associated cycle is

$$[x_1 x_2 \cdots x_n \bar{x}_1 \bar{x}_2 \cdots \bar{x}_n] = [X \bar{X}].$$

Moreover, when $S = [X \bar{X}]$ we have that $\bar{S} = [\bar{X} X]$, i.e., $\bar{S} \simeq S$ and hence S is a self-dual sequence. Combining this observation and using a proof similar to the one of Lemma 3.3 we have the following result.

Lemma 3.5. *A sequence S is a self-dual sequence if and only if S can be represented as $S = [X \bar{X}]$.*

Corollary 3.3. *A sequence S of length n is a self-dual sequence if it can be represented as $S = [X \bar{X} X \bar{X} \cdots X \bar{X}]$, where $[X \bar{X}]$ is a sequence of full-order.*

Corollary 3.4. *The period (and the length) of a self-dual sequence is an even integer.*

Lemma 3.6. *All the cycles of the CCR_n are self-dual cycles. A self-dual cycle from the CCR_n has period d, where d is an even divisor of $2n$ and d does not divide n.*

Proof. By the definition of the feedback function of the CCR_n we have that if $x = (x_1, x_2, \ldots, x_n)$, then it forms the cycle $[x_1 x_2 \cdots x_n \bar{x}_1 \bar{x}_2 \cdot \bar{x}_n]$ and hence by Lemma 3.5 all the cycles of the CCR_n are self-dual cycles.

Assume now that a cycle from the CCR_n has the form $[c_1 c_2 \cdots c_d]$.

By Corollary 3.1 the period d of such a cycle is a divisor of $2n$. Assume first, on the contrary, that d is odd. Hence, d divides n, i.e., $n = \mu d$, $2n = 2\mu d$, and the extended representation of the associated cycle (sequence) from the CCR_n can be written as

$$[\overbrace{c_1 c_2 \cdots c_d}^{\mu \text{ times}} \cdots \cdots c_1 c_2 \cdots c_d \overbrace{c_1 c_2 \cdots c_d}^{\mu \text{ times}} \cdots \cdots c_1 c_2 \cdots c_d].$$

However, this sequence is not of the form $[X\bar{X}]$, a contradiction. Hence, d is an even divisor of $2n$.

Assume now that d divides n. It follows that $n = \mu d$, i.e., $2n = 2\mu d$ and with the same argument, the associated extended representation of the cycle of length $2n$ is again

$$[\overbrace{c_1 c_2 \cdots c_d}^{\mu \text{ times}} \cdots \cdots c_1 c_2 \cdots c_d \overbrace{c_1 c_2 \cdots c_d}^{\mu \text{ times}} \cdots \cdots c_1 c_2 \cdots c_d].$$

However, again, this sequence is not of the form $[X\bar{X}]$, a contradiction. Hence, d is an even divisor of $2n$ and d does not divide n. □

Let $SD(n)$ denote the number of self-dual cycles of period n.

Lemma 3.7. *For each positive integer n*

$$Z^*(n) = \sum_{\substack{d|2n \\ d \nmid n}} SD(d).$$

Proof. By Lemma 3.6 we have that all the CCR_n cycles are self-dual cycles with period $d = 2k$, where d divides $2n$ and d does not divide n.

Assume now that $\mathcal{C} = [X\bar{X}]$ is a self-dual cycle of length and period d, where d divides $2n$, but d does not divide n. This implies that $2n = \mu d$, where μ is an odd integer, i.e., $\mu = 2r + 1$ and $2n = (2r + 1)d$. Hence, $\mathcal{C}$ has an extended representation of length $2n$ given by

$$[\overbrace{X\bar{X} \cdots \cdots X\bar{X}}^{r \text{ times}} X\bar{X} \overbrace{X\bar{X} \cdots \cdots X\bar{X}}^{r \text{ times}}].$$

This structure is associated with an extended representation $E(\mathcal{C})$ of a CCR_n cycle $\mathcal{C}$ since the $(n+i)$th bit is the complement of the ith bit for each $1 \leq i \leq n$.

Thus by the analysis of the CCR_n cycles and of the structure of self-dual sequences we have that

$$Z^*(n) = \sum_{\substack{d|2n \\ d \nmid n}} SD(d).$$ □

Corollary 3.5. *The cycles of length d in the state diagram of the* CCR_n *are all the self-dual cycles of period d.*

Theorem 3.2. *The weight of the feedback function of a span n de Bruijn sequence with the maximum weight function is* $2^{n-1} - Z^*(n) + 1$.

Proof. Assume that S is a de Bruijn sequence of order n with weight function $\omega(n)$. The associated truth table (top half) has $2^{n-1} - \omega(n)$ zeros. By Lemma 1.20 we have that changing a *zero* to a *one* in the truth table of the state diagram either splits one cycle into two cycles or merges two cycles into one cycle. Therefore changing all the $2^{n-1} - \omega(n)$ zeros of the truth table of S to *ones* yields a state diagram with at most $2^{n-1} - \omega(n) + 1$ cycles. However, this truth table of the state diagram that was obtained is the truth table of the CCR_n since it has weight function 2^{n-1}. Therefore $2^{n-1} - \omega(n) + 1 \geq Z^*(n)$, i.e., $2^{n-1} - Z^*(n) + 1 \geq \omega(n)$

To complete the proof we have to show that there exists a span n de Bruijn sequence S whose weight function is at least $2^{n-1} - Z^*(n) + 1$. Applying appropriately the merge-or-split method $Z^*(n) - 1$ times on the cycles of the CCR_n by using Theorem 1.20 yields a span n de Bruijn sequence with weight function $2^{n-1} - Z^*(n) + 1$.

The two parts of the proof imply that the weight of the feedback function of a de Bruijn sequence of length 2^n and the maximum weight function is $2^{n-1} - Z^*(n) + 1$. □

The third FSR_n, which will be considered, is called the PSR_n for the **pure summing register** of order n. Its linear feedback function is

$$f(x_1, x_2, \ldots, x_n) = \sum_{i=1}^{n} x_i.$$

Similarly to Lemmas 3.1 and 3.4 that give the weight functions for the PCR_n and the CCR_n, respectively, we can easily calculate the weight function of the PSR_n.

Lemma 3.8. *The weight function of the* PSR_n *is* 2^{n-2}.

Proof. The value of $f(x_1, x_2, \ldots, x_n)$ is $\sum_{i=1}^{n} x_i$, i.e., *zero* if the weight of $(x_1, x_2, \ldots, x_n)$ is even, and *one* if the weight of $(x_1, x_2, \ldots, x_n)$ is odd. Since $x_1 = 0$ for calculating the weight function, it follows that the number of *ones*, in the top half of the truth table, equals the number of distinct $(n-1)$-tuples $(x_2, \ldots, x_n)$ of odd weight. Exactly half of the $(n-1)$-tuples have odd weight and hence the weight function of the PSR_n is 2^{n-2}. □

The **extended representation** $E(C)$ of a cycle C from the PSR_n is given by an $(n+1)$-tuple $E(C) \triangleq [x_0 x_1 \cdots x_{n-1} x_n]$, where $(x_0, x_1, \ldots, x_{n-1})$ is a state on C and $x_n = \sum_{i=0}^{n-1} x_i$. Clearly, for each i, $0 \leq i \leq n$, the ith bit x_i in the extended representation is the sum modulo 2 of the other n bits, which form (when read cyclically) a state on the cycle $\mathcal{C}$. Note further that the sum of the $n + 1$ bits in this extended representation is even and all the possible extended representations are cyclic shifts of the others. Hence, by Corollary 3.1 the length of the cycles contained in the state diagram of the PSR$_n$ is a divisor of $n + 1$. Let $S(n)$ be the number of cycles in the state diagram of the PSR$_n$. The computation for the exact value of $S(n)$ will be presented in the next section.

The fourth FSR$_n$ is called the CSR$_n$ for the *complemented summing register* of order n, and its nonlinear feedback function is

$$f(x_1, x_2, \ldots, x_n) = \sum_{i=1}^{n} x_i + 1.$$

The truth table for the function of the CSR$_n$ is the complement of the truth table for the function of the PSR$_n$ and hence we have the following lemma.

Lemma 3.9. *The weight function of the* CSR$_n$ *is* 2^{n-2}.

The *extended representation* $E(\mathcal{C})$ of a cycle $\mathcal{C}$ from the CSR$_n$ is given by an $(n + 1)$-tuple $E(\mathcal{C}) \triangleq [x_0 x_1 \cdots x_{n-1} x_n]$, where $(x_0, x_1, \ldots, x_{n-1})$ is a state on $\mathcal{C}$ and $x_n = 1 + \sum_{i=0}^{n-1} x_i$. Clearly, for each i, $0 \leq i \leq n$, the ith bit x_i in the extended representation is the complement of the sum modulo 2 of the other n bits, which form (when read cyclically) a state on the cycle $\mathcal{C}$. This implies that the sum of the $n + 1$ bits in this extended representation is odd. Note further that all the extended representations are cyclic shifts of the other. Hence, by Corollary 3.1, the length of the cycles contained in the state diagram of the CSR$_n$ is a divisor of $n + 1$. Let $S^*(n)$ be the number of cycles in the state diagram of the CSR$_n$. The computation for the exact value of $S^*(n)$ will be presented in the next section.

Lemma 3.10. *All the cycles in the state diagram of the* CSR$_n$ *have odd weights.*

Proof. The sum of the $n + 1$ bits in this extended representation of a cycle of the CSR$_n$ is odd and hence the weight w of the extended representation is odd. The length of a cycle from the CSR$_n$ is a divisor d of $n + 1$ and the weight of such a cycle is $w\frac{d}{n+1}$. Since w is an odd integer and d divides $n + 1$, it follows that $w\frac{d}{n+1}$ is also an odd integer. Therefore all the cycles in the state diagram of the CSR$_n$ have odd weights. □

It is important to note that for the PSR$_n$ no lemma is analogous to Lemma 3.10. A cycle of the PSR$_n$ can be of either even weight or odd weight, although the extended representation for a cycle of the PSR$_n$ is always of even weight. A slightly restricted analog to Lemma 3.10 is the following lemma.

Lemma 3.11. *When n is even all the cycles in the state diagram of the* PSR$_n$ *have even weights.*

Proof. If n is even, then the extended representation of a cycle from the PSR_n has $n + 1$ bits, where $n + 1$ is odd and hence by Corollary 3.1 the length of the cycle in the state diagram is an odd d that divides $n + 1$. Therefore $\frac{n+1}{d}$ is odd. If w is the weight of the extended representation, then $w\frac{d}{n+1}$ is the weight of the cycle in the state diagram. Since w is even and $\frac{n+1}{d}$ is odd, it follows that $w\frac{d}{n+1}$ is even, and hence all the cycles in the state diagram of the PSR_n have even weight.   □

Example 3.1. Consider $n = 9$ and the state $x = (001000110)$. The extended representation of its cycle C in the PSR_9 is $E(C) = [0010001101]$ whose weight is 4 and it contains 10 states, i.e., its period is 10. The state $y = (001110011)$ is on a cycle C whose extended representation is $E(C) = [0011100111]$ and it has period 5 and hence as a cycle it is represented by $[00111]$, a cycle of length 5 with 5 states and weight 3.

Consider now $n = 8$ and the state $x = (01101011)$. The extended representation of its cycle C in the PSR_8 is $E(C) = [011010111]$ whose weight is 6 and it contains 9 states, i.e., its period is 9. The state $y = (01101101)$ is on a cycle C whose extended representation is $E(C) = [011011011]$ and it has period 3 and hence as a cycle it is represented by $[011]$, a cycle of length 3 with 3 states and weight 2.   ■

Lemma 3.12. *When n is even, the cycles of the CSR_n are the complements of the cycles from the PSR_n.*

Proof. Each $n + 1$ bits have either an odd weight or an even weight and can serve as the extended representation of a cycle. If these $n + 1$ bits are of odd weight, then they are associated with a cycle of the CSR_n and if they are of even weight, they are associated with a cycle of the PSR_n. When $n + 1$ is odd, the complement of each $(n + 1)$-tuple of odd weight is an $(n + 1)$-tuple of even weight. Hence, when n is even, the cycles of the CSR_n are the complements of the cycles from the PSR_n.   □

The values of $Z(n)$, $Z^*(n)$, $S(n)$, and $S^*(n)$, are highly related and their exact computation will be done in the next section. Now, some of the connections between these values will be proved.

Lemma 3.13. *For every $n \geq 1$ we have that $Z(n + 1) = S(n) + S^*(n)$.*

Proof. Each extended representation $E(C)$ of either the PSR_n or CSR_n forms a cycle of PCR_{n+1} of either even extended weight or odd extended weight, respectively. Similarly, each extended representation $E(C)$ of a cycle C from the PCR_{n+1} forms an extended representation of a cycle from either the PSR_n or the CSR_n, depending on whether the (extended) weight of $E(C)$ is even or odd, respectively.   □

For the next theorem the operator **D**, which was defined and partially analyzed in Section 1.2, will be required. This operator will be analyzed in detail in Section 4.2 and Chapters 5 and 6.

Lemma 3.14. *For every $n \geq 1$ we have that $Z^*(n) = S^*(n-1)$.*

Proof. Let C be a cycle from the CCR_n. By Lemma 3.6, all the cycles of the CCR_n are self-dual cycles, where the structure of such a cycle C of period $2d$ is $[X\bar{X}]$, $X = x_1 x_2 \cdots x_{d-1} x_d$, and d divides n. By the definition of the operator **D** we have that $\mathbf{D}C = [b_1 b_2 \cdots b_{d-1} b_d b_{d+1} b_{d+2} \cdots b_{2d-1} b_{2d}]$, where $b_i = x_i + x_{i+1}$, $b_{d+i} = \bar{x}_i + \bar{x}_{i+1} = b_i$ for $1 \leq i \leq d-1$, $b_d = x_d + \bar{x}_1$, and $b_{2d} = \bar{x}_d + x_1 = b_d$. Hence, $b_{i+d} = b_i$ for each $1 \leq i \leq d$, which implies that the period of $\mathbf{D}C$ is half of the period of C, i.e., d. Moreover, $\sum_{i=1}^{d} b_i = \sum_{i=1}^{d-1}(x_i + x_{i+1}) + x_d + \bar{x}_1 = x_1 + \bar{x}_1 = 1$ and hence the weight of $\mathbf{D}C$ is odd, which implies that $\mathbf{D}C$ is a cycle of the CSR_{n-1}.

Furthermore, the mapping implies that each two complement n-tuples are mapped to the same $(n-1)$-tuple. Therefore the mapping on the CCR_n cycles maps all the n-tuples (states) of the CCR_n cycles onto all the $(n-1)$-tuples (states) of the CSR_{n-1}. Moreover, given a cycle C of the CSR_{n-1} whose period is d and its weight is odd, we can find exactly one cycle C' for which $\mathbf{D}C' = C$ and this cycle of period $2d$ has the form $C' = [X\bar{X}]$, where X is a sequence of length d. Now, it is easy to verify that this is a one-to-one mapping between the cycles of CCR_n and the cycles of the CSR_{n-1}, i.e., $Z^*(n) = S^*(n-1)$. $\square$

Corollary 3.6. *If n is a positive integer, then $S(n) = Z(n+1) - Z^*(n+1)$.*

Proof. By Lemma 3.13, we have that $S(n) = Z(n+1) - S^*(n)$. By Lemma 3.14, we have $S^*(n) = Z^*(n+1)$ and hence $S(n) = Z(n+1) - Z^*(n+1)$. $\square$

Lemma 3.15. *If n is even, then $S(n) = S^*(n)$.*

Proof. By Lemma 3.12, if n is even, then $n+1$ is odd, and there is a one-to-one correspondence between the cycles of the PSR_n and the cycles of the CSR_n, under which an extended representation $E(C)$ of a PSR_n (CSR_n, respectively) cycle C is mapped into its complement that is an extended representation of the cycle $\bar{C}$ of the CSR_n (PSR_n, respectively). This implies that $S(n) = S^*(n)$. $\square$

By Corollary 1.13, the number of FSR_n is $2^{2^{n-1}}$. By Corollary 2.1, the number of LFSR_n is 2^{n-1}. Hence, it is obvious that most shift registers are nonlinear and their representation can be done either by the associated truth table or by the function derived from this truth table based on basic methods of digital systems for reducing the number of terms in a function. Span n de Bruijn sequences that form factors with one cycle in G_n are all nonlinear and we are interested in their feedback functions. By Theorem 1.29, there are $2^{2^{n-1}-n}$ span n de Bruijn sequences. This is still a small fraction of the total number of FSR_n. Constructions of de Bruijn sequences will be given in the next chapter by construction algorithms rather than their function. It is still interesting to find feedback functions that form interesting shift registers in terms of their state diagrams, i.e., the sequences that they form.

The cycles of some nonlinear FSR_n ($NLFSR_n$ in short) can be of increased interest based on the discussion that we had so far. Given an $LFSR_n$ with a feedback function $f(x_1, x_2, \ldots, x_n)$, we already know how to find the number of cycles of each length in its state diagram (see Lemma 2.5 and Theorem 2.5). Consider now the simple modification of the feedback function to

$$h(x_1, x_2, \ldots, x_n) = f(x_1, x_2, \ldots, x_n) + 1.$$

If $h(x_1, x_2, \ldots, x_n) = x_1 + 1$ or $h(x_1, x_2, \ldots, x_n) = \sum_{i=1}^{n} x_i + 1$, then the length of the cycles and their number are discussed in this section and the following one. Can we say something about other functions? There are some cases where the answer is very simple.

Lemma 3.16. *If $f(x_1, x_2, \ldots, x_n)$ is the characteristic (feedback) function for an M-sequence S, then the state diagram of the feedback function*

$$h(x_1, x_2, \ldots, x_n) = f(x_1, x_2, \ldots, x_n) + 1$$

has two sequences, $\bar{S}$ and $[1]$.

Proof. Let $(y_1, y_2, \ldots, y_n)$ be an arbitrary state. If $f(y_1, y_2, \ldots, y_n) = b$, then by using Lemma 2.1 we have

$$
\begin{aligned}
h(\bar{y}_1, \bar{y}_2, \ldots, \bar{y}_n) &= f(\bar{y}_1, \bar{y}_2, \ldots, \bar{y}_n) + 1 = f(y_1 + 1, y_2 + 1, \ldots, y_n + 1) + 1 \\
&= f(y_1, y_2, \ldots, y_n) + f(1, 1, \ldots, 1) + 1 \\
&= b + 0 + 1 = \bar{b}.
\end{aligned}
$$

Therefore since the function $f(x_1, x_2, \ldots, x_n)$ generates the sequences S and $[0]$, it follows that the function $h(x_1, x_2, \ldots, x_n)$ generates the sequences $\bar{S}$ and $[1]$. $\qquad\square$

For a factor $\mathcal{F}$ in the de Bruijn graph G_n, the **complement factor** $\mathcal{F}^c$ is defined as $\mathcal{F}^c \triangleq \{e : e \in E, e \notin \mathcal{F}\}$, i.e., $\mathcal{F}^c$ contains all the edges of G_n that are not contained in $\mathcal{F}$. The **binary complement factor** $\bar{\mathcal{F}}$ of $\mathcal{F}$ is defined by $\bar{\mathcal{F}} \triangleq \{C : \bar{C} \in \mathcal{F}\}$. These definitions are associated with state diagrams of an FSR. The following observation can be easily verified.

Lemma 3.17. *If $f(x_1, x_2, \ldots, x_n)$ form a factor $\mathcal{F}$, then $h(x_1, x_2, \ldots, x_n) = f(x_1, x_2, \ldots, x_n) + 1$ if the feedback function of the FSR_n whose state diagram is the factor $\mathcal{F}^c$. Moreover, the sequences of their truth tables are complements.*

By Lemma 3.6, all the cycles in the state diagram formed from the function $h(x_1, x_2, \ldots, x_n) = x_1 + 1$ are self-dual. If $h(x_1, x_2, \ldots, x_n) = \sum_{i=1}^{n} x_i + 1$, then by Lemma 3.10 all the cycles of the state diagram are of odd weight. If n is even, then by Lemma 3.12 they are the complements of the cycles generated by the function $f(x_1, x_2, \ldots, x_n) = \sum_{i=1}^{n} x_i$. If n is odd, then by the discussion on the CSR_n we have that for each cycle C generated by $h(x_1, x_2, \ldots, x_n) = \sum_{i=1}^{n} x_i + 1$, also its complement $\bar{C}$ is generated by the same function since both extended representations have even length $n + 1$ and odd weight. This implies that a general characterization, for the function $h(x_1, x_2, \ldots, x_n) = f(x_1, x_2, \ldots, x_n) + 1$ based on the function $f(x_1, x_2, \ldots, x_n)$, will be difficult to obtain. Consider the associated shift register for these two feedback functions. There is another interesting property that is a generalization of Lemma 3.16.

Theorem 3.3. *Let $f(x_1, x_2, \ldots, x_n)$ be a linear function of an $LFSR_n$. The sequence defined by the truth table (top half) of f is a CR sequence if and only if the sequences generated by the feedback function*

$$h(x_1, x_2, \ldots, x_n) = f(x_1, x_2, \ldots, x_n) + 1$$

are the complements of the sequences that are generated by the feedback function $f(x_1, x_2, \ldots, x_n)$.

Proof. Note first that if $(0, x_2, \ldots, x_n)$ is in the ith row of the truth table, then $(0, \bar{x}_2, \ldots, \bar{x}_n)$ is in the ith row from the bottom (the $(2^n - i + 1)$th row) of the top half of the truth table. If the sequence defined by the top half of the truth table is a CR sequence, then $f(0, x_2, \ldots, x_n)$ and $f(0, \bar{x}_2, \ldots, \bar{x}_n)$ must have different values. Therefore $f(0, x_2, \ldots, x_n)$ and $f(1, \bar{x}_2, \ldots, \bar{x}_n)$ have the same values for all n-tuples of the form $(0, x_2, \ldots, x_n)$ if and only if the truth-table sequence defined by the function f is a CR sequence.

Assume first that the sequence, defined by the top half of the truth table of the feedback function f, is a CR sequence. Let $f(0, x_2, \ldots, x_n) = b$ for some $x_i \in \{0, 1\}$, $2 \le i \le n$, and $b \in \{0, 1\}$ and hence by the definition of h we have that $h(0, x_2, \ldots, x_n) = \bar{b}$. Since the sequence defined by the truth table of the function f is a CR sequence, it follows that $f(0, \bar{x}_2, \ldots, \bar{x}_n) = \bar{b}$ and therefore $f(1, \bar{x}_2, \ldots, \bar{x}_n) = b$ and $h(1, \bar{x}_2, \ldots, \bar{x}_n) = \bar{b}$.

Thus we have that $f(0, x_2, \ldots, x_n) = b$ implies that $h(1, \bar{x}_2, \ldots, \bar{x}_n) = \bar{b}$, $f(1, \bar{x}_2, \ldots, \bar{x}_n) = b$, and $h(0, x_2, \ldots, x_n) = \bar{b}$. Hence, the cycles of the state diagram defined by the feedback function h are the complements of the cycles of the state diagram defined by the feedback function f.

Now, assume that the sequences generated by the feedback function h are the complements of the sequences generated by the feedback function f. Let $f(0, x_2, \ldots, x_n) = b$ for some $x_i \in \{0, 1\}$, $2 \le i \le n$, and $b \in \{0, 1\}$. This implies that $f(1, x_2, \ldots, x_n) = \bar{b}$, $h(0, x_2, \ldots, x_n) = \bar{b}$, $h(1, x_2, \ldots, x_n) = b$, and since the sequences defined by h are the complements of those defined by f and $f(1, x_2, \ldots, x_n) = \bar{b}$, it follows that $h(0, \bar{x}_2, \ldots, \bar{x}_n) = b$. Hence, by the definition of the function h we have that $f(0, \bar{x}_2, \ldots, \bar{x}_n) = \bar{b}$ and since

$f(0, x_2, \ldots, x_n) = b$, it follows that the sequence defined by the truth table of the feedback function f is a CR sequence. $\qquad\square$

Corollary 3.7. *The sequence defined by the truth table of an M-sequence is a CR sequence.*

Proof. This is an immediate consequence from Theorem 3.3 and Lemma 3.16. $\qquad\square$

Problem 3.1. Given a factor $\mathcal{F}$ in G_n, can the number of cycles of the complement factor $\mathcal{F}^c$ and their length be given as a function of the cycles of $\mathcal{F}$ and their length? What if $\mathcal{F}$ is the state diagram of an LFSR_n?

3.2 Enumeration methods for polynomials and cycles

This section is devoted to the enumeration of sequences having certain properties. We enumerate the number of irreducible polynomials and the number of primitive polynomials of any given degree. In particular, we concentrate on counting the number of cycles in the state diagrams of the PCR_n, the CCR_n, the PSR_n, and the CSR_n, i.e., finding the values of $Z(n)$, $Z^*(n)$, $S(n)$, and $S^*(n)$, respectively. These enumerations are based on the Euler function $\phi(\cdot)$ and the Möbius function $\mu(\cdot)$.

As we saw in Theorem 2.5, the length of a sequence obtained from an LFSR_n depends on the factorization of its characteristic polynomial. Moreover, sequences of maximum length $q^n - 1$, over $\mathbb{F}_q$, are obtained from primitive polynomials that form a subset of the set of irreducible polynomials. How many irreducible polynomials of degree n over $\mathbb{F}_q$ exist? How many of them are primitive polynomials? This section will be devoted first to answering these two questions and related ones.

Theorem 3.4. *The number of distinct primitive polynomials of degree n over $\mathbb{F}_q$ is*

$$\frac{\phi(q^n - 1)}{n}.$$

Proof. Let α be a primitive element in $\mathbb{F}_{q^n}$. If g.c.d.$(k, q^n - 1) = 1$, then α^k has order $q^n - 1$ in $\mathbb{F}_{q^n}$, i.e., α^k is also a primitive element in $\mathbb{F}_{q^n}$. Moreover, α^k is not a primitive element in $\mathbb{F}_{q^n}$ if and only if g.c.d.$(k, q^n - 1) > 1$. Hence, the number of primitive elements in $\mathbb{F}_{q^n}$ is $\phi(q^n - 1)$. As a consequence of Theorem 1.24, we have that each primitive polynomial of degree n over $\mathbb{F}_q$ has n distinct primitive elements, which are the roots of the polynomial. Hence, the number of distinct primitive polynomials of degree n over $\mathbb{F}_q$ is

$$\frac{\phi(q^n - 1)}{n}.$$

$\qquad\square$

We are now going to provide a few applications of the Möbius inversion formula and in particular, several such applications that are associated with the computation of the number of cycles in the state diagram of some FSR_ns or the number of cycles with some given properties. The first application is to find the number of irreducible polynomials of degree n over the field $\mathbb{F}_q$.

Lemma 3.18. *The number of distinct irreducible polynomials of degree n over $\mathbb{F}_q$ is*

$$\frac{1}{n} \sum_{d \mid n} \mu(d) \cdot q^{n/d}.$$

Proof. Let $I_q(n)$ be the number of distinct irreducible polynomials of degree n over $\mathbb{F}_q$. By Theorem 1.24, each element of $\mathbb{F}_{q^n}$ is a root of an irreducible polynomial whose degree divides n. Hence, since an irreducible polynomial of degree d has d distinct roots, we have that

$$q^n = \sum_{d \mid n} d \cdot I_q(d).$$

Now, we apply the Möbius inversion formula (see Theorem 1.10) and obtain

$$n \cdot I_q(n) = \sum_{d \mid n} \mu(d) \cdot q^{n/d}$$

and the claim of the lemma follows. $\qquad\qquad\qquad\qquad\qquad\square$

Theorem 3.5. *The number of cycles of length n in the state diagram of the PCR_n is*

$$\frac{1}{n} \sum_{d \mid n} \mu(d) \cdot 2^{n/d}.$$

Proof. Let F_n be the number of PCR_n cycles of length n. Counting the number of states in the state diagram of the PCR_n in two different ways implies that

$$2^n = \sum_{d \mid n} d \cdot F_d.$$

Now, we apply the Möbius inversion formula and obtain

$$n \cdot F_n = \sum_{d \mid n} \mu(d) \cdot 2^{n/d}$$

and the claim of the theorem follows. $\qquad\qquad\qquad\qquad\qquad\square$

Theorem 3.5 can be generalized for any necklaces factor with σ beads, i.e., σ alphabet letters, using the same proof.

Theorem 3.6. *The number of necklaces of length n over σ beads is*

$$\frac{1}{n} \sum_{d|n} \mu(d) \cdot \sigma^{n/d}.$$

Each necklace of length k, $k \le n$, having beads with σ colors, is also a cycle in $G_{\sigma,n}$, and therefore we have the following consequence.

Corollary 3.8. *The number of cycles of period k, $k \le n$ in $G_{\sigma,n}$ is*

$$\frac{1}{k} \sum_{d|k} \mu(d) \cdot \sigma^{k/d}.$$

The connection between Lemma 3.18 and Theorem 3.5 is readily verified. There is a one-to-one correspondence between the cycles of length n in the state diagram of the PCR_n and the number of binary irreducible polynomials of degree n. Having found the number of PCR_n cycles with length n, we would like to find the total number of PCR_n cycles. To this end, we will use Burnside's lemma (see Theorem 1.2).

Theorem 3.7. *For each positive integer n, the number of cycles in the state diagram of the PCR_n is*

$$Z(n) = \frac{1}{n} \sum_{d|n} \phi(d) \cdot 2^{n/d}.$$

Proof. Let U be the set of all binary words of length n and let G be a finite group decomposed by the n cyclic permutations on the coordinates of the words in U. First, one should note that an n-tuple remains unchanged under a cyclic permutation if the n-tuple has period p and the cyclic permutation is a multiple of p. Therefore the number of n-tuples that remained unchanged by a cyclic shift of i positions is $2^{\text{g.c.d.}(n,i)}$. If we consider two words that are different only by a cyclic shift as equivalent, then the number of equivalence classes of this relation is $Z(n)$. Hence, by Burnside's lemma, we have

$$Z(n) = \frac{1}{n} \sum_{i=1}^{n} 2^{\text{g.c.d.}(n,i)} = \frac{1}{n} \sum_{d|n} \sum_{\substack{\text{g.c.d.}(n,i)=d \\ 1 \le i \le n}} 2^d = \frac{1}{n} \sum_{d|n} 2^d \sum_{\substack{\text{g.c.d.}(n,i)=d \\ 1 \le i \le n}} 1$$

$$= \frac{1}{n} \sum_{d|n} 2^d \sum_{\substack{\text{g.c.d.}(n/d,i)=1 \\ 1 \le i \le n/d}} 1 = \frac{1}{n} \sum_{d|n} \phi\left(\frac{n}{d}\right) \cdot 2^d = \frac{1}{n} \sum_{d|n} \phi(d) \cdot 2^{n/d}. \quad \square$$

By Theorem 3.1 we have that the minimum weight function of a span n de Bruijn sequence is $Z(n) - 1$. The parity of $Z(n)$ that will be discussed next is important not only for its theoretical value, it will be used later in the proof of Theorem 5.6.

Lemma 3.19. *The value of $Z(n)$ is an even integer for each $n \geq 3$.*

Proof. We distinguish between three cases depending on whether n is odd, a power of 2 greater than 2, or another even integer.

Case 1: If n is odd, then dividing $\sum_{d|n} \phi(d) \cdot 2^{n/d}$ by n does not change the parity. Since all the powers of 2 in $\sum_{d|n} \phi(d) \cdot 2^{n/d}$ are even integers, it follows that $Z(n)$ is even.

Case 2: If $n = 2^k$, where $k > 1$, then since $\phi(1) = 1$ and $\phi(2^\ell) = 2^{\ell-1}$ for $\ell > 0$, we have that

$$Z(n) = \frac{1}{n} \sum_{d|n} \phi(d) \cdot 2^{n/d} = \frac{1}{2^k} \sum_{i=0}^{k} \phi(2^i) \cdot 2^{2^{k-i}} = 2^{2^k - k} + \sum_{i=1}^{k} 2^{-k+(i-1)+2^{k-i}}.$$

Clearly, $2^{2^k - k}$ is even and $-k + (i-1) + 2^{k-i} > 0$, unless $i = k-1$ or $i = k$. If $i = k-1$, then $-k + (i-1) + 2^{k-i} = 0$ and if $i = k$, then $-k + (i-1) + 2^{k-i} = 0$. Hence, in both cases ($i \in \{k-1, k\}$) we have that $2^{-k+(i-1)+2^{k-i}} = 1$. However, since both $i = k-1$ and $i = k$ are in the summation they add up to 2 and the whole expression of $Z(n)$ is even.

Case 3: n is not odd and not a power of 2 greater than 2, i.e., n can be written as $n = 2^k m$, where $k > 0$ and $m > 1$ is an odd integer. The divisors of $n = 2^k m$ are the divisors of m multiplied by 2^i for each $0 \leq i \leq k$. Hence, we have (using the divisors of $\phi(d)$ as given in Lemma 1.4) that

$$Z(n) = \frac{1}{n} \sum_{d|n} \phi(d) \cdot 2^{n/d}$$

$$= \frac{1}{2^k m} \sum_{d|2^k m} \phi(d) \cdot 2^{n/d}$$

$$= \frac{1}{2^k m} \sum_{d'|m} \sum_{i=0}^{k} \phi(2^i d') \cdot 2^{2^k m/(2^i d')}$$

$$= \frac{1}{m} \sum_{d|m} \phi(d) \cdot \frac{1}{2^k} \sum_{i=0}^{k} \phi(2^i) \cdot 2^{2^{k-i}(m/d)}$$

$$= \frac{1}{m} \sum_{d|m} \phi(d) \cdot \left(2^{(2^k \cdot m/d) - k} + \sum_{i=1}^{k} 2^{-k+(i-1)+2^{k-i}(m/d)} \right).$$

As in Case 1, dividing by the odd integer m does not change the parity. If $d > 1$ and since d is odd, it follows that $\phi(d)$ is even, and hence we only have to show that when d is odd and $d > 1$ the expression in the parenthesis is an integer.

Exactly as in Case 2 we have that

$$2^{(2^k \cdot m/d)-k} + \sum_{i=1}^{k} 2^{-k+(i-1)+2^{k-i}(m/d)}$$

is an integer and hence

$$\phi(d) \cdot \left(2^{(2^k \cdot m/d)-k} + \sum_{i=1}^{k} 2^{-k+(i-1)+2^{k-i}(m/d)} \right)$$

is even. If $d = 1$, then $m/d = m$ and as in Case 2 we have that

$$2^{(2^k \cdot m)-k} + \sum_{i=1}^{k} 2^{-k+(i-1)+2^{k-i}m}$$

is an integer and this integer is even since $m > 1$. $\square$

Corollary 3.9. *For $n \geq 3$, the weight function of an FSR_n with a function f is even or odd, respectively, if and only if the number of cycles in the state diagram of f is even or odd, respectively.*

Proof. By Lemma 3.19 the number of cycles in the state diagram of the PCR_n, where $n \geq 3$, is even and the weight function is 0, i.e., an even integer. Assume that the weight function of some function f of an FSR_n is k. This implies that the truth table of the function f is obtained from the truth table of the PCR_n by changing k *zeros* to *ones*. By Lemma 1.20 and Corollary 1.15, each change of one *zero* to a *one* either increases the number of cycles in the state diagram by 1 or decreases this number by 1. Therefore the number of cycles in the state diagram of f is even if k is even and odd if k is odd. $\square$

Corollary 3.10. *The weight of the feedback function of any span n de Bruijn sequence is odd when $n \geq 3$.*

Proof. By Corollary 3.9 we have that all span n de Bruijn sequences have the same parity for the weight of their feedback functions. By Theorem 3.1 the minimum weight function of a span n de Bruijn sequence is $Z(n) - 1$. By Lemma 3.19, this number is odd and the claim follows. $\square$

Corollary 3.11. *The value of $Z^*(n)$ is an even integer for each $n \geq 3$.*

Proof. By Lemma 3.2 the maximum function weight of a span n de Bruijn sequence is $2^{n-1} - Z^*(n) + 1$ and by Corollary 3.10 this number is odd for $n \geq 3$. Hence, $Z^*(n)$ is even for $n \geq 3$. $\square$

Corollary 3.11 and Lemma 3.14 imply the following consequence.

Corollary 3.12. *The value of $S^*(n)$ is an even integer for each $n \geq 2$.*

Corollaries 3.6 and 3.12 and Lemma 3.15 imply the following consequence.

Corollary 3.13. *The value of $S(n)$ is an even integer for each $n \geq 2$.*

Proof. By Corollary 3.6 we have that $S(n) = Z(n+1) - Z^*(n+1)$. By Corollary 3.11, we have that $Z^*(n)$ is even for $n \geq 3$. By Lemma 3.19, we have that $Z(n)$ is even for $n \geq 3$. Thus we have that $S(n)$ is even for each $n \geq 2$. □

Finally, we want to make an asymptotic computation of the number of cycles in the PCR_n. Recall that each cycle of the PCR_n is a necklace with two beads.

Lemma 3.20. *For $N = 2^n$, there are $O\left(\sqrt{N}\right)$ states in the degenerate necklaces of G_n and $\frac{N}{\log N} - O\left(\frac{\sqrt{N}}{\log N}\right)$ full-order necklaces.*

Proof. By Corollary 3.1 each degenerate necklace has a length that is a divisor $d < n$ of n and hence the number of states in the degenerate necklaces is at most

$$\sum_{\substack{d|n \\ d<n}} 2^d \leq 2^{\frac{n}{2}+1} = O\left(\sqrt{N}\right).$$

The remaining $N - O\left(\sqrt{N}\right)$ states are on full-order necklaces. Each full-order necklace contains $\log N$ states and therefore there are $\frac{N}{\log N} - O\left(\frac{\sqrt{N}}{\log N}\right)$ full-order necklaces. □

Theorem 3.7 can be generalized easily using the same proof for the necklaces factor with σ beads. Similarly to Theorem 3.7 we can prove the following claim.

Theorem 3.8. *For each positive integer n, the number of cycles in the necklaces factor of order n, over σ beads, is*

$$\frac{1}{n}\sum_{d|n} \phi(d) \cdot \sigma^{n/d}.$$

Theorem 3.8 indicates that we can handle factors of $G_{\sigma,n}$ in the same way that we handle state diagrams of FSR_n. This is left as an exercise for the reader.

Theorem 3.9. *For each positive integer n, the number of cycles in the state diagram of the CSR_n is*

$$S^*(n) = \frac{1}{2(n+1)} \sum_{\substack{d|n+1 \\ d \text{ odd}}} \phi(d) \cdot 2^{(n+1)/d}.$$

Proof. Assume first that n is even, i.e., $n + 1$ is odd. By Lemma 3.15 we have that $S(n) = S^*(n)$ and by Lemma 3.13 we have that $Z(n + 1) = S(n) + S^*(n)$, i.e., $S^*(n) = \frac{1}{2}Z(n + 1)$. We also have by Theorem 3.7 that

$$Z(n + 1) = \frac{1}{n + 1} \sum_{d|n+1} \phi(d) \cdot 2^{(n+1)/d}.$$

Since $n + 1$ is odd, it follows that all the divisors of $n + 1$ are odd, and therefore

$$S^*(n) = \frac{1}{2(n + 1)} \sum_{d|n+1} \phi(d) \cdot 2^{(n+1)/d}.$$

Assume now that n is odd, i.e., $n + 1$ is even. Let m be a positive integer, A_m be the set of words of length m and even weight, and let B_m be the set of words of length m and odd weight. It is easy to verify that $|A_m| = |B_m|$. Consider now the extended representation of the cycles in the CSR_n and the words of length $n + 1$ obtained by the $n + 1$ cyclic shift of each cycle. For a cycle C in the CSR_n, the first n bits of each cyclic shift of $E(C)$ is a state on C. Note also that if $n + 1$ is even, then each cycle of the CSR_n has period $\frac{n+1}{d}$ for some divisor d of $n + 1$. Since $w = \text{wt}(E(C))$ is odd, it follows that also $\frac{w}{d}$ is odd, which implies that d is odd and hence the period $\frac{n+1}{d}$ is even. Let U be the set of binary words of length $n + 1$ and odd weight, i.e., B_{n+1}. Let G be the finite group decomposed of the $n + 1$ cyclic permutations on the coordinates of words in U. By Burnside's lemma, we have that

$$S^*(n) = \frac{1}{n + 1} \sum_{\substack{i=1 \\ \frac{n+1}{\text{g.c.d.}(n+1,i)} \text{ odd}}}^{n+1} 2^{\text{g.c.d.}(n+1,i)-1} = \frac{1}{n + 1} \sum_{\substack{d|n+1 \\ \frac{n+1}{d} \text{ odd}}} \sum_{\substack{\text{g.c.d.}(n+1,i)=d \\ 1 \le i \le n+1}} 2^{d-1}$$

$$= \frac{1}{2(n + 1)} \sum_{\substack{d|n+1 \\ \frac{n+1}{d} \text{ odd}}} 2^d \sum_{\substack{\text{g.c.d.}(\frac{n+1}{d},i)=1 \\ 1 \le i \le n+1}} 1 = \frac{1}{2(n + 1)} \sum_{\substack{d|n+1 \\ \frac{n+1}{d} \text{ odd}}} \phi\left(\frac{n + 1}{d}\right) \cdot 2^d$$

$$= \frac{1}{2(n + 1)} \sum_{\text{odd } d|n+1} \phi(d) \cdot 2^{(n+1)/d},$$

where the division by 2 in the first equation is because $|A_m| = |B_m|$, where $m = \text{g.c.d.}(n + 1, i)$. Note that $m = \text{g.c.d.}(n + 1, i)$ is an even integer, but $d = \frac{n+1}{m}$ is an odd integer. The number of words of length $n + 1$ that remains unchanged under a cyclic permutation by i positions is 2^m, but only half of them have extended representations with odd weights.

This completes the proof of the theorem. $\square$

Corollary 3.14. *For each positive integer n, the number of cycles in the state diagram of the* CCR_n *is*

$$Z^*(n) = \frac{1}{2n} \sum_{\substack{d \mid n \\ d \text{ odd}}} \phi(d) \cdot 2^{n/d}.$$

Theorem 3.10. *For each positive even integer n, the number of cycles in the state diagram of the* PSR_n *is*

$$S(n) = \frac{1}{2(n+1)} \sum_{d \mid n+1} \phi(d) \cdot 2^{(n+1)/d}.$$

For each positive odd integer n, the number of cycles in the state diagram of the PSR_n *is*

$$S(n) = \frac{1}{2(n+1)} \sum_{d \mid n+1} \phi(d) \cdot 2^{(n+1)/d} + \frac{1}{2(n+1)} \sum_{\substack{d \mid n+1 \\ d \text{ even}}} \phi(d) \cdot 2^{(n+1)/d}$$

$$= \frac{1}{2(n+1)} \sum_{\substack{d \mid n+1 \\ d \text{ odd}}} \phi(d) \cdot 2^{(n+1)/d} + \frac{1}{n+1} \sum_{\substack{d \mid n+1 \\ d \text{ even}}} \phi(d) \cdot 2^{(n+1)/d}.$$

Proof. Assume first that n is even. By Lemma 3.15 we have that $S(n) = S^*(n)$ and hence by Theorem 3.9 we have that

$$S(n) = \frac{1}{2(n+1)} \sum_{d \mid n+1} \phi(d) \cdot 2^{(n+1)/d}.$$

Assume now that n is odd. By Lemma 3.13, $S(n) = Z(n+1) - S^*(n)$ and together with the value of $Z(n)$ provided in Theorem 3.7 and the value of $S^*(n)$ provided in Theorem 3.9 we have that

$$S(n) = \frac{1}{n+1} \sum_{d \mid n+1} \phi(d) \cdot 2^{(n+1)/d} - \frac{1}{2(n+1)} \sum_{\substack{d \mid n+1 \\ d \text{ odd}}} \phi(d) \cdot 2^{(n+1)/d}$$

$$= \frac{1}{2(n+1)} \sum_{d \mid n+1} \phi(d) \cdot 2^{(n+1)/d} + \frac{1}{2(n+1)} \sum_{\substack{d \mid n+1 \\ d \text{ even}}} \phi(d) \cdot 2^{(n+1)/d}. \qquad \square$$

Since for even n, $n + 1$ has only odd divisors, we have the following consequence.

Corollary 3.15. *For each positive integer n, the number of cycles in the state diagram of the* PSR$_n$ *is*

$$S(n) = \frac{1}{2(n+1)} \sum_{d|n+1} \phi(d) \cdot 2^{(n+1)/d} + \frac{1}{2(n+1)} \sum_{\substack{d|n+1 \\ d \text{ even}}} \phi(d) \cdot 2^{(n+1)/d}.$$

Let $N(n, \epsilon)$ be the number of PCR$_n$ cycles of length n and weight ϵn, where $0 \le \epsilon \le 1$ (ϵn is not necessarily an integer).

Lemma 3.21. *If we define $\binom{\beta}{\alpha} = 0$ when α is not an integer, then*

$$N(n, \epsilon) = \frac{1}{n} \sum_{d|n} \binom{d}{\epsilon d} \cdot \mu\left(\frac{n}{d}\right).$$

Proof. The number of states of length n with weight ϵn is $\binom{n}{\epsilon n}$. This number of states is also equal to $\sum_{d|n} d \cdot N(d, \epsilon)$. Therefore

$$\binom{n}{\epsilon n} = \sum_{d|n} d \cdot N(d, \epsilon)$$

and hence, by the Möbius inversion formula (Theorem 1.10) we have that

$$N(n, \epsilon) = \frac{1}{n} \sum_{d|n} \binom{d}{\epsilon d} \cdot \mu\left(\frac{n}{d}\right). \qquad \square$$

Lemma 3.22. *The number of* PCR$_n$ *cycles of length n and odd weight is*

$$\frac{1}{n} \sum_{d|n} \mu\left(\frac{n}{d}\right) \cdot Q(n, d),$$

where $Q(n, d) = 2^{d-1}$ if $\frac{n}{d}$ is odd and $Q(n, d) = 0$ if $\frac{n}{d}$ is even.

Proof. The number of PCR$_n$ cycles of length n and odd weight is equal to the number of CSR$_{n-1}$ cycles with an extended representation of length n. This number is equal to

$$\sum_{k=0}^{\lfloor n/2 \rfloor} N\left(n, \frac{2k+1}{n}\right).$$

By Lemma 3.21, we have that

$$\sum_{k=0}^{\lfloor n/2 \rfloor} N\left(n, \frac{2k+1}{n}\right) = \sum_{k=0}^{\lfloor n/2 \rfloor} \frac{1}{n} \sum_{d|n} \binom{d}{\frac{(2k+1)d}{n}} \cdot \mu\left(\frac{n}{d}\right).$$

By changing the order of the summation we have

$$\sum_{k=0}^{\lfloor n/2 \rfloor} \frac{1}{n} \sum_{d|n} \binom{d}{\frac{(2k+1)d}{n}} \cdot \mu\left(\frac{n}{d}\right) = \frac{1}{n} \sum_{d|n} \mu\left(\frac{n}{d}\right) \sum_{k=0}^{\lfloor n/2 \rfloor} \binom{d}{\frac{(2k+1)d}{n}}.$$

If $\frac{n}{d}$ is even, then for every k, $\frac{(2k+1)d}{n}$ is not an integer and hence

$$\sum_{k=0}^{\lfloor n/2 \rfloor} \binom{d}{\frac{(2k+1)d}{n}} = 0.$$

If $\frac{n}{d}$ is odd, then $\frac{(2k+1)d}{n}$ is either odd or not an integer. Moreover, all the odd integers between 0 and d have exactly one representation as $\frac{(2k+1)d}{n}$. Hence,

$$\sum_{k=0}^{\lfloor n/2 \rfloor} \binom{d}{\frac{(2k+1)d}{n}} = \sum_{\text{odd } k} \binom{d}{k} = 2^{d-1}.$$

Therefore we have

$$\frac{1}{n} \sum_{d|n} \mu\left(\frac{n}{d}\right) \sum_{k=0}^{\lfloor n/2 \rfloor} \binom{d}{\frac{(2k+1)d}{n}} = \frac{1}{n} \sum_{d|n} \mu\left(\frac{n}{d}\right) \cdot Q(n, d). \qquad \square$$

Theorem 3.11. *For every positive integer n, the number of* CCR_n *cycles of length (period) 2n is*

$$\frac{1}{n} \sum_{d|n} \mu\left(\frac{n}{d}\right) \cdot Q(n, d).$$

Proof. By Lemmas 3.6 and 3.7 and by Corollary 3.5, the self-dual sequences of period $2n$ are exactly the CCR_n cycles of length $2n$. Similarly to the proof of Lemma 3.14 we have that the number of CCR_n cycles of length $2n$ is equal to the number of CSR_{n-1} cycles of length n (and odd weight). This number is equal, as explained in the proof of Lemma 3.13, to the number of PCR_n cycles of length n and odd weight that by Lemma 3.22 equals

$$\frac{1}{n} \sum_{d|n} \mu\left(\frac{n}{d}\right) \cdot Q(n, d). \qquad \square$$

By Lemma 3.6 all the self-dual sequences have even lengths. Now, we can also calculate their number as an immediate consequence from Theorem 3.11.

Corollary 3.16. *The number of self-dual sequences of period 2n is*

$$SD(2n) = \frac{1}{n} \sum_{d|n} \mu\left(\frac{n}{d}\right) \cdot Q(n, d).$$

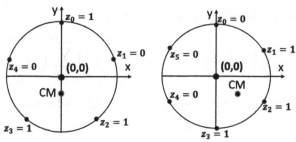

FIGURE 3.1 A state of G_5, colored with an I, on the left and a state of G_6, colored with an R, on the right.

3.3 Maximum number of cycles in a state diagram

Which factor of G_n has the largest number of cycles, i.e., which FSR_n has a state diagram with the largest number of cycles, and what is the number of cycles in the associated factor? Is the factor that attains the largest number of cycles unique? All these questions will be considered and answered in this section.

So far, states were denoted with lowercase letters, and their entries with lowercase letters with subscripts. Sometimes, as in this section, subscripts will be required also for states and they will be denoted by uppercase letters. There will be other places where uppercase letters will be required to denote states and this should be understood from the context.

We will now present a proof that the number of cycles in the FSR_n which has the largest number of cycles in its state diagram is $Z(n)$. The proof will be obtained by coloring the vertices in G_n. Each digit of any given state $Z = (z_0, z_1, \cdots, z_{n-1})$ is placed on the unit circle of the two-dimensional plane, where z_0 is placed on the point $(x, y) = (0, 1)$ and the other digits are evenly spaced on the unit circle written in increasing order of indices clockwise. The coordinates of the digit z_i are

$$(x, y) = \left(\sin \frac{2\pi i}{n}, \cos \frac{2\pi i}{n} \right), \quad i = 0, 1, \ldots n - 1.$$

Consider each z_i to be a mass of weight z_i, which is either 0 or 1, and find the **center of mass** (denoted by CM) of these weights. The state Z is colored with an L if its CM is to the left of the y-axis, colored with an I if its CM is on the y-axis, and colored with an R if its CM is to the right of the y-axis. Examples of states from G_5 and G_6 are given in Fig. 3.1.

The next few lemmas prove a few properties of this coloring associated with the cycles in G_n and in particular with the PCR_n cycles.

Lemma 3.23. *Either all the states on a cycle from the PCR_n have color I, or the cycle contains exactly one block of consecutive Ls and one block of consecutive Rs, and at most two Is separating these two blocks, i.e., each color I that exists is exactly after one of these two blocks.*

Proof. The colors of the states associated with a cycle C from the PCR_n are obtained by rotating the circle that represents the states of C in a way that the first bit of each state will be in the coordinates $(x, y) = (0, 1)$. This implies that the CM will move around in a circle with its center at the origin. If the CM in one of the states is at the origin $(0, 0)$, then by this rotation the CM of all the states is at the origin and the colors of all the states will be I. If the CM is not at the origin, then the CM will move in a circle and will cross the y-axis exactly twice, once from right to left and once from left to right. Only in these two times can the CM be on the y-axis (either once or twice or none at all) and the associated states will have the color I. In all the other rotations the same color L or R is kept, as long the y-axis is not crossed. □

Corollary 3.17. *If one state, of a PCR_n cycle C, has a CM at the origin, then all the states of C have CM at the origin.*

Lemma 3.24. *Let C be a cycle from G_n that contains at least one state colored with L, and at least one state that is not colored with L. Let Z be a state on C that is colored by L, and that has a predecessor on C that is not colored by L. Suppose that Z is on a cycle C' from PCR_n. Then, Z is the first state in the block of states colored with L on C'.*

Proof. Two conjugate states differ in their first bit and therefore they differ only in the weight that is placed on the point $(x, y) = (0, 1)$ that is on the y-axis. This implies that the CM of both conjugate states are either on the y-axis or on the same side of the y-axis. Therefore two such conjugate states have the same color. The two predecessors of Z are conjugate states and hence they have the same color. The color of these two predecessors of Z is not L by the assumption. Thus by Lemma 3.23 the state Z is the first state in a cycle C' of the PCR_n in the block of states that are colored by L. □

Lemma 3.25. *If C is a cycle of G_n that does not contain both colors L and R (it might contain a state colored by either L or R), then C is a cycle from the PCR_n for which all the states have a CM at the origin.*

Proof. Define the **moment** M_Z of a state $Z = (z_0, z_1, \ldots, z_{n-1})$ to be the total moment about the y-axis of the weights z_is. In other words

$$M_Z = \sum_{i=0}^{n-1} z_i \cdot \sin \frac{2\pi i}{n}.$$

The moment M_Z of the state Z is negative, zero, or positive, respectively, according to how the state Z is colored with L, I, or R, respectively. Since the sum of all the n-roots of unity is 0, it follows that for a cycle $C = [c_0 c_1 \ldots c_{\ell-1}]$ whose states are $Z_0, Z_1, \ldots, Z_{\ell-1}$ we have that

$$\sum_{t=0}^{\ell-1} M_{Z_t} = \sum_{t=0}^{\ell-1} \sum_{t'=0}^{n-1} c_{t+t'} \cdot \sin \frac{2\pi t'}{n} = \sum_{t=0}^{\ell-1} c_t \sum_{t'=0}^{n-1} \sin \frac{2\pi t'}{n} = 0.$$

Hence, C contains a state with a positive moment if and only if C contains a state with a negative moment, i.e., C contains a state with the color R if and only if C contains a state with the color L. However, since the cycle C does not contain both colors L and R, it follows that all the states on C are colored with the color I.

If one of the states X on the cycle C, whose states are colored with an I, is not on a PCR_n cycle for which the CM of its states is at the origin (see Corollary 3.17), then by Lemma 3.23 one of the predecessors of X is colored by either L or R and its second predecessor has the same color. However, all the states on C are colored with I and hence this contradicts the fact that its predecessor on C is colored with an I. Thus all the states of C are on PCR_n cycles with CM at the origin.

To complete the proof it is required to show that all the states of C are on one PCR_n cycle. If C is not one of the cycles from the PCR_n with CM of all its states at the origin, then there exists a τ and two cycles C_1 and C_2 from the PCR_n such that Z_τ is a state on C_1 and $Z_{\tau+1}$ is a state on C_2, where both states are colored with an I and their CM is at the origin. This implies, by Corollary 3.17, that all the states of C_1 and C_2 have CM at the origin. The companion $Z'_{\tau+1}$ of $Z_{\tau+1}$ must be on C_1, and it is colored with an I, but its CM cannot be at the origin. This contradicts the fact that if a PCR_n cycle has one state with CM at the origin, then all its states have CM at the origin (see Corollary 3.17).

Thus we have that C is a PCR_n cycle for which all its states have CM at the origin. $\qquad\square$

A **V-set** in a directed graph G is a set of vertices that will leave the graph with no cycles when removed (with their incident edges) from G.

The definition of a V-set immediately implies the following result.

Lemma 3.26. *A V-set contains at least one vertex from each cycle of any arbitrary factor.*

Let $v(n)$ be the minimum number of vertices that should be removed from G_n (with their incident edges) to leave the graph with no cycles. Let $\eta(n)$ be the maximum number of cycles in a factor of G_n.

Lemma 3.27. *For each $n \geq 2$ we have that $v(n) \geq \eta(n)$.*

Proof. Let Γ be a set with the minimum number of vertices in G_n that leave G_n with no cycles when removed from the graph and let $\mathcal{F}$ be the factor in G_n with the maximum number of cycles.

Clearly, Γ must contain a vertex from each cycle of $\mathcal{F}$ as otherwise if Γ is removed from G_n, then at least one cycle of $\mathcal{F}$ will remain untouched. Therefore $v(n) \geq \eta(n)$. $\qquad\square$

Theorem 3.12. *The minimum number of states that, if removed from G_n will leave G_n with no cycles is at most $Z(n)$, i.e., $Z(n) \geq v(n)$.*

Proof. To leave G_n with no cycles it is required to remove at least one state from each cycle of the PCR_n. Let Γ be a set of $Z(n)$ states that contains:

(1) An arbitrary state from each cycle in the PCR_n whose states have CM at the origin.

(2) The first state in the block of states colored with L from each other cycle of the PCR_n (see Lemma 3.23).

Let C be an arbitrary cycle in G_n.

If C does not contain a state colored by L, then by Lemma 3.25 we have that C is a cycle from the PCR_n whose states have CM at the origin. Therefore by (1), we have that Γ contains a state from C.

If C contains a state colored by L, then by Lemma 3.24 and (2), we have that Γ contains a state from C.

Thus Γ is a V-set and the claim of the theorem follows. □

Corollary 3.18. *The maximum number of cycles in a factor of G_n (a state diagram of an FSR_n) is $Z(n)$, i.e., $\eta(n) = Z(n)$.*

Proof. In the PCR_n there are $Z(n)$ cycles and hence by definition, we have that $Z(n) \leq \eta(n)$. By Lemma 3.27 and Theorem 3.12 we have that

$$\eta(n) \leq v(n) \leq Z(n).$$

This implies that $Z(n) \geq \eta(n)$. Thus we have that $Z(n) = \eta(n)$. □

Corollary 3.19. *For each $n \geq 2$ we have that $v(n) = \eta(n) = Z(n)$.*

The coloring technique used in this section is unique and we have the following question.

Problem 3.2. Can the coloring technique used in this section be used to solve other problems associated with the de Bruijn graph?

Are there other FSR_ns, except for the PCR_n, whose state diagrams (factors) have $Z(n)$ cycles? The answer is that there exist many such factors for each $n > 3$. For $n = 3$, the PCR_n has four cycles [0], [1], [001], and [011]. The following four cycles of the PSR_n, [0], [1], [01], and [0011], form the only other another state diagram in G_3 with 4 cycles as in PCR_3. Such factors can be found for every $n \geq 3$, by considering two PCR_n cycles C_1 and C_2 in which there are two pairs $\{x, x'\}$ and $\{y, y'\}$ of companion states, where $x, y \in C_1$ and $x', y' \in C_2$. Two such cycles, not necessarily in the PCR_n, with two pairs of companion pairs, are called an **adjacency pair**. By applying the merge-or-split method on $\{x, x'\}$ the two cycles C_1 and C_2 are merged into one cycle C. By applying the merge-or-split method on $\{y, y'\}$, which are now two states on C, the cycle C is split into two cycles C_3 and C_4, which are different from C_1 and C_2. The newly generated factor has $Z(n)$ cycles and this factor is not the state diagram of the PCR_n.

Example 3.2. Assume $n \geq 3$ and the two PCR_n cycles $[0^{n-1}1]$ and $[0^{n-2}11]$, are both of length n. These two cycles have two pairs of companion states

$$\{(0^{n-2}10), (0^{n-2}11)\} \text{ and } \{(10^{n-1}), (10^{n-2}1)\}.$$

Now, if the merge-or-split method is applied on the pair of companion states $\{(0^{n-2}10), (0^{n-2}11)\}$, then the cycle $[0^{n-1}110^{n-2}1]$ of length $2n$ is generated. This cycle is split into two cycles by applying the merge-or-split method on the pair of companion states $\{(10^{n-1}), (10^{n-2}1)\}$. These two cycles are $[0^{n-1}11]$ of length $n + 1$ and $[10^{n-2}]$ of length $n - 1$. Note that if we start by applying the merge-or-split method on the pair of companion states $\{(10^{n-1}), (10^{n-2}1)\}$, the cycle $[0^{n-2}110^{n-1}1]$ is generated. ∎

3.4 Notes

Although there has been more research on $LFSR_n$s, the research on $NLFSR_n$s was carried out in parallel. $NLFSR_n$s were used to form span n de Bruijn sequences that are always nonlinear. Each sequence is generated by some $LFSR_n$ and this will be considered in Chapter 5. The current chapter considered the nonlinear theory only in three directions, properties of four interesting nonlinear FSR_ns and in particular their cycle structure and the number of cycles in their state diagrams, which is the second direction. Finally, the intriguing problem on the state diagram with the maximum number of cycles was solved. More analysis from a few different directions of $NLFSR_n$s can be found in the book of Golomb [12].

Section 3.1. Analysis of the cycles from the PCR_n, the CCR_n, the PSR_n, and the CSR_n, was given first in the book of Golomb [12] who developed the various formulas and was the first to consider these enumerations. An analysis of the number of cycles in the state diagrams of these shift registers was presented also in Etzion [8]. In particular, this work considered the enumeration of self-dual sequences and their connection to the cycles of the PCR_n, the CCR_n, the PSR_n, and the CSR_n. Self-dual sequences will appear also in other chapters of the book, where some more properties of these sequences will be proved. Sloane [36] observed a connection between the enumerations for $Z(n)$, $Z^*(n)$, $S(n)$, $S^*(n)$ and single-deletion-correcting codes.

Necklaces were considered and used under different types of structures. They were classified and categorized in different ways. For many applications, it is required to choose one representative from each full-order necklace. This will be done in some chapters of the book. One important set of representatives was suggested by Lyndon [22] who took the word that is the least one in lexicographic order. These words are called by his name, Lyndon words. Properties and construction methods for all the necklaces and some of their classes were extensively studied. In particular, one would like to have efficient algorithms for their construction. For some work on such construction methods see Cattell,

Ruskey, Sawada, and Serra [2], Fredricksen and Kessler [11], Ruskey, Miers, and Sawada [31], Ruskey and Sawada [32], Sawada [33,34], and Sawada and Ruskey [35].

Other nonlinear shift registers were also observed. For example, Coven and Hedlund [5] determined all the possible periods of shift-register sequences for nonlinear feedback functions of the form

$$f(x_1, x_2, \ldots, x_n) = x_1 + \prod_{i=1}^{k}(x_i + b_i), \quad m - 1 \geq k \geq 2, \ b_i \in \{0, 1\},$$

where the least period of $[b_1 \ \cdots \ b_k]$ is k, i.e., $[b_1 \ \cdots \ b_k]$ is a full-order necklace. Other papers that consider the structure of sequences of nonlinear shift-register sequences and also their decomposition into other cycles are, for example, Cheng [3], Cohn and Lempel [4], Kjeldsen [16], Ma, Qi, and Tian [23], Mykkeltveit, Siu, and Tong [29], Søreng [37], Søreng [38], and Zhang, Qi, Tian, and Wang [41].

Problem 3.3. Find more types of nonlinear feedback functions for which the number of generated cycles or their periods can be analyzed.

Self-dual sequences were defined only for binary sequences and it is natural to ask whether there is a generalization for these sequences over larger alphabets. Assume S is a sequence over $\mathbb{F}_q$, q a prime power. We say that S is a self-dual sequence over $\mathbb{F}_q$ if $(\alpha, \alpha, \ldots, \alpha) + S = S$ for each element $\alpha \in \mathbb{F}_q^*$. The same definition applies also if we take $\mathbb{Z}_m$ instead of $\mathbb{F}_q$ for every positive integer $m \geq 2$. The theory for non-binary self-dual sequences is less developed, but such sequences will appear again in the book. Similarly, we can defined non-binary shift registers over $\mathbb{F}_q$ and analog registers for the PCR$_n$, the CCR$_n$, the PSR$_n$, and the CSR$_n$ as follows:

$$f(x_1, x_2, \ldots, x_n) = x_1,$$
$$f(x_1, x_2, \ldots, x_n) = x_1 + \alpha, \quad \alpha \in \mathbb{F}_q^*,$$
$$f(x_1, x_2, \ldots, x_n) = \sum_{i=1}^{n} x_i$$

and

$$f(x_1, x_2, \ldots, x_n) = \alpha - \sum_{i=1}^{n} x_i, \quad \alpha \in \mathbb{F}_q^*,$$

respectively, where all computations are done in $\mathbb{F}_q$.

Problem 3.4. Continue to develop the nonlinear theory of non-binary shift registers with comparison to the theory of the related binary shift registers.

The structures of cycles from NLFSR$_n$s and in particular those that yield complement sequences, reverse sequences, and self-dual sequences were discussed by Walker [39]. It was also proved in that paper that a de Bruijn sequence S cannot be a self-dual sequence. This will be also proved using a simpler proof in Chapter 5. The parity of the number of cycles of length k in G_n was given by Duvall and Kibler [6].

Theorem 3.13. *If $n \geq 2$, then the number of simple cycles of length k in G_n is odd if and only if*

$$k = \frac{1}{3} \left(2^{n+1} \pm 3 + (-1)^n \right).$$

Section 3.2. Some of the enumeration results in this section can be found in the book of Golomb [12] and also in Etzion [8]. The enumeration for the number of full-order necklaces in Theorem 3.5 (Theorem 3.6 for the binary case) is much older and can be attributed to MacMahon [24]. Some of the enumerations are also associated with elements in coding theory and can be found in the excellent book of MacWilliams and Sloane [25] on coding theory. In this section, we enumerated the number of cycles in some FSR$_n$s and also the number of cycles of specific structures like self-dual sequences. A different enumeration problem is to find the number of cycles of any given length in $G_{\sigma,n}$. The number of cycles of short length in G_n was computed by Bryant and Christensen [1] and Wan, Xiong, and Yu [40]. Asymptotic results on the number of cycles of any given length were presented by Maurer in [26] who also provided many tables and computer-search enumerations. Li, Jiang, and Lin [21] studied the properties of all the cycles in the PCR$_n$ which contain only states of weight at most k. They also considered cycles of larger length that contain these states. Constructions of cycles that contain these states were considered before by Fredricksen [10]. Asymptotic enumerations for the number of full-length necklaces and the number of degenerate necklaces were carried out by Prasad and Iyengar [30]. Finally, the function $N(n, \epsilon)$ and Lemma 3.21 were defined and proved by Etzion [7]. Further analysis of the number of cycles associated with this function was presented in Etzion [8]. As noted in Section 2.5 it was proved by Lempel [19] that in $G_{\sigma,n}$ there exists a cycle of length k for each $1 \leq k \leq \sigma^n$. Hemmati and Costello [14] suggested an algebraic construction for such sequences when σ is a power of a prime. Etzion [7] gave an efficient algorithm to construct such binary sequences. The same ideas can be used to construct such sequences over any alphabet.

Section 3.3. The proof that $Z(n)$ is the maximum number of cycles in a state diagram of an FSR$_n$ was presented by Mykkeltveit in [27]. The idea to use the concept of V-set in the proof was suggested by Lempel [18]. Adjacency pairs in the PCR$_n$ were considered by van Lantschoot [17] and Mykkeltveit [28]. They illustrated many pairs of PCR$_n$ cycles with two pairs of companion states. Each such pair of PCR$_n$ cycles can be used to form new factors of G_n with $Z(n)$

cycles. In many cases, a few pairs of companion states in pairs of PCR_n cycles can be used to form different factors of G_n with $Z(n)$ cycles. Adjacency pairs in other states diagrams were considered, for example, by Hauge [13], Hemmati, Schilling, and Eichmann [15], and Li, Jiang, and Lin [20]. Adjacency pairs such as those considered in Hemmati, Schilling, and Eichmann [15] were used and considered in parallel by Etzion and Lempel [9].

References

[1] P.R. Bryant, J. Christensen, The enumeration of shift register sequences, J. Comb. Theory, Ser. A 35 (1983) 154–172.

[2] K. Cattell, F. Ruskey, J. Sawada, M. Serra, Fast algorithms to generate necklaces, unlabeled necklaces, and irreducible polynomials over GF(2), J. Algorithms 37 (2000) 267–282.

[3] U. Cheng, On the cycle structure of certain classes of nonlinear shift registers, J. Comb. Theory, Ser. A 37 (1984) 61–68.

[4] M. Cohn, A. Lempel, Cycle decomposition by disjoint transpositions, J. Comb. Theory, Ser. A 13 (1972) 83–89.

[5] E.M. Coven, G.A. Hedlund, Periods of some nonlinear shift registers, J. Comb. Theory, Ser. A 27 (1979) 186–197.

[6] P.F. Duvall Jr., R.E. Kibler, On the parity of the frequency of cycle lengths of shift register sequences, J. Comb. Theory, Ser. A 18 (1975) 357–361.

[7] T. Etzion, An algorithm for generating shift-register cycles, Theor. Comput. Sci. 44 (1986) 209–224.

[8] T. Etzion, Self-dual sequences, J. Comb. Theory, Ser. A 44 (1987) 288–298.

[9] T. Etzion, A. Lempel, Construction of de Bruijn sequences of minimal complexity, IEEE Trans. Inf. Theory 30 (1984) 705–709.

[10] H. Fredricksen, The number of nonlinear shift registers that produce all vectors of weight $\leq t$, IEEE Trans. Inf. Theory 39 (1993) 1989–1990.

[11] H. Fredricksen, I.J. Kessler, An algorithm for generating necklaces of beads in two colors, Discrete Math. 61 (1986) 181–188.

[12] S.W. Golomb, Shift Register Sequences, Holden Day, San Francisco, 1967, Aegean Park, Laguna Hills, CA, 1980, World Scientific, Singapore, 2017.

[13] E.R. Hauge, On the cycles and adjacencies in the complementary circulating register, Discrete Math. 145 (1995) 105–132.

[14] F. Hemmati, D.J. Costello Jr., An algebraic construction for q-ary shift register sequences, IEEE Trans. Comput. 27 (1978) 1192–1195.

[15] F. Hemmati, D.L. Schilling, G. Eichmann, Adjacencies between the cycles of a shift register with characteristic polynomial $(1 + x)^n$, IEEE Trans. Comput. 33 (1984) 675–677.

[16] K. Kjeldsen, On the cycle structure of a set of nonlinear shift registers with symmetric feedback functions, J. Comb. Theory, Ser. A 20 (1976) 154–169.

[17] E.J. van Lantschoot, Double adjacencies between cycles of a circulating shift register, IEEE Trans. Comput. 22 (1973) 944–955.

[18] A. Lempel, On the extremal factors of the de Bruijn graph, J. Comb. Theory, Ser. B 11 (1971) 17–27.

[19] A. Lempel, m-ary closed sequences, J. Comb. Theory 10 (1971) 253–258.

[20] M. Li, Y. Jiang, D. Lin, The adjacency graphs of some feedback shift registers, Des. Codes Cryptogr. 82 (2017) 695–713.

[21] M. Li, Y. Jiang, D. Lin, Properties of the cycles that contain all vectors of weight $\leq k$, Des. Codes Cryptogr. 91 (2023) 221–239.

[22] R.C. Lyndon, On Burnside problem I, Trans. Am. Math. Soc. 77 (1954) 202–215.

[23] Z. Ma, W.-F. Qi, T. Tian, On the decomposition of an NFSR into the cascade connection of an NFSR into an LFSR, J. Complex. 29 (2013) 173–181.

[24] P.A. MacMahon, Application of a theory of permutations in circular procession to the theory of numbers, Proc. Lond. Math. Soc. 23 (1892) 305–313.

[25] F.J. MacWilliams, N.J.A. Sloane, The Theory of Error-Correcting Codes, North-Holland, Amsterdam, 1977.

[26] U.M. Maurer, Asymptotically-tight bounds on the number of cycles in generalized de Bruijn-Good graphs, Discrete Appl. Math. 37/38 (1992) 421–436.

[27] J. Mykkeltveit, A proof of Golomb's conjecture for the de Bruijn graph, J. Comb. Theory, Ser. B 13 (1972) 40–45.

[28] J. Mykkelveit, Generating and counting the double adjacencies in a pure circulating shift register, IEEE Trans. Comput. 24 (1975) 299–304.

[29] J. Mykkeltveit, M.-K. Siu, P. Tong, On the cycle structure of some nonlinear shift-register sequences, Inf. Control 43 (1979) 202–215.

[30] L. Prasad, S.S. Iyengar, An asymptotic equality for the number of necklaces in a shuffle-exchange network, Theor. Comput. Sci. 102 (1992) 355–365.

[31] F. Ruskey, C.R. Miers, J. Sawada, The number of irreducible polynomials and Lyndon words with given trace, SIAM J. Discrete Math. 14 (2001) 240–245.

[32] F. Ruskey, J. Sawada, An efficient algorithm for generating necklaces with fixed density, SIAM J. Comput. 29 (1999) 671–684.

[33] J. Sawada, Generating bracelets in constant amortized time, SIAM J. Comput. 31 (2001) 259–268.

[34] J. Sawada, A fast algorithm to generate necklaces with fixed content, Theor. Comput. Sci. 301 (2003) 477–489.

[35] J. Sawada, F. Ruskey, Generating Lyndon brackets. An addendum to: fast algorithms to generate necklaces, unlabeled necklaces and irreducible polynomials over GF(2), J. Algorithms 46 (2003) 21–26.

[36] N.J.A. Sloane, On single-deletion-correcting codes, Codes Des. 10 (2002) 273–291.

[37] J. Søreng, The periods of the sequences generated by some symmetric shift registers, J. Comb. Theory, Ser. A 21 (1976) 164–187.

[38] J. Søreng, Symmetric shift registers, Pac. J. Math. 85 (1979) 201–229.

[39] E.A. Walker, Non-linear recursive sequences, Can. J. Math. 11 (1959) 370–378.

[40] Z.-X. Wan, R.-H. Xiong, M.-A. Yu, On the number of cycles of short length in the de Bruijn-Good graph G_n, Discrete Math. 62 (1986) 85–98.

[41] J.-M. Zhang, W.-F. Qi, T. Tian, Z.-X. Wang, Further results on the decomposition of an NFSR into the cascade connection of an NFSR into an LFSR, IEEE Trans. Inf. Theory 61 (2015) 645–654.

Chapter 4

Constructions of full cycles

Full cycles, enumeration, **D**-morphism, algorithms

Full cycles, known also as de Bruijn cycles, are the most important family of cycles in the de Bruijn graph. During the years since de Bruijn found their number, many algorithms were developed to construct either all these cycles or a large subset of these cycles. There were a few goals in mind when algorithms to construct these cycles were developed. The first one is to construct all of them, the second one was to construct these cycles with an efficient algorithm. Next, it was required to construct a subset of these cycles with desired properties. Algorithms are usually implemented via associated software, but it was also important to construct these cycles via hardware. Span n M-sequences can be constructed with $LFSR_n$ via their feedback function associated with a corresponding primitive polynomial. However, also to find a primitive polynomial of a large degree is not an easy task. Hence, designing span n full cycles via their associated FSR_n is also a goal in the construction of full cycles. All these goals will be addressed in this chapter.

Constructions of all span n de Bruijn sequences as Eulerian cycles can be carried out with a reverse spanning tree algorithm that can be used to construct all Eulerian cycles in any directed graph that satisfies the Euler criteria (see Theorem 1.15). This algorithm is discussed in Section 4.1. This algorithm, i.e., implementing the one-to-one correspondence between the number of reverse-spanning trees and the number of Eulerian cycles in G_n implies an enumeration method to find the number of span $n + 1$ de Bruijn sequences. This method will be considered in this section and will be also used in some later chapters. In this section, we will also discuss the method of de Bruijn to find the number of full cycles in G_n. His method is more general to find the number of Eulerian cycles in a connected directed graph, where each vertex has in-degree 2 and out-degree 2. It can be generalized to a directed graph in which the in-degree equals the out-degree for each vertex. Finally, the first known algorithm to construct full cycles will be presented.

Section 4.2 is devoted to recursive constructions for de Bruijn sequences. An important tool in these constructions is the operator **D**, which will be called the **D**-morphism in this section. This operator induces a homomorphism between de Bruijn graphs of consecutive orders. This operator was already mentioned in Section 1.2, where it is also called the derivative. It was also used in Sections 2.3

Sequences and the de Bruijn Graph. https://doi.org/10.1016/B978-0-44-313517-0.00010-X

and 3.1. It will be also used in various chapters of our exposition and it has an important role in analyzing the properties of words, sequences, and codes.

The merge-or-split method introduced in Section 1.3 is the main tool in efficient constructions of many full cycles in G_n. The method requires a factor (a state diagram) with a set of cycles and a method to find bridging states to apply the merge-or-split method to merge all the cycles of the factor. In Section 4.3 the state diagrams of the most celebrated FSR_n discussed in Chapter 3, the PCR_n, the CCR_n, the PSR_n, and the CSR_n, will be considered for this purpose.

4.1 Enumeration of all Eulerian cycles

Theorems 1.17 and 1.18 imply that an Eulerian cycle in G_n is associated with a span $n+1$ de Bruijn sequence and a Hamiltonian cycle in G_n is associated with a span n de Bruijn sequence. While in general there are efficient algorithms to generate Eulerian cycles in a graph and finding a Hamiltonian cycle is a difficult problem, in specific graphs, like the de Bruijn graph, it is not difficult to find an efficient algorithm to generate a Hamiltonian cycle. It is more difficult to generate all the de Bruijn sequences of order n efficiently. In the following sections, we will exhibit several efficient algorithms to generate a large class of de Bruijn sequences. The most interesting method to generate all of them is based on the reverse spanning tree algorithm that in general yields Eulerian cycles in each directed graph in which for each vertex the in-degree is equal to the out-degree. In this section, we concentrate on the algorithm only for in-degree 2 and out-degree 2 for each vertex.

A *reverse spanning tree* in a directed graph $G = (V, E)$ is a subgraph of G whose set of vertices is V, one vertex r with out-degree *zero* and from each other vertex $v \in V$ there is a unique directed path from v to r. Note that T is a reverse spanning tree in G if and only if T^R is a tree in G^R. The following lemma is a simple observation.

Lemma 4.1. *The out-degree of each vertex, excluding the root, in a reverse spanning tree is one.*

By Lemma 1.17, G_n and G_n^R are isomorphic graphs. This implies the following theorem that is proved again for completeness.

Theorem 4.1. *The number of reverse spanning trees in $G_n = (V_n, E_n)$ whose root r is the all-zeros word is equal to the number of spanning trees in G_n whose root r is the all-zeros word.*

Proof. Assume $T = (V_n, E)$ is a spanning tree in G_n whose root r is the all-zeros word. We construct the following subgraph $T' = (V_n, E')$ of G_n, where

$$E' \triangleq \{x \to y : y^R \to x^R \in E\}$$
$$= \{(x_0, \ldots, x_{n-1}) \to (x_1, \ldots, x_n) : (x_n, \ldots, x_1) \to (x_{n-1}, \ldots, x_0) \in E\}.$$

By the definition of the edges in G_n we have that

$$(x_0, x_1, \ldots, x_{n-1}) \to (x_1, \ldots, x_{n-1}, x_n) \in E_n$$

if and only if

$$(x_n, x_{n-1}, \ldots, x_1) \to (x_{n-1}, \ldots, x_1, x_0) \in E_n$$

and hence $E' \subset E_n$. This also implies that the number of edges in E is equal to the number of edges in E'. Therefore to complete the proof it is sufficient to show that from each vertex $v \in V_n$, $v \neq r$, there is a directed path from v to r in T'. Let

$$r \to v_1 \to v_2 \to \cdots \to v_{\ell-1} \to v_\ell \to v^R$$

be the path from r to v^R in T. Then, the reverse path

$$v \to v_\ell^R \to v_{\ell-1}^R \to \cdots \to v_2^R \to v_1^R \to r$$

is a path from v to r is T'. $\qquad\square$

A reverse spanning tree in G_4 and its associated spanning tree in G_4 are depicted in Fig. 4.1.

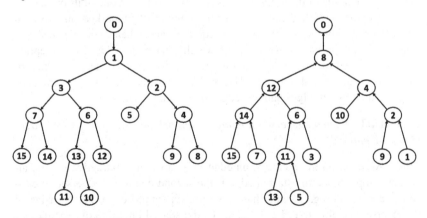

FIGURE 4.1   A tree and its associated reverse tree in G_4.

For a connected directed graph $G = (V, E)$, in which the in-degree is equal to the out-degree at each vertex, by Theorem 1.15 we have that there exists an Eulerian cycle in G. From each reverse spanning tree of G, one can construct some Eulerian cycles. For the de Bruijn graph G_n, it can be proven that there exists a one-to-one correspondence between the set of reverse spanning trees rooted at one of the self-loops (e.g., (0^n)) and the set of de Bruijn sequences of length 2^{n+1} (Eulerian cycles in G_n). We will first show that each reverse spanning tree in G_n yields a distinct span $n + 1$ de Bruijn sequence.

The reverse spanning tree algorithm:

Let T be a reverse spanning tree in G_n rooted at the vertex $r = (0^n)$. Each edge of T is starred in G_n. All the edges of G_n are set to be unmarked. Set the current vertex v to be r. Mark the self-loop edge $r \to r$.

(T1) If the only unmarked out-edge of v is the starred edge $v \to u_1$, then mark the edge $v \to u_1$ and set the current vertex to $v := u_1$; otherwise, mark the un-starred edge $v \to u_2$ and set $v := u_2$.

(T2) If $v \neq r$ then go to **(T1)**; otherwise stop.   ∎

Starting with n consecutive *zeros* and continuing with the sequence of (last) bits in the traversed edges (in the order in which they were marked) we obtain a sequence S from the algorithm. Similarly, we can consider the consecutive edges represented as $(n + 1)$-tuples as a path $\mathcal{P}$ in G_n. Every $n + 1$ consecutive bits of S are associated with an edge in G_n. Every $n + 2$ consecutive bits are associated with two consecutive edges being traversed by the algorithm. These two edges are connected by a vertex whose n bits form the suffix of length n of the first edge and the prefix of length n of the second edge.

Lemma 4.2. *When the reverse spanning tree algorithm terminates, the current vertex is $v = r$ and all the in-edges and out-edges of r were traversed.*

Proof. At each step of the algorithm when a vertex is left on one of its out-edges, the number of unmarked in-edges and the number of unmarked out-edges at each vertex, except for the self-loop r, is equal. Hence, when a vertex v, $v \neq r$, is visited there are unmarked out-edges on which it can be left. The only exception is the root r that immediately after the first step has exactly one unmarked in-edge and no unmarked out-edge. Hence, the algorithm terminates at the self-loop r and all the in-edges and out-edges of r were traversed.   □

Lemma 4.3. *When the reverse spanning tree algorithm terminates, all the edges of G_n are marked.*

Proof. Assume, on the contrary, an edge $v_1 \to v_2$ is not marked when the algorithm stops. Since the starred edge is the second one to be traversed at each vertex $v \neq r$, w.l.o.g. assume that $v_1 \to v_2$ is a starred edge, i.e., an edge of the reverse spanning tree. Let $v_2 \to v_3$ be the starred out-edge of v_2. Since v_2 was visited at most once and $v_2 \to v_3$ is a starred edge, it follows that also $v_2 \to v_3$ is unmarked. Continue by induction implies that the starred edge $u \to r$ is unmarked too, contradicting Lemma 4.2 that the algorithm terminates at the self-loop r and all the in-edges and out-edges of r were traversed.   □

Theorem 4.2. *The sequence S obtained by the reverse spanning tree algorithm is a span $n + 1$ de Bruijn sequence.*

Proof. By Lemma 4.3, when the algorithm terminates all the edges were visited, and hence the sequence of the visited edges forms an Eulerian cycle. By the discussion in Section 1.3 the written sequence is a span $n + 1$ de Bruijn sequence. □

Lemma 4.4. *Any two distinct reverse spanning trees yield two distinct de Bruijn sequences.*

Proof. Let T_1 and T_2 be two distinct reverse spanning trees of G_n. Two copies of the algorithm can be applied; one on T_1 and the second on T_2. As long as the two copies of that algorithm visit the same vertices and the same out-edges, from G_n, of these visited vertices, the generated subsequences are the same. Since T_1 and T_2 are distinct spanning trees, it follows that at one point the applied copies of the algorithm arrive for the first time at a vertex v that has different out-edges that are not edges of the tree, $v \to v_1$ for T_1 and $v \to v_2$ for T_2. At this point, the algorithm generates a different bit (the last bit of either v_1 or v_2, respectively) for the two sequences and hence the two generated de Bruijn sequences are distinct. □

Lemma 4.5. *Each span $n + 1$ de Bruijn sequence is generated by the algorithm using exactly one reverse spanning tree of G_n.*

Proof. By Lemma 4.4 any span $n + 1$ de Bruijn sequence can be generated by at most one reverse spanning tree.

Let S be a span $n + 1$ de Bruijn sequence, where the $(n + 1)$-tuples are ordered in a way that S starts with $n + 1$ *zeros*. Let $T = (V_n, E)$ be a graph defined as follows. Each nonzero n-tuple $x = (x_1, x_2, \ldots, x_n)$ is contained exactly twice in S. If $(x_1, x_2, \ldots, x_n, x_{n+1})$ is the second $(n + 1)$-tuple whose prefix is x, then $(x_1, x_2, \ldots, x_n) \to (x_2, \ldots, x_n, x_{n+1})$ is an edge in E. Clearly, by this definition, the number of edges in T is $2^n - 1$ and each vertex in T, except the all-zeros vertex, has an out-degree equal to one.

Now, we exhibit a path from each vertex to the all-zeros vertex. Assume, on the contrary, that from some vertex $y = (y_1, y_2, \ldots, y_n)$ there is no directed path in T to the all-zeros vertex. Assume further that y is the last n-tuple in S with this property. Let $(y_1, y_2, \ldots, y_n, y_{n+1})$ be the second appearance of y as a prefix of an $(n + 1)$-tuple in S. By definition, $y \to (y_2, \ldots, y_n, y_{n+1})$ is an edge in T. Hence, since there is no path from y to the all-zeros vertex, it follows that there is no path from $(y_2, \ldots, y_n, y_{n+1})$ to the all-zeros vertex, a contradiction to the assumption that y was the last n-tuple in S with this property. Therefore the definition of T implies that the algorithm generates S using the reverse spanning tree T.

Thus each span $n + 1$ de Bruijn sequence is generated by the algorithm using exactly one reverse spanning tree of G_n. □

Corollary 4.1. *There is a one-to-one correspondence between the Eulerian cycles in G_n and the reverse spanning trees in G_n.* ■

Theorem 4.3. *Let G be a strongly connected directed graph with m vertices for which each vertex has in-degree 2 and out-degree 2. If G has Δ Eulerian cycles, then its line graph $L(G)$ has $2^{m-1}\Delta$ Eulerian cycles, where it is assumed that $\Delta = 1$ for a graph with one vertex.*

Proof. The proof is by induction on m.

The basis for the induction is $m = 1$. In this case, the graph G has one vertex v and two self-loops e_1 and e_2 from v to v. The line graph $L(G)$ has two vertices e_1 and e_2 and four edges (e_1, e_1), (e_1, e_2), (e_2, e_1), and (e_2, e_2), i.e., it is the de Bruijn graph of order 1, G_1. This graph has exactly one Eulerian cycle as in G and the proof of the basis is complete.

Assume now that the claim is true for all strongly connected directed graphs with $m - 1$ vertices and in-degree 2 and out-degree 2 for each vertex, i.e., for such a strongly directed graph G with $m - 1$ vertices and Δ distinct Eulerian cycles, in $L(G)$ there are $2^{m-2}\Delta$ distinct Eulerian cycles.

For the induction step, we are given a strongly connected graph G with m vertices, in-degree 2 and out-degree 2 for each vertex, for which G has Δ distinct Eulerian cycles.

Assume first that G has m vertices and all of them are self-loops. Since the graph G is strongly connected and except for the self-loop each vertex has one out-edge and one in-edge, it follows that there is a cycle going through the m vertices,

$$v_1 \to v_2 \to \cdots \to v_m \to v_1.$$

Let a_i be the vertex in $L(G)$ that represents the edge (self-loop) $v_i \to v_i$ and b_i be the vertex in $L(G)$ that represents the edge $v_i \to v_{i+1}$. By the definition of the line graph, $L(G)$ is as depicted in Fig. 4.2.

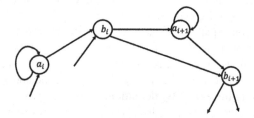

FIGURE 4.2 Vertices and edges in $L(G)$.

A cycle in $L(G)$ that passes through b_i must continue and pass through b_{i+1}. An Eulerian cycle in $L(G)$ will start at b_0, continue to b_1 and after that to b_2 until it reaches b_0 and one round has finished. It will continue through all these vertices of $L(G)$ in the same order and when the second round is finished an Eulerian cycle will be formed. Such a cycle has two ways of going from b_i to b_{i+1}, either directly or through a_{i+1}. It should be chosen, for each b_i, which option is taken in the first round and which one is taken in the second round.

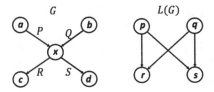

FIGURE 4.3 A local scenario in G and $L(G)$.

The first edge in this process can be chosen arbitrarily and for the other edges we have in total 2^{m-1} choices for 2^{m-1} Eulerian cycles in $L(G)$ compared to one Eulerian cycle in G. Therefore the number of distinct Eulerian cycles is $2^{m-1}\Delta$ in this case.

Assume now that the graph G with m vertices has a vertex x that is not a self-loop. The scenario is depicted in Fig. 4.3, where P, Q, R, S are different edges of G (although some of the vertices a, b, c, and d may coincide).

From G we form another two graphs with $m-1$ vertices, in which each vertex has in-degree 2 and out-degree 2, where the vertex x is removed from the graph G with all its incident edges. These two graphs are constructed as follows:

(G1) The graph $G(1)$ is obtained by replacing the two edges P and R with one edge from a to c, and replacing the two edges Q and S with one edge from b to d.

(G2) The graph $G(2)$ is obtained by replacing the two edges P and S with one edge from a to d, and replacing the two edges Q and R with one edge from b to c.

It is easy to verify that the number of Eulerian cycles in G is the sum of the number of Eulerian cycles in $G(1)$ and $G(2)$.

By the induction hypothesis, the theorem applies to $G(1)$ and also to $G(2)$. There are three different types of Eulerian cycles in $L(G)$, depending on whether the two paths, on the Eulerian cycle, leaving r and returning to p or q, both go to p, both go to q, or one goes to p and one goes to q. This also determines the structure of the paths from s to p and to q. We distinguish between these three cases to be analyzed. For each case, the associated figure will be depicted.

Case 1: Assume first that the first path, $\mathcal{P}_1$, goes from r to p, the second path, $\mathcal{P}_2$, goes from s to q, the third path, $\mathcal{P}_3$, goes from s to p, and the fourth path, $\mathcal{P}_4$, goes from r to q. This scenario is depicted in Fig. 4.4.

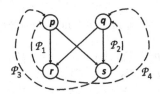

FIGURE 4.4 The scenario of the paths in $L(G)$ for Case 1.

These paths yield the following four Eulerian cycles in $L(G)$:

$$
\begin{array}{llllllll}
\mathcal{P}_1 & pr & \mathcal{P}_4 & qs & \mathcal{P}_3 & ps & \mathcal{P}_2 & qr \\
\mathcal{P}_1 & ps & \mathcal{P}_2 & qr & \mathcal{P}_4 & qs & \mathcal{P}_3 & pr \\
\mathcal{P}_1 & ps & \mathcal{P}_3 & pr & \mathcal{P}_4 & qs & \mathcal{P}_2 & qr \\
\mathcal{P}_1 & ps & \mathcal{P}_2 & qs & \mathcal{P}_3 & pr & \mathcal{P}_4 & qr
\end{array}
$$

In $L(G(1))$ and $L(G(2))$ these four paths are reduced to the scenario as depicted in Fig. 4.5.

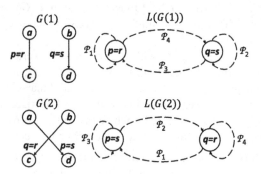

FIGURE 4.5 The scenario for the four paths of $L(G(1))$ and $L(G(2))$ in Case 1.

In $L(G(1))$, there is exactly one Eulerian cycle that will use the paths $\mathcal{P}_1$, $\mathcal{P}_2$, $\mathcal{P}_3$, and $\mathcal{P}_4$. Similarly, in $L(G(2))$, there is exactly one Eulerian cycle that will use the paths $\mathcal{P}_1$, $\mathcal{P}_2$, $\mathcal{P}_3$, and $\mathcal{P}_4$. These two Eulerian cycles are associated with the four Eulerian cycles in $L(G)$.

Case 2: Assume now that the first path, $\mathcal{P}_5$, goes from r to p, the second path, $\mathcal{P}_6$, goes from r to p, the third path, $\mathcal{P}_7$, goes from s to q, and the fourth path, $\mathcal{P}_8$, goes from s to q. This scenario is depicted in Fig. 4.6.

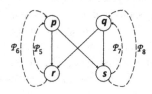

FIGURE 4.6 The scenario of the paths in $L(G)$ for Case 2.

These paths yield the following four Eulerian cycles in $L(G)$:

$$
\begin{array}{llllllll}
\mathcal{P}_5 & pr & \mathcal{P}_6 & ps & \mathcal{P}_7 & qs & \mathcal{P}_8 & qr \\
\mathcal{P}_5 & pr & \mathcal{P}_6 & ps & \mathcal{P}_8 & qs & \mathcal{P}_7 & qr \\
\mathcal{P}_5 & ps & \mathcal{P}_7 & qs & \mathcal{P}_8 & qr & \mathcal{P}_6 & pr \\
\mathcal{P}_5 & ps & \mathcal{P}_8 & qs & \mathcal{P}_7 & qr & \mathcal{P}_6 & pr
\end{array}
$$

In $L(G(1))$ and $L(G(2))$ these four paths are reduced to the scenario as depicted in Fig. 4.7.

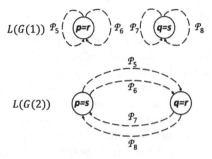

FIGURE 4.7 The scenario for the four paths in $L(G(1))$ and $L(G(2))$ in Case 2.

In $L(G(1))$ there is no Eulerian cycle that will use the paths $\mathcal{P}_5$, $\mathcal{P}_6$, $\mathcal{P}_7$, and $\mathcal{P}_8$. In $L(G(2))$ there are exactly two Eulerian cycles that will use the paths $\mathcal{P}_5$, $\mathcal{P}_6$, $\mathcal{P}_7$, and $\mathcal{P}_8$. These two Eulerian cycles in $L(G(1))$ and $L(G(2))$ are associated with the four Eulerian cycles in $L(G)$.

Case 3: Assume for the last scenario that the first path, $\mathcal{P}_9$, goes from r to q, the second path, $\mathcal{P}_{10}$, goes from r to q, the third path, $\mathcal{P}_{11}$, goes from s to p, and the fourth path, $\mathcal{P}_{12}$, goes from s to p. This is depicted in Fig. 4.8 and is handled exactly as in Case 2, when the roles of r and s are exchanged.

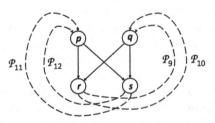

FIGURE 4.8 The scenario of the paths in $L(G)$ for Case 3.

Therefore the number of Eulerian cycles in $L(G)$ (12 cycles in Cases 1, 2, and 3) is twice the sum of the number of Eulerian cycles in $L(G(1))$ and $L(G(2))$ (6 cycles in Cases 1, 2, and 3). On the other hand, the number of Eulerian cycles in G is equal to the sum of the associated numbers of Eulerian cycles in $G(1)$ and $G(2)$.

Thus the claim of the theorem is implied by the induction hypothesis. □

Corollary 4.2. *The number of Eulerian cycles in G_{n-1} is $2^{2^{n-1}-n}$.*

Proof. The proof is again by simple induction, with the basis being the unique Eulerian cycle in G_1. For the induction step we observe that by Theorem 1.16 we have that $G_n = L(G_{n-1})$ and hence we can apply Theorem 4.3 to obtain the required result (stated also in Theorem 1.29). □

Corollary 4.3. *The number of span n de Bruijn sequences is $2^{2^{n-1}-n}$.*

Corollary 4.4. *The number of reverse spanning trees in G_n (and the number of spanning trees in G_n) is $2^{2^{n-1}-n}$.*

After having enumerated the number of span n de Bruijn sequences, we turn to constructions of such sequences. The reverse spanning tree algorithm can be used to generate all span n de Bruijn sequences, but it is an inefficient algorithm. First, it should find all the spanning trees of G_n and the algorithm also has to store each one of them. Therefore we would like to have an efficient algorithm that constructs a subset of these sequences. We start with the most celebrated algorithm to construct one such sequence. The first known algorithm to form a span n de Bruijn sequence is known as the **prefer one** algorithm. We start with the all-zeros n-tuple as the first n bits (initial state) of the sequence. Given the last n bits of the current sequence (current state), which was already generated, $(x_1, x_2, \ldots, x_n)$, examine whether the n-tuple $(x_2, \ldots, x_n, 1)$ appeared as a window of length n in the sequence generated so far. If it did not appear, then add 1 to the current sequence; otherwise, add 0 to the current sequence. There is no need to check if $(x_2, \ldots, x_n, 0)$ already appeared in the sequence as it will be proved in the following lemma that such a scenario cannot happen. Finally, the algorithm stops when the $n-1$ consecutive *zeros* appear again.

Lemma 4.6. *The prefer one algorithm generates a span n de Bruijn sequence.*

Proof. Assume first, on the contrary, that an n-tuple $x_1 x_2 \cdots x_{n-1} x_n$ appears twice in the sequence. Clearly, $x_n = 0$ (since an n-tuple that ends with a *one* cannot appear twice by the definition) and it also implies that an n-tuple $z x_1 x_2 \cdots x_{n-1}$ also appears twice in the sequence (since $x_1 x_2 \cdots x_{n-1}$ appears at least three times), and using induction we have that the n-tuple 0^n appears twice in the sequence, a contradiction. Therefore each n-tuple is contained at most once in the sequence. Next, we have to prove that the last n-tuple in the sequence is $10 \cdots 0$. When a given nonzero $(n-1)$-tuple $x_1 x_2 \cdots x_{n-1}$ appears in the sequence for the first time it is associated with the edge

$$(b, x_1, x_2, \ldots, x_{n-1}) \rightarrow (x_1, x_2, \ldots, x_{n-1} 1)$$

in G_n. It can appear a second time when it is associated with the edge $(\bar{b}, x_1, x_2, \ldots, x_{n-1}) \rightarrow (x_1, x_2, \ldots, x_{n-1} 0)$ in G_n. It cannot appear a third time since both its predecessors $b x_1 x_2 \cdots x_{n-1}$ and $\bar{b} x_1 x_2 \cdots x_{n-1}$ already appeared in the sequence. This implies that for any nonzero n-tuple that does not end with $n-1$ *zeros* we can add either 0 or 1 at any step of the algorithm. Hence, the only n-tuple that can appear, for which we cannot add a 0 nor a 1, is $(1, 0, \ldots, 0)$ since the edge $(0, 0, \ldots, 0) \rightarrow (0, \ldots, 0, 1)$ already appeared (the first edge after the self-loop at the all-zeros vertex), but the edge $(1, 0, \ldots, 0) \rightarrow (0, \ldots, 0, 0)$ (the last edge in the sequence) did not appear. Hence, the last n-tuple in the sequence is $10 \cdots 0$ and the algorithm stops.

Assume now that the n-tuple $x_1 x_2 \cdots x_n$ does not appear in the sequence, which implies that also the n-tuples $x_2 \cdots x_n 0$ and $x_3 \cdots x_n 00$ were not constructed and by induction, we will have that the n-tuple $10 \cdots 0$ was not constructed, a contradiction.

Thus the prefer one algorithm generates a span n de Bruijn sequence. $\quad\square$

Remark. There is a very clear similarity between the proof of Lemma 4.6 and the proofs of Lemmas 4.2 and 4.3.

There are two main disadvantages of the prefer one algorithm. The first one is that we have to store the whole sequence and search all of it to compute the next bit of the sequence. The second one is that the algorithm generates exactly one span n de Bruijn sequence. In Section 4.3 we will generate sequences that overcome these two disadvantages.

4.2 The D-morphism and recursive constructions

One of the most important operators on sequences is the ***derivative***, which was already defined in Section 1.2, and was also analyzed in Sections 2.3 and 3.1. This operator $\mathbf{D}$, when applied on acyclic sequences is a mapping $\mathbf{D} : \mathbb{F}_2^n \rightarrow \mathbb{F}_2^{n-1}$. In its definition, we distinguished between its application on cyclic sequences and acyclic sequences. In this section, we start by using its definition on acyclic sequences, continue with its application on cyclic sequences associated with cycles in G_n, and conclude with its applications to generate de Bruijn sequences and factors in G_n. For a binary word $X = (x_1, x_2, \ldots, x_{n-1}, x_n)$, the derivative (operator) $\mathbf{D}X$ was defined by

$$\mathbf{D}X \triangleq (x_1 + x_2, x_2 + x_3, \ldots, x_{n-1} + x_n).$$

The following lemma is an immediate observation from the definition.

Lemma 4.7. *The mapping* $\mathbf{D}$ *is a two-to-one mapping from* $\mathbb{F}_2^n$ *to* $\mathbb{F}_2^{n-1}$ *under which* $\mathbf{D}X = \mathbf{D}\bar{X}$.

An equivalent statement of that in Lemma 4.7 is the following simple lemma.

Lemma 4.8. *For* $X, Y \in \mathbb{F}_2^n$, $\mathbf{D}X = \mathbf{D}Y$ *if and only if* $Y = X$ *or* $Y = \bar{X}$.

As a consequence of these properties, the mapping $\mathbf{D}$ will be also called the **D-*morphism***. The mapping $\mathbf{D}$ has many more interesting properties that will be discussed in this section and throughout the book. The following result is also very easy to verify from the definition of the operator $\mathbf{D}$.

Lemma 4.9. *If* $X \rightarrow Y$ *is an edge in* G_n, *then* $\mathbf{D}X \rightarrow \mathbf{D}Y$ *is an edge in* G_{n-1}.

Lemma 4.9 induces a mapping from cycles of G_n to cycles of G_{n-1}. If $\mathcal{C} = [X_1, X_2, \ldots, X_k]$ is a cycle in G_n, where X_i, $1 \leq i \leq k$, is a vertex in G_n, then

$$\mathbf{D}\mathcal{C} \triangleq [\mathbf{D}X_1, \mathbf{D}X_2, \ldots, \mathbf{D}X_k],$$

is a cycle in G_{n-1}, where $\mathbf{D}X_i$ is a vertex in G_{n-1}, $\mathbf{D}X_i \rightarrow \mathbf{D}X_{i+1}$, $1 \leq i \leq k - 1$, and $\mathbf{D}X_k \rightarrow \mathbf{D}X_1$ are edges in G_{n-1}.

If $\mathbf{D}\mathcal{C}$ is a simple cycle, then also $\mathcal{C}$ is a simple cycle. For the other direction, $\mathcal{C}$ might contain a pair of vertices X and $\bar{X}$, and hence $\mathbf{D}\mathcal{C}$ might not be a simple cycle. Hence, if $\mathcal{C}$ is a simple cycle, we have a slightly different result, which is also easy to verify by using Lemma 4.8.

Lemma 4.10.

- If $\mathbf{D}\mathcal{C}$ is a simple cycle in G_{n-1}, then $\mathcal{C}$ is a simple cycle and for each state (vertex) X in $\mathcal{C}$ we have that $\bar{X} \notin \mathcal{C}$.
- If the cycle $\mathcal{C}$ is a simple cycle and for each $X \in \mathcal{C}$ we have that $\bar{X} \notin \mathcal{C}$, then $\mathbf{D}\mathcal{C}$ is also a simple cycle.

Consider again the cycle $\mathcal{C} = [X_1, X_2, \ldots, X_k]$, where the X_is are the consecutive vertices of the cycle and each vertex is a binary word of length n. The cycle $\mathbf{D}\mathcal{C}$ can be described also by using the representation of the cycle as a binary sequence. If $\mathcal{C} = [c_1, c_2, \ldots, c_k]$, where c_i is the first bit of X_i, $1 \leq i \leq k$, then $\mathbf{D}\mathcal{C} = [c_1 + c_2, c_2 + c_3, \ldots, c_{k-1} + c_k, c_k + c_1]$ is the representation of $\mathbf{D}\mathcal{C}$ as a sequence. In this representation, each n consecutive bits of $\mathcal{C} = [c_1, c_2, \ldots, c_k]$ is a corresponding vertex of $\mathcal{C}$ in G_n that is represented by n bits. Each $n - 1$ consecutive bits of $\mathbf{D}\mathcal{C} = [c_1 + c_2, c_2 + c_3, \ldots, c_{k-1} + c_k, c_k + c_1]$, which is a cycle of length k, have an associated vertex of $\mathbf{D}\mathcal{C}$ in G_{n-1} which is represented by $n - 1$ bits.

For the mapping $\mathbf{D}$ from the vertices of G_n to the vertices of G_{n-1}, i.e., from $\mathbb{F}_2^n$ to $\mathbb{F}_2^{n-1}$, we can also define an inverse mapping $\mathbf{D}^{-1}$ from $\mathbb{F}_2^{n-1}$ to $\mathbb{F}_2^n$ as follows:

$$\mathbf{D}_b^{-1}(x_1, x_2, \ldots, x_{n-1}) \triangleq (b, b + x_1, b + x_1 + x_2, \ldots, b + \sum_{i=1}^{n-1} x_i),$$

where b can be chosen as 0 or as 1. Clearly this is a one-to-two mapping, where $\mathbf{D}_0^{-1} X$ is the complement of $\mathbf{D}_1^{-1} X$ for any $X \in \mathbb{F}_2^{n-1}$. It is now interesting to see how this inverse mapping works on cycles and their associated sequences. First, we would like to extend the definition $\mathbf{D}^{-1}$ for cycles and sequences. If the cycle $\mathcal{C}$ is represented by its consecutive vertices (states), i.e., $\mathcal{C} = [X_1, X_2, \ldots, X_k]$ then

$$\mathbf{D}_{b_1}^{-1}(\mathcal{C}) = [\mathbf{D}_{b_1}^{-1} X_1, \mathbf{D}_{b_2}^{-1} X_2, \ldots \mathbf{D}_{b_m}^{-1} X_m], \quad b_i \in \{0, 1\},$$

where $\mathbf{D}_{b_i}^{-1} X_i \rightarrow \mathbf{D}_{b_{i+1}}^{-1} X_{i+1}$, $1 \leq i \leq m - 1$, $\mathbf{D}_{b_m}^{-1} X_m \rightarrow \mathbf{D}_{b_1}^{-1} X_1$, and either $m = k$ or $m = 2k$ (as we will see later). If $\mathcal{C}$ is represented by a sequence, i.e., $\mathcal{C} = [c_1, c_2, \cdots, c_k]$, then for $b \in \{0, 1\}$, we have

$$\mathbf{D}_b^{-1}(\mathcal{C}) = [b, b_1 + c_1, b + c_1 + c_2, \ldots, b + \sum_{i=1}^{k-1} c_i] \quad \text{if} \quad \sum_{i=1}^{k} c_i \equiv 0 \pmod 2$$

since $b + \sum_{i=1}^{k} c_i = b$; if $\sum_{i=1}^{k} c_i \equiv 1 \pmod 2$, then $b + \sum_{i=1}^{k} c_i = \bar{b}$ and

$$\mathbf{D}_b^{-1}(C) = [b, b + c_1, \ldots, b + \sum_{i=1}^{k-1} c_i, \bar{b}, \bar{b} + c_1, \ldots, \bar{b} + \sum_{i=1}^{k-1} c_i].$$

The justification for these representations of $\mathbf{D}_b^{-1}(C)$ is also a consequence of the next few results.

Definition 4.1. A *primitive* cycle C in G_n is a simple cycle that does not contain two complementary vertices.

Theorem 4.4. *A simple k-cycle $\Upsilon = [\varsigma_1, \varsigma_2, \cdots, \varsigma_k]$ in G_{n-1} is the $\mathbf{D}$-morphic image of a primitive k-cycle $C = [c_1, c_2, \ldots, c_k]$ in G_n if and only if wt(Υ) is even.*

Proof. If C is a primitive cycle in G_n, then by Definition 4.1 and Lemma 4.10 we have that $\Upsilon = \mathbf{D}C$ is a simple cycle in G_{n-1} and

$$\sum_{i=1}^{k} \varsigma_i = \sum_{i=1}^{k-1} (c_i + c_{i+1}) + (c_k + c_1) = 0.$$

Thus the modulo 2 sum of the ς_is is zero and, hence, wt(Υ) is even.

Now, let $\Upsilon = [\varsigma_1, \varsigma_2, \cdots, \varsigma_k]$ be a simple k-cycle of even weight in G_{n-1}. Let $c_1 \triangleq b$, where b can be either 0 or 1, and let

$$c_i = b + \sum_{j=1}^{i-1} \varsigma_j, \quad i = 2, 3, \ldots, k.$$

This definition implies that $\varsigma_i = c_i + c_{i+1}$ for $1 \le i \le k - 1$ and since C is a cycle of length k, it follows that we must have $\varsigma_k = c_k + c_1$. Since wt$(\Upsilon) = \sum_{i=1}^{k} \varsigma_i$ is even, it follows that

$$c_k = b + \sum_{i=1}^{k-1} \varsigma_i = c_1 + \varsigma_k \quad \text{and} \quad b + \sum_{i=1}^{k} \varsigma_i = b = c_{k+1} = c_1,$$

which implies that the definition of C is consistent and C has period k. Finally, since $\Upsilon = \mathbf{D}C$ is a simple k-cycle, then by Lemma 4.8 there are no complementary states in C, and hence C is primitive. $\square$

Corollary 4.5. *There exists a one-to-one correspondence between the simple k-cycles Υ of even weight in G_{n-1} and the primitive pairs of k-cycles C and $\bar{C}$ in G_n under which $\Upsilon = \mathbf{D}C = \mathbf{D}\bar{C}$.*

The next result provides a different proof of Lemma 3.5.

Lemma 4.11. *A simple cycle* $C = [c_1, c_2, \ldots, c_k]$ *in* G_n *is self-dual if and only if it is of even period* $k = 2\pi$ *and*

$$c_{i+\pi} = \bar{c}_i, \quad i = 1, 2, \ldots, \pi.$$

Proof. Clearly, if C is self-dual then for each state X on C also $\bar{X}$ is on C and $X \to Y$ is an edge on C if and only if $\bar{X} \to \bar{Y}$ is an edge on C. Hence, each such pair of vertices $\{X, \bar{X}\}$ must be at the same distance on C, i.e., C is of even period $k = 2\pi$ and $c_{i+\pi} = \bar{c}_i$ for $1 \le i \le \pi$.

If C is of even period $k = 2\pi$ and $c_{i+\pi} = \bar{c}_i$ for $1 \le i \le \pi$, then this definition immediately implies that C is a self-dual cycle. □

The next lemma is also an immediate consequence of the definitions.

Lemma 4.12. *The* **D**-*morphic image of a self-dual* (2π)-*cycle (not necessarily simple) is a* π-*cycle. If* $C = [c_1, c_2, \ldots, c_{2\pi}]$ *is a self-dual* (2π)-*cycle in* G_n, *then the* π-*cycle* $\mathbf{D}C = \Upsilon = [\varsigma_1, \varsigma_2, \cdots, \varsigma_\pi]$ *in* G_{n-1} *is given by*

$$\varsigma_i = c_i + c_{i+1}, \quad i = 1, 2, \ldots, \pi.$$

Theorem 4.5. *A* π-*cycle* Υ *in* G_{n-1} *is the* **D**-*morphic image of a self-dual* (2π)-*cycle* C *in* G_n *if and only if* $\mathrm{wt}(\Upsilon)$ *is odd.*

Proof. The proof proceeds along the same line as the proof of Theorem 4.4, where we have in the proof that since $\mathrm{wt}(\Upsilon)$ is odd, it follows that $c_{\pi+i} = \bar{c}_i$ for $1 \le i \le \pi$. □

Corollary 4.6. *There exists a one-to-one correspondence between the* k-*cycles* Υ *of odd weight in* G_{n-1} *and the self-dual* $(2k)$-*cycles* C *in* G_n *under which* $\Upsilon = \mathbf{D}C$.

Corollary 4.6 implies the claim in Lemma 3.14.

Corollary 4.7. *The* **D**-*morphism forms a one-to-one mapping from the cycles of the* CCR_{n+1} *to the cycles of the* CSR_n.

Given a simple cycle C we define $\mathbf{D}^{-1}C$ as the two cycles $\mathbf{D}_0^{-1}C$ and $\mathbf{D}_1^{-1}C$ if C has even weight. If C has odd weight, then $\mathbf{D}^{-1}C$ is either $\mathbf{D}_0^{-1}C$ or $\mathbf{D}_1^{-1}C$ (these two cycles are equivalent).

Finally, the inverse mapping can be applied to the cycles of a factor in G_n to obtain the following result.

Theorem 4.6. *If* $\mathcal{F}$ *is a factor in* G_n, *then* $\mathbf{D}^{-1}\mathcal{F}$ *is a factor in* G_{n+1}, *where*

$$\mathbf{D}^{-1}\mathcal{F} \triangleq \{\mathcal{D}^{-1}C : C \in \mathcal{F}\}.$$

This is the place to pay attention to some differences between sequences and cycles in a graph. While in a cycle we are interested that the period of the cycle will be also its length (it still might not be a simple cycle), in sequences sometimes it is worthwhile to take a sequence of length $k \cdot \pi$ whose period is π. For example, to take the sequence [101010] rather than the sequence [10]. It should be understood from the context which kind of sequence is considered.

The following lemma can be proved by several methods with the exposition we had so far.

Lemma 4.13. *If $n > 1$ and S is a span n de Bruijn sequence, then $\mathbf{D}_0^{-1}S$ and $\mathbf{D}_1^{-1}S$ are two complementary sequences that form a factor in G_{n+1} with exactly two cycles.*

Proof. A span n de Bruijn sequence is associated with a simple cycle in G_n and it has even weight. Therefore the lemma follows directly from Corollary 4.5. $\square$

Note that in Lemma 4.13 we have that $n > 1$ since the span 1 de Bruijn sequence is [01] and $\mathbf{D}_0^{-1}[01] = [0011]$. [01] is the only de Bruijn sequence of odd weight.

Corollary 4.8. *Let $n > 1$, let C be a span n de Bruijn cycle, let $X = (010101 \cdots)$ be a state in G_{n+1}, and let X' be its companion state. The cycles $\mathbf{D}_0^{-1}C$ and $\mathbf{D}_1^{-1}C$ contain all the states of G_{n+1}. If X is contained in $\mathbf{D}_b^{-1}C$, then X' is contained in $\mathbf{D}_{\bar{b}}^{-1}C$.*

Proof. Clearly by Lemma 4.13 since $\mathbf{D}_0^{-1}C$ and $\mathbf{D}_1^{-1}C$ are two complementary sequences in G_{n+1}, if follows that X and $\bar{X}$ are not on the same cycle, i.e., $X \in \mathbf{D}_0^{-1}C$ if and only if $\bar{X} \in \mathbf{D}_1^{-1}C$. W.l.o.g. assume that $X \in \mathbf{D}_0^{-1}C$. Moreover, both X and X' are successors of $\bar{X} = (101010 \cdots)$ in G_{n+1} and therefore since $X \in \mathbf{D}_0^{-1}C$, it follows that X' is the successor of $\bar{X} = (101010 \cdots)$ on $\mathbf{D}_1^{-1}C$. $\square$

Now, we can apply the merge-or-split method (see Lemma 1.19) to merge the two cycles $\mathbf{D}_0^{-1}C$ and $\mathbf{D}_1^{-1}C$ into a span $n + 1$ de Bruijn cycle using the companion states X and X' (as defined in Corollary 4.8) as bridging states. Assume first that C is generated via the feedback function

$$f(x_1, x_2, \ldots, x_n) = x_1 + g(x_2, \ldots, x_n).$$

Theorem 4.7. *Let the cycle C and the state X be as in Corollary 4.8. After merging the two cycles $\mathbf{D}_0^{-1}C$ and $\mathbf{D}_1^{-1}C$ via the companion states X and X' the feedback function h of the constructed span $n + 1$ de Bruijn sequence is*

$$h(x_0, x_1, \ldots, x_n)$$
$$= f(x_0 + x_1, x_1 + x_2, \ldots, x_{n-1} + x_n) + x_n + \bar{x}_1 x_2 \bar{x}_3 x_4 \cdots \bar{x}_{n-1} x_n$$

when n is even and

$$h(x_0, x_1, \ldots, x_n)$$
$$= f(x_0 + x_1, x_1 + x_2, \ldots, x_{n-1} + x_n) + x_n + \bar{x}_1 x_2 \bar{x}_3 x_4 \cdots \bar{x}_{n-2} x_{n-1} \bar{x}_n$$

when n is odd.

Proof. Given an $(n + 1)$-tuple $X = (x_0, x_1, \ldots, x_n)$, either X is a state on the cycle $\mathbf{D}_0^{-1} \mathcal{C}$ or X is a state on the cycle $\mathbf{D}_1^{-1} \mathcal{C}$. We also have that

$$\mathbf{D} X = (x_0 + x_1, x_1 + x_2, \ldots, x_{n-1} + x_n)$$

is a state on $\mathcal{C}$. Since $\mathcal{C}$ was generated by the feedback function f, it follows that the next state after $\mathbf{D} X$ on $\mathcal{C}$ is

$$\mathbf{D} Z = (x_1 + x_2, \ldots, x_{n-1} + x_n, f(x_0 + x_1, x_1 + x_2, \ldots, x_{n-1} + x_n)).$$

Therefore by the definition of the operator $\mathbf{D}$, the next state after X on either the cycle $\mathbf{D}_0^{-1} \mathcal{C}$ or the cycle $\mathbf{D}_1^{-1} \mathcal{C}$ is

$$Z = (x_1, \ldots, x_n, y),$$

where $x_n + y = f(x_0 + x_1, x_1 + x_2, \ldots, x_{n-1} + x_n)$. This implies that the feedback function $\gamma(x_0, x_1, \ldots, x_n)$ of the state diagram whose cycles are $\mathbf{D}_0^{-1} \mathcal{C}$ and $\mathbf{D}_1^{-1} \mathcal{C}$ is

$$\gamma(x_0, x_1, \ldots, x_n) = f(x_0 + x_1, x_1 + x_2, \ldots, x_{n-1} + x_n) + x_n.$$

It remains to show that the defined function $h(x_0, x_1, \ldots, x_n)$ is indeed the function of the feedback shift register that generates the de Bruijn sequence after merging the cycles $\mathbf{D}_0^{-1} \mathcal{C}$ and $\mathbf{D}_1^{-1} \mathcal{C}$ via the bridging states X and X'. This is an immediate consequence of Eq. (1.8), as explained for the merge-or-split method in Section 1.3. $\qquad \square$

There are more pairs of companion states, one in $\mathbf{D}_0^{-1} \mathcal{C}$ and its companion pair in $\mathbf{D}_1^{-1} \mathcal{C}$. This implies that there are more ways to merge these cycles. Other ways to merge cycles of a state diagram to form a de Bruijn sequence will be demonstrated in the next section and also in Chapter 5.

Example 4.1. Consider the span 3 de Bruijn sequence $S = [00011101]$. Applying the operator $\mathbf{D}^{-1}$ on S we obtain the two sequences

$$\mathbf{D}_0^{-1} S = [00001011] \quad \text{and} \quad \mathbf{D}_1^{-1} S = [11110100].$$

Merging $\mathbf{D}_0^{-1} S$ and $\mathbf{D}_1^{-1} S$ via the companion pair (0101) and (0100) we obtain the span 4 de Bruijn sequence

$$[1111010110000100].$$

Merging $\mathbf{D}_0^{-1} S$ and $\mathbf{D}_1^{-1} S$ via the companion pair (0010) and (0011) we obtain the span 4 de Bruijn sequence

$$[1110100101100001].$$

■

We will end this section with another class of sequences in G_n.

Definition 4.2. A span n *half de Bruijn sequence* is a (cyclic) sequence of period 2^{n-1} that has the property that for each possible n-tuple X, either X or $\bar{X}$ appears in the sequence exactly once as a subsequence.

There are many different ways to construct half de Bruijn sequences. Some of the approaches to construct de Bruijn sequences can be used to construct half de Bruijn sequences. There is a simple method, in which a half de Bruijn sequence of length 2^{n-1} is generated from a span $n - 1$ de Bruijn sequence S. The sequences $\mathbf{D}_0^{-1} S$ and $\mathbf{D}_1^{-1} S$ are half de Bruijn sequences of order n. Another method is based on M-sequences, as given in the following theorem.

Theorem 4.8. *If S is an M-sequence of order $n - 1$, then for each pair of n-tuples X and $\bar{X}$ either X or $\bar{X}$ appears in S, except for the pair that consists of the all-zeros and all-ones n-tuples.*

Proof. Let $g(x)$ be the characteristic polynomial of order $n - 1$ that generates M-sequence S. By Lemma 2.5 the polynomial $g(x)(x + 1)$ generates the sequences S and $\bar{S}$. □

Corollary 4.9. *Let S be an M-sequence of order $n - 1$ and let S' be the sequence obtained from S by adding another one to the unique run with $n - 1$ ones. Then, S' is a span n half de Bruijn sequence.*

4.3   Merging cycles of large factors

One of the main disadvantages of the prefer one algorithm is that its straightforward implementation requires one to remember the subsequence that was generated so far and this is highly inefficient. However, we can generate de Bruijn sequences with more efficient algorithms.

A factor in G_n with exactly one cycle is associated with a span n de Bruijn sequence. The unique cycle is a Hamiltonian cycle and hence a span n de Bruijn sequence. Any factor in G_n can be used to form a span n de Bruijn sequence by merging the cycles of the factor into one cycle using the merge-or-split method. If the factor has Δ cycles, then by Theorem 1.20, the merge-or-split method can be applied $\Delta - 1$ times to generate a de Bruijn cycle. The merge-or-split method, described in Section 1.3, is a basic method that can be used to merge all the cycles of the factor into one cycle. It will be demonstrated how to produce efficiently a large set of de Bruijn cycles from the state diagrams of the most celebrated feedback shift registers that were considered in Chapter 3, namely, the PCR_n, the CCR_n, the PSR_n, and the CSR_n. For this purpose, we store some bits that indicate which full cycle will be constructed. These bits are associated with the bridging states for the merge-or-split method. To have an efficient procedure, it will be required that at each step of the algorithm, only the last n bits (or slightly more) of the cycle will be known and the procedure will use these n bits and the stored bits to generate the next bit of the cycle.

We will now describe a general method to combine the cycles of a state diagram for any FSR_n whose state diagram has Δ cycles. We choose a set of $\Delta - 1$ pairs of bridging states. From each pair, only one bridging state has to be stored and w.l.o.g. assume this bridging state is a state whose last bit is a *one*. If we form an undirected graph in which each vertex represents a cycle of the state diagram and an edge between two vertices (cycles of the state diagram) that share a pair of bridging states, then we can form a full cycle if the graph is connected. Since $\Delta - 1$ pairs of bridging states were taken and the graph has Δ vertices, it follows that when the graph is connected it is a tree. We order the cycles of the state diagram such that the first two cycles share a pair of bridging states and each other cycle in this order shares a pair of bridging states with a cycle that comes before it in this order. Moreover, we have to choose many different sets with $\Delta - 1$ pairs of bridging states that satisfy these requirements. This will enable us to form many full cycles from the same state diagram. At each step of the algorithm, we will have a main cycle that was created by a set of say the first $k - 1$ cycles from this order. The kth cycle in this order is merged with the main cycle to form a new main cycle. This is possible since it has a bridging state whose pair is on the main cycle, as implied by this order. Initially, the main cycle is the first cycle in this order.

We will apply now this idea to the specific FSR_ns that were mentioned in Chapter 3.

Merging cycles from the PCR_n:

We will describe a method to combine all the $Z(n)$ PCR_n cycles using $Z(n) - 1$ bridging states. By Theorem 3.1, the constructed full cycle by such a method is of a minimum weight function.

The following simple lemma will be used to form a set of bridging states.

Lemma 4.14. *Each cycle of (extended) weight $i > 0$ in the state diagram of the PCR_n has a state ending with a* one *whose companion is a state ending with a zero on a cycle of weigh $i - 1$ in the state diagram of the PCR_n.*

At each step, we have a *main cycle* obtained by joining a subset of the PCR_n cycles, and the remaining PCR_n cycles. Initially, the main cycle is chosen to be the unique PCR_n cycle of weight zero. Next, the main cycle is extended by joining to it the (unique) cycle of weight one. In general step i, we extend the main cycle by joining to it all the PCR_n cycles of weight i (in arbitrary order).

This is always possible because the current main cycle contains all the states whose weight is less than i and since each PCR_n cycle of weight $i > 1$ has a state ending in a *one*, it can be joined to the current main cycle as its companion ends with a *zero* and has weight $i - 1$ and hence it is contained in the current main cycle. This procedure ends when all the PCR_n cycles have been joined together.

We proceed now to a precise and detailed description of the proposed construction.

Consider an ordered set $V = \{V(i)\}_{i=0}^{k-1}$ of k states, $1 \leq k \leq 2^{(n-4)/2}$, constructed as follows (all the logarithms in this section are in base-2):

1. The first $\lceil \log k \rceil + 1$ bits of each $V(i)$ form the binary representation of i. (note that the first bit in such a representation is always a *zero*).
2. The last $\lfloor \log k \rfloor + 1$ bits of each $V(i)$ are *ones*. Before these bits there is a single *zero*.
3. In positions $\lceil \log k \rceil + 2 + (\lfloor \log k \rfloor + 1)j$, for integers j satisfying

$$0 \leq j < \left\lceil \frac{n - \lceil \log k \rceil - \lfloor \log k \rfloor - 3}{\lfloor \log k \rfloor + 1} \right\rceil$$

each $V(i)$ has a *zero*.
4. The remaining bits of each $V(i)$ are chosen arbitrarily.

Example 4.2. Let $n = 16$ and $k = 8$. The set V for these values of n and k takes the form

$$00000x_1^{(0)}x_2^{(0)}x_3^{(0)}0x_4^{(0)}x_5^{(0)}01111$$
$$00010x_1^{(1)}x_2^{(1)}x_3^{(1)}0x_4^{(1)}x_5^{(1)}01111$$
$$00100x_1^{(2)}x_2^{(2)}x_3^{(2)}0x_4^{(2)}x_5^{(2)}01111$$
$$00110x_1^{(3)}x_2^{(3)}x_3^{(3)}0x_4^{(3)}x_5^{(3)}01111$$
$$01000x_1^{(4)}x_2^{(4)}x_3^{(4)}0x_4^{(4)}x_5^{(4)}01111$$
$$01010x_1^{(5)}x_2^{(5)}x_3^{(5)}0x_4^{(5)}x_5^{(5)}01111$$
$$01100x_1^{(6)}x_2^{(6)}x_3^{(6)}0x_4^{(6)}x_5^{(6)}01111$$
$$01110x_1^{(7)}x_2^{(7)}x_3^{(7)}0x_4^{(7)}x_5^{(7)}01111$$

where the $x_j^{(i)}$ are free parameters. ∎

It can be easily verified that the right-hand block of $\lfloor \log k \rfloor + 1$ *ones*, in each state of V, forms the unique largest run of *ones* in each $V(i)$, and that every pair of states differ in their first $\lceil \log k \rceil + 1$ bits. Therefore we have

Lemma 4.15. *No two states of V belong to the same cycle of the PCR_n.*

The construction of a span n de Bruijn cycle from the PCR_n cycles proceeds by a sequence of joins where at each step a cycle of least weight among the remaining PCR_n cycles is joined to the current main cycle. A join is performed using a pair of companion states X and X', with X on the next PCR_n cycle C in line and X' on the current main cycle. The states X and X' are the bridging states of the join. The bridging state X on C is determined as follows: if C contains a state from V, then it is chosen as the bridging state of C. Otherwise, the choice of X is as follows. Let M be the state on C whose value $|M|$ when viewed as a number in base 2 is maximal. If $|M| = \ell \cdot 2^r$, where ℓ is an odd integer and $r \geq 0$, then the state X such that $|X| = \ell$ is also on C, and X is taken to be the bridging state of C.

In any case, the chosen bridging state X for the current PCR_n cycle C always ends in a *one*. By Lemma 4.14, its companion X' belongs to a PCR_n cycle whose weight is smaller than that of C. Therefore X' must be on the current main cycle. By Lemma 1.19, interchanging the predecessors of X and X' will create the new next main cycle by joining C to the current one.

A span n de Bruijn cycle obtained by joining the PCR_n cycles as described above can be generated bit-by-bit following a procedure based on the underline rules for the joining of cycles. In this procedure, the $(i+n)$th bit b_{i+n} of the full cycle is determined from the preceding n-bit state $\beta_i = (b_i, b_{i+1}, \ldots, b_{i+n-1})$. If β_i served as a predecessor of a bridging state (X or X'), then $b_{i+n} = b_i + 1$; otherwise, $b_{i+n} = b_i$. The formal steps for determining b_{i+n} are presented in the following algorithm.

Algorithm merge PCR_n:

Choose a constant k such that $1 \leq k \leq 2^{(n-4)/2}$. Choose and store an ordered set of bridging states $V = \{V(i)\}_{i=0}^{k-1}$. Initially, $\beta_0 = (0, 0, \ldots, 0) = 0^n$. Given $\beta_i = (b_i, b_{i+1}, \ldots, b_{i+n-1})$, proceed to produce $\beta_{i+1} = (b_{i+1}, \ldots, b_{i+n-1}, b_{i+n})$.

(A1) Examine the cyclic shifts of $\beta_i^* = (b_{i+1}, \ldots, b_{i+n-1}, 1)$ for the existence of a shift X that begins with a *zero* and ends with $1 + \lfloor \log k \rfloor$ ones. If no such X exists, then go to (A3).

(A2) Let X^* be the first $1 + \lceil \log k \rceil$ bits of X and let $|X^*| = t$ be the binary value of X^*. If $t > k - 1$, then go to (A3); otherwise, if $X = V(t) = \beta_i^*$, then go to (A5); if $X = V(t) \neq \beta_i^*$, then go to (A4).

(A3) Let M be the cyclic shift of β_i^* with the largest binary value $|M| = \ell \cdot 2^r$, where ℓ is odd and $r \geq 0$. Let Y be the shift of β_i^* such that $|Y| = \ell$. If $Y = \beta_i^*$, then go to (A5).

(A4) Set $b_{i+n} := b_i$ and stop.

(A5) Set $b_{i+n} := b_i + 1$.

Theorem 4.9.

(a) *For every choice of k, where $1 \leq k \leq 2^{(n-4)/2}$, and of the set V, algorithm merge PCR_n produces a full cycle of length 2^n.*

(b) *For a given choice of k there are $2^{k \cdot g(n,k)}$ distinct choices for the set V, where*

$$g(n,k) = n - 3 - \lceil \log k \rceil - \lfloor \log k \rfloor - \left\lceil \frac{n - \lceil \log k \rceil - \lfloor \log k \rfloor - 3}{\lfloor \log k \rfloor + 1} \right\rceil;$$

thus algorithm merge PCR_n can be used to produce $2^{k \cdot g(n,k)}$ distinct full cycles.

(c) *The working space that algorithm merge PCR_n requires to produce a full cycle is $3n + k \cdot g(n,k)$ bits and the work required to produce the next bit is at most $2n$ cyclic shifts and about the same number of n-bits comparisons.*

Proof.

(a) This follows from the discussion preceding algorithm merge PCR_n.

(b) It is due to the fact that each $V(i)$ is specified up to exactly $g(n,k)$ free parameters and that no state except for $(0^{n-\alpha}1^\alpha)$, where $\alpha = 1 + \lfloor \log k \rfloor$, may serve as a bridging state via both of the two criteria: (i) by being a member of the set V; (ii) representing the odd part of a maximal shift. This, together with Lemma 4.15, imply that distinct choices for the set V correspond to distinct sets of bridging states and, hence, to distinct full cycles.

(c) It follows directly from algorithm merge PCR_n. Note that only information about members of the set V has to be stored and, there, only the $g(n,k)$ free bit-values of each $V(i)$ require storage. □

The big advantage of the method that we have presented for constructing full cycles by merging the cycles of the PCR_n is that we can control the number of full cycles generated by the algorithm. If we store Δ bits for the bridging states, then we generate 2^Δ full cycles, which is the largest number of full cycles that can be generated with storage of Δ bits. The disadvantage is that for a large Δ, the value of n should be very large. The next construction will yield a much larger number of full cycles for a relatively smaller n. However, for a very large n the number of full cycles generated by algorithm merge PCR_n is much larger.

Merging cycles from the CSR_n:

We continue and apply the merge-or-split method on the CSR_n cycles. The following lemma is an immediate result of the definition of extended weight.

Lemma 4.16. *For a cycle C from the CSR_n we have that $wt_E(C) = 2k + 1$ for some $0 \le k \le \lfloor n/2 \rfloor$ and for each state X of C we have $2k \le wt(X) \le 2k + 1$.*

A CSR_n cycle C is called a **run-cycle** if all the *ones* in C form a cyclic run.

For each cycle C of the CSR_n, with $wt_E(C) = 2k + 1 < n + 1$, we define a unique **preferred state** $\mathcal{P}(C)$. For a run-cycle $\mathcal{P}(C) = (1^{2k+1}0^{n-2k-1})$; for a cycle with more than one (cyclic) run of *ones* the preferred state is defined as follows.

Let $E^*(C) = [0^r 1^t 0 b_1 \cdots b_{n-t-r-2} 1 0]$ be the unique extended representation of C that satisfies the following properties:

1. $r \geq 0$;
2. t is the longest run of *ones*;
3. among all the extended representations of this form, with the same maximal t, $E^*(C)$ is the largest when viewed as a binary number.

Then, the preferred state of C is $\mathcal{P}(C) = (0^r 1^t 0 b_1 \cdots b_{n-t-r-2} 1)$.

Lemma 4.17. *Let C_1 be a nonrun-cycle from the* CSR_n *and with a preferred state* $\mathcal{P}(C_1) = (0^r 1^{t_1} 0 b_1 \cdots b_{n-t_1-r-2} 1)$. *This implies that the state* $B = (10^r 1^{t_1} 0 b_1 \cdots b_{n-t_1-r-2})$ *and the companion of* $\mathcal{P}(C_1)$ *are two states on the same cycle* $C_2 \neq C_1$, *with* $\mathrm{wt}_E(C_2) = \mathrm{wt}_E(C_1)$. *Furthermore, if t_2 is the longest run of* ones *in* $\mathcal{P}(C_2)$, *then either* $t_2 = t_1 + 1$ *or* $t_2 = t_1$ *and* $|\mathcal{P}(C_2)| > |\mathcal{P}(C_1)|$.

Proof. Clearly, $\mathrm{wt}(B) = \mathrm{wt}(\mathcal{P}(C_1)) = \mathrm{wt}_E(C_1) = 2k + 1$ for some k. Hence, by Lemma 4.16 we have that $\mathrm{wt}_E(C_2) = 2k + 1 = \mathrm{wt}_E(C_1)$. It is also clear that $E(C_2) = [10^r 1^{t_1} 0 b_1 \cdots b_{n-t_1-r-2} 0]$. Hence, if $r = 0$, then $t_2 = t_1 + 1$; and if $r > 0$, then an alternate extended representation of C_2 is given by $E'(C_2) = [0^{r-1} 1^{t_1} 0 b_1 \cdots b_{n-t_1-r-2} 010]$.
 This implies that

$$|\mathcal{P}(C_2)| \geq \left| (0^{r-1} 1^{t_1} 0 b_1 \cdots b_{n-t_1-r-2} 01) \right| > |\mathcal{P}(C_1)|.$$

Thus in any case $C_1 \neq C_2$, and, since the two possible successors of B are $\mathcal{P}(C_1)$ and the companion of $\mathcal{P}(C_1)$, it follows that the companion of $\mathcal{P}(C_1)$ is the successor of B on C_2. □

Lemma 4.18. *Let $U = (u_1, \ldots, u_{n-1}, 1)$ be a state on a cycle C_1 of the* CSR_n *with* $\mathrm{wt}(U) + 1 = \mathrm{wt}_E(C_1) = 2k + 1$ *for some $k \geq 1$. Then, the companion U' of U is on a* CSR_n *cycle C_2 with* $\mathrm{wt}_E(C_2) = 2k - 1$.

Proof. The claim of the lemma is an immediate consequence of the definition of a companion state and from Lemma 4.16. □

Lemmas 4.16, 4.17, and 4.18 lead to a construction of a large class of full cycles by joining the cycles of the CSR_n. Lemma 4.17 suggests a way of joining all the CSR_n cycles with the same extended weight. For each extended weight of $2k + 1$, we start with the run-cycle of this weight as the main cycle. In each step, the current main cycle is expanded by joining to it the CSR_n cycle of extended weight $2k + 1$ with the longest run of *ones*; if there are two or more cycles with the same longest run of *ones*, join the one with the largest preferred state. Once all the CSR_n cycles of extended weight $2k + 1$ are joined together into a corresponding main cycle MC_k, $0 \leq k \leq \lfloor n/2 \rfloor$, we apply Lemma 4.18 to join these $\lfloor n/2 \rfloor + 1$ MC_k cycles, in order of increasing k, to form a full cycle.
 We proceed now to describe an algorithm for producing the $(i + n)$th bit, b_{i+n}, of the resulting full cycle from the following inputs:

1. the preceding n-bit state $\beta_i = (b_i, b_{i+1}, \ldots, b_{i+n-1})$;

2. the parity p_i of β_i, i.e., $p_i = \sum_{j=0}^{n-1} b_{i+j}$;

3. the weight of β_i, $\mathrm{wt}(\beta_i)$.

The production of b_{i+n} from the above inputs is based on the fact that when $(x_1, \ldots, x_{n-1}, x_n)$ is the successor of $(x_0, x_1, \ldots, x_{n-1})$ then $\sum_{i=0}^{n} x_i$ is odd if and only if both states are on the same CSR_n cycle.

In the algorithm, for merging the CSR_n cycles, given below, we first check whether the state β_i serves as a predecessor of a bridging state (X or X'). If it does, then we set $b_{i+n} := p_i$ and $p_{i+1} := b_i$. If β_i is not a predecessor of a bridging, then we set $b_{i+n} := p_i \oplus 1$ and $p_{i+1} := b_i \oplus 1$. In both cases we set $w_{i+1} := w_i - b_i + b_{i+n}$

Algorithm merge CSR_n:

For every k such that $1 \le k \le \frac{n}{2}$ choose and store a bridging state $U^{(2k)}$ of the form $U^{(2k)} = (u_1^k, u_2^k, \cdots, u_{n-1}^k, 1)$ with $\mathrm{wt}(U^{(2k)}) = 2k$. Initially, set $\beta_0 = (0, 0, \ldots, 0) = 0^n$, $p_0 = 0$, and $w_0 = \mathrm{wt}(\beta_0) = 0$. Given the current state $\beta_i = (b_i, b_{i+1}, \ldots, b_{i+n-1})$, p_i, and $w_i = \mathrm{wt}(\beta_i)$, proceed to produce the next state $\beta_{i+1} = (b_{i+1}, \ldots, b_{i+n-1}, b_{i+n})$, p_{i+1}, and w_{i+1}, as follows:

(B1) If $p_i + b_i = 0$, then go to **(B3)**.

(B2) If $(b_{i+1}, \ldots, b_{i+n-1}, 1) = U^{(w_i - b_i + 1)}$, then go to **(B6)**; otherwise go to **(B5)**.

(B3) If $\beta_i^* = [b_{i+1}, \ldots, b_{i+n-1} 1 0]$ is a run-cycle, then go to **(B5)**; otherwise, find the cyclic shift $E_i^* = [0^r 1^t 0 b_s \cdots b_{n-t-r+s-3} 1 0]$ of β_i^* whose first n bits form a preferred state.

(B4) If $E_i^* = \beta_i^*$, then go to **(B6)**.

(B5) Set $b_{i+n} := p_i \oplus 1$, $p_{i+1} := b_i \oplus 1$, $w_{i+1} := w_i - b_i + b_{i+n}$, and stop.

(B6) Set $b_{i+n} := p_i$, $p_{i+1} := b_i$, $w_{i+1} := w_i - b_i + b_{i+n}$.

Theorem 4.10.

(a) *For every choice of the set of states $\{U^{(2k)}\}_{k=1}^{\lfloor n/2 \rfloor}$ algorithm merge CSR_n produces a full cycle of length 2^n.*

(b) *Algorithm merge CSR_n can be used to produce*

$$\prod_{k=1}^{\lfloor n/2 \rfloor} \binom{n-1}{2k-1}$$

distinct full cycles.

(c) *The working space that algorithm merge CSR_n requires to produce a full cycle is about $\frac{n^2}{2}$ bits and the work required to produce the next bit is n cyclic shifts and about the same number of n-bits comparisons.*

Proof.

(a) It follows directly from the discussion preceding algorithm merge CSR_n.

(b) It is because different sets of bridging states produce different full cycles. The number of ways to choose the set $\{U^{(2k)}\}$ is

$$\prod_{k=1}^{\lfloor n/2 \rfloor} \binom{n-1}{2k-1}.$$

(c) It follows directly from algorithm merge CSR_n. It is clear that most of the work consists of finding the preferred state and

$$\beta_i^* = [b_{i+1}, \ldots, b_{i+n-1}10] = [0^{r_1}1^{t_1}X_110]$$

in **(B3)**. Let E_i^* be the shift of β_i^* whose first n bits form a preferred state. Initially, $E_i^* = \beta_i^*$. Given $E_i^* = [0^{r_2}1^{t_2}X_210]$ and a shift $E_i^+ = [0^{r_3}1^{t_3}X_310]$ of β_i^*, set $E_i^* = E_i^+$ if either $t_3 > t_2$, or $t_2 = t_3$ and $\left|E_i^+\right| > \left|E_i^*\right|$. After n shifts E_i^* will have the required form. $\qquad\square$

Merging the cycles of the PSR_n is carried out in the same way as the procedure for the cycles of the CSR_n with the only exception that the extended weight of any cycle will be even instead of odd for the CSR_n. This distinction will dictate appropriate changes in the algorithm.

Merging cycles from the CCR_n:

The last FSR_n that will be considered for the construction of span n de Bruijn sequences is the CCR_n. We will describe a method to combine all the $Z^*(n)$ CCR_n cycles using $Z^*(n) - 1$ bridging states. By Theorem 3.2 the constructed full cycle by such a method is of a maximum weight function.

Recall that by Corollary 4.7, $\mathbf{D}$ is a one-to-one mapping from the cycles of the set of CCR_{n+1} cycles to the cycles of the set of CSR_n cycles. If the states X and X' are bridging states for the construction of the full cycle from the CSR_n cycles, then we can choose $\mathbf{D}_0^{-1}X$ (or $\mathbf{D}_1^{-1}X$) and its companion as bridging states for the construction of a full cycle $\mathcal{C}$ from the CCR_{n+1} cycles. If $\delta_i = (d_i, d_{i+1}, \ldots, d_{i+n})$ is a state $\mathcal{C}$, then $(d_{i+1}, \ldots, d_{i+n}, 1)$ serves as a bridging state for the CCR_{n+1} cycles if and only if $d_{i+1} = 0$ and $(b_{i+1}, \ldots, b_{i+n-1}, 1)$, for $b_{i+j} = d_{i+j} + d_{i+j+1}$, $1 \leq j \leq n - 1$, serves as a bridging state for the CSR_n cycles. If δ_i serves as a bridging state, then $d_{i+n+1} = d_i$; otherwise $d_{i+n+1} = d_i + 1$. The formal steps for the production of the next bit in the full cycle of span $n + 1$, d_{i+n+1}, are given in algorithm merge CCR_{n+1}.

Algorithm merge CCR_{n+1}:

For every k such that $1 \leq k \leq \frac{n}{2}$ choose and store a bridging state $U^{(2k)}$ (for the CSR_n) of the form $U^{(2k)} = (u_1^k, u_2^k, \cdots, u_{n-1}^k, 1)$ with $\mathrm{wt}(U^{(2k)}) = 2k$. Initially, set $\delta_0 = (0, 0, \ldots, 0) = 0^{n+1}$, $\beta_0 = (0, 0, \ldots, 0) = 0^n$, $p_0 = 0$, $w_0 = 0$. Given $\delta_i = (d_i, d_{i+1}, \ldots, d_{i+n})$ $\beta_i = (b_i, b_{i+1}, \ldots, b_{i+n-1})$, the parity p_i of β_i, and $w_i = \mathrm{wt}(\beta_i)$ proceed to produce $\delta_{i+1} = (d_{i+1}, \ldots, d_{i+n}, d_{i+n+1})$, $\beta_{i+1} = (b_{i+1}, \ldots, b_{i+n-1}, b_{i+n})$, p_{i+1}, and w_{i+1}, as follows:

(**C1**) If $d_{i+1} = 1$, then go to (**C6**).

(**C2**) If $p_i + b_i = 0$, then go to (**C4**).

(**C3**) If $(b_{i+1}, \ldots, b_{i+n-1}, 1) = U^{(w_i - b_i + 1)}$, then go to (**C7**); otherwise go to (**C6**).

(**C4**) If $\beta_i^* = [b_{i+1}, \ldots, b_{i+n-1} 1 0]$ is a run-cycle, then go to (**C6**); otherwise, find the cyclic shift $E_i^* = [0^r 1^t 0 b_s \cdots b_{n-t-r+s-3} 1 0]$ of β_i^* whose first n bits form a preferred state.

(**C5**) If $E_i^* = \beta_i^*$, then go to (**C7**).

(**C6**) Set $d_{i+n+1} := d_i \oplus 1$ and go to (**C8**).

(**C7**) Set $d_{i+n+1} := d_i$.

(**C8**) Set the following: $b_{i+n} := d_{i+n} \oplus d_{i+n+1}$, $p_{i+1} := p_i \oplus b_i \oplus b_{i+n}$, and $w_{i+1} := w_i - b_i + b_{i+n}$.

Theorem 4.11. *Algorithm merge* CCR_{n+1} *produces the same number of full cycles of span* $n + 1$ *as the number of full cycles of span* n *that algorithm merge* CSR_n *produces. The working space and the time complexity of algorithm merge* CCR_{n+1} *is about the same as those of algorithm merge* CSR_n.

Proof. The claim follows directly from the discussion preceding algorithm merge CCR_{n+1} since there is only some more constant number of additions in algorithm merge CCR_{n+1} compared to those in algorithm merge CSR_n. $\square$

Let $FC(FSR_n)$ be the number of full cycles obtained by merging all the Δ cycles of a given FSR_n, using exactly $\Delta - 1$ steps of the merge-or-split method.

Lemma 4.19. *For each positive integer* n,

$$FC(CCR_{n+1}) = FC(CSR_n) \cdot 2^{Z^*(n+1)-1} = FC(CSR_n) \cdot 2^{S^*(n)-1}.$$

Proof. By Corollary 4.7, $\mathbf{D}$ is a one-to-one mapping from the cycles of the set of CCR_{n+1} cycles to the cycles of the set of CSR_n cycles. Given a set of bridging states for joining the CSR_n cycles, then since for each pair of bridging states X and X' in $\mathbb{F}_2^n$, $\mathbf{D}_b^{-1} X$ and $\mathbf{D}_b^{-1} X'$, where $b \in \{0, 1\}$, yield two pairs of companion states, it follows that we can choose one of the two pairs of states ($\{\mathbf{D}_0^{-1} X, \mathbf{D}_0^{-1} X'\}$ or $\{\mathbf{D}_1^{-1} X, \mathbf{D}_1^{-1} X'\}$) as bridging states for the join of the CCR_{n+1} cycles. Also, since there are $S^*(n) = Z^*(n + 1)$ CSR_n cycles we have that there are $Z^*(n + 1) - 1$ states in the set of bridging states (pairs). Hence, $FC(CCR_{n+1}) \geq FC(CSR_n) \cdot 2^{Z^*(n+1)-1}$.

Assume now that we have a set of $Z^*(n + 1) - 1$ bridging states (pairs) from the CCR_{n+1}, where the representatives of each pair come from a different cycle of the $Z^*(n + 1)$ cycles of the CCR_{n+1}. In other words, only one CCR_{n+1} cycle does not have a representative in this set (note that there must be at least one bridging state from this cycle, but it is not taken as the representative). The complement pair $\{\bar{X}, \bar{X}'\}$ of each pair $\{X, X'\}$ of bridging states can be used as a pair of bridging states and since for a CCR_{n+1} cycle $\mathcal{C}$, $X \in \mathcal{C}$ if and only if $\bar{X} \in \mathcal{C}$, the two pairs join the same two cycles (but differently) during the merging process. Each such two pairs $\{\bar{X}, \bar{X}'\}$ and $\{X, X'\}$ are associated with one pair of bridging states $\{\mathbf{D}X, \mathbf{D}X'\}$ for joining the cycles of the CSR_n. Therefore $\mathrm{FC}(\mathrm{CCR}_{n+1}) \leq \mathrm{FC}(\mathrm{CSR}_n) \cdot 2^{Z^*(n+1)-1}$.

Thus $\mathrm{FC}(\mathrm{CCR}_{n+1}) = \mathrm{FC}(\mathrm{CSR}_n) \cdot 2^{Z^*(n+1)-1} = \mathrm{FC}(\mathrm{CSR}_n) \cdot 2^{S^*(n)-1}$. $\square$

Lemma 4.19 can be also used to construct a larger set of full cycles from the CCR_n, where the number of such constructed cycles can be chosen in advance.

The full cycles produced in the algorithm merge PCR_n have minimum weight function, while the full cycles produced by the algorithm merge CCR_{n+1} have maximum weight function. It is also important to note that the order that was suggested to merge cycles might be different. For example, concerning the PCR_n cycles, it is possible that a cycle with extended weight i would be merged to the main cycle only after some cycle of extended weight $i + 1$ was joined. However, this different order will not change the minimum weight function of the full cycle.

4.4   Notes

Constructing de Bruijn sequences and various algorithms for their construction is probably the most researched topic on de Bruijn sequences.

Section 4.1. Finding the number of Eulerian cycles in a directed graph for which each vertex has the same in-degree as out-degree is known as the BEST theorem for Nicholas de Bruijn and Tatyana van Aardenne-Ehrenfest [1], and Cedric Smith, and Bill Tutte [63]. The first work to find the number of Eulerian cycles in G_n was done by de Bruijn [10]. He proved Theorem 4.3 and Corollary 4.2 in his paper. The proof that was given in this section for Theorem 4.3 is based on the proof given by de Bruijn [10]. It was not known that the number of cycles in $G_{\sigma,n}$ was found much earlier in the work of Flye-Sainte Marie [25]. Flye-Sainte Marie solved in his work a question that was asked by Rivière [60]. van Aardenne-Ehrenfest and de Bruijn [1] proved the following result.

Theorem 4.12. *The number of Eulerian cycles in $G_{\sigma,n}$ is $(\sigma - 1)!^{\sigma^{n-1}} \cdot \sigma^{\sigma^{n-1}-n}$.*

The reverse spanning algorithm and generating the de Bruijn sequences with this algorithm was presented by Mowle [53,54]. The reverse spanning tree algorithm can be adapted with almost no change to $G_{\sigma,n}$ and Corollary 4.1. The only change in the algorithm is that the order in which the $\sigma - 1$ un-starred out-edges of a vertex v are chosen is arbitrary. This will be also done in another variant of the algorithm that will be presented in Section 6.3.

Corollary 4.10. *There is a one-to-one correspondence between the Eulerian cycles in $G_{\sigma,n}$ and the reverse spanning trees in $G_{\sigma,n}$.*

Section 4.2. The $\mathbf{D}$-morphism and its properties when applied to the de Bruijn graph, its cycles, factors, and words, was considered first in Lempel [44]. Merging the two sequences $\mathbf{D}_0^{-1}S$ and $\mathbf{D}_1^{-1}S$ in various ways, where S is a span n

de Bruijn sequence, was considered and analyzed by Games [37]. The operator was defined in parallel by Goka [38] as an operator on binary sequences. It was applied also on infinite sequences by Nathanson [56] who called it the derivative. The operator was used later for analyzing words and sequences, but most of the time these two papers were not credited for their original first work. In connection to de Bruijn sequences the **D**-morphism and its inverse were heavily used, as will be mentioned in the following chapters. Theorem 4.8 was also proved differently by Arazi [7].

An efficient way to implement the **D**-morphism for the construction of 2^{n-2} span n de Bruijn sequences was given by Annexstein [6]. Another efficient way to implement the **D**-morphism for the construction of de Bruijn sequences was presented by Chang, Park, Kim, and Song [13].

In general, the operator **D** was used in many applications, some of which will be demonstrated in the following chapters throughout the whole book. There are applications for the operator that were not mentioned, e.g., see Repke and Rytter [59].

The operator **D** was generalized also to non-binary words and sequences. Given a word $w = (w_1, w_2, \ldots, w_n)$ over $\mathbb{F}_q$ or $\mathbb{Z}_m$, its derivative **D**w was defined by $\mathbf{D}w \triangleq (w_2 - w_1, w_3 - w_2, \ldots, w_n - w_{n-1})$. For two words X and Y of length n we have that $\mathbf{D}X = \mathbf{D}Y$ if and only if $Y = X + (\gamma^n)$, where γ is an element either in $\mathbb{F}_q$ or $\mathbb{Z}_m$.

Alhakim and Akinwande [3] generalized the **D**-morphism to a non-binary alphabet (not necessarily a prime power) and show how to use it to obtain span n de Bruijn sequences from a span $n-1$ de Bruijn sequence. They also demonstrated how to generalize the efficient implementation given by Annexstein [6] for the non-binary case. The **D**-morphism was further generalized and developed to obtain a much more efficient recursive construction by Alhakim and Nouiehed [4]. The operator **D** is associated with the polynomial $x-1$ (or $\mathbf{E}-1$) since we have that for a cyclic sequence S, $\mathbf{D}S = (\mathbf{E}-1)S = \mathbf{E}S - S$. In the same way, we defined in Section 2.3 the operator $\mathbf{D}_g$ on a span n sequence S for each irreducible polynomial $g(x)$, over $\mathbb{F}_q$, of degree k less than n. The operator $\mathbf{D}_g$ is a q^k-to-one mapping from $\mathbb{F}_q^n$ onto $\mathbb{F}_q^{n-k}$, where it operates on q-ary n-tuples. In the same way that $\mathbf{D}^{-1}$ is a mapping of the n-tuple states of G_n onto the $(n+1)$-tuple states of G_{n+1}, we have that $\mathbf{D}_g^{-1}$ is a mapping from the n-tuple states of $G_{q,n}$ onto the $(n+k)$-tuple states of $G_{q,n+k}$. Similarly, $\mathbf{D}_g^{-1}$ maps cyclic sequences of $G_{q,n}$ into cyclic sequences of $G_{q,n+k}$. If S is a span n de Bruijn sequence over $\mathbb{F}_q$, then $\mathbf{D}_g^{-1}S$ yields some sequences in $G_{q,n+k}$. Alhakim and Nouiehed [4] presented a method to merge these sequences into a span $n+k$ de Bruijn sequence. This makes the recursive construction much more efficient when we want to obtain de Bruijn sequences with a large span. Other recursive constructions for de Bruijn sequences were given, for example, by Zhao, Tian, and Qi [66]. The operator $\mathbf{D}_g$, where $g(x) = x-1$, was heavily used, as will be demonstrated in some of the following chapters. Much less is known for other polynomials. Hence, we have the following problem.

Problem 4.1. Develop the theory for the operator $\mathbf{D}_g$, over $\mathbb{F}_2$, and its inverse, where $g(x) = x^k - 1$, $k \neq 2^m$, $m \geq 0$. Find applications for the theory. The same is asked for other polynomials and the first one over $\mathbb{F}_2$ that is of special interest is $g(x) = x^2 + x + 1$. Extend the theory for the operator $\mathbf{D}_g$ over other finite fields.

Section 4.3. The prefer one algorithm was suggested first by Martin [50]. It was rediscovered several more times, e.g., by Ungar [64] and by Ford [26], and its sequence is sometimes called the Ford sequence. The behavior of the Ford sequence was examined many times. It was shown that the sequence is generated by merging the cycles of the PCR_n by Fredricksen [27] who called this sequence the lexicographically least de Bruijn cycle. The truth table of the sequence was analyzed by Mossige [52]. Fredricksen presented an efficient algorithm to generate this sequence in [28] and generalized it to construct 2^{2n-5} full cycles in [29]. The construction was further generalized in this section to obtain more full cycles from the PCR_n and it was presented by Etzion and Lempel [23]. The efficiency of the construction that requires the minimum amount of space per number of generated sequences and having a minimum required time to compute each bit, makes this algorithm of practical use. The method was generalized to m-ary de Bruijn sequences in another paper by Etzion [21]. The construction of full cycles from the CSR_n was introduced by Etzion [22] based on the ideas of their construction from the PSR_n presented in Etzion and Lempel [23]. The algorithm to form full cycles from the CCR_n was given by Etzion [22]. An old survey for constructions and properties of full cycles was given by Fredricksen [30]. At the same time, a survey for the algorithmic approach to the construction problem was presented by Ralston [58].

As mentioned by Etzion and Lempel [23], any choice of a state ending in a *one* and its companion as bridging states will lead to another de Bruijn sequence. The states that were chosen, for this purpose, in the prefer one algorithm are the same as those chosen in step **(A4)** of algorithm merge PCR_n. In order of consistency between the algorithms, this was also the choice in the algorithm merge PCR_n, although a choice like the Lyndon word for each PCR_n cycle would have made it more efficient. Sawada, Williams, and Wong [61] chose the states with the least value in binary representation for each PCR_n cycle (Lyndon words) to be the states ending in a *one*. This implies a simpler algorithm for implementation than the algorithm for the prefer one method, i.e., a simpler rule to check the next bit, given the current n bits of the sequence. Sawada, Williams, and Wong [62] generalized this rule for m-ary de Bruijn sequences.

There are other beautiful methods to construct de Bruijn sequences, some of which look rather surprising. If we order the necklaces (PCR_n cycles) with beads having m colors in a way that the first state is the maximum one as a number in base m, and concatenate them by the order of this state from the largest one to the smallest one, then the sequence obtained will be an m-ary de Bruijn sequence. This construction was suggested by Fredricksen and Maiorana [32], see also

Moreno [51]. Using Lyndon words Sawada, Williams, and Wong [61] also used the same idea. Since the Lyndon words have the smallest value in the necklace, the sequence was called the least lexicographic de Bruijn sequence.

Example 4.3. If $n = 5$ and $m = 2$, then the span 5 binary de Bruijn sequence is

$$1, 11110, 11100, 11010, 11000, 10100, 10000, 0,$$

where the commas separate between the different necklaces.

If $n = 4$ and $m = 3$, then the span 4 ternary de Bruijn sequence is

$$2, 2221, 2220, 2211, 2210, 2201, 2200, 21, 2120, 2111, 2110,$$

$$2101, 2100, 20, 2011, 2010, 2001, 2000, 1, 1110, 1100, 10, 1000, 0.$$

The last 16 bits of this sequence form an associated span 4 binary de Bruijn sequence. ■

Similarly, each span n de Bruijn sequence, over an alphabet of size less than m, is a suffix of the sequence over an alphabet of size m. Moreover, this sequence is identical to the prefer one sequence for the binary alphabet and also to the sequence obtained by preferring always the largest possible symbol in the alphabet of size m.

Fredricksen and Kessler [31] suggested considering all partitions of an integer n and their lexicographic ordering to form the necklaces with beads in two colors and use their concatenation to form a span $n + 1$ binary de Bruijn sequence. Fredricksen [30] suggested a prefer same algorithm that prefers to repeat the previous symbol subject to some conditions. Alhakim [2] suggested a prefer opposite algorithm, where if possible you add to the current sequence the complement of the last bit unless it creates an n-tuple that already appeared. Alhakim, Sala, and Sawada [5] analyzed the sequences obtained from the prefer same and prefer opposite algorithms and found that the resulting sequences have some unique extremal properties. The prefer opposite is associated with the cycles of the CCR_n. Gabric, Sawada, Williams, and Wong [35] suggested a framework for a successor rule to generate de Bruijn sequences. Their method leads to a few constructions that can be applied to various state diagrams of FSR_n. Their framework was generalized in Gabric, Sawada, Williams, and Wong [36] for a larger alphabet. It should be noted that the proof for the correctness of all these algorithms is very similar.

There are many interesting strategies to combine cycles from different FSR_n. However, there are other different interesting methods to construct span n de Bruijn sequences. The merge-or-split method is usually used by starting with a state diagram and merging all its cycles. Hence, there is no split in the method. A completely different approach is to start with a known de Bruijn sequence S. This de Bruijn sequence can be further modified by first splitting the sequence S into two sequences S_1 and S_2 using a pair P_1 of bridging states and after that merging S_1 and S_2 into a span n de Bruijn sequence S' via another pair P_2 of bridging states. The pair $\{P_1, P_2\}$ is called a *cross-join pair*. The method of splitting the de Bruijn sequences into two cycles and merging these two cycles again by using a cross-join pair is called the *cross-join method*. It is straightforward to see that adding a *zero* to the longest run of *zeros* in a span n M-sequence yields a span n de Bruijn sequence S. However, it was conjectured by Chang, Song, Cho [14] that for each span n M-sequence, there exists $\frac{(2^{n-1}-1)(2^{n-1}-2)}{6}$ distinct cross-join pairs. The conjecture was proved by Helleseth and Kløve [40]. Constructions of de Bruijn sequences by using the cross-join method based on such M-sequences were given by Dubrova [19]. Mykkeltveit and Szmidt [55] proved that any span n de Bruijn sequence can be obtained from any other span n de Bruijn sequence by a sequence of cross-join operations.

Merging cycles of the state diagram of an FSR_n using bridging states (either a pair of conjugate states or a pair of companion states) is the most common method to generate a full cycle. Therefore it is desirable to learn the structure of the cycles and to find the possible bridging states. For this purpose, we can define the *adjacency graph* of the state diagram. The adjacency graph is an undirected graph whose vertices are the cycles of the state diagram. Two vertices are connected by an edge if the associated cycles have a pair of companion states. The tree (using the bridging states for edges) that was described at the beginning of the section is a subgraph of the adjacency graph. Li, Zeng, Li, and Helleseth [48] studied the cycle structure of the state diagram obtained from the characteristic polynomial $(1 + x^3)g(x)$, where $g(x)$ is a primitive polynomial of degree at least 3. Based on the cycle structure they have generated a class of de Bruijn sequences. Li, Zeng, Li, Helleseth, and Li [49] considered a characteristic polynomial that is a multiplication of distinct primitive polynomials with distinct degrees that are pairwise co-primes. They studied the structure of the state diagram and the adjacency graph for the associated LFSR. When one of the polynomials is $x + 1$ they gave an efficient time and space algorithm to generate many de Bruijn sequences by merging the cycles of the associated state diagram. Chang, Ezerman, Ling, and Wang [11] studied the adjacency graph of the $LFSR_n$ whose characteristic polynomial is $g(x)f(x)$, where $g(x)$ and $f(x)$ are distinct irreducible polynomials. Using the structure of the adjacency graph they constructed a class of de Bruijn cycles. Chang, Ezerman, Ling, and Wang [12] suggested an efficient way to find the conjugate pairs between any two cycles of an $LFSR_n$ that induces the adjacency graph. Using this method they designed an efficient algorithm to produce de Bruijn cycles.

Hauge and Helleseth [39] suggested, using the merge-or-split method, how to generate de Bruijn sequences from irreducible cyclic codes. The number of sequences obtained by their method is related to the cyclotomic numbers. More constructions of de Bruijn sequences (each construction has some specific uniqueness) are given by Cooper and Heitsch [15,16], Dong and Pei [17], Dragon, Hernandez, Sawada, Williams, and Wong [18], Dubrova [20], Gabric and Sawada [33,34], Hsieh, Sohn, and Bricker [41], Hsieh, Sohn, and

Bricker [41], Jansen, Franx, and Boekee [42], Li and Lin [45,46], Li, Zeng, Helleseth, Li, and Hu [47], Ralston [57], and Yang, Mandal, Aagaard, and Gong [65].

There are many other construction problems. For example, one might ask if a span n de Bruijn sequence can be extended to a span $n + 1$ de Bruijn sequence. In other words, can we write the entries of the span n de Bruijn sequence and continue it to obtain a span $n + 1$ de Bruijn sequence? The problem was considered by Flaxman, Harrow, and Sorkin [24] and Leach [43]. When $\sigma > 2$, a span n de Bruijn sequence is a Hamiltonian cycle in $G_{\sigma,n}$. If the edges of the cycle are removed from the graph, we obtain a new strongly connected graph G' in which each vertex has in-degree $\sigma - 1 \geq 2$ and out-degree $\sigma - 1 \geq 2$. Hence, G' has an Eulerian cycle that can be attached to the Hamiltonian cycle (the last vertex of the Hamiltonian cycle coincides with the first vertex of the Eulerian cycle). The new sequence contains each edge of $G_{\sigma,n}$ exactly once, and hence it is a span $n + 1$ de Bruijn sequence over an alphabet of size $\sigma > 2$. For $\sigma = 2$ this process does not work as G' will not be a connected graph. These ideas were used by Becher and Heiber [9] who also proved that a span n binary de Bruijn sequence can be extended to a span $n + 2$ binary de Bruijn sequence. Another way to extend a de Bruijn sequence is by adding a symbol to the alphabet. This question is solved with the construction by Fredricksen and Maiorana [32] that was mentioned earlier. Another construction was given by Becher and Cortés [8] who used graph theory and a network flow problem.

Since the number of span n de Bruijn sequences is super-exponential, it follows that we cannot design an efficient algorithm to generate all the sequences. However, we might be able to do something in this direction i.e., enumerate these sequences. For this, we end this chapter with the following research problems.

Problem 4.2. Enumerate all span n de Bruijn sequences in some order. Design an efficient algorithm to generate the kth sequence in this order. Design an efficient algorithm that generates the $(k + 1)$th de Bruijn sequence from the kth sequence.

Problem 4.2 seems to be extremely difficult, so we ease this research problem with the following problem.

Problem 4.3. Partition the set of span n de Bruijn sequences into equivalence classes, possibly exponential (or even super-exponential, but as small as possible) number of classes. Design an efficient algorithm to generate the sequences of each class.

References

[1] T. van AArdenne-Ehrenfest, N.G. de Bruijn, Circuit and trees in ordered linear graphs, Simon Steven 28 (1951) 203–217.

[2] A. Alhakim, A simple combinatorial algorithm for de Bruijn sequences, Am. Math. Mon. 117 (2010) 728–732.

[3] A. Alhakim, M. Akinwande, A recursive construction of nonbinary de Bruijn sequences, Des. Codes Cryptogr. 60 (2011) 155–169.

[4] A. Alhakim, M. Nouiehed, Stretching de Bruijn sequences, Des. Codes Cryptogr. 85 (2017) 381–394.

[5] A. Alhakim, E. Sala, J. Sawada, Revisiting the prefer-same and prefer-opposite de Bruijn sequence constructions, Theor. Comput. Sci. 852 (2021) 73–77.

[6] F.S. Annexstein, Generating de Bruijn sequences: an efficient implementation, IEEE Trans. Comput. 46 (1997) 198–200.

[7] B. Arazi, Method of constructing de Bruijn sequences, Electron. Lett. 12 (1976) 858–859.

[8] V. Becher, L. Cortés, Extending de Bruijn sequences to larger alphabets, Inf. Process. Lett. 168 (2021) 106085.

[9] V. Becher, P.A. Heiber, On extending de Bruijn sequences, Inf. Process. Lett. 111 (2011) 930–932.

[10] N.G. de Bruijn, A combinatorial problem, Ned. Akad. Wet. 49 (1946) 758–764.

[11] Z. Chang, M.F. Ezerman, S. Ling, H. Wang, Construction of de Bruijn sequences from product of two irreducible polynomials, Cryptogr. Commun. 10 (2018) 251–275.

[12] Z. Chang, M.F. Ezerman, S. Ling, H. Wang, On binary de Bruijn sequences from LFSRs with arbitrary characteristic polynomials, Des. Codes Cryptogr. 87 (2019) 1137–1160.

[13] T. Chang, B. Park, Y.H. Kim, I. Song, An efficient implementation of the D-homomorphism for generation of de Bruijn sequences, IEEE Trans. Inf. Theory 45 (1999) 1280–1283.

[14] T. Chang, I. Song, S.H. Cho, Some properties of cross-join pairs in maximum length linear sequences, in: Proc. ISITA 90, Honolulu, Hawaii, 1990, pp. 1077–1079.

[15] J. Cooper, C. Heitsch, The discrepancy of the lex-least de Bruijn sequence, Discrete Math. 310 (2010) 1152–1159.

[16] J. Cooper, C.E. Heitsch, Generalized Fibonacci recurrences and the lex-least de Bruijn sequence, Adv. Appl. Math. 50 (2013) 465–473.

[17] J. Dong, D. Pei, Construction for de Bruijn sequences with large stage, Des. Codes Cryptogr. 87 (2015) 343–458.

[18] P.B. Dragon, O.I. Hernandez, J. Sawada, A. Williams, D. Wong, Constructing de Bruijn sequences with co-lexicographic order: the k-ary grandmama sequence, Eur. J. Comb. 72 (2018) 1–11.

[19] E. Dubrova, A scalable method for constructing Galois NLFSRs with period $2^n - 1$ using cross-join pairs, IEEE Trans. Inf. Theory 59 (2013) 703–709.

[20] E. Dubrova, Generation of full cycles by a composition of NLFSRs, Des. Codes Cryptogr. 73 (2014) 469–486.

[21] T. Etzion, An algorithm for constructing m-ary de Bruijn sequences, J. Algorithms 7 (1986) 331–340.

[22] T. Etzion, Self-dual sequences, J. Comb. Theory, Ser. A 44 (1987) 288–298.

[23] T. Etzion, A. Lempel, Algorithms for the generation of full-length shift-register cycles, IEEE Trans. Inf. Theory 30 (1984) 480–484.

[24] A. Flaxman, A. Harrow, G. Sorkin, Strings with maximally many distinct subsequences and substrings, Electron. J. Comb. 11 (2004) #R8.

[25] C. Flye-Sainte Marie, Solution to problem number 58, l'Intermediare Math. 1 (1894) 107–110.

[26] L.R. Ford, A cyclic arrangement of m-tuples, Rand Corp. Reports (1957) P-1071.

[27] H.M. Fredricksen, The lexicographically least de Bruijn cycle, J. Comb. Theory 9 (1970) 1–5.

[28] H.M. Fredricksen, Generation of the Ford sequence of length 2^n, n large, J. Comb. Theory, Ser. A 12 (1972) 153–154.

[29] H.M. Fredricksen, A class of nonlinear de Bruijn cycles, J. Comb. Theory, Ser. A 19 (1975) 192–199.

[30] H. Fredricksen, A survey of full length nonlinear shift register cycle algorithms, SIAM Rev. 24 (1982) 195–221.

[31] H. Fredricksen, I. Kessler, Lexicographic compositions and de Bruijn sequences, J. Comb. Theory, Ser. A 22 (1977) 17–30.

[32] H. Fredricksen, J. Maiorana, Necklaces of beads in k colors and k-ary de Bruijn sequences, Discrete Math. 23 (1978) 207–210.

[33] D. Gabric, J. Sawada, Constructing de Bruijn sequences by concatenating smaller universal cycles, Theor. Comput. Sci. 743 (2018) 12–22.

[34] D. Gabric, J. Sawada, Investigating the discrepancy property of de Bruijn sequences, Discrete Math. 345 (2022) 112780.

[35] D. Gabric, J. Sawada, A. Williams, D. Wong, A framework for constructing de Bruijn sequences via simple successor rule, Discrete Math. 341 (2018) 2977–2987.

[36] D. Gabric, J. Sawada, A. Williams, D. Wong, A successor rule framework for constructing k-ary de Bruijn sequences and universal cycles, IEEE Trans. Inf. Theory 66 (2020) 679–687.

[37] R.A. Games, A generalized recursive construction for de Bruijn sequences, IEEE Trans. Inf. Theory 29 (1983) 843–850.

[38] T. Goka, An operator on binary sequences, SIAM Rev. 12 (1970) 264–266.

[39] E.R. Hauge, T. Helleseth, de Bruijn sequences, irreducible codes, and cyclotomy, Discrete Math. 159 (1996) 143–154.

[40] T. Helleseth, T. Kløve, The number of cross-join pairs in maximum length linear sequences, IEEE Trans. Inf. Theory 37 (1991) 1731–1733.

[41] Y.-C. Hsieh, H.-S. Sohn, D.L. Bricker, Generating $(n, 2)$ de Bruijn sequences with some balance and uniformity properties, Ars Comb. 72 (2004) 277–286.

[42] C.J.A. Jansen, W.G. Franx, D.E. Boekee, An efficient algorithm for the generation of deBruijn cycles, IEEE Trans. Inf. Theory 37 (1991) 1475–1478.

[43] E.B. Leach, Regular sequences and frequency distributions, Proc. Am. Math. Soc. 11 (1960) 566–574.

[44] A. Lempel, On a homomorphism of the de Bruijn graph and its applications to the design of feedback shift registers, IEEE Trans. Comput. 19 (1970) 1204–1209.

[45] M. Li, D. Lin, de Bruijn sequences, adjacency graphs, and cyclotomy, IEEE Trans. Inf. Theory 64 (2018) 2941–2952.

[46] M. Li, D. Lin, Partial cycle structure of FSRs and its applications in searching de Bruijn sequences, IEEE Trans. Inf. Theory 69 (2023) 598–609.

[47] C. Li, X. Zeng, T. Helleseth, C. Li, L. Hu, The properties of a class of linear FSRs and their applications to the construction of nonlinear FSRs, IEEE Trans. Inf. Theory 60 (2014) 3052–3061.

[48] C. Li, X. Zeng, C. Li, T. Helleseth, A class of de Bruijn sequences, IEEE Trans. Inf. Theory 60 (2014) 7955–7969.

[49] C. Li, X. Zeng, C. Li, T. Helleseth, M. Li, Construction of de Bruijn sequences from LFSRs with reducible characteristic polynomials, IEEE Trans. Inf. Theory 62 (2016) 610–624.

[50] M.H. Martin, A problem in arrangements, Bull. Am. Math. Soc. 40 (1934) 859–864.

[51] E. Moreno, On the theorem of Fredricksen and Maiorana about de Bruijn sequences, Adv. Appl. Math. 33 (2004) 413–415.

[52] S. Mossige, Constructive theorems for the truth table of the Ford sequence, J. Comb. Theory 11 (1971) 106–110.

[53] F.J. Mowle, Relations between P_n cycles and stable feedback shift registers, IEEE Trans. Comput. 15 (1966) 375–378.

[54] F.J. Mowle, An algorithm for generating stable feedback shift registers of order n, J. Assoc. Comput. Mach. 14 (1967) 529–542.

[55] J. Mykkeltveit, J. Szmidt, On cross joining de Bruijn sequences, Contemp. Math. 632 (2015) 335–346.

[56] M.B. Nathanson, Derivatives of binary sequences, SIAM J. Appl. Math. 21 (1971) 407–412.

[57] A. Ralston, A new memoryless algorithm for de Bruijn sequences, J. Algorithms 2 (1981) 50–62.

[58] A. Ralston, de Bruijn sequences - a model example of the interaction of discrete mathematics and computer science, Math. Mag. 55 (1982) 131–143.

[59] D. Repke, W. Rytter, On semi-perfect de Bruijn words, Theor. Comput. Sci. 720 (2018) 55–63.

[60] A. de Rivière, Question number 58, l'Intermediare Math. 1 (1894) 19–20.

[61] J. Sawada, A. Williams, D. Wong, A surprisingly simple de Bruijn sequence construction, Discrete Math. 339 (2016) 127–131.

[62] J. Sawada, A. Williams, D. Wong, A simple shift rule for k-ary de Bruijn sequences, Discrete Math. 340 (2017) 524–531.

[63] C.A.B. Smith, W.T. Tutte, On unicursal paths in a network of degree 4, Am. Math. Mon. 48 (1941) 233–237.

[64] P. Ungar, Problem number 4385, Am. Math. Mon. (1950) 188.

[65] B. Yang, K. Mandal, M.D. Aagaard, G. Gong, Efficient composited de Bruijn sequence generators, IEEE Trans. Comput. 66 (2017) 1354–1368.

[66] X.-X. Zhao, T. Tian, W.-F. Qi, An interleaved method for constructing de Bruijn sequences, Discrete Appl. Math. 254 (2019) 234–245.

Chapter 5

Linear complexity of sequences

Algorithms, complexity of de Bruijn sequences

The linear complexity $c(S)$ of a sequence S, is the minimum degree n of an LFSR$_n$ (equivalently, a linear recurrence) that generates S. The linear complexity of S is one of the measures of its predictability – S is completely determined by $2c(S)$ consecutive bits. Writing $c(S)$ equations, each one based on $c(S) + 1$ consecutive bits, for the linear recurrence is enough to solve it, i.e., recover the $c(S)$ coefficients of the linear recurrence. Although high complexity does not necessarily mean low predictability, the converse is always true: low complexity implies high predictability. In many applications, it is therefore important to know the linear complexity and to use appropriate sequences with high linear complexity. In this chapter, we are mainly interested in the linear complexity of sequences whose length and alphabet size are powers of the same prime. In particular, we are interested in the linear complexity of de Bruijn sequences.

Section 5.1 starts with some general definitions and results on the complexity of general sequences. It continues with properties of binary sequences of length 2^n with some prescribed linear complexities. An algorithm to compute the linear complexity of such sequences is also given. Finally, the set $\Omega_g(n)$ for $g(x) = x + 1$ is analyzed (see Section 2.3).

Section 5.2 is devoted to the linear complexities of binary de Bruijn sequences. If S is a span n de Bruijn sequence, then $2^{n-1} + n \leq c(S) \leq 2^n - 1$. The lower and the upper bounds are attained and constructions for sequences that attain these bounds are presented. It is also proved that there is no span n de Bruijn sequence whose linear complexity is $2^{n-1} + n + 1$.

Section 5.3 considers the linear complexities of sequences whose length and alphabet size (that is larger than 2) are powers of the same prime. Several properties of these sequences are discussed and an efficient algorithm is given to compute their linear complexities. Note that we regard the field $\mathbb{F}_{2^m}$ as non-binary although the characteristic of the field is 2.

Section 5.4 is devoted to the linear complexity of non-binary de Bruijn sequences. Surprisingly, the bounds that will be derived are not generalizations for the bounds of the binary case.

Sequences and the de Bruijn Graph. https://doi.org/10.1016/B978-0-44-313517-0.00011-1

5.1 Binary sequences whose length is 2^n

Recall that every sequence $S = [s_0, s_1, \ldots, s_{k-1}]$, over $\mathbb{F}_q$, satisfies a *linear recursion*

$$s_{i+m} + \sum_{j=1}^{m} a_j s_{i+m-j} = 0, \quad i \geq 0, \quad a_j \in \mathbb{F}_q,$$

where m, the degree of the recursion, is less than or equal to k.

In terms of the shift operator $\mathbf{E}$ ($\mathbf{E}s_i = s_{i+1}$) the linear recursion takes the form

$$f(\mathbf{E})S = \left(\mathbf{E}^m + \sum_{j=1}^{m} a_j \mathbf{E}^{m-j} \right) S = [0^k].$$

In other words, $\left(\mathbf{E}^m + \sum_{j=1}^{m} a_j \mathbf{E}^{m-j} \right) s_i = 0$ for all $i \geq 0$.

Definition 5.1. The *linear complexity* (or just *complexity*) $c(S)$ of a sequence $S = [s_0, s_1, \ldots, s_{k-1}]$ is the least integer m for which there exists a polynomial $f(\mathbf{E})$ of degree m such that $f(\mathbf{E})S = [0^k]$. It is also the linear recurrence of the smallest degree that generates the sequence, i.e., if the linear complexity is c, then there exist c coefficients, $b_i \in \mathbb{F}_q$, $0 \leq i \leq c - 1$, such that $s_{j+c} = \sum_{i=0}^{c-1} b_i s_{j+i}$.

Definition 5.1 implies that the linear complexity of a sequence S is the degree of the minimal polynomial that generates the sequence as defined in Definition 2.3. The following lemma is an immediate result.

Lemma 5.1. *If S is a sequence of length k, over $\mathbb{F}_q$, then $c(S) \leq k$.*

Lemma 5.2. *Let S be a sequence of length k and $f(\mathbf{E})$ be the polynomial with the least degree such that $f(\mathbf{E})S = [0^k]$. If $g(\mathbf{E})S = [0^k]$, then $f(\mathbf{E})$ divides $g(\mathbf{E})$.*

Proof. We can write $g(\mathbf{E}) = \alpha(\mathbf{E})f(\mathbf{E}) + \beta(\mathbf{E})$, where $\deg \beta(\mathbf{E}) < \deg f(\mathbf{E})$. Hence, we have

$$[0^k] = g(\mathbf{E})S = (\alpha(\mathbf{E})f(\mathbf{E}) + \beta(\mathbf{E}))S = \alpha(\mathbf{E})f(\mathbf{E})S + \beta(\mathbf{E})S = \beta(\mathbf{E})S.$$

Since $\beta(\mathbf{E})S = [0^k]$ and $\deg \beta(\mathbf{E}) < \deg f(\mathbf{E})$, it follows that $\beta(\mathbf{E})$ is the zero polynomial and hence $g(\mathbf{E}) = \alpha(\mathbf{E})f(\mathbf{E})$, i.e., $f(\mathbf{E})$ divides $g(\mathbf{E})$. $\square$

Corollary 5.1. *The polynomial $g(x)$ of the smallest degree that generates a sequence $S = [s_0, s_1, \ldots, s_{k-1}]$ divides the polynomial $x^k - 1$.*

This leads to the following two observations.

Lemma 5.3. *Let* $S = [s_0 s_1 \cdots s_{k-1}]$ *be a sequence of length k, over* $\mathbb{F}_q$, *and let*

$$S(x) = \sum_{i=0}^{k-1} s_i x^i$$

be its generating function. Then, the linear complexity $c(S)$ *of S is*

$$\min\{deg\ f(x) \ : \ f(x) \not\equiv 0\ (\text{mod } x^k - 1),\ f(x) \cdot S(x) \equiv 0\ (\text{mod } x^k - 1)\}.$$

Corollary 5.2. *Let S be a sequence of length* $k = p^{\ell_1}$ *over* $\mathbb{F}_q$, *where* $q = p^{\ell_2}$ *and p is a prime. The linear complexity of S is c if and only if*

$$(x - 1)^{c-1} S(x) \equiv \gamma (1 + x + x^2 + \cdots x^{k-1})\ (\text{mod } x^k - 1) \qquad (5.1)$$

for some $\gamma \in \mathbb{F}_q \setminus \{0\}$.

Lemma 5.3 implies another equivalent definition for a sequence's linear complexity, which will be useful in some applications.

Definition 5.2. Let S be a sequence of length k over $\mathbb{F}_q$ and $S(x)$ be its generating function. The ***linear complexity*** of S is defined as

$$c(S) \triangleq \min\{\deg f(x) \ : \ f(x) \not\equiv 0\ (\text{mod } x^k - 1),\ f(x) \cdot S(x) \equiv 0\ (\text{mod } x^k - 1)\}.$$

We continue to examine now only binary sequences whose length (and as a consequence also their period) is a power of two. We make use of the simple fact that in the binary case $\mathbf{E} - 1 = \mathbf{E} + 1$. The following lemma is the binary analog to Corollary 5.2.

Lemma 5.4. *A binary sequence S of length* 2^n *has linear complexity c if and only if* $(\mathbf{E} + 1)^{c-1} S = [1^{2^n}]$.

Proof. First, note that

$$(\mathbf{E} + 1)^{2^n} S = (\mathbf{E}^{2^n} + 1) S = \mathbf{E}^{2^n} S + S = [0^{2^n}].$$

Therefore by Lemma 5.2 we have that $c(S) = c$ if and only if c is the least integer such that $(\mathbf{E} + 1)^c S = [0^{2^n}]$ and hence $(\mathbf{E} + 1)^{c-1} S = [1^{2^n}]$ if and only if $c(S) = c$. $\square$

Corollary 5.3. *A sequence S, whose length is a power of 2, has least period* 2^n *if and only if* $2^{n-1} + 1 \leq c(S) \leq 2^n$.

Corollary 5.4.
- *For a nonzero sequence S of length* 2^n *we have that* $c(\mathbf{D}S) = c(S) - 1$.
- *For a nonzero sequence S of length* 2^n *we have that* $c(\mathbf{D}_0^{-1} S) = c(S) + 1$ *and* $c(\mathbf{D}_1^{-1} S) = c(S) + 1$.

Corollary 5.5. *Let S be a sequence of least period 2^n and linear complexity c.*
- $\mathbf{D}^{-1}S$ *contains two sequences,* $\mathbf{D}_0^{-1}$ *and* $\mathbf{D}_1^{-1}$ *of least period 2^n with complexity $c + 1$ if and only if $2^{n-1} < c < 2^n$.*
- $\mathbf{D}^{-1}S$ *contains one sequence of least period 2^{n+1} with complexity $c + 1$ if and only if $c = 2^n$.*

Lemma 5.5. *Let S_1 and S_2 be two binary sequences of length 2^n.*
- *If $c(S_1) = c(S_2)$, then $c(S_1 + S_2) < c(S_1)$.*
- *If $c(S_1) < c(S_2)$, then $c(S_1 + S_2) = c(S_2)$.*

Proof. Let S_1 and S_2 be two binary sequences of length 2^n.
- If $c = c(S_1) = c(S_2)$, then $(\mathbf{E} + \mathbf{1})^{c-1}S_1 = (\mathbf{E} + \mathbf{1})^{c-1}S_2 = [1^{2^n}]$, which implies that

$$(\mathbf{E} + \mathbf{1})^{c-1}(S_1 + S_2) = (\mathbf{E} + \mathbf{1})^{c-1}S_1 + (\mathbf{E} + \mathbf{1})^{c-1}S_2 = [0^{2^n}].$$

Thus $c(S_1 + S_2) \leq c - 1 < c(S_1)$.
- If $c_1 = C(S_1) < c_2 = C(S_2)$, then $(\mathbf{E} + \mathbf{1})^{c_1-1}S_1 = (\mathbf{E} + \mathbf{1})^{c_2-1}S_2 = [1^{2^n}]$ and $(\mathbf{E} + \mathbf{1})^{c_2-1}S_1 = [0^{2^n}]$, which implies that

$$(\mathbf{E}+\mathbf{1})^{c_2-1}(S_1+S_2) = (\mathbf{E}+\mathbf{1})^{c_2-1}S_1 + (\mathbf{E}+\mathbf{1})^{c_2-1}S_2 = [0^{2^n}] + [1^{2^n}] = [1^{2^n}].$$

Thus $c(S_1 + S_2) = c_2 = c(S_2)$. $\qquad\qquad\square$

Lemma 5.6. *A binary sequence S of length 2^n has linear complexity 2^n if and only if the weight of S is odd.*

Proof. By Lemma 5.4 we have that $c(S) = c$ if and only if $(\mathbf{E}+\mathbf{1})^{c-1}S = [1^{2^n}]$. Clearly,

$$(\mathbf{E} + \mathbf{1})^{2^n-1} = \frac{(\mathbf{E}+\mathbf{1})^{2^n}}{\mathbf{E}+\mathbf{1}} = \frac{\mathbf{E}^{2^n}+\mathbf{1}}{\mathbf{E}+\mathbf{1}} = \mathbf{E}^{2^n-1} + \cdots \mathbf{E}^2 + \mathbf{E} + \mathbf{1} = \sum_{i=0}^{2^n-1} \mathbf{E}^i$$

and hence $c(S) = 2^n$ if and only is

$$\sum_{i=0}^{2^n-1} \mathbf{E}^i S = [1^{2^n}] \quad \Leftrightarrow \quad \sum_{i=0}^{2^n-1} s_i = 1,$$

i.e., $c(S) = 2^n$ if and only if the weight of S is odd. $\qquad\qquad\square$

Lemma 5.7. *A binary sequence S of length 2^n has linear complexity $2^{n-1} + 1$ if and only if $S = [X\bar{X}]$, i.e., S is a self-dual sequence.*

Proof. If $c(S) = 2^{n-1} + 1$, then by Corollary 5.4 we have that $c(\mathbf{D}S) = 2^{n-1}$ and hence since the length of S is 2^n, it follows by Corollary 5.3 that $\mathbf{D}S$ has period 2^{n-1}. In other words, $\mathbf{D}S = [Y\ Y]$, which by Lemma 5.6 implies that $\text{wt}(Y)$ is odd. Therefore by Corollary 4.6 we have that $S = \mathbf{D}^{-1}\mathbf{D}S$ is a self-dual sequence, i.e., $S = [X\bar{X}]$.

If $S = [X\bar{X}]$, then $(\mathbf{E} + 1)^{2^{n-1}} S = (\mathbf{E}^{2^{n-1}} + 1)S = \mathbf{E}^{2^{n-1}}S + S = [1^{2^n}]$ and hence, by Lemma 5.4, we have that $c(S) = 2^{n-1} + 1$. $\qquad\square$

An important factor in computing the linear complexity of a sequence is an efficient algorithm that computes the linear complexity of a given sequence S of length 2^n. This is done in the following algorithm.

The Games–Chan algorithm:

The input for the algorithm is a binary sequence S of length 2^n. If $S = [0^{2^n}]$, then $c(S) = 0$. If $S \neq [0^{2^n}]$, then set $c_n := 0$ and $A_n := S$. The algorithm starts at step n.

At general step k of the algorithm there is a sequence A_k of length 2^k and the accumulated linear complexity is c_k. The sequence A_k is partitioned into its left half $L(A_k) = [a_0, \ldots, a_{2^{k-1}-1}]$ and its right half $R(A_k) = [a_{2^{k-1}}, \ldots, a_{2^k-1}]$. Let $B_k := L(A_k) + R(A_k)$.

1. If $B_k = [0^{2^{k-1}}]$, then set $A_{k-1} := L(A_k)$ and $c_{k-1} := c_k$.
2. If $B_k \neq [0^{2^{k-1}}]$, then set $A_{k-1} := B_k$ and $c_{k-1} := c_k + 2^{k-1}$.

We continue with step $k - 1$ and stop at step 0 with c_0. The output is $c(S) = c_0 + 1$. $\qquad\blacksquare$

Theorem 5.1. *Given a sequence S of length 2^n, the Games–Chan algorithm finds the linear complexity of S, $c(S)$.*

Proof. Let $c = c(S)$ and assume $c - 1 = \sum_{i=0}^{n-1} a_i 2^i$, where $a_i \in \{0, 1\}$. Clearly,

$$(\mathbf{E} + 1)^{c-1} S = \left(\prod_{a_i=1} (\mathbf{E} + 1)^{2^i} \right) S = \left(\prod_{a_i=1} (\mathbf{E}^{2^i} + 1) \right) S.$$

The Games–Chan algorithm terminates when the final sequence consists only of *ones*. This relates to the equation $(\mathbf{E} + 1)^{c-1} S = [1^{2^n}]$, i.e., $c - 1$ consecutive applications of $\mathbf{E}S + S$, but since $(\mathbf{E} + 1)^{2^k} = \mathbf{E}^{2^k} + 1$, we can speed the process, as is done in the algorithm. Moreover, the algorithm only performs $\left(\prod_{a_i=1} (\mathbf{E}^{2^i} + 1) \right) S$, since whenever the two halves of the sequence are equal we have $a_i = 0$, we continue with one half of the sequence, and the accumulated linear complexity remained unchanged. $\qquad\square$

Consider now the set of all sequences whose period is a power of 2. Let $\Omega(n)$ denote the subset of these sequences that are generated by the polynomial

$(x + 1)^{n+1}$ (i.e., $(\mathbf{E} + 1)^{n+1} S = \mathbf{0}$) and are not generated by the polynomial $(x + 1)^n$ (i.e., $(\mathbf{E} + 1)^n S \neq \mathbf{0}$), where $n \geq -1$. This implies that these sequences have linear complexity $n + 1$. Note that this definition is slightly different from that of $\Omega_g(n)$, where $\Omega(n) = \Omega_g(n + 1)$ when $g(x) = x + 1$.

Example 5.1. The following sets contain the sequences in $\Omega(i)$, for $-1 \leq i \leq 7$:

$\Omega(-1) = \{[0]\}$,

$\Omega(0) = \{[1]\}$,

$\Omega(1) = \{[01]\}$,

$\Omega(2) = \{[0011]\}$,

$\Omega(3) = \{[0001], [0111]\}$,

$\Omega(4) = \{[00001111], [00101101]\}$,

$\Omega(5) = \{[00000101], [11111010], [00011011], [11100100]\}$,

$\Omega(6) = \{[00000011], [01010110], [00001001], [01011100]$
$\quad\quad\quad [11111100], [10101001], [11110110], [10100011]\}$,

$\Omega(7) = \{[00000001], [00110010], [00000111], [00110100]$
$\quad\quad\quad [01010100], [01100111], [01010010], [01100001]$
$\quad\quad\quad [11111110], [11001101], [11111000], [11001011]$
$\quad\quad\quad [10101011], [10011000], [10101101], [10011110]\}$

and $\Omega(8)$ contains exactly all the 16 distinct self-dual sequences of period 16.
∎

By the observations we have made so far, the following claims can be easily verified.

Lemma 5.8. *The set $\Omega(n)$ can be defined recursively by the operator $\mathbf{D}$ as follows:*

$$\Omega(n) \triangleq \{S : S = \mathbf{D}_0^{-1} S', \ S' \in \Omega(n - 1)\} \bigcup \{S : S = \mathbf{D}_1^{-1} S', \ S' \in \Omega(n - 1)\},$$

where $\Omega_g(-1) = \{[0]\}$.

Lemma 5.9. *The sequences in $\bigcup_{i=-1}^{n-1} \Omega(i)$ are formed by a linear feedback shift register whose characteristic polynomial is $(x + 1)^n$.*

Lemma 5.10. *The sequences in $\bigcup_{i=-1}^{n-1} \Omega(i)$, where $n \geq 1$, contain each n-tuple exactly once as a window in one of the sequences.*

Lemma 5.11. *The sequences in $\Omega(n)$, where $n \geq 1$, contain each n-tuple exactly once.*

Proof. This follows immediately from the following two facts:

- The sequences in $\bigcup_{i=-1}^{n-1} \Omega(i)$ contain each n-tuple exactly once.
- The sequences in $\bigcup_{i=-1}^{n} \Omega(i)$ contain each $(n+1)$-tuple exactly once. $\square$

Corollary 5.6. *If $f(x_1, x_2, \ldots, x_n)$ is the feedback function for the state diagram of the sequences in $\bigcup_{i=-1}^{n-1} \Omega(i)$, then*

$$h(x_1, x_2, \ldots, x_n) = f(x_1, x_2, \ldots, x_n) + 1$$

is the feedback function for the state diagram of the sequences in $\Omega(n)$.

Corollary 2.25 can be adapted for the polynomial $g(x) = x + 1$ as follows.

Lemma 5.12. *The number of sequences in $\Omega(n)$, $n \geq 1$, is $2^{n - \lfloor \log n \rfloor - 1}$ and the length of each sequence in $\Omega(n)$ is $2^{\lfloor \log n \rfloor + 1}$.*

Proof. First, note that Lemma 5.11 implies that if all the sequences in $\Omega(n)$ are of length $2^{\lfloor \log n \rfloor + 1}$, then the number of sequences in $\Omega(n)$ is $2^{n - \lfloor \log n \rfloor - 1}$. Hence, we only have to prove the claim for the length of the sequences.

The proof is by induction. The basis is $n = 1, 2$, and 3, with trivial claims. Assume the claim is true for n, where the length of a sequence is $2^{\lfloor \log n \rfloor + 1}$.

As for the step of the induction, we distinguish between two cases depending on whether n is one less than a power of 2 or not.

Case 1: If n is one less than a power of 2, then by the definition of $\Omega(n)$ the linear complexity of all the sequences in $\Omega(n)$ is a power of 2. This implies by Lemma 5.6 that the weight of all the sequences in $\Omega(n)$ is odd. Therefore by Corollary 4.6 we have that $\Omega(n + 1)$ contains the same number of sequences as in $\Omega(n)$ whose length is doubled. Thus the claim is proved.

Case 2: If n is not one less than a power of 2, then again by the definition of $\Omega(n)$ the weight of all the sequences in $\Omega(n)$ is even. Therefore by Corollary 4.5 we have that $|\Omega(n + 1)| = 2 \cdot |\Omega(n)|$ and the length of the sequences in $\Omega(n)$ is the same as their length in $\Omega(n)$. Thus the claim is proved. $\square$

Corollary 5.7. *The factor $\Omega(n)$, of G_n, contains 2^{n-k} cycles of length 2^k for each n, where $2^{k-1} \leq n < 2^k$.*

5.2 Complexity of binary de Bruijn sequences

de Bruijn sequences are of special interest throughout the book; hence, we would like to find the possible linear complexities of these sequences. The answer for binary de Bruijn sequences will be given in this section.

Theorem 5.2. *If S is a span n de Bruijn sequence, $n \geq 3$, then*

$$2^{n-1} + n \leq c(S) \leq 2^n - 1 .$$

Proof. Let S be a de Bruijn sequence of length 2^n.

Since the length of S is 2^n, it follows by Lemma 5.1 that $c(S) \leq 2^n$. The weight of S is 2^{n-1} and hence by Lemma 5.6 we have that $c(S) \neq 2^n$, which implies that $c(S) \leq 2^n - 1$.

Since S is a span n de Bruijn sequence of length 2^n, it follows that the period of S is 2^n and hence $2^{n-1} < c(S)$.

Since each n-tuple is contained exactly once in S, it follows by Lemma 4.7 that each $(n-1)$-tuple is contained exactly twice in $\mathbf{D}S$. Similarly, in $\mathbf{D}^2 S$ each $(n-2)$-tuple is contained exactly 2^2 times. If we continue by induction, then we have that in $\mathbf{D}^i S$, $1 \leq i \leq n-1$, each $(n-i)$-tuple is contained exactly 2^i times. Hence, the weight of $\mathbf{D}^i S$, $1 \leq i \leq n-1$, is 2^{n-1}. Assume now, on the contrary, that $2^{n-1} + 1 \leq c(S) = 2^{n-1} + r < 2^{n-1} + n$. Since $r < n$, it follows that $\text{wt}(\mathbf{D}^r S) = 2^{n-1}$.

Now, we also use the fact obtained in Corollary 5.4 that $c(\mathbf{D}S) = c(S) - 1$. This implies that $c(\mathbf{D}^r S) = 2^{n-1}$. However, by Corollary 5.3 we have that the period of $\mathbf{D}^r S$ is 2^{n-1}, i.e., $\mathbf{D}^r S = [XX]$, and by Lemma 5.6 we have that the weight of X is odd, i.e., $\text{wt}(\mathbf{D}^r S)$ is twice an odd integer, a contradiction to the fact that $\mathbf{D}^r S$ is of weight 2^{n-1}. Thus $r \geq n$, $c(S) \geq 2^{n-1} + n$, and hence for $n \geq 3$ we have that

$$2^{n-1} + n \leq c(S) \leq 2^n - 1 . \qquad \square$$

Up to a certain complexity, the distribution of the linear complexities among de Bruijn sequences can be relatively easily found by a computer search. Some of the results of such a search are as follows. There are two de Bruijn sequences of length 8 derived from the two M-sequences of length 7. By Theorem 5.2 their linear complexity is 7. There are sixteen de Bruijn sequences of length 16. Four of these sequences have linear complexity 12, four have complexity 14, and eight have complexity 15. Let $\gamma(c, n)$ be the number of span n de Bruijn sequences with linear complexity c. The complexity distribution of span 5 de Bruijn sequences is given in the following table:

c	$\gamma(c, 5)$	c	$\gamma(c, 5)$	c	$\gamma(c, 5)$
21	8	25	32	29	224
22	0	26	36	30	448
23	12	27	64	31	1024
24	20	28	180		

The complexity distribution of span 6 de Bruijn sequences is given in the following table:

c	$\gamma(c,6)$	c	$\gamma(c,6)$	c	$\gamma(c,6)$
38	448	47	1168	56	259320
39	0	48	2772	57	519752
40	32	49	2352	58	1041252
41	96	50	5224	59	2090716
42	160	51	8704	60	4162352
43	80	52	18096	61	8342176
44	432	53	34224	62	16692832
45	288	54	67700	63	33731200
46	896	55	126592		

The number of span 7 de Bruijn sequences is 2^{57}. This number is too large, which makes it impossible to compute the linear complexity of all span 7 de Bruijn sequences. Nevertheless, it was possible to find the number of span 7 de Bruijn sequences that have relatively low linear complexity, close to the minimal linear complexity $2^{n-1} + n$. The first few values for span 7 de Bruijn sequences are $\gamma(71,7) = 477240$, $\gamma(72,7) = 0$, $\gamma(73,7) = 688$, $\gamma(74,7) = 696$, $\gamma(75,7) = 5760$, $\gamma(76,7) = 1232$, $\gamma(77,7) = 12432$, $\gamma(78,7) = 4868$, $\gamma(79,7) = 10040$, $\gamma(80,7) = 7764$, $\gamma(81,7) = 8276$, $\gamma(82,7) = 7496$, $\gamma(83,7) = 18840$, and $\gamma(84,7) = 26964$.

Problem 5.1. Continue with the computation of $\gamma(c,7)$ for $c > 84$. Find consequences from the current data regarding the linear complexity of de Bruijn sequences and their complexity distribution.

The complexity distribution that was found raises several conjectures and questions. First, the number of span n de Bruijn sequences of minimal complexity is much higher than the number of sequences with even relatively much higher complexity. This phenomenon is not difficult to explain and constructions for a large number of such sequences will be given. Next, we have that it appears that there are no span n de Bruijn sequences with linear complexity $2^{n-1} + n + 1$. This fact will be proved later for all $n \geq 3$. The number of span n de Bruijn sequences with complexity $c > 7$ is divisible by 4. This is not correct for some larger complexities, as will be explained in Section 5.5. Finally, we ask the following question.

Problem 5.2. Is the number of span n de Bruijn sequences with linear complexity $2^n - 1$ at least half of the total number of span n de Bruijn sequences?

We continue and provide proof that the lower bound on the minimum complexity of de Bruijn sequences is attained for all parameters.

Theorem 5.3. *If the sufficient condition, stated below, holds for a given n, then there exists a de Bruijn sequence S of order n with $c(S) = 2^{n-1} + n$.*
<u>**The sufficient condition:**</u>

   Consider the set $\Omega(n)$. Then, it is possible to choose one state (of size n) in each of the sequences of $\Omega(n)$, designated as the first state of the sequence, and it is possible to arrange the members of $\Omega(n)$ in pairs $P_i = (A_i, B_i)$, $1 \le i \le 2^{n-\lfloor \log n \rfloor - 2}$, so that Property 1 *through* Property 4 *hold.*
Property 1: *For each pair P_i, the first state of A_i is the companion of the first state of B_i.*
Property 2: *For each i, $A_i + B_i = A_1 + B_1$.*
Property 3: *$c(A_1 + B_1) = n$.*
Property 4: *The graph $(V(n), E(n))$, where $V(n) = \{v_i \ : \ 1 \le i \le 2^{n-\lfloor \log n \rfloor - 2}\}$ and $\{v_i, v_j\} \in E(n)$ if and only if A_i and A_j have a pair of companion states in the same position (relative to their respective first states), is a connected graph.*

Proof. Given an arrangement of the set $\Omega(n)$ that satisfies Property 1 through Property 4, let $(V(n), T)$ denote a spanning tree of $(V(n), E(n))$. We join the members of $\Omega(n)$ to form a single sequence S by applying Lemma 1.19 as follows.

   First, we form S_1, by joining all the A_i sequences via the companion pairs that define the edges of $(V(n), T)$. Then, we form S_2, by joining all the B_i sequences via the corresponding companion pairs whose existence is guaranteed by Property 2. We designate the first states of A_1 and B_1 to be the first states of S_1 and S_2, respectively. It is easy to verify that under this convention the following hold:

1. Two states occupying the same position in an (A_i, B_i) pair are also located opposite each other (same position modulo 2^{n-1}) in S_1 and S_2.
2. The position of each state in S_1 (S_2, respectively) is congruent to its original A_i-position (B_i-position, respectively) modulo $2^{\lfloor \log n \rfloor + 1}$ (which is the length of the A_is and the length of the B_is).

   As a result, it follows that

$$S_1 + S_2 = (A_1 + B_1)^k,$$

where $k = 2^{n-\lfloor \log n \rfloor - 2}$ and S^k is a concatenation of k occurrences of S. Also, by Property 3, $c(S_1 + S_2) = n$. Finally, we join S_1 and S_2 via their respective first states to form a de Bruijn sequence

$$S = [S_1 \ S_2].$$

Because of this form of S and the fact that $c(S_1 + S_2) = n$, it follows directly from the Games–Chan algorithm that $c(S) = 2^{n-1} + n$. $\square$

   Note that we can choose many companion pairs to merge the sequences S_1 and S_2 in Theorem 5.3 (and not just the first states of A_1 and B_1) and obtain a span n de Bruijn sequence with minimal complexity. This will imply the existence of many de Bruijn sequences with minimal complexity. Moreover, as n becomes larger, there will be many choices of edges for the tree $(V(n), T)$ that will imply more such de Bruijn sequences.

Example 5.2. Let $n = 8$, where $|\Omega(8)| = 16$, so that there are 16 inequivalent sequences of length 16 with complexity 9. These sequences are listed below in eight pairs that satisfy Property 1 through Property 4. It is easy to verify that this arrangement satisfies Property 1 through Property 3. To check Property 4, let $POC(i, j)$ denote a position in A_i and A_j that implies $\{v_i, v_j\} \in E(n)$ according to Property 4. It is easy to see now that the seven edges implied by $POC(1, 2) = 4$, $POC(2, 3) = 1$, $POC(3, 4) = 5$, $POC(2, 5) = 3$, $POC(5, 6) = 6$, $POC(3, 7) = 2$, and $POC(5, 8) = 7$ form a tree of $(V(8), E(8))$, thus validating Property 4.

$$(A_1, B_1) = ([0111111110000000], [0111111010000001]),$$
$$(A_2, B_2) = ([0110111110010000], [0110111010010001]),$$
$$(A_3, B_3) = ([1110111100010000], [1110111000010001]),$$
$$(A_4, B_4) = ([1110011100011000], [1110011000011001]),$$
$$(A_5, B_5) = ([0100111110110000], [0100111010110001]),$$
$$(A_6, B_6) = ([0100101110110100], [0100101010110101]),$$
$$(A_7, B_7) = ([1010111101010000], [1010111001010001]),$$
$$(A_8, B_8) = ([0100110110110010], [0100110010110011]).$$

We continue and merge the sequences A_1 and A_2 via the companion states $\{(00000110), (00000111)\}$ defined by $POC(1, 2) = 4$ and similarly we merge the sequences B_1 and B_2 via the associated companions $\{(00010110), (00010111)\}$ guaranteed by Property 2. The merged sequences are

[0110111110010000   [0110111010010001
0111111110000000]   0111111010000001].

We continue and merge the sequences A_3 and B_3 via the two pairs of companion states $\{(00100000), (001000001)\}$ and $\{(00100010), (001000011)\}$, respectively, defined by $POC(2, 3) = 1$. The merged sequences are

[0111111110000000   [0111111010000001
0110111110010000   0110111010010001
1110111100010000]   1110111000010001].

We continue and merge the sequences A_4 and B_4 via the two pairs of companion states $\{(00011100), (00011101)\}$ and $\{(00111100), (00111101)\}$, respectively, defined by $POC(3, 4) = 5$. The merged sequences are

[0111111110000000   [0111111010000001

0110111110010000   0110111010010001

1110011100011000   1110011000011001

1110111100010000]   1110111000010001].

We continue and merge the sequences A_5 and B_5 via the two pairs of companion states $\{(10000010), (10000011)\}$ and $\{(10001010), (10010011)\}$, respectively, defined by $POC(2, 5) = 3$. The merged sequences are

[0100111110010000   [0100111010010001

1110011100011000   1110011000011001

1110111100010000   1110111000010001

0111111110000000   0111111010000001

0110111110110000]   0110111010110001].

We continue and merge the sequences A_6 and B_6 via the two pairs of companion states $\{(00010010), (00010011)\}$ and $\{(01010010), (01010011)\}$, respectively, defined by $POC(5, 6) = 6$. The merged sequences are

[0100101110110000   [0100101010110001

0100111110010000   0100111010010001

1110011100011000   1110011000011001

1110111100010000   1110111000010001

0111111110000000   0111111010000001

0110111110110100]   0110111010110101].

We continue and merge the sequences A_7 and B_7 via the two pairs of companion states $\{(01000010), (01000011)\}$ and $\{(01000110), (01000111)\}$, respectively, defined by $POC(3, 7) = 2$. The merged sequences are

[0100101110110000   [0100101010110001

0100111110010000   0100111010010001

1110011100011000   1110011000011001

1110111100010000   1110111001010001

1010111101010000   1010111000010001

0111111110000000   0111111010000001

0110111110110100]   0110111010110101].

We continue and merge the sequences A_8 and B_8 via the two pairs of companion states $\{(00100110), (00100111)\}$ and $\{(10100110), (10100111)\}$, respectively, defined by $POC(5, 8) = 7$, to obtain the merged sequences S_1 and S_2 as follows:

$S_1 =$ [0100110110110000 $S_2 =$ [0100110010110001

      0100111110010000   0100111010010001

      1110011100011000   1110011000011001

      1110111100010000   1110111001010001

      1010111101010000   1010111000010001

      0111111110000000   0111111010000001

      0110111110110100   0110111010110101

      0100101110110010]   0100101010110011].

Finally, we merge the sequences S_1 and S_2 via one of the 16 companion pairs defined by Property 1 through Property 3. From these pairs, we chose the first pair to obtain the following span 8 de Bruijn sequence with minimal linear complexity 136.

[0100110110110000 0100111110010000 1110011100011000 1110111100010000

  1010111101010000 0111111110000000 0110111110110100 0100101110110010

  0100110010110001 0100111010010001 1110011000011001 1110111001010001

  1010111000010001 0111111010000001 0110111010110101 0100101010110011].

∎

We continue with the construction of de Bruijn sequences with minimal complexity $2^{n-1} + n$, where $n = 2^m$ and $m \geq 3$. For this purpose, we will order the sequences in the set $\Omega(2^m)$ and show that they can be arranged in a way that can satisfy Property 1 through Property 4.

Construction 5.1. *Starting with the set $\Omega(8)$ arranged in* Example 5.2. *Given such an arrangement for $\Omega(2^m)$, construct a set $\Omega'(2^{m+1})$. Let $Y(m)$ denote the set of 2^{2^m-1} elements consisting of the $2^{2^m-1} - 1$ elements of the form $(0, y_1, y_2, \ldots, y_{2^m-1})$, where at least one of the y_is is not* zero, *and the all-ones word of length 2^m. For each $S = [X \ \bar{X}] \in \Omega(2^m)$ and for every $Y \in Y(m)$, let*

$$S_Y = [Y \quad X+Y \quad \bar{Y} \quad X+\bar{Y}].$$

Let

$$\Omega'(2^{m+1}) = \bigcup_{S \in \Omega(2^m)} S(m),$$

where

$$S(m) \triangleq \bigcup_{Y \in Y(m)} S_Y.$$

∎

By the properties of the operator **D** we have the following immediate consequence.

Lemma 5.13. *The set* $\Omega'(2^{m+1})$ *obtained from* $\Omega(2^m)$ *as defined in* Construction 5.1 *is the set* $\Omega(2^{m+1})$.

Proof. The proof will be done iteratively. Assume we are given the set $\Omega(2^m)$ and apply Construction 5.1. Let $S = [X \ \bar{X}]$ be a sequence in $\Omega(2^m)$ whose length is 2^{m+1}.

By the definition, we have that $S_Y = [Y \ \ X + Y \ \ \bar{Y} \ \ X + \bar{Y}]$. Now, we apply $\mathbf{D}^{2^m}$ on S_Y and obtain

$$\mathbf{D}^{2^m} S_Y = (\mathbf{E} + 1)^{2^m} S_Y = (\mathbf{E}^{2^m} + 1) S_Y = (\mathbf{E}^{2^m} + 1)[Y \ \ X + Y \ \ \bar{Y} \ \ X + \bar{Y}]$$

$$= [Y + X + Y \ \ X + Y + \bar{Y} \ \ \bar{Y} + X + \bar{Y} \ \ X + \bar{Y} + Y] = [X \ \ \bar{X} \ \ X \ \ \bar{X}].$$

This implies that $S_Y \in D^{-2^m} S$. Since all the sequences in $\Omega(2^m)$ were inequivalent, it follows that the sequences in $\Omega'(2^{m+1})$ are also inequivalent.

By Lemma 5.12 we have that $|\Omega(2^m)| = 2^{2^m - m - 1}$ and by the construction of $S(m)$ we have that $|S(m)| = |Y(m)| = 2^{2^m - 1}$ and therefore we have that $|\Omega'(2^{m+1})| = 2^{2^m - m - 1} \cdot 2^{2^m - 1} = 2^{2^{m+1} - (m+1) - 1}$, which implies that $\Omega'(2^{m+1})$ have the same number of sequences as in $\Omega(2^{m+1})$.

Thus the set $\Omega'(2^{m+1})$ defined in Construction 5.1 is the set $\Omega(2^{m+1})$. $\square$

We proceed to show that the defined set $\Omega'(2^{m+1})$ can be arranged so that Property 1 through Property 4 are satisfied. To this end, we add Property 5 to be satisfied only when $n = 2^m$.

For an arrangement of the set $\Omega(2^m)$ and $A_i, A_j \in \Omega(2^m)$, let $d(A_i, A_j)$ denote the first position in which A_i differs from A_j, and let

$$d_m = \{d(A_i, A_j) = \text{POC}(i, j) - 1 \ : \ \{v_i, v_j\} \in T(2^m)\},$$

where $T(2^m)$ is a spanning tree of $(V(2^m), E(2^m))$ for $\Omega(2^m)$.
Property 5: $d_m = \{0, 1, 2, \ldots, 2^m - 2\}$.

Example 5.3. For the set of Example 5.2 we have

$$d(A_1, A_2) = 3, \quad d(A_2, A_3) = 0, \quad d(A_3, A_4) = 4, \quad d(A_2, A_5) = 2,$$
$$d(A_5, A_6) = 5, \quad d(A_3, A_7) = 1, \quad d(A_5, A_8) = 6.$$

Hence, $d_3 = \{0, 1, 2, 3, 4, 5, 6\}$.   ∎

Lemma 5.14. *If* $\Omega(2^{m+1}) = \Omega'(2^{m+1})$ *is obtained via* Construction 5.1 *from an ordered set* $\Omega(2^m)$ *satisfying* Property 1 *through* Property 5, *then* $\Omega(2^{m+1})$ *can be arranged to satisfy* Property 1 *through* Property 5.

Proof. Consider the pairs $P_i = (A_i, B_i)$ of $\Omega(2^m)$ and let $A_i = [X_i \ \bar{X}_i]$. Since X_i is the first state of A_i, it follows from Property 1 that $B_i = [X'_i \ \bar{X}'_i]$ (X' is the companion of X) and therefore $A_i + B_i = (0^{2^m-1} 1 0^{2^m-1} 1)$. For each P_i of $\Omega(2^m)$ and for every $Y \in Y(m)$, we form the pair $P_{iY} = (A_{iY}, B_{iY})$ as described in Construction 5.1. Therefore, $A_{iY} + B_{iY} = (0^{2^{m+1}-1} 1 0^{2^{m+1}-1} 1)$, which immediately implies Property 1 through Property 3 for $\Omega(2^{m+1})$.

To complete the proof we have to show that given a spanning tree $T(2^m)$ of the set $\Omega(2^m)$ with $d_m = \{0, 1, 2, \ldots, 2^m - 2\}$, the graph $(V(2^{m+1}), E(2^{m+1}))$ for the set $\Omega(2^{m+1})$ is connected and it has a spanning tree $T(2^{m+1})$ for which $d_{m+1} = \{0, 1, 2, \ldots, 2^{m+1} - 2\}$.

Consider two sequences $A_i, A_j \in \Omega(2^m)$ such that $\{v_i, v_j\} \in T(2^m)$. Since both A_i and A_j are self-dual sequences and $A_j \neq A_i$, it follows from Property 4 that $A_i + A_j = (0^k 1 0^{2^m-1} 1 0^{2^{m-1-k}})$, where $k = d(A_i, A_j) = \text{POC}(i, j) - 1$ (see Example 5.2).

For every $Y \in Y(m)$ let $G(Y)$ be the subgraph of $(V(2^{m+1}), E(2^{m+1}))$ that is spanned by the vertices v_{iY} corresponding to the sequences A_{iY}, where $1 \leq i \leq 2^{2^m - m - 2}$. It can be easily verified that $\{v_i, v_j\} \in T(2^m)$ implies that $\{v_{iY}, v_{jY}\} \in E(2^{m+1})$ with $d(A_{iY}, A_{jY}) = 2^m + d(A_i, A_j)$ and therefore $\text{POC}(iY, jY) = 2^m + \text{POC}(i, j)$. Hence, the set

$$T_{m+1}(Y) \triangleq \{\{v_{iY}, v_{jY}\} : \{v_i, v_j\} \in T(2^m)\}$$

forms a tree of $G(Y)$, isomorphic to $T(2^m)$, and

$$d_{m+1}(Y) \triangleq \{d(A_{iY}, A_{jY}) : \{v_i, v_j\} \in T(2^m)\} = \{2^m, 2^m + 1, \ldots, 2^{m+1} - 2\}.$$

We now show that the 2^{2^m-1} trees, $T_{m+1}(Y)$, $Y \in Y(m)$, can be embedded in a tree $T(2^{m+1})$ of $(V(2^{m+1}), E(2^{m+1}))$ so that the corresponding set d_{m+1} will include the set $\{0, 1, \ldots, 2^m - 1\}$ along with $d_{m+1}(Y)$. To this end, consider $2^m - 1$ pairs $(A_i, A_j)_k$, one for each $k \in d_m$ such that $A_i, A_j \in \Omega(2^m)$, $\{v_i, v_j\} \in T(2^m)$ and $d(A_i, A_j) = k$. For each such pair (A_i, A_j) we form the pair $(A_{iY_k}, A_{jY_{k+1}})$, where $Y_r = (0^r 1 2^{2^m-r})$, $r \in d_m$. As before, it is easy to see that $\{v_i, v_j\} \in T(2^m)$ implies that $\{v_{iY_k}, v_{jY_{k+1}}\} \in E(2^{m+1})$ with $d(A_{iY_k}, A_{jY_{k+1}}) = k$. Moreover, the union G^*_{m+1} of these $2^m - 1$ members of $E(2^{m+1})$ and of the 2^{2^m-1} trees $T_{m+1}(Y)$, $Y \in Y(m)$ contains no cycle.

For $k = 2^m - 1$ we take any pair $P_i = (A_i, B_i)$ and we form the pair $(A_{iY}, B_{iY'})$, where $A_{iY}, B_{iY'} \in \Omega(2^{m+1})$, $Y = (0^{2^m-2} 1 1)$, and $Y' = (0^{2^m-2} 1 0)$. Since we have that $A_i + B_i = (0^{2^m-1} 1 0^{2^m-1} 1)$, it follows that we also have $A_{iY} + B_{iY'} = (0^{2^m-1} 1 0^{2^{m+1}-1} 1 0^{2^m})$, which implies $\{v_{iY}, v_{iY'}\} \in E(2^{m+1})$ with $d(A_{iY}, B_{iY'}) = 2^m - 1$. Adding $\{v_{iY}, v_{iY'}\}$ to G^*_{m+1} creates the cycle-free graph G^* and completes the construction of $d_{m+1} = \{0, 1, \ldots, 2^{m+1} - 2\}$. Since we have used a B-sequence (rather than an A-sequence) with Y', we have to interchange $A_{jY'}$ with $B_{jY'}$ for every $j = 1, \ldots, 2^{2^m-m-2}$. That is, the original pairs $(A_{1Y'}, B_{1Y'})$, $(A_{2Y'}, B_{2Y'})$, etc. associated with Y' now become the pairs $(B_{1Y'}, A_{1Y'})$, $(B_{2Y'}, A_{2Y'})$, etc. (It is easy to verify that the graph $G(Y')$ obtained from the $B_{jY'}$s is isomorphic to the one obtained from the $A_{jY'}$s.)

So far, we have shown that $G(Y')$ and $G(Y_k)$, for each $k \in d_m$, form a connected graph G^*. To see that the rest of the $G(Y)$s are connected to G^*, consider the subgraph $G(Y_1)$ of G^* and any maximum weight Y (starts with a zero and weight $2^m - 2$) such that $G(Y)$ is not in G^*. Then, wt$(Y + Y_1) = 1$, i.e., $Y + Y_1 = (0^r 10^{2^m - r - 1})$ for some $1 \le r \le 2^m - 1$. Let $A_i, A_j \in \Omega(2^m)$ be two sequences such that $\{v_i, v_j\} \in T(2^m)$ and $d(A_i, A_j) = r$. Then, as before, it follows that $\{v_{iY}, v_{jY_1}\} \in E(2^{m+1})$. This procedure can be repeated until all of the vertices of $G(Y)$s are shown to be connected by considering at each step, the current proven connected piece G^* and a maximum weight Y such that $G(Y)$ is not in G^*. For any such Y there exists $\hat{Y}$ such that $G(\hat{Y})$ is in G^* and wt$(Y + \hat{Y}) = 1$. As before, this proves that $G(Y)$ is also connected to G^*. $\square$

We continue to show that the sufficient condition is satisfied for any given n if it is satisfied for $2^{\lfloor \log n \rfloor}$ and $2^{\lceil \log n \rceil}$, as was already proved in Lemma 5.14.

Construction 5.2. *Given a positive integer $n \ge 8$ that is not a power of 2, construct the set $\Omega(n)$ by repeatedly applying the recursion*

$$\Omega'(k + 1) = \mathbf{D}_0^{-1}\Omega(k) \cup \mathbf{D}_1^{-1}\Omega(k),$$

where

$$\mathbf{D}_b^{-1}\Omega(k) = \bigcup_{S \in \Omega(k)} \mathbf{D}_b^{-1}S, \quad b = 0, 1,$$

beginning with the set $\Omega(2^{\lfloor \log n \rfloor})$ obtained by Construction 5.1. ∎

By Lemma 5.8 we have the following conclusion.

Lemma 5.15. Construction 5.2 *yields the set $\Omega(k + 1)$ (defined as $\Omega'(k + 1)$) from the set $\Omega(k)$.*

The validity of the following three lemmas can be easily verified.

Lemma 5.16. *If the first states of S_1 and S_2 are companions, then the first states of $\mathbf{D}_b^{-1}S_1$ and $\mathbf{D}_b^{-1}S_1$ are companions, where $b \in \{0, 1\}$.*

Lemma 5.17. *The mth state of $\mathbf{D}_b^{-1}S_1$ is the companion of either the mth state of $\mathbf{D}_b^{-1}S_2$ or the mth state of $\mathbf{D}_{1-b}^{-1}S_2$, $b \in \{0, 1\}$ if and only if the mth states of S_1 and S_2 are companions.*

Lemma 5.18. *If S_1 and S_2 are sequences of the same length and the same parity, then $\mathbf{D}_0^{-1}(S_1 + S_2) = \mathbf{D}_b^{-1}S_1 + \mathbf{D}_b^{-1}S_2$, where $b \in \{0, 1\}$.*

Given the pair $P_i = (A_i, B_i)$ of $\Omega(k-1)$, it follows, by Lemma 5.16, that the pairs $P_{i0} = (\mathbf{D}_0^{-1}A_i, \mathbf{D}_0^{-1}B_i)$ and $P_{i1} = (\mathbf{D}_1^{-1}A_i, \mathbf{D}_1^{-1}B_i)$ of $\Omega(k)$, $2^{\lfloor \log n \rfloor} < k \le n$, satisfy Property 1. By Lemma 5.18, Construction 5.2 preserves Property 2 and, by Corollary 5.4 and Lemma 5.18, we have that Construction 5.2 preserves Property 3.

By Lemma 5.17 and Construction 5.2, the existence of a tree $T(k-1)$ for $\Omega(k-1)$, where $2^{\lfloor \log n \rfloor} < k \le n$, implies the existence of a corresponding pair of trees $T_1(k)$ and $T_2(k)$, isomorphic to $T(k-1)$, which form subtrees of the graph $(V(k), E(k))$ for $\Omega(k)$. $T_1(k)$ and $T_2(k)$ are disjoint and together include every element of $V(k)$. Thus all we need to complete the proof that Construction 5.2 yields a set $\Omega(k)$ that satisfies Properties 1 through 4 is to demonstrate the existence of an edge in $E(k)$ that connects $T_1(k)$ with $T_2(k)$, $2^{\lfloor \log n \rfloor} < k \le n$. That is, we have to show that it is possible to find two sequences, $\mathbf{D}_i^{-1}A_r$ and $\mathbf{D}_j^{-1}A_s$, $i, j \in \{0, 1\}$, such that $\mathbf{D}_i^{-1}A_r$, corresponds to a vertex of $T_1(k)$, $\mathbf{D}_j^{-1}A_s$ corresponds to a vertex of $T_2(k)$, and the mth state of $\mathbf{D}_i^{-1}A_r$ is the companion of the mth state of $\mathbf{D}_j^{-1}A_s$, for some positive integer m. Assume, on the contrary, that $k_0 \ge 2^{\lfloor \log n \rfloor} + 1$ is the least integer for which there is no edge in $E(k_0)$ that connects $T_1(k)$ and $T_2(k)$. Then, by Lemma 5.17, none of the sets $\Omega(k_0)$ through $\Omega(2^{\lceil \log n \rceil} - 1)$, obtained from $\Omega(k_0 - 1)$ via Construction 5.2, correspond to a connected graph. Consider the set $\Omega \triangleq \Omega(2^{\lceil \log n \rceil} - 1)$. This set can be partitioned into Ω_0 and Ω^*, where $\Omega_0 = \mathbf{D}_0^{-(2^{\lfloor \log n \rfloor}-1)}\Omega(2^{\lfloor \log n \rfloor})$ (applying $\mathbf{D}_0^{-1}$ on the sequences $2^{\lfloor \log n \rfloor} - 1$ times, where $\mathbf{D}_0^{-i}S = \mathbf{D}_0^{-1}(\mathbf{D}_0^{-i+1}S)$) and $\Omega^* = \Omega \setminus \Omega_0$. Given that the graph (V, E) for Ω is not connected, it follows that the graph $(\hat{V}, \hat{E})$ for $\hat{\Omega} \triangleq \mathbf{D}_0^{-1}\Omega_0 \cup \mathbf{D}_0^{-1}\Omega^*$ is not connected. However, it is easy to verify that the set $\hat{\Omega}$ is identical to the set $\Omega(2^{m+1})$ of Construction 5.1 for $m = \lfloor \log n \rfloor$. Since by Lemma 5.14 the graph $(\hat{V}, \hat{E})$ of $\Omega(2^{m+1}) = \hat{\Omega}$ is connected (the commutation of some of the (A_i, B_i) pairs in the proof of Lemma 5.14 does not affect connectivity), we have a contradiction. This invalidates our assumption regarding k_0 and completes the proof of the following result.

Theorem 5.4. *For every $n \ge 3$ there exists a span n de Bruijn sequence whose linear complexity is $2^{n-1} + n$.*

Theorem 5.4 implies that the lower bound of Theorem 5.2 is attained for all $n \ge 3$. Now, we will prove that the upper bound is also attained for any $n \ge 3$.

Theorem 5.5. *For every $n \ge 2$ there exists a span n de Bruijn sequence with linear complexity $2^n - 1$.*

Proof. The proof will be given by induction. The basis is $n = 2$, where the only de Bruijn sequence of length 4, [0011] has linear complexity 3. Assume that the claim holds for some $n \ge 2$.

Let S be a span n de Bruijn sequence with linear complexity $2^n - 1$. Consider the two sequences $\mathbf{D}_0^{-1}S$ and $\mathbf{D}_1^{-1}S$. By Corollary 5.4 we have that the linear complexity of $\mathbf{D}_0^{-1}S$ is 2^n and the linear complexity of $\mathbf{D}_1^{-1}S$ is 2^n. Let R be the sequence in $\{\mathbf{D}_0^{-1}S, \mathbf{D}_1^{-1}S\}$ that contains the n-tuple (state) $X = (0101 \cdots)$ and w.l.o.g. assume it is the first n-tuple in R. The second sequence in $\{\mathbf{D}_0^{-1}S, \mathbf{D}_1^{-1}S\}$ is $\bar{R}$ and it starts with the n-tuple $\bar{X} = (1010 \cdots)$. The next n-tuple of $\bar{R}$ is the companion of X, X', and we can apply Lemma 1.19 on the cycles R and $\bar{R}$ with the pair of bridging states X and X' and obtain the span $n + 1$ de Bruijn sequence $[R \mathbf{E}\bar{R}]$. If

$$R = [x_1, x_2, x_3, \ldots, x_{2^n}],$$

then

$$R + \mathbf{E}\bar{R} = [x_1 + \bar{x}_2, x_2 + \bar{x}_3, \cdots, x_{2^n} + \bar{x}_1] = [x_1 + x_2 + 1, x_2 + x_3 + 1, \cdots, x_{2^n} + x_1 + 1].$$

This implies that

$$c(R + \mathbf{E}\bar{R}) = c(R + \mathbf{E}R + [1^{2^n}]) = c((\mathbf{D}R) + [1^{2^n}]) = c(\mathbf{D}R) = c(S) = 2^n - 1.$$

Now, by the Games–Chan algorithm, we have that

$$c([R \mathbf{E}\bar{R}]) = 2^n + c(R + \mathbf{E}\bar{R}) = 2^n + 2^n - 1 = 2^{n+1} - 1,$$

which completes the proof of the theorem. □

After we saw that the lower bound of Theorem 5.2 is attained and also that the upper bound of Theorem 5.2 is attained, we can observe from the complexity distribution of span n de Bruijn sequences, where $n \leq 7$, that there are no span n de Bruijn sequences with linear complexity $2^{n-1} + n + 1$. It will be proved now that indeed there are no span n de Bruijn sequences with this linear complexity for every $n \geq 3$.

Theorem 5.6. *For $n \geq 3$, there is no span n de Bruijn sequence with linear complexity $2^{n-1} + n + 1$.*

Proof. By Corollary 3.10, a Boolean function of a span n de Bruijn sequence, where $n \geq 3$, has an odd weight.

Assume, on the contrary, that S is a de Bruijn with linear complexity $2^{n-1} + n + 1$. As in the proof of Theorem 5.2, we apply the operator $\mathbf{D}$ on S a few times, i.e., we consider $\mathbf{D}^i S$, $1 \leq i \leq n$. By Corollary 5.4, the linear complexity of $\mathbf{D}^n S$ is $2^{n-1} + 1$. By Lemma 5.7 this implies that $\mathbf{D}^n S$ is a self-dual sequence. Since $\mathbf{D} = \mathbf{E} + 1$, it follows that $\mathbf{D}^n S = (\mathbf{E} + 1)^n S = (\mathbf{E}^n + g(\mathbf{E}) + 1)S$, where $g(\mathbf{E})$ is a polynomial of degree at most $n - 1$ whose smallest power is at least *one*. By rearranging we have

$$(\mathbf{E}^n + g(\mathbf{E}) + 1)S = \mathbf{E}^n S + S + g(\mathbf{E})S$$

and we can write it as

$$\mathbf{E}^n s_i + s_i + g(\mathbf{E})s_i = s_i + s_{i+n} + g(\mathbf{E})s_i$$

for each i. It is important to note that the sequence $(\mathbf{E}^n + g(\mathbf{E}) + 1)S$ has the same weight as the function that is associated with $\mathbf{E}^n s_i + s_i + g(\mathbf{E})s_i$ computed for each row of the truth table. The sequence $g(\mathbf{E})S$ has weight 2^{n-1} as it sums some columns from columns 2 through n, of the truth table, for the values of the variables in the truth table (for each row). Performing the addition $s_i + s_{i+n}$ (or equivalently $\mathbf{E}^n S + S$) on all the associated bits is equivalent to adding the first column with the function column in the truth table. The addition of these two columns is a column of length 2^n and weight 2δ, where δ is the weight of the function associated with the de Bruijn sequence S. By Corollary 3.10 we have that δ is an odd integer. Moreover, the two halves of the column in this summation are identical. The summation of some columns from columns 2 through n yields a column whose weight is 2^{n-1}, with identical top and bottom halves. Therefore the addition of this subset of $n + 1$ columns is a column of length 2^n whose two identical halves have odd weight γ. This implies that the weight of $\mathbf{D}^n S$ is 2γ, where γ is an odd integer. This is a contradiction to the fact that $\mathbf{D}^n S$ is a self-dual sequence whose weight is 2^{n-1}.

Thus for $n \geq 3$, there is no span n de Bruijn sequence with linear complexity $2^{n-1} + n + 1$. □

5.3 Sequences over p^m whose length is p^n, p prime

We continue now with the non-binary case and with the same direction as in the previous sections. First, we want to know what is the linear complexity (minimum degree polynomial) that generates a sequence of length p^n over $\mathbb{F}_{p^m}$, where p is a prime. In particular, we are interested in de Bruijn sequences of length p^n over $\mathbb{F}_{p^m}$ and they will be the topic of the next section. Many claims regarding non-binary sequences are straightforward generalizations of the binary case. However, some results are quite different. Moreover, the proofs for the binary case are less complicated. For these reasons, we make almost a complete separation between the two cases. Nevertheless, some of the results in this section hold also in the binary case.

Theorem 5.7. *A sequence S over $\mathbb{F}_{p^m}$ has period p^n for some n and linear complexity $c(S)$ if and only if the minimum polynomial of S is $(\mathbf{E} - 1)^{c(S)}$. Furthermore, if S is a nonzero sequence that has period p^n, then*

$$p^{n-1} + 1 \leq c(S) \leq p^n,$$

unless $n = 0$, in which case $c(S) = 1$.

Proof. Suppose S is a sequence of period p^n over $\mathbb{F}_{p^m}$, with minimal polynomial $g(x)$. Then, $g(\mathbf{E})S$ is the all-zeros sequence and from Lemma 5.2 and Corollary 5.1, $g(x)$ divides $x^{p^n} - 1$. Over a field with characteristic p, we have $x^{p^n} - 1 = (x - 1)^{p^n}$ since the binomial coefficients $\binom{p^n}{i}$ are *zero* for $1 \leq i \leq p^n - 1$. Thus the minimal polynomial of S is simply $(x - 1)^{c(S)}$ and

S satisfies the linear recurrence

$$(\mathbf{E} - \mathbf{1})^{c(S)} S = [0^{p^n}].$$

This implies that

$$(\mathbf{E}^{p^n} - \mathbf{1}) S = (\mathbf{E} - \mathbf{1})^{p^n - c(S)} (\mathbf{E} - \mathbf{1})^{c(S)} S = [0^{p^n}]$$

for any n such that $p^n \geq c(S)$.

Suppose now that S has a period that is a power of p and suppose further that S is a nonzero sequence so that S has minimal polynomial $(x - 1)^{c(S)}$, where $c(S) \geq 1$. The case $c(S) = 1$ is trivial, so we may assume that $c = c(S) \geq 2$. Then, there is a unique integer n such that $p^{n-1} \leq c(S) \leq p^n$. Now,

$$(\mathbf{E}^{p^n} - \mathbf{1}) S = (\mathbf{E} - \mathbf{1})^{p^n} S = (\mathbf{E} - \mathbf{1})^{p^n - c} (\mathbf{E} - \mathbf{1})^c S = (\mathbf{E} - \mathbf{1})^{p^n - c} [0^{p^n}] = [0^{p^n}],$$

while

$$(\mathbf{E}^{p^{n-1}} - \mathbf{1}) S = (\mathbf{E} - \mathbf{1})^{p^{n-1}} S \neq [0^{p^n}],$$

for otherwise S would have linear complexity at most p^{n-1}. Hence, S has period p^n. $\qquad\square$

The following lemma is the analog of Lemma 5.4. It was also given with the polynomial definition of linear complexity (see Corollary 5.2) and it is proved similarly.

Lemma 5.19. *A sequence S of length p^n over $\mathbb{F}_{p^m}$ has linear complexity c if and only if $(\mathbf{E} - \mathbf{1})^{c-1} S = [\gamma^{p^n}]$, where γ is a nonzero element in $\mathbb{F}_{p^m}$*

Similarly, the following lemma is the analog of Corollary 5.4 and it is observed from the proof of Theorem 5.7.

Corollary 5.8. *The linear complexity of sequence S of length p^n over $\mathbb{F}_{p^m}$ is c if and only if $c((\mathbf{E} - \mathbf{1}) S) = c - 1$.*

As in the binary case, to compute the linear complexity of a sequence whose period is p^n we used and will use the following simple equation:

$$(\mathbf{E} - \mathbf{1})^{p^r} = \mathbf{E}^{p^r} - \mathbf{1}. \tag{5.2}$$

Again, this equation follows immediately from the fact that all the binomial coefficients $\binom{p^r}{i}$, $1 \leq i \leq p^r - 1$, are equal to 0 over $\mathbb{F}_{p^m}$. Similarly to the Games–Chan algorithm, the idea in computing the linear complexity is to reduce the linear complexity of the given sequence S by the largest possible p^r until we reach a sequence whose linear complexity is 1. The following algorithm is a generalization of the Games–Chan algorithm.

The non-binary linear complexity algorithm:

The input for the algorithm is a sequence $S = [A_1 \ A_2 \ \cdots \ A_p]$, over $\mathbb{F}_{p^m}$, of length p^n, where each A_i has length p^{n-1}. If $S = [0^{2^n}]$, then $c(S) = 0$. If $S \neq [0^{2^n}]$, then set $c_n := 0$ and $B_n := S$. The algorithm starts at step n.

At general step k of the algorithm there is a sequence B_k of length p^k and the accumulated linear complexity is c_k. The sequence B_k is partitioned into p equal parts $A_i(B_k)$, $1 \leq i \leq p$, i.e., $B_k = [A_1(B_k) \ A_2(B_k) \ \cdots \ A_p(B_k)]$, and let

$$D_k = [A_2(B_k) - A_1(B_k), \ \cdots \ , A_p(B_k) - A_{p-1}(B_k), A_1(B_k) - A_p(B_k)].$$

1. If $D_k = [0^{p^k}]$, then set $B_{k-1} := A_1(B_k)$ and $c_{k-1} := c_k$; continue to step $k - 1$.
2. If $D_k \neq [0^{p^{k-1}}]$, then set $B_k := D_k$ and $c_k := c_k + p^{k-1}$; apply step k again.

We stop at step 0 with c_0. The output is $c(S) = c_0 + 1$. ∎

The correctness of the algorithm is implied by the simple observation that

$$D_k = \mathbf{E}^{p^{k-1}} B_k - B_k = (\mathbf{E} - 1)^{p^{k-1}} B_k$$

and hence each iteration at step k, which does not result in the all-zeros sequence, reduces the linear complexity of the sequence by p^{k-1}, and therefore we add p^{k-1} to the accumulated linear complexity. If the iteration results in the all-zeros sequence, then B_k is periodic and we continue with a sequence of length p^{k-1}, whose linear complexity is at most p^{k-1}, rather than a sequence of length p^k. This implies the following theorem.

Theorem 5.8. *Given a sequence S, over $\mathbb{F}_{p^m}$, of length p^n, the non-binary linear complexity algorithm finds the linear complexity of S, $c(S)$.*

Lemma 5.20. *A sequence S of length p^n, $n \geq 2$, over $\mathbb{F}_{p^m}$ has linear complexity $p^{n-1} + 1$ if and only if $S = [X \quad X + Y \quad X + 2Y \ \cdots \ X + (p - 1)Y]$, where Y is a nonzero constant sequence, i.e., $Y = \gamma^{p^{n-1}}$ and $\gamma \in \mathbb{F}_{p^m}^*$.*

Proof. Assume first that

$$S = [X \quad X + Y \quad X + 2Y \ \cdots \ X + (p - 1)Y],$$

where $Y = \gamma^{p^{n-1}}$. Hence,

$$(\mathbf{E} - 1)^{p^{n-1}} S = (\mathbf{E}^{p^{n-1}} - 1)[X \quad X + Y \quad X + 2Y \ \cdots \ X + (p - 1)Y] = [Y \ Y \ Y \ \cdots \ Y].$$

Since the complexity of $[Y \ Y \ \cdots \ Y]$ is 1, it follows that $c(S) = p^{n-1} + 1$.

On the other hand, assume that $c(S) = p^{m-1} + 1$. The only sequence with complexity 1 is a nonzero constant sequence Y. Assume further that $S = [X_1 \ X_2 \ X_3 \ \cdots \ X_p]$, where X_i is a sequence of length p^{n-1}. Since $c(S) = p^{n-1} + 1$, it follows that

$$(\mathbf{E} - 1)^{p^{n-1}} S = [X_2 - X_1 \ X_3 - X_2 \ \cdots \ X_1 - X_p] = [Y \ Y \ \cdots \ Y],$$

which implies that $S = [X \ X+Y \ X+2Y \ \cdots \ X+(p-1)Y]$, i.e., $X_i = X + (i-1)Y$, where $1 \le i \le p$. $\qquad\qquad\square$

Sequences over $\mathbb{F}_{p^m}$ and period p^n, whose sum of elements is zero or nonzero over $\mathbb{F}_{p^m}$, can be distinguished by their linear complexities.

Lemma 5.21. *If a sequence* $S = [s_0, s_1, \ldots, s_{p^n-1}]$ *over* $\mathbb{F}_{p^m}$ *has linear complexity between* $p^{n-1}+1$ *and* p^n-1, *then*

$$\sum_{i=0}^{p^n-1} s_i = 0.$$

Proof. Let S be a sequence over $\mathbb{F}_{p^m}$ of length p^n. By Theorem 5.7 we have that $p^{n-1}+1 \le c(S) \le p^n$.

If $p^{n-1}+1 \le c(S) \le p^n-1$, then there exists a sequence T for which $c(T) = c(S)+1$ such that $S = (\mathbf{E}-1)T$. Therefore $p^{n-1}+2 \le c(T) \le p^n$ and hence the period of T is p^n. Let $T = [t_0, t_1, \ldots, t_{p^n-1}]$, which implies that $S = [t_1 - t_0, t_2 - t_1, \ldots, t_{p^n-1} - t_{p^n-2}, t_0 - t_{p^n-1}]$, where the computation is performed in $\mathbb{F}_{p^m}$. It is easy to verify that the sum of the elements in S is 0 in $\mathbb{F}_{p^m}$. $\qquad\qquad\square$

Corollary 5.9. *For every sequence over* $\mathbb{F}_{p^m}$ *of length* p^n *and complexity less than* p^n *the sum of its elements is 0 in* $\mathbb{F}_{p^m}$.

By carefully considering the sequence $(\mathbf{E}-1)S$, where $c(S) = p^n+1$, we will infer the following consequence.

Corollary 5.10. *If* $S = [s_0, s_1, \ldots, s_{p^n-1}]$ *is a sequence over* $\mathbb{F}_{p^m}$ *of length* p^n *and linear complexity* p^n, *then*

$$\sum_{i=0}^{p^m-1} s_i \ne 0.$$

Proof. If S is a sequence over $\mathbb{F}_{p^m}$ of length p^n and complexity p^n, then there exists a sequence T such that $S = (\mathbf{E}-1)T$ and $c(T) = p^n+1$. The sequence T must have the following form:

$$T = [X \ X+Y \ X+2Y \ \cdots \ X+(p-1)Y],$$

where X is any sequence of length p^n over $\mathbb{F}_{p^m}$ and Y is a constant nonzero sequence over $\mathbb{F}_{p^m}$, i.e., $Y = \gamma \ \cdots \ \gamma$, where $\gamma \in \mathbb{F}_{p^m}^*$. This implies by Lemma 5.20 that $c(T) = p^n+1$ and hence by Corollary 5.8 $c((\mathbf{E}-1)T) = p^n$. By the structure of T, we have that

$$S = (\mathbf{E}-1)T = [(\mathbf{E}-1)X + 0^{p^n-1}\gamma, (\mathbf{E}-1)X + 0^{p^n-1}\gamma, \ldots, (\mathbf{E}-1)X + 0^{p^n-1}\gamma].$$

Therefore the period of S is p^n and $S = [(\mathbf{E}-1)X + 0^{p^n-1}\gamma]$. The sequence $(\mathbf{E}-1)X$ is of length p^n and complexity less than p^n and hence by Corollary 5.9 the sum of its elements is 0 in $\mathbb{F}_{p^m}$. This implies that the sum of elements in $S = (\mathbf{E}-1)T = [(\mathbf{E}-1)X + 0^{p^n-1}\gamma]$ is not 0 in $\mathbb{F}_{p^m}$. $\qquad\square$

Corollary 5.10 can be proved by enumeration of the number of sequences of a given period and complexity with even weights and those with odd weights. Let $\mathcal{N}(p^m, c)$ denote the number of sequences over $\mathbb{F}_{p^m}$ whose period is a power of p and linear complexity c, where each sequence is counted for all its cyclic shifts.

Lemma 5.22. *For every prime p, a positive integer m, and $c \geq 1$, we have*

$$\mathcal{N}(p^m, c) = (p^m - 1)p^{m(c-1)}.$$

Proof. Consider the set of sequences having a period that is a power of p and linear complexity at most $c \geq 1$. It follows from Theorem 5.7 that a sequence S is in this set if and only if $(\mathbf{E}-1)^c$ is the minimal polynomial of S. Thus each of these sequences is uniquely determined by its first c terms, using the recursion implied by $(\mathbf{E}-1)^c s_i = 0$ for each i of a sequence $s_0 s_1 \cdots s_{p^k-1}$. Hence, there are exactly p^{mc} sequences in this set and $p^{m(c-1)}$ sequences that satisfy the recursion $(\mathbf{E}-1)^{c-1} s_i = 0$. Hence, there are $p^{mc} - p^{m(c-1)} = (p^m - 1)p^{m(c-1)}$ sequences of linear complexity exactly c and having a period that is a power of p. It follows that

$$\mathcal{N}(p^m, c) = (p^m - 1)p^{m(c-1)}. \qquad\square$$

Corollary 5.11. *For a prime p and $k > 0$, let $p^{k-1} + 1 \leq c \leq p^k$. Then, there are exactly $(p^m - 1)p^{mc-m-k}$ nonequivalent sequences over $\mathbb{F}_{p^m}$ with period p^k and linear complexity c.*

Proof. If the linear complexity of a sequence is between $p^{k-1} + 1$ and p^k, then by Theorem 5.7 the period of the sequence is p^k. Each nonequivalent sequence in Lemma 5.22 was counted p^k times, once for any initial c terms. Hence, the claim follows. $\qquad\square$

Note that the proofs of Lemma 5.22 and Corollary 5.11 can be used as alternatives for the computation of the number of sequences of $\Omega(n)$ for each positive integer n.

5.4 Complexity of non-binary de Bruijn sequences

After analyzing the linear complexity and complexity distribution of non-binary sequences whose length is a power of a prime p and its alphabet size is a power of the same prime, we turn our attention to the related de Bruijn sequences. As in the binary case, we provide some computer-search results for some of the complexity distribution of non-binary de Bruijn sequences. Let $\gamma_q(c, n)$ be the number of span n de Bruijn sequences over $\mathbb{F}_q$ with linear complexity c. For example $\gamma_3(7, 2) = 12$ and $\gamma_3(8, 2) = 12$. The following tables provide some of the complexity distributions.

The first table is for span 2 de Bruijn sequences over $\mathbb{F}_4$:

c	$\gamma_4(c, 2)$	c	$\gamma_4(c, 2)$	c	$\gamma_4(c, 2)$
10	96	12	336	14	3312
11	144	13	1200	15	15648

The second table is for span 2 de Bruijn sequences over $\mathbb{F}_5$:

c	$\gamma_5(c, 2)$	c	$\gamma_5(c, 2)$	c	$\gamma_5(c, 2)$
11	240	16	1920	21	6430280
12	0	17	10080	22	31677520
13	0	18	54800	23	159523800
14	0	19	256360	24	796064720
15	760	20	1307520		

The third table is for span 3 de Bruijn sequences over $\mathbb{F}_3$:

c	$\gamma_3(c, 3)$	c	$\gamma_3(c, 3)$	c	$\gamma_3(c, 3)$
17	48	21	1620	25	82920
18	60	22	3096	26	246144
19	60	23	9240		
20	504	24	29556		

As we can see from the tables, in these examples, the maximum complexity of sequences of length q^n, where q is a prime power, is $q^n - 1$. As for a lower bound, no general formula is seen from these examples. Some of the bounds for the minimum complexity will be discussed in the rest of this section.

Theorem 5.9. *If S is a span n de Bruijn sequence over $\mathbb{F}_{p^m}$, where $m > 1$ if $p = 2$, then*

$$p^{mn-1} + n \leq c(S) \leq p^{mn} - 1.$$

Proof. The upper bound is an immediate consequence of Corollary 5.10.

The lower bound is proved similarly to the proof of Theorem 5.2. By Theorem 5.7 we have that $p^{mn-1} + 1 \leq c(S)$. If S is a span n de Bruijn sequence over $\mathbb{F}_{p^m}$, then each n-tuple over $\mathbb{F}_{p^m}$ is contained exactly once in S. In the sequence $(\mathbf{E} - \mathbf{1})S$ we have that each $(n - 1)$-tuple over $\mathbb{F}_{p^m}$ is contained exactly p^m times. Similarly, in the sequence $(\mathbf{E} - \mathbf{1})^i S$, $1 \leq i \leq n - 1$, we have that each $(n - i)$-tuple over $\mathbb{F}_{p^m}$ is contained exactly p^{mi} times. Assume now, on the contrary, that $p^{mn-1} + 1 \leq c(S) = p^{mn-1} + r < p^{mn-1} + n$. Since $r < n$, it follows that each element of $\mathbb{F}_{p^m}$ is contained the same number of times in $(\mathbf{E} - 1)^r S$.

Now, we also use the fact that $c((\mathbf{E} - 1)S) = c(S) - 1$. This implies that $c((\mathbf{E} - 1)^r S) = p^{mn-1}$. However, by Theorem 5.7 we have that the period of $(\mathbf{E} - 1)^r S$ is p^{mn-1}, i.e., $(\mathbf{E} - 1)^r S = [X \; X \; \cdots \; X]$, and by Corollary 5.10 we have that the sum of the elements in X is not 0 in $\mathbb{F}_{p^m}$, i.e., each element of $\mathbb{F}_{p^m}$ is not contained the same number of times in X. The sequence $(\mathbf{E} - 1)^r S = [X \; X \; \cdots \; X]$ has the same property, a contradiction to the fact that we proved in the previous paragraph that in $(\mathbf{E} - 1)^r S$ each element of $\mathbb{F}_{p^m}$ is contained the same number of times. Thus $r \geq n$, $c(S) \geq p^{mn-1} + n$, and hence the claim of the theorem follows. $\qquad\qquad\square$

Theorem 5.10. *If S is a span 2 de Bruijn sequence over $\mathbb{F}_p$, p an odd prime, then*

$$c(S) \geq 2p + 1.$$

Proof. Let S be a span 2 de Bruijn sequence over $\mathbb{F}_p$ given as

$$S = [s_0 s_1 \cdots s_{p^2-1}].$$

W.l.o.g., we can assume that $s_0 = 0$ (adding a constant sequence of length p^2 will not change the linear complexity). By Theorem 5.7, Lemma 5.20, and Corollary 5.10, we have that $p + 2 \leq c(S) \leq p^2 - 1$. Assume, on the contrary, that $p + 2 \leq c(S) \leq 2p$. By Corollary 5.8 this implies that $2 \leq c((\mathbf{E} - 1)^p S) \leq p$ and hence by Theorem 5.7, $(\mathbf{E} - 1)^p S$ has period p. If $X = (\mathbf{E} - 1)^p S$, we can write $X = [x_0 x_1 \cdots x_{p-1}]$ by considering the period p of X. Since $(\mathbf{E}^p - 1) S = (\mathbf{E} - 1)^p S = X$, it follows that

$$S = [\hat{S}, \hat{S} + X, \hat{S} + 2X, \ldots, \hat{S} + (p-1)X]$$

for $\hat{S} = s_0 s_1 \cdots s_{p-1}$. Note that $s_p = x_0$ since the first bit of $\hat{S}$ is $s_0 = 0$. Let

$$d_i = \begin{cases} s_{i+1} - s_i & \text{for } 0 \leq i \leq p - 2 \\ x_0 - s_{p-1} & \text{for } i = p - 1 \end{cases}$$

and

$$e_i = x_{i+1} - x_i \quad \text{for } 0 \leq i \leq p - 1,$$

where $x_p = x_0$. In the sequence

$$T = [(s_0, s_1 - s_0), (s_1, s_2 - s_1), \ldots, (s_{p^2-1}, s_0 - s_{p^2-1})]$$

every ordered pair of elements of $\mathbb{F}_p$ appears exactly once since every ordered pair of $\mathbb{F}_p$ appears exactly once in the span 2 de Bruijn sequence S (note that

T is combined from S and $(\mathbf{E} - 1)S$. However, T may be written in the form

$$[(s_0, d_0), (s_1, d_1), \ldots, (s_{p-1}, d_{p-1}),$$
$$(s_0 + x_0, d_0 + e_0), (s_1 + x_1, d_1 + e_1), \ldots, (s_{p-1} + x_{p-1}, d_{p-1} + e_{p-1}),$$
$$(s_0 + 2x_0, d_0 + 2e_0), (s_1 + 2x_1, d_1 + 2e_1), \ldots, (s_{p-1} + 2x_{p-1}, d_{p-1} + 2e_{p-1}),$$
$$\vdots$$
$$(s_0 + (p-1)x_0, d_0 + (p-1)e_0), (s_1 + (p-1)x_1, d_1 + (p-1)e_1), \ldots,$$
$$(s_{p-1} + (p-1)x_{p-1}, d_{p-1} + (p-1)e_{p-1})].$$

Note that since S has period p^2, it follows that not all the terms of X are *zeros* (for otherwise we will have that $c(X) = 0$ and $c(S) = p$). We claim that no x_i is *zero*. Suppose, on the contrary, that $x_k = 0$ and $x_r \neq 0$. Since $X = (\mathbf{E}^p - 1)S$, it implies that $s_k, s_{k+p}, \ldots, s_{k+(p-1)p}$ are all equal but $s_r, s_{r+p}, s_{r+2p}, \ldots, s_{r+(p-1)p}$ are all distinct, so that some element of $\mathbb{F}_p$ appears more than p times in a period of s. This contradicts the fact that by the span 2 property, each element of $\mathbb{F}_p$ appears p times in the span 2 de Bruijn sequence S. Hence, there are two terms of X that are equal, say $x_k = x_r$. All the $2p$ pairs

$$(s_k, d_k), (s_k + x_k, d_k + e_k), \ldots, (s_k + (p-1)x_k, d_k + (p-1)e_k),$$
$$(s_r, d_r), (s_r + x_r, d_r + e_r), \ldots, (s_r + (p-1)x_r, d_k + (p-1)e_r)$$

are distinct. Hence, the set of the following two equations

$$s_k + i x_k = s_r + j x_r, \qquad d_k + i e_k = d_r + j e_r,$$

does not have a solution. However, since $x_k = x_r$ and all the other variables (except for i and j) are given, it follows that $e_k = e_r$. However, from the fact that $e_k = e_r$ and since $x_k = x_r$, it follows that $x_{k+1} = x_{r+1}$. Repeatedly applying the argument above, we quickly find that all the x_is are equal. However, this implies that $c(X) = 1$ and consequently $c(S) = p + 1$, a contradiction. Thus $c(s) \geq 2p + 1$. $\qquad\qquad\square$

The computational results presented in the tables (such as $\gamma_3(c, 3) = 0$ for $c < 17$) suggest that the lower bounds presented in Theorems 5.9 and 5.10 can be improved in some cases. Moreover, other computational results (such as $\gamma_3(12, 2) = \gamma_3(13, 2) = \gamma_3(14, 2) = 0$) suggest there might be a related result to that of Theorem 5.6. Therefore we suggest the following research problem.

Problem 5.3. Improve the lower bounds on the minimal complexity of span n de Bruijn sequences over $\mathbb{F}_q$. This is most appealing when q is a prime and $n > 2$.

Problem 5.4. Find more values of γ, c, and n, for which $\gamma_q(c, n) = 0$

Problem 5.5. Extend the complexity-distribution tables and try to find more interesting properties associated with the complexity distribution of non-binary de Bruijn sequences.

Construction 5.3. *Let p be an odd prime. We generate a sequence S over $\mathbb{F}_p$ as follows. For $0 \le i, j \le p - 1$, we define*

$$s_{jp+i} = \begin{cases} i + \frac{1}{2}(j - 1)j & \text{for } i \text{ even} \\ i + \frac{1}{2}(j + 1)j & \text{for } i \text{ odd,} \end{cases}$$

where the computation is performed modulo p, and let $S = [s_0 s_1 \cdots s_{p^2-1}]$. ■

Example 5.4. For $p = 3$, the sequence S is [012022110].
  For $p = 5$, the sequence S is [01234022441143103204211330]. ■

Theorem 5.11. Construction 5.3 *generates a span 2 de Bruijn sequence of linear complexity $2p + 1$.*

Proof. We start by calculating $c(S)$. Let

$$T = (\mathbf{E} - 1)^p S = (\mathbf{E}^p - 1) = [t_0 t_1 \cdots t_{p^2-1}]$$

and by simple algebraic calculations, we have that for $0 \le i, j \le p - 1$,

$$t_{jp+i} = \begin{cases} j & \text{for } i \text{ even} \\ j + 1 & \text{for } i \text{ odd.} \end{cases}$$

For example, when $p = 5$, $T = [0101012121232323434340404]$. From this, it is easy to see that $(\mathbf{E} - 1)^{2p} S = (\mathbf{E}^p - 1)(\mathbf{E}^p - 1)S = [1^{p^2}]$ and hence by the linear complexity algorithm, we have that $c(S) = 2p + 1$.

  Next, we show that for all $k, d \in \mathbb{F}_p$, the pair $(k, k + d)$ appears as a pair of consecutive elements in S, so that S is a span 2 de Bruijn sequence. We consider a period of S as being built up from p blocks, the elements in the jth block being $s_{jp}, s_{jp+1}, \dots, s_{jp+p-1}$. It is easy to check from the definition of S that if i is even, then the difference between s_{jp+i} and s_{jp+i+1} is $1 + j$, while if i is odd, then the difference is $1 - j$, where the arithmetic is done modulo p. It follows that to find all pairs $(k, k + d)$ in S, we need to show that the elements $s_{(d-1)p+i}$ in block $d - 1$ with i even and the elements $s_{(1-d)p+i}$ in block $1 - d$ with i odd together comprise $\mathbb{F}_p$. From the definition of S, these elements are

$$i + \frac{1}{2}(d - 2)(d - 1) \pmod{p}, \ i \text{ even}$$

and

$$i + \frac{1}{2}(2 - d)(1 - d) \pmod{p}, \ i \text{ odd}.$$

These p elements have the desired property and so the theorem is proved. □

It is interesting to note that Construction 5.3 and the proof of Theorem 5.11 on the span of the sequences are still valid if p is any odd integer $p > 1$, so that Construction 5.3 still produces p-ary span 2 de Bruijn sequences. However, of course, the linear complexity is defined only when p is prime for this construction.

Problem 5.6. Define a measure for the linear complexity of de Bruijn sequences over the ring $\mathbb{Z}_m$ and perform a comprehensive analysis for this topic.

Construction 5.4. *Let $m \geq 2$ and let S be a span 2 de Bruijn sequence of length p^{2m-2} over $\mathbb{F}_{p^{m-1}}$. Let T be the following sequence of p^{2m-1} elements:*

$$T \triangleq [\overbrace{0, 0, \ldots, 0}^{p^{2m-2} \text{ times}}, \overbrace{1, 1, \ldots, 1}^{p^{2m-2} \text{ times}}, \ldots, \overbrace{p-1, p-1, \ldots, p-1}^{p^{2m-2} \text{ times}}],$$

consisting of p^{2m-2} copies of each element of $\mathbb{F}_p$. Let A be the sequence of length p^{2m-1}:

$$A \triangleq [0, 1, \ldots, p-1, 0, 1, \ldots, p-1, \ldots, 0, 1, \ldots, p-1],$$

consisting of p^{2m-2} copies of the sequence $0, 1, \ldots, p-1$. Furthermore, define

$$Q = [T, T + A, T + 2A, \ldots, T + (p-1)A],$$

so that Q is a sequence over $\mathbb{F}_p$ with period p^{2m}. Let R be the sequence of period p^{2m} over $\mathbb{F}_{p^m}$, where the elements of $\mathbb{F}_{p^m}$ are represented by m-tuples over $\mathbb{F}_p$. The first $m - 1$ components of each m-tuple are the components of an element of S and the last component of each m-tuple is the associated element of Q. In other words, let S' be a sequence of length p^{2m} that consists of p^2 consecutive copies of S. If $S' = [s'_1, s'_2, \ldots, s'_{p^{2m}}]$ and $Q = [q_1, q_2, \ldots, q_{p^{2m}}]$, then define

$$R \triangleq [(s'_1, q_1), (s'_2, q_2), \ldots, (s'_{p^{2m}}, q_{p^{2m}})].$$

■

Theorem 5.12. *The sequence R constructed in Construction 5.4 is a span 2 de Bruijn sequence over $\mathbb{F}_{p^m}$ with linear complexity $p^{2m-1} + 2$, where $m \geq 2$.*

Proof. We begin by calculating the linear complexities of the component sequences of R. The first $m - 1$ components in an element from R come from an element of S, a sequence of period p^{2m-2}. Hence, these components (ignoring the elements from Q) have linear complexity at most $p^{2m-2} - 1$. The last component sequence of R is Q, and it is easy to see that $(\mathbf{E} - 1)^{p^{2m-1}} Q = [A]$, a sequence of linear complexity 2. Hence, $c(Q) = p^{2m-1} + 2$ and by using Lemma 5.19 we have that $c(R) = p^{2m-1} + 2$.

Next, we show that R is a span 2 de Bruijn sequence. Let α and β be two arbitrary elements of $\mathbb{F}_{p^m}$. We can write $\alpha = (\alpha_0, \alpha_1)$ and $\beta = (\beta_0, \beta_1)$, where $\alpha_0, \beta_0 \in \mathbb{F}_{p^{m-1}}$ and $\alpha_1, \beta_1 \in \mathbb{F}_p$. We will show that α and β appear consecutively as terms in the sequence R. First, since S is a span 2 de Bruijn sequence over $\mathbb{F}_{p^{m-1}}$, it follows that there exists a unique j with $0 \le j < p^{2m-2}$ such that $(s'_{j+kp^{2m-2}}, s'_{j+1+kp^{2m-2}}) = (\alpha_0, \beta_0)$ for every k. Therefore we need only show that for some k, we have $(q_{j+kp^{2m-2}}, q_{j+1+kp^{2m-2}}) = (\alpha_1, \beta_1)$. This claim is a simple consequence of the construction of Q from T and A. $\qquad\Box$

5.5 Notes

The linear complexity of sequences is important in the design of stream ciphers. Some important research in this direction was reported in Menezes, van Oorschot, and Vanstone [35] and Rueppel [39]. Computing the linear complexity of a sequence, i.e., the length of the shortest linear shift register that generates the sequence is performed by the well-known Berlkamp–Massey algorithm [1,31]. This algorithm is better known as the algorithm for the decoding of BCH codes that form an important family of error-correcting codes.

Section 5.1. The Games–Chan algorithm was presented first in [24]. The proof that was given in this section was presented by Etzion, Kalouptsidis, Kolokotronis, Limniotis, and Paterson [17]. Another algorithm to compute the linear complexity of a binary sequence whose period is 2^n was presented by Robshaw [38]. While in the Games–Chan algorithm, the linear complexity of the sequence is accumulated in steps, in the Robshaw algorithm the amount that has to be subtracted from 2^n to obtain the linear complexity is calculated.

Implementation of the Games–Chan algorithm requires $N = 2^n$ bit operations on the sequence S and another n bit operations reduced from integer operations to compute the complexity c. In the following two decades, a few algorithms were suggested to generalize this algorithm for binary sequences with other periods and also for periodic sequences over $\mathbb{F}_q$. The complexity of these algorithms for sequences with period N was kept as low as βN for some constant β, but relatively much higher than the $N + \log N$ bit operations required for The Games–Chan algorithm. The generalization for sequence with period p^n over $\mathbb{F}_{p^t}$ was given by Ding [12]. Wei, Xiao, and Chen [43] and Xiao, Wei, Lam, and Imamura [44] gave an algorithm to compute the linear complexity of sequences with period $N \in \{p^t, 2p^t\}$ over $\mathbb{F}_q$, when q is a primitive root modulo p^2. Chen [9] gave an algorithm for sequences over $\mathbb{F}_{p^t}$ with period $\ell \cdot 2^n$, where 2^n divides $p^t - 1$ and g.c.d.$(\ell, p^t - 1) = 1$. Chen [10] generalized this algorithm to determine the linear complexity of sequences with period $\ell \cdot n$ over $\mathbb{F}_{p^t}$, where ℓ divides $p^t - 1$ and g.c.d.$(n, p^t - 1) = 1$. The main idea in Chen [10] is to reduce the calculation for the linear complexity of a sequence with period $\ell \cdot n$ over $\mathbb{F}_{p^t}$ to the calculation for the linear complexity of ℓ sequences with period n over $\mathbb{F}_{p^t}$. The algorithms in Chen [9,10], Wei, Xiao, and Chen [43], and Xiao, Wei, Lam, and Imamura [44] are designed for sequences over a field of odd order. The ideas given by Chen [9,10] are generalized in Meidl [33] for binary sequences. In [33] Meidl presents the most efficient algorithm for computing the linear complexities of binary sequences of period $\ell \cdot 2^n$. To apply Meidl's algorithm on a sequence S, one forms a family of sequences of length 2^n from S and applies the Games–Chan algorithm to each of these sequences. Then, for specific values of ℓ, Meidl showed that the algorithm requires βN bit operation, where β is a small constant. The algorithm in Meidl [33] is of interest for large N, a small odd integer ℓ such that the smallest k for which ℓ divides $2^k - 1$ is not large. Chee, Chrisnata, Etzion, and Kiah [8] also generalized the Games–Chan algorithm and applied it efficiently for several families of binary sequences. Their algorithm is very simple and requires βN bit operations for a small constant β, where N is the period of the sequence.

There are related concepts for linear complexity. These concepts were mainly introduced as part of the importance of high linear complexity for cryptographic applications. If a sequence has large linear complexity, and during transmission a small number of changes to its terms greatly reduces its linear complexity, then the resulting keystream is also cryptographically weak: knowledge of the first few bits allows the efficient generation of a sequence that closely approximates the original one. Hence, the linear complexity of a sequence S should also remain high even if some of its terms are altered. This observation led to the definition of the *k-error linear complexity* $c_k(S)$ given by Stamp and Martin [41] that was first introduced by Ding [11,12] as *sphere complexity*. The *error linear complexity spectrum*, defined by analogy to the linear complexity spectrum, indicates how linear complexity decreases as the number k of bits allowed to be modified increases. It is called the *k-error linear complexity profile* in Stamp and Martin [41] by analogy with the linear complexity profile introduced by Rueppel [39]. We note that Niederreiter in [36] has given an alternative definition of the k-error linear complexity profile: it is defined there as a measure of how the linear complexity of S changes when considering an increasing number of initial bits of S but a fixed number of errors. An efficient algorithm to compute, for fixed k, the value of $c_k(S)$ for binary sequences with period 2^n was presented by Stamp and Martin [41]. Lauder and Paterson [30] generalized this algorithm to compute the entire error linear complexity spectrum of such sequences. A formula relating the minimum number of bits that need to be altered in a sequence S to reduce the linear complexity of a sequence S to a given value $c(S) - \epsilon$ was given by Kurosawa, Sato, Sakata, and Kishimoto [29]. Sălăgean [40] presented an algorithmic method, based on the Lauder–Paterson algorithm [30], which computes the minimum number of bits, as well as their positions, that should be modified to reduce the linear complexity below any given constant c. Furthermore, exact formulas for the counting function and the expected value for the 1-error linear complexity of binary sequences that have period 2^n, as well as the corresponding bounds for the expected value for the k-error linear complexity for $k \geq 2$, were given by Meidl [32]. Generalization of these results to sequences that have period p^m over the finite field $\mathbb{F}_p$, where

p is prime, is presented by Meidl and Venkateswarlu [34]. The case of binary sequences that have period p^n is studied by Han, Chung, and Yang [25], whereas the 1-error linear complexity of binary sequences with period $2^n - 1$ was treated by Kolokotronis, Rizomiliotis, and Kalouptsidis [28]. Finally, Etzion, Kalouptsidis, Kolokotronis, Limniotis, and Paterson [17] examined the points in which the linear complexity of the sequence decreases and categorized the sequences that have exactly two such distinct points in their spectrums. They analyzed the related sequences and also sequences that have more than two such points.

Section 5.2. A comprehensive study on the complexity of de Bruijn sequences was first carried out by Chan, Games, and Key [7]. They proved the lower and upper bounds on the linear complexity of span n de Bruijn sequence. Based on the Berlkamp–Massey algorithm to find the linear complexity of general binary sequence [1,31] they found that when a *zero* is added to the run with $n - 1$ *zeros* in a span n M-sequence, the outcome is a span n de Bruijn sequence with maximum linear complexity $2^n - 1$. They also computed the complexity distribution of de Bruijn sequences up to span 6, which are tabled in this section, and based on the obtained data they offered Problem 5.2. Games [23] considered the linear complexity of the sequence obtained by merging the two sequences $\mathbf{D}_0^{-1} S$ and $\mathbf{D}_1^{-1} S$, where S is a span n de Bruijn sequence. The nonexistence of span n de Bruijn sequences with linear complexity $2^{n-1} + n + 1$ was proved by Games [22]. The linear complexities of span n de Bruijn sequences with linear complexity between $2^{n-1} + n$ and $2^{n-1} + 2^{n-2}$ were considered by Etzion [13]. The construction of span n de Bruijn sequences with minimal complexity $2^{n-1} + n$ was given in detail by Etzion and Lempel [19]. These sequences have importance not only in their theoretical value but they are also used in constructions of a two-dimensional de Bruijn array by Etzion [15] that will be discussed in Section 9.1 (see Theorem 9.2). The complexity of de Bruijn sequences is also important in recursive constructions of such arrays presented by Fan, Fan, Ma, and Siu [20] and analyzed by Paterson [37] (see Section 9.4).

All the span 6 de Bruijn sequences with minimal complexity 38 can be constructed via the construction using the sufficient condition of Theorem 5.3. For span 7, 447,184 of the de Bruijn sequences with minimal complexity 71 have the form $S = [S_1 \ S_2]$, where $c(S_1 + S_2) = 7$ and $(S_1 + S_2) = (00000011)^8$. They can be constructed using the sufficient condition. However, there exist 56 span 7 de Bruijn sequences for which $(S_1 + S_2) = (11111100)^8$ and they cannot be constructed using the sufficient condition.

Problem 5.7. What are the possible sequences obtained by $S_1 + S_2$, where $[S_1 \ S_2]$ is a de Bruijn sequence of minimal complexity?

The construction of de Bruijn sequences with minimal complexity was generalized by Etzion [16] where the following theorem is proved.

Theorem 5.13. *For every $n \geq 4$ there exists a de Bruijn sequence of order n and linear complexity*

$$2^{n-1} + n + \sum_{i=\lfloor \log n \rfloor + 1}^{n-3} \alpha_i 2^i$$

for any selected $\alpha_i \in \{0, 1\}$, $1 \leq i \leq \lfloor \log n \rfloor + 1$.

Problem 5.8. For $n \geq 7$, are all the linear complexities between $2^{n-1} + n + 2$ and $2^n - 2$ attainable by span n de Bruijn sequences?

It seems by the search results that for $n \geq 4$, the number of span n de Bruijn sequences with linear complexity c is divisible 4. A possible proof was to consider each span n de Bruijn sequence $S = [s_1, s_2, \ldots, s_{2^n}]$, its complement $\bar{S}$, and its reverse S^R. The fourth sequence to be considered is the reverse of the complement of S, i.e., $\bar{S}^R$. These four sequences have the same linear complexity. If n is even or the complexity of S is even, then the four sequences S, $\bar{S}$, S^R, and $\bar{S}^R$ are distinct sequences and hence in these cases we have that $\gamma(c, n) \equiv 0 \pmod{4}$ [18]. However, when n is odd and c is odd, S might be a CR sequence, i.e., $S = \bar{S}^R$ and hence $\bar{S} = S^R$. Let $\delta(c, n)$ be the number of span n de Bruijn CR sequences. The complexity distribution of these sequences for $n = 7$ that was computed by Etzion [14] is presented in the following table. Clearly, $\delta(c, n)$ is not divisible by 4 for $c = 99, 101, 105, 109, 117, 119$, and 121. For all these complexities we will have that $\gamma(c, n)$ is not divisible by 4.

c	$\delta(c, 7)$	c	$\delta(c, 7)$	c	$\delta(c, 7)$
71	448	91	1620	111	559216
73	8	93	2560	113	1102220
75	168	95	6424	115	2116456
77	24	97	7488	117	4210074
79	88	99	11802	119	8328830
81	40	101	20258	121	16875998
83	224	103	31144	123	33706580
85	326	105	72250	125	67480984
87	284	107	143238	127	131815424
89	844	109	285742		

Section 5.3. As was mentioned before, the linear complexity of a sequence S of length N over a finite field $\mathbb{F}_q$ can be determined with the well-known Berlkamp–Massey algorithm, see Berlkamp [1] and Massey [31], in $O(N^2)$ symbol field operations. This algorithm was implemented over the years in various ways, e.g., see Fitzpatrick [21] and Sugiyama, Kasahara, Hirasawa, and Namekawa [42]. The complexity of this algorithm was improved to

$O(N(\log N)^2 \log\log N)$ and this can be read in the papers of Blackburn [2,3] and the books of Blahut [5,6]. All these algorithms are considered acyclic sequences of any length over any finite field $\mathbb{F}_q$. However, in many applications, only periodic sequences are considered and hence the algorithm to find the linear complexity of such sequences can be considerably improved. Another measure that is similar to the linear complexity, called the depth of a sequence, will be discussed in Section 6.4 when this concept will be used to classify words and codes.

Section 5.4. A comprehensive study on the complexity of non-binary de Bruijn sequences was carried out by Blackburn, Etzion, and Paterson [4]. They also considered complexities of span 1 de Bruijn sequences over $\mathbb{F}_p$, p prime. These de Bruijn sequences are just permutations of the elements in $\mathbb{F}_p$. Nevertheless, they yield some interesting properties. They are associated with permutation polynomials and these properties can be applied to obtain results on the linear complexities of de Bruijn sequences with higher spans. The work of Blackburn, Etzion, and Paterson [4] was largely devoted to the complexity of permutation polynomials. Follow-up research on the permutation polynomial was performed by Hines [26,27]. Blackburn, Etzion, and Paterson [4] obtained Theorem 5.9 using permutation polynomials. The lower bound of Theorem 5.9 was attained for $\mathbb{F}_{p^m}$, where $m \geq 2$, in Theorem 5.12 for $n = 2$. The lower bound was also attained in other cases and the upper bound of Theorem 5.9 can be also attained. These results and also the following theorem can be found in Blackburn, Etzion, and Paterson [4].

Theorem 5.14. *Let S be an M-sequences of length $p^{mn} - 1$ over $\mathbb{F}_{p^m}$ and suppose that $s_0 = s_1 = \cdots = s_{n-2} = 0$. Let T denote the span n de Bruijn sequence attained by adding a* zero *into this run of $n - 1$ zeros. Then, T is a span n de Bruijn sequence, where $c(T) = p^{mn} - 1$, unless $p = 2$, $m = 1$, and $n = 1$, in which case $c(T) = 2$.*

Similarly to the binary case, we can consider now the set of all sequences whose period is a power of a prime p. Let $\Omega(p, n)$ denote the subset of these sequences whose linear complexity is $n + 1$, where $n \geq -1$. This definition is a generalization of the definition of $\Omega(n)$. There are many properties similar for all primes, but there are also many essential differences.

Problem 5.9. Can the sets in the set $\{\Omega(p, n)\}$ be used to form span n de Bruijn sequences over $\mathbb{F}_p$ as the sets in the set $\{\Omega(n)\}$ in the binary case?

References

[1] E.R. Berlkamp, Algebraic Coding Theory, McGraw-Hill, New York, 1968.
[2] S.R. Blackburn, A generalization of the discrete Fourier transform: determining the minimal polynomial of a periodic sequence, IEEE Trans. Inf. Theory 40 (1994) 1702–1704.
[3] S.R. Blackburn, Fast rational interpolation, Reed-Solomon decoding and the linear complexity profiles of sequences, IEEE Trans. Inf. Theory 43 (1997) 537–548.

[4] S.R. Blackburn, T. Etzion, K.G. Paterson, Permutation polynomials, de Bruijn sequences, and linear complexity, J. Comb. Theory, Ser. A 76 (1996) 55–82.

[5] R.E. Blahut, Theory and Practice of Error Control Codes, Addison-Wesley, Reading, MA, 1983.

[6] R.E. Blahut, Fast Algorithms for Digital Signal Processing, Addison-Wesley, Reading, MA, 1985.

[7] A.H. Chan, R.A. Games, E.L. Key, On the complexities of de Bruijn sequences, J. Comb. Theory, Ser. A 33 (1982) 233–246.

[8] Y.M. Chee, J. Chrisnata, T. Etzion, H.M. Kiah, Efficient algorithm for the linear complexity of sequences and some related consequences, in: Proc. IEEE Int. Symp. on Infor. Theory (ISIT), 2020, pp. 2897–2902.

[9] H. Chen, Fast algorithms for determining the linear complexity of sequences over $GF(p^m)$ with period $2^t n$, IEEE Trans. Inf. Theory 51 (2005) 1854–1856.

[10] H. Chen, Reducing the computation of linear complexities of periodic sequences over $GF(p^m)$, IEEE Trans. Inf. Theory 52 (2006) 5537–5539.

[11] C. Ding, Lower bounds on the weight complexities of cascaded binary sequences, Lect. Notes Comput. Sci. 453 (1990) 39–43.

[12] C. Ding, A fast algorithm for the determination of linear complexity of sequences over $GF(p^m)$ with period p^n, in: C. Ding, G. Xiao, W. Shan (Eds.), The Stability Theory of Stream Ciphers, in: Lecture Notes in Computer Science, vol. 56, Springer-Verlag, Berlin-Heidelberg, 1991, pp. 141–144.

[13] T. Etzion, On the distribution of de Bruijn sequences of low complexity, J. Comb. Theory, Ser. A 38 (1985) 241–253.

[14] T. Etzion, On the distribution of de Bruijn CR-sequences, IEEE Trans. Inf. Theory 32 (1986) 422–423.

[15] T. Etzion, Constructions for perfect maps and pseudorandom arrays, IEEE Trans. Inf. Theory 34 (1988) 1308–1316.

[16] T. Etzion, Linear complexity of de Bruijn sequences – old and new results, IEEE Trans. Inf. Theory 45 (1999) 693–698.

[17] T. Etzion, N. Kalouptsidis, N. Kolokotronis, K. Limniotis, K.G. Paterson, Properties of the error linear complexity spectrum, IEEE Trans. Inf. Theory 55 (2009) 4681–4686.

[18] T. Etzion, A. Lempel, On the distribution of de Bruijn sequences of given complexity, IEEE Trans. Inf. Theory 30 (1984) 611–614.

[19] T. Etzion, A. Lempel, Construction of de Bruijn sequences of minimal complexity, IEEE Trans. Inf. Theory 30 (1984) 705–709.

[20] C.T. Fan, S.M. Fan, S.L. Ma, M.K. Siu, On de Bruijn arrays, Ars Comb. 19A (1985) 205–213.

[21] P. Fitzpatrick, On the key equation, IEEE Trans. Inf. Theory 41 (1995) 1290–1302.

[22] R.A. Games, There are no de Bruijn sequences of span n with complexity $2^{n-1} + n + 1$, J. Comb. Theory, Ser. A 34 (1983) 248–251.

[23] R.A. Games, A generalized recursive construction for de Bruijn sequences, IEEE Trans. Inf. Theory 29 (1983) 843–850.

[24] R.A. Games, A.H. Chan, A fast algorithm for determining the complexity of a binary sequence of period 2^n, IEEE Trans. Inf. Theory 29 (1983) 144–146.

[25] Y.K. Han, J.-H. Chung, K. Yang, On the k-error linear complexity of p^m-periodic binary sequences, IEEE Trans. Inf. Theory 53 (2007) 2297–2304.

[26] P.A. Hines, Characterising the linear complexity of span 1 de Bruijn sequences over finite fields, J. Comb. Theory, Ser. A 81 (1998) 140–148.

[27] P.A. Hines, On the minimum linear complexity of de Bruijn sequences over non-prime finite fields, J. Comb. Theory, Ser. A 86 (1999) 127–139.

[28] N. Kolokotronis, P. Rizomiliotis, N. Kalouptsidis, Minimum linear span approximation of binary sequences, IEEE Trans. Inf. Theory 48 (2002) 2758–2764.

[29] K. Kurosawa, F. Sato, T. Sakata, W. Kishimoto, A relationship between linear complexity and k-error linear complexity, IEEE Trans. Inf. Theory 46 (2000) 694–698.

[30] A.G.B. Lauder, K.G. Paterson, Computing the error linear complexity spectrum of a binary sequence with period 2^n, IEEE Trans. Inf. Theory 49 (2003) 273–280.
[31] J.L. Massey, Shift-registers synthesis and BCH decoding, IEEE Trans. Inf. Theory 15 (1969) 122–127.
[32] W. Meidl, On the stability of 2^n-periodic binary sequences, IEEE Trans. Inf. Theory 51 (2005) 1151–1155.
[33] W. Meidl, Reducing the calculation of the linear complexity of $u2^v$-periodic binary sequences to Games-Chan algorithm, Des. Codes Cryptogr. 46 (2008) 57–65.
[34] W. Meidl, A. Venkateswarlu, Remarks on the k-error linear complexity of p^n-periodic sequences, Des. Codes Cryptogr. 42 (2007) 181–193.
[35] A.J. Menezes, P.C. van Oorschot, S.A. Vanstone, Handbook of Applied Cryptography, CRC Press, Boca Raton, FL, 1997.
[36] H. Niederreiter, Some computable complexity measures for binary sequences, in: Proc. Sequences and Their Applications (SETA), 1999, pp. 67–78.
[37] K.G. Paterson, Perfect maps, IEEE Trans. Inf. Theory 40 (1994) 743–753.
[38] M.J.B. Robshaw, On evaluating the linear complexity of a sequence of least period 2^n, Des. Codes Cryptogr. 4 (1994) 263–269.
[39] R.A. Rueppel, Analysis and Design of Stream Ciphers, Springer-Verlag, Berlin, 1986.
[40] A. Sălăgean, On the computation of the linear complexity and the k-error linear complexity of binary sequences with period a power of two, IEEE Trans. Inf. Theory 51 (2005) 1145–1150.
[41] M. Stamp, F.Y. Martin, An algorithm for the k-error linear complexity of binary sequences with period 2^n, IEEE Trans. Inf. Theory 39 (1993) 1398–1401.
[42] Y. Sugiyama, M. Kasahara, S. Hirasawa, T.N. Namekawa, A method for solving key equation for solving Goppa codes, Inf. Control 27 (1975) 87–99.
[43] S. Wei, G. Xiao, Z. Chen, A fast algorithm for determining the minimal polynomial of a sequence with period $2p^n$ over $GF(q)$, IEEE Trans. Inf. Theory 48 (2002) 2754–2758.
[44] G. Xiao, S. Wei, K.Y. Lam, K. Imamura, A fast algorithm for determining the linear complexity of a sequence with period p^n over $GF(q)$, IEEE Trans. Inf. Theory 46 (2000) 2203–2206.

Chapter 6

Classification of sequences

Balanced sequences, the depth of a sequence

This chapter is devoted to several methods to classify sequences of the same length, particularly binary sequences of the same length n and weight $\lceil \frac{n}{2} \rceil$. These sequences are called balanced sequences. The families of de Bruijn sequences and M-sequences form the most important classes of sequences in these classifications and they will deserve special consideration. The linear complexity of a sequence defined and discussed in Chapter 5 can be a criterion to distinguish between sequences of the same length over the same alphabet. The importance of this classification will be demonstrated for example in the rotating-table game presented in Section 7.4. However, of course, this is not the only motivation to classify sequences. Classification of elements is a very natural requirement in many areas of our life. In this chapter, we will discuss several methods to classify sequences and in particular binary sequences, sequences of length $2^n - 1$ or 2^n, balanced sequences, and de Bruijn sequences.

Section 6.1 is devoted to a hierarchy of inclusions of binary sequences of length $2^n - 1$ with exactly 2^{n-1} *ones*. This hierarchy is presented by a lattice, where $A \longrightarrow B$ in the lattice implies that the set A contains the set B. At the top of the hierarchy, we have all these sequences. At the bottom of this hierarchy, we have the M-sequences of this length. Sequences satisfying property **R-2** or property **R-3** (see Section 2.2) form two important classes of the hierarchy.

In Section 6.2 we will concentrate only on the classification of de Bruijn sequences or shortened de Bruijn sequences. These classifications will be based on the linear complexity of the shortened de Bruijn sequences or the weight of the truth table of the de Bruijn sequences.

Section 6.3 is devoted to binary sequences of even length n in which exactly half of the entries are *ones*. Two types of hierarchies will be presented in this set of sequences. In the first hierarchy, all these sequences are contained at the top of the hierarchy. If $n = \mu \cdot 2^\ell$, where μ is an odd integer and $\ell \geq 1$, then at the bottom of the hierarchy we have all the sequences of length n in which each ℓ-tuple appears exactly μ times as a window of length ℓ in the sequence. When n is a power of two, de Bruijn sequences form the bottom of this hierarchy. The second hierarchy is based on the derivatives of the sequence, where in the ℓth level of the hierarchy we have those sequences for which all their first ℓ derivatives are balanced.

Sequences and the de Bruijn Graph. https://doi.org/10.1016/B978-0-44-313517-0.00012-3

The classification in Section 6.4 is very similar to classification by the linear complexity for binary sequences and coincides with it for sequences whose length is a power of 2. It will be interesting to note that this classification, called the depth, in which the sequences are acyclic, can be also used to classify and characterize linear codes.

6.1   Classification of sequences with length $2^n - 1$

Let U be the set of all binary sequences of period $\pi = 2^n - 1$ that contains 2^{n-1} *ones* and $2^{n-1} - 1$ *zeros* (property **R-1**). Let PN be the subset of U that contains the M-sequences of length $2^n - 1$. Let R be the subset of U with the "run property", i.e., in each sequence, there are 2^{n-i-2} runs of exactly i *ones* and 2^{n-i-2} runs of exactly i *zeros*, $1 \le i \le n - 2$, one run of exactly $n - 1$ *zeros* and one run of exactly n *ones* (property **R-2**).

Let S be the subset of R consisting of those sequences that have span n. In other words, in these sequences each nonzero n-tuple appears exactly once in a window of length n in the sequence, i.e., these are the shortened de Bruijn sequences.

Let C be the subset of U consisting of the sequences that satisfy the auto-correlation property (property **R-3**). Denote a sequence by $A_0 = (a_0, a_1, \ldots, a_{\pi-1})$, and its kth cyclic shifts by $A_k = (a_k, a_{k+1}, \ldots, a_{\pi-1}, a_0, \ldots, a_{k-1})$, then $A_0 \in C$ if and only if $A_i + A_j \in U$ for all $i \ne j$. The sequences in C are in natural correspondence with Hadamard difference sets (see Section 2.4).

Let M be a subset of U consisting of those sequences that have 2 as a "multiplier". In other words, for some cyclic shift $A' = (a_0', a_1', \ldots, a_{\pi-1}')$ of $A = (a_0, a_1, \ldots, a_{\pi-1})$, we have $a_{2i}' = a_i'$ for each i, $0 \le i \le \pi - 1$, where all the subscripts are taken modulo π. Since, by Theorem 2.17, 2 is a multiplier for all the Hadamard difference sets we have that $C \subset M$.

The hierarchy of inclusions among these sets of sequences is shown in lattice form and as a Venn diagram in Fig. 6.1. The intersection between the different classes is investigated for the rest of this section.

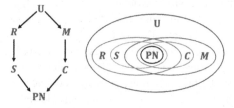

FIGURE 6.1   A lattice of the hierarchy on the left and its Venn diagram on the right.

We would like first to consider whether $R \cap M$ contains sequences that are not contained in $S \cup C$. Such sequences have the run property and the multiplier by 2 property, but do not have the span n property and do not have the correlation property. If $\pi = 31$, then by using a computer search it was found that there are eight sequences in $R \cap M$ as follows:

> 111111011100010101101000011100100
> 111011001110000110101001000101011
> 111010001001010110000111001101 1
> 100101100111110001101110101 0000
> 10000101011101100011111 00110100
> 100100110000101101010001110 1111
> 11111010100111001100011010 10000
> 10000101011000110011100101 01111

The first six sequences are the only M-sequences of order 5. The last two sequences are neither in S nor in C. Specifically, the span of these sequences is 9.

There are also sequences in $S \cap M$ that are not contained in C. The first such example is of period 127 and it is constructed in the following way for every odd n for which there exists a primitive polynomial of the form $x^n + x + 1$ (a polynomial of the form $x^n + x^k + 1$ is called a ***trinomial***). Let $f(x) = x^n + x + 1$ be a primitive trinomial over $\mathbb{F}_2$, and let $\pi = 2^n - 1$, where $n > 4$ is odd. Consider the associated M-sequence, having the linear recurrence $x_n = x_{n-1} + x_0$. If the sequence starts with $n - 1$ *zeros* followed by a *one*, then the following $n - 1$ bits are *ones*. Hence, we can consider a cyclic shift of the M-sequence as starting with the longest run of *ones* (n *ones*) and ending with the longest run of *zeros* ($n - 1$ *zeros*). As an M-sequence, the sequence has the multiplier property and the span n property. If we now complement all the terms of the sequence except for the term at position 0, we preserve the multiplier property and the span n property. However, the correlation property of the set C is no longer maintained. Since the reverse of the sequence $\{b_i\} = \{a_{\pi-i}\}$ preserves all the discussed properties, it follows that there are at least two examples of sequences in $S \cap M$ that are not in C for each period $\pi = 2^n - 1$ corresponding to the existence of primitive trinomials $f(x) = x^n + x + 1$, where n is odd.

Problem 6.1. Are there infinitely many primitive polynomials (primitive trinomials) of the form $x^n + x^k + 1$, and in particular infinitely many such polynomials where n is odd, and the same question for odd n and $k = 1$?

Next, we ask whether $R \cap C =$ PN. A counterexample has been found with $n = 7$, i.e., with period 127. The related sequence is

> 1111101111001111111001001011101010101110001100000100110111001 1000
> 11011011101001000110100001010100110100101000111011000010 1000000

Based on the examples that were given in this section we have a more delicate hierarchy than the one depicted in Fig. 6.1. This delicate hierarchy is depicted in Fig. 6.2.

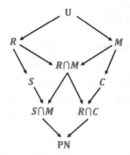

FIGURE 6.2 A lattice of the updated hierarchy.

It should be clear that the sequences that were found in $R \cap M$, in $S \cap M$, and in $R \cap C$, which are not M-sequences in PN, were not found for all periods $2^n - 1$, and hence we have the following problems.

Problem 6.2.

- Characterize all values of n for which $R \cap M \neq$ PN and find constructions for sequences of length $2^n - 1$ in $R \cap M$ that are not M-sequences.
- Characterize all values of n for which $S \cap M \neq$ PN and find constructions for sequences of length $2^n - 1$ in $S \cap M$ that are not M-sequences.
- Characterize all values of n for which $R \cap C \neq$ PN and find constructions for sequences of length $2^n - 1$ in $R \cap C$ that are not M-sequences.

Finally, up to $n = 8$, it was found by a computer search that all sequences in $S \cap C$ are M-sequences, which are all the sequences in PN. However, in general, we have the following problem.

Problem 6.3. Does $S \cap C = $ PN?

6.2 Classification of de Bruijn sequences

In this section, we will concentrate only on de Bruijn sequences and shortened de Bruijn sequences. We have already classified de Bruijn sequences by their linear complexity in Chapter 5. Other measures can classify de Bruijn sequences, like the weight of their FSR_n function or the maximum number of terms of a product in their feedback function represented by a sum of products.

What is the linear complexity distribution of shortened de Bruijn sequences that form the set S defined in the hierarchy of the previous section? Using a computer search the complexity distribution of spans 4, 5, and 6, shortened de Bruijn sequences can be found. There are two span 4 shortened de Bruijn sequences with linear complexity 4 (the M-sequences), four with linear complexity 12, and ten with linear complexity 14. For span 5, there are six M-sequences, ten shortened de Bruijn sequences with complexity 15, four with complexity 20, 306 with complexity 25, and 1722 with complexity 30. The distribution for span 6 shortened de Bruijn sequences is given in the following table:

complexity	number of sequences	complexity	number of sequences
6	6	45	28702
27	10	47	86056
30	8	48	134290
32	12	50	401102
33	8	51	453734
35	62	53	1364978
36	152	54	1819148
38	478	56	5453680
39	1036	57	3190982
41	3572	59	9555084
42	6100	60	11148860
44	17240	62	33441564

We will try now to explain the attainable values for the linear complexities of span n shortened de Bruijn sequences as were found by a computer search. Since any sequence of period π is generated by the LFSR_π, where $\pi = 2^n - 1$, whose feedback function is $x_\pi = x_0$, which is associated with the polynomial $f(x) = x^\pi - 1$, i.e., the PCR_π, it follows that the minimal polynomial that generates the sequence is a factor of $x^\pi - 1$. Since a shortened de Bruijn sequence S has span n, it follows that its linear complexity is at least n. Note further that since the weight of S is even, it follows that $(\mathbf{E}^{\pi-1} + \mathbf{E}^{\pi-2} + \cdots + \mathbf{E} + 1)S$ is the all-zeros sequence and hence its linear complexity is at most $\pi - 1$. Therefore we have the following theorem.

Theorem 6.1. *For $n \geq 4$, the linear complexity L of a span n shortened de Bruijn sequence satisfies*

$$n \leq L \leq 2^n - 2.$$

The number of irreducible polynomials of degree n over $\mathbb{F}_q$, $I_q(n)$, computed in Lemma 3.18 is used in the attainable linear complexities of span n shortened de Bruijn sequences.

Theorem 6.2. *For $n \geq 4$, a linear complexity L attained by span n shortened de Bruijn sequences satisfies*

$$L = \sum_{\substack{d \mid n \\ d \neq 1 \\ a_d \in \{0, 1, \dots, I_2(d)\}}} a_d \cdot d.$$

Proof.* A span n shortened de Bruijn sequence S has period $\pi = 2^n - 1$, and hence its minimal polynomial divides the polynomial $f(x) = x^\pi - 1$. By Theorem 1.23, the irreducible factors of $f(x)$ are all the irreducible polynomials over $\mathbb{F}_2$ of degree d, where d divides n. Hence, the minimal polynomial that generates S is formed from a multiplication of a subset of the irreducible polynomials that divides $f(x)$, except for $x + 1$ (since its associated sequence [1] will complement all the entries of the sequence and will make it an odd-weight sequence). Therefore the sum of the degrees of each such subset of irreducible polynomials can be the linear complexity of a shortened de Bruijn sequence. Thus the claim of the theorem follows. □

Corollary 6.1. *For $n \geq 4$, the linear complexity L attained by a span n shortened de Bruijn sequence, where n is a prime, satisfies $L \equiv 0$ (mod n).*

Proof. This is an immediate consequence from Theorem 6.2 since the only divisor of n different from 1 is n. □

Not all the linear complexities that are allowed by Theorem 6.2 occur, i.e., Theorem 6.2 is a necessary condition, but not a sufficient condition on the attainable linear complexities of shortened de Bruijn sequences. This is immediately observed from the computer search that was detailed. For example, for span 6 shortened de Bruijn sequences, all the values that are allowed by Theorems 6.1 and 6.2 starting from linear complexity 27, except for linear complexity 28, are attained. We expect a similar phenomenon for higher spans.

Problem 6.4. Analyze the attainable values for the linear complexities of shortened de Bruijn sequences and improve on the bounds of Theorems 6.1 and 6.2.

Problem 6.5. Prove (or disprove) that for $n > 3$, the linear complexity of most shortened span n de Bruijn sequences is greater than 2^{n-1}.

We turn now to consider the classification of de Bruijn sequences by the weights of their truth table. The two span 3 de Bruijn sequences have weight function 3. Twelve of the span 4 de Bruijn sequences have weight function 5 and four have weight function 7. For span 5, 576 sequences have weight function 7, 960 sequences have weight function 9, 448 have weight function 11, and 64 sequences have weight function 13. A table for the span 6 de Bruijn sequences is given below:

weight	number of sequences	weight	number of sequences
13	2211840	21	9912320
15	11059200	23	2637824
17	21086208	25	344064
19	19841024	27	16384

It should be noted that the number of sequences with minimum weight function and with maximum weight function, which are generated by merging the PCR_n

cycles and the CCR$_n$ cycles, respectively, are relatively small compared to the sequences with the other weight functions.

Some preliminary analysis of the weight functions of de Bruijn sequences will be provided now. For the next theorem, we will generalize the definition of the reversed function that was presented for linear FSRs in Section 2.1. The *reversed function* $f^R(x_1, x_2, \ldots, x_n)$ of a feedback function

$$f(x_1, x_2, \ldots, x_n) = x_1 + g(x_2, \ldots, x_n)$$

is the function defined by

$$f^R(x_1, x_2, \ldots, x_n) \triangleq x_1 + g(x_n, \ldots, x_2).$$

Theorem 2.6 is generalized in the next theorem, whose proof is identical to that of Theorem 2.6.

Theorem 6.3. *The sequences generated by the reversed function $f^R(x_1, \ldots, x_n)$ of a feedback function $f(x_1, x_2, \ldots, x_n)$ are the reverse sequences of those generated by the feedback function f.*

Theorem 6.4. *A de Bruijn sequence and its reverse sequence belong to the same weight class.*

Proof. Assume that S is a span n de Bruijn sequence with feedback function f and weight function w. Let f^R be the feedback function of the reverse sequence S^R. We have to consider two cases.

1. If $f(0, x_2, \ldots, x_n) = 0$, then $f^R(0, x_n, \ldots, x_2) = 0$.
2. If $f(0, x_2, \ldots, x_n) = 1$, then $f(1, x_2, \ldots, x_n) = 0$, i.e., $f^R(0, x_n, \ldots, x_2) = 1$.

These two cases imply that the weight function of f^R is also w. $\qquad\square$

Theorem 6.5. *A de Bruijn sequence and its complement sequence belong to the same weight class.*

Proof. Assume S is a span n de Bruijn sequence with a feedback function f and weight function w, in other words, the top half of the truth table has weight w and the bottom half has weight $2^{n-1} - w$. We also have that $f(x_1, x_2, \cdots, x_n) = b$ if and only if $f(\bar{x}_1, x_2, \cdots, x_n) = \bar{b}$. For the function $\bar{f}$ of the complement sequence $\bar{S}$ we have that if $f(x_1, x_2, \cdots, x_n) = b$, then $\bar{f}(\bar{x}_1, \bar{x}_2, \cdots, \bar{x}_n) = \bar{b}$. This implies that the weight of the bottom half of the truth table for $\bar{f}$ is the same as the weight of the bottom half of the truth table for f. Thus the weight function of $\bar{f}$ is w. $\qquad\square$

We already mentioned in Section 5.5 that for odd n there exist de Bruijn sequences whose reverses equal their complement (CR sequences) and hence we can only say that the number of sequences in each weight class is divisible by 2. When n is even, CR sequences cannot exist and hence the number of sequences in each class is divisible by four when n is even. Nevertheless, the experimental results show that the divisibility is by a much higher power of 2 and it is intriguing to find an explanation for this phenomenon that does not happen in the distribution of linear complexities where similar sequences were examined (see Section 5.5).

6.3 Classification of balanced sequences

In Section 6.1, the discussion was on sequences of length $2^n - 1$ that have 2^{n-1} *ones* and $2^{n-1} - 1$ *zeros*, where at the top of the hierarchy we had all the $\binom{2^n-1}{2^{n-1}-1}$ such sequences and at the bottom of the hierarchy we have the $\frac{\phi(2^n-1)}{n}$ M-sequences. In this section, we consider sequences of even length with half *ones* and half *zeros*. If the length of the sequences is 2^n, then the $\binom{2^n}{2^{n-1}}$ such sequences are at the top of the hierarchy and the $2^{2^{n-1}-n}$ span n de Bruijn sequences will be at the bottom of the hierarchy. If the length of the sequences is $\mu \cdot 2^\ell$, where μ is an odd integer and $\ell \geq 1$, then at the top of the hierarchy, we have all the $\binom{\mu \cdot 2^\ell}{\mu \cdot 2^{\ell-1}}$ such sequences and at the bottom of the hierarchy will be all the sequences in which each ℓ-tuple appears exactly μ times in a window of length ℓ. This type of sequence forms a natural generalization to the de Bruijn sequences.

We start by considering algorithms to generate the set of $\binom{2n}{n}$ binary sequences of length $2n$ and weight n. There is a simple encoding algorithm to order this set of sequences. Given a sequence $S = (s_1, s_2, \ldots, s_{2n})$, the next sequence $T = (t_1, t_2, \ldots, t_{2n})$ is computed as follows:

(E1) Find the largest ℓ, $1 \leq \ell \leq 2n - 1$, such that $s_\ell = 0$ and $s_{\ell+1} = 1$; if no such ℓ exists, then S is the last sequence.

(E2) Set $t_i := s_i$ for $1 \leq i \leq \ell - 1$; set $t_\ell := 1$; for the rest of the sequence $t_{\ell+1} \cdots t_n$ set the entries such that the total weight of the sequence is n and all the *ones* are consecutive at the end of the sequence.

The words of length $2n$ and weight n are generated with this algorithm in lexicographic order. The method used for this lexicographic order is called **enumerative coding**. This generation of all the binary sequences of length $2n$ and weight n is very efficient in obtaining the next sequence from the current one. This procedure is used to order all the $\binom{2n}{n}$ balanced words of length $2n$, but it can be also used to order the set of words of length n and any given weight w, $1 \leq w \leq n$. We can also find the ith sequence in this ordering and also compute the place of each sequence in this ordering. These procedures are also called **combinatorial index systems**. However, the procedures to find the ith sequence or to find the place of a sequence in this ordering are not efficient enough.

The following algorithm, known as Knuth's algorithm, generates balanced words (not all of them) more efficiently and can be used to generate the kth sequence in the list or to find the position of a given word.

Knuth's algorithm:

Given an arbitrary binary sequence $S = (s_1 s_2 \cdots s_{2n})$ of length $2n$.

(K1) If the sequence S is balanced, then set S_1 to be S and go to **(K3)**.

(K2) Starting from $i = 1$ flip the values of the bits from s_i to $\bar{s}_i$ until the sequence is balanced. Let S_1 be the generated sequence.

(K3) Let k be the last index for which s_k was flipped, where $k = 0$ if no bit was flipped. Encode k into a balanced sequence S_2, such that each k is encoded into a different balanced sequence S_2 of even length $n' \approx \log n$.

(K4) The output is the sequence $S_1 S_2$. ■

Theorem 6.6. Knuth's algorithm generates a balanced sequence of length $2n + n'$.

Proof. It is easily verified that to prove the claim of the theorem it is sufficient to show that if S is not a balanced sequence, then there exists a k such that the sequence $\bar{s}_1, \bar{s}_2, \ldots, \bar{s}_k, s_{k+1}, s_{k+2}, \ldots, s_{2n}$ is a balanced sequence.

Let $N_1(S)$ denote the number of *ones* in S and $N_0(S)$ denote the number of *zeros* in S. If S is not balanced, then either $N_0(S) > N_1(S)$ or $N_1(S) > N_0(S)$.

Assume w.l.o.g. that $N_1(S) < n$, i.e., $N_0(S) > N_1(S)$. Hence, for the complement sequence $\bar{S}$, $N_1(\bar{S}) = N_0(S) > N_1(S) = N_0(\bar{S})$. Flipping all the bits changes the number of *ones* from $N_1(S)$ in S to $2n - N_1(S) = N_0(S) = N_1(\bar{S})$. Each flip of a bit either increases the number of *ones* by one (and decreases the number of *zeros* by one) or decreases the number of *ones* by one (and increases the number of *zeros* by one). Since the process starts when $N_1(S) < n$ and ends when $N_1(\bar{S}) > n$, it follows that there exists a certain point, say k, when s_k is flipped and the sequence

$$\bar{s}_1, \bar{s}_2, \ldots, \bar{s}_k, s_{k+1}, s_{k+2}, \ldots, s_{2n}$$

is balanced. □

Theorem 6.7. *Two different sequences S and S' of length $2n$ are mapped by Knuth's algorithm to two different sequences of length $2n + n'$.*

Proof. If the last bit that is flipped by the algorithm is in a different entry, then by **(K3)** we have that S and S' are mapped into different sequences.

If the last bit that is mapped in both S and S' is k, then clearly the sequences obtained from S and S' in **(K2)** are different and hence S and S' are mapped into different sequences. □

Corollary 6.2. *Knuth's algorithm induces a one-to-one mapping from $\mathbb{F}_2^{2n}$ into a subset of $\mathbb{F}_2^{2n+n'}$ whose size is 2^{2n}.*

Definition 6.1. A binary cyclic sequence S of length $n = \mu \cdot 2^{\ell}$ is an (ℓ, μ)-BdB (for balanced de Bruijn) sequence, if each binary ℓ-tuple is contained exactly μ times as a window of length ℓ in the sequence.

An $(\ell, 1)$-BdB sequence is a span ℓ de Bruijn sequence and the set of all balanced sequences of even length n is the set of $(1, n/2)$-BdB sequences. Let $\mathcal{B}(\ell, \mu)$ be the set of all inequivalent (ℓ, μ)-BdB sequences.

Lemma 6.1. *A sequence S of length n is a $(2, \frac{n}{4})$-BdB sequence if and only if S is balanced and its derivative $\mathbf{D}S$ is balanced.*

Proof. Assume first that $S \in \mathcal{B}(2, \frac{n}{4})$. Since each 2-tuple is contained $\frac{n}{4}$ times in S, it follows that S is balanced. Derivating each 2-tuple of S yields $\frac{n}{2}$ occurrences of *zeros* and $\frac{n}{2}$ occurrences of *ones* in $\mathbf{D}S$. Hence, $\mathbf{D}S$ is balanced.

Assume now that both S and $\mathbf{D}S$ are balanced. Let x_{ij} be the number of pairs (i, j) in S, where $i, j \in \{0, 1\}$. Clearly, after an appearance of a pair $(0, 1)$ in S, a pair $(1, 0)$ appears in S before the next appearance of another pair $(0, 1)$ in S. Hence, we have

$$x_{01} = x_{10} \tag{6.1}$$

for any cyclic sequence. Since $\mathbf{D}S$ is balanced, it follows by counting the number of *ones* in $\mathbf{D}S$ and the number of *zeros* in $\mathbf{D}S$ that we have

$$x_{00} + x_{11} = x_{01} + x_{10}. \tag{6.2}$$

Since S is balanced, it follows by counting the number of *ones* in S and the number of *zeros* in S that we have

$$2x_{00} + x_{01} + x_{10} = 2x_{11} + x_{01} + x_{10}. \tag{6.3}$$

Note that each *zero* and each *one* is counted twice in (6.3). Solving the three equalities in Eqs. (6.1), (6.2), and (6.3), implies that

$$x_{00} = x_{11} = x_{01} = x_{10}. \qquad \square$$

Let $\beta(\ell, \mu)$ be the set of all (ℓ, μ)-BdB words, i.e., the distinct words of length $\mu \cdot 2^{\ell}$ contained in the cyclic sequences of $\mathcal{B}(\ell, \mu)$

Example 6.1. For $\ell = 3$ and $\mu = 2$, the following three words are contained in $\beta(3, 2)$:

$$(0001110100011101)$$
$$(0000100111101101) \;.$$
$$(0010011110110100)$$

However, these three words are associated with only two sequences of $\mathcal{B}(3, 2)$,

$$[0001110100011101]$$
$$[0000100111101101]$$

since the third word is a cyclic shift of the second word. ∎

It should be noted that in contrast to de Bruijn sequences, i.e., $(\ell, 1)$-BdB sequences and $(\ell, 1)$-BdB words where the difference in their number is a factor of 2^ℓ (the number of different cyclic shifts), the difference when $\mu > 1$ is slightly more difficult to compute. The number of sequences in $\mathcal{B}(\ell, \mu)$, i.e., (ℓ, μ)-BdB sequences and the number of words in $\beta(\ell, \mu)$, i.e., (ℓ, μ)-BdB words, can be found using a graph in a similar way to the computation of the number of de Bruijn sequences from the de Bruijn graph.

A *generalized de Bruijn graph* $\Gamma_{\ell,\mu}$ is a directed graph with $2^{\ell-1}$ vertices, which are the same vertices as in $G_{\ell-1}$ and $\mu \cdot 2^\ell$ edges, where each edge of $G_{\ell-1}$ is duplicated μ times to form μ parallel edges in $\Gamma_{\ell,\mu}$. In $\Gamma_{\ell,\mu}$ a Hamiltonian cycle is a span $\ell - 1$ de Bruijn sequence the same as in $G_{\ell-1}$, but Eulerian cycles in $\Gamma_{\ell,\mu}$ are different from those in $G_{\ell-1}$. The enumeration of the words in $\beta(\ell, \mu)$ and the sequences in $\mathcal{B}(\ell, \mu)$ will be done using the reverse spanning tree algorithm to form Eulerian cycles in $\Gamma_{\ell,\mu}$. Although the algorithm is not deterministic we can count the number of distinct words formed by the algorithm based on the distinct choices it has to perform. The reverse spanning tree algorithm, to generate sequences in $B(\ell, \mu)$, is similar to the one introduced in Section 4.1, to find the number of Eulerian cycles in a graph. The algorithm is written the same as the one written in Section 4.1 with two exceptions. Let r be the all-zeros vertex. The edge $r \to r$ is not marked when the algorithm starts. If $\mu > 1$, then the algorithm usually has many choices to choose which edge to traverse next, while previously the choice was deterministic throughout its execution. The algorithm is a very important tool in this section and hence it will be presented again. Moreover, some of the associated proofs given in Section 4.1 should be slightly modified.

The reverse spanning tree algorithm:

Let T be a reverse spanning tree in $\Gamma_{\ell,\mu}$ rooted at the vertex $r = (0^{\ell-1})$. Each edge of T is starred in $\Gamma_{\ell,\mu}$. All the edges of $\Gamma_{\ell,\mu}$ are set to be unmarked. Set the current vertex v to be r.

(T1) If the only unmarked out-edge of v is the starred edge $v \to u_1$, then set the current vertex $v := u_1$, and mark the edge $v \to u_1$; otherwise, mark an edge, either $v \to u_1$ (if one of the associated $\mu - 1$ un-starred edges is unmarked) or $v \to u_2$ (if one of these μ edges is unmarked) and set $v := u_1$ or $v := u_2$, respectively.

(T2) If there are unmarked out-edges of v, then go to **(T1)**; otherwise stop.

The output sequence is generated by considering the first bit of the consecutive edges on the generated path. ∎

Example 6.2. The graph $\Gamma_{3,2}$ is depicted in Fig. 6.3, where the starred edges correspond to the reverse spanning tree T with the set of edges.

$$\{\, 010,\ 110,\ 100 \,\}$$

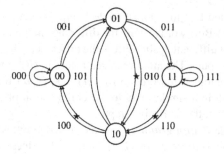

FIGURE 6.3 The graph $\Gamma_{3,2}$.

One of the paths generated by the reverse spanning tree algorithm when invoked with T is

$$00 \xrightarrow{000} 00 \xrightarrow{000} 00 \xrightarrow{001} 01 \xrightarrow{010} 10 \xrightarrow{100} 00 \xrightarrow{001} 01 \xrightarrow{011} 11 \xrightarrow{111} 11$$

$$\xrightarrow{111} 11 \xrightarrow{110} 10 \xrightarrow{101} 01 \xrightarrow{011} 11 \xrightarrow{110*} 10 \xrightarrow{101} 01 \xrightarrow{010*} 10 \xrightarrow{100*} 00,$$

which corresponds to the $(3, 2)$-BdB word (0000100111101101). ∎

We will prove the basic features of the algorithms as was proved in Section 4.1. The proof of the first lemma is identical to the proof of Lemma 4.2.

Lemma 6.2. *The path generated by the reverse spanning tree algorithm ends at the root r and all the in-edges and out-edges of r are traversed.*

The next lemma is the analog of Lemma 4.3 and its proof is essentially the same.

Lemma 6.3. *When the reverse spanning tree algorithm terminates, all the edges of the graph are traversed, i.e., all the edges of $\Gamma_{\ell,\mu}$ are marked.*

It follows from Lemma 6.3 that each reverse spanning tree T corresponds to some sequences of length $\mu \cdot 2^\ell$ in which each ℓ-tuple appears exactly μ times, i.e., an (ℓ, μ)-BdB word. Since the chosen root is $r = (0^{\ell-1})$, it follows that each such sequence starts with $\ell - 1$ consecutive *zeros*.

Lemma 6.4. *Each sequence of $\beta(\ell, \mu)$ that starts with $\ell - 1$ consecutive* zeros *is obtained by an algorithm from exactly one reverse spanning tree.*

Proof. Let $S \in \beta(\ell, \mu)$ be a sequence that starts with $\ell - 1$ *zeros*, and observe the cycle in $\Gamma_{\ell,\mu}$ that is associated with S. For each vertex $v \neq r$, its last appearance on the sequence S as an $(\ell - 1)$-tuple, is associated with an edge $v \rightarrow u$ in the graph $\Gamma_{\ell,\mu}$. Let E' denote the set of these edges; we will prove next that E' stands for a reverse spanning tree of $\Gamma_{\ell,\mu}$. Let $v_1, \ldots, v_{2^{\ell-1}-1}$ denote the start points of these edges, which do no not include the root r, ordered by the position of their last appearance in S, and denote $v_{2^{\ell-1}} = r$. It follows from the definition of E', the ordering of the vertices, and from the fact that the path associated with S ends at the root r (since it starts with $\ell - 1$ *zeros*), that E' contains only edges of the form $v_i \to v_j$, where $i < j$, which implies that E' contains no cycle. Thus given any vertex v_i, we can construct a unique path $v_i \to v_{j_1} \to v_{j_2} \to \cdots \to v_{j_{m-1}} \to v_{j_m}$, where $i < j_1 < j_2 < \cdots < j_{m-1} < j_m$, by iteratively traversing the single outgoing edge of each vertex. Since E' contains no cycles, this path must end at the root r.

Thus S can be obtained by the algorithm using the reverse spanning tree whose set of edges is E'. This tree is uniquely defined by the sequence and therefore S is obtained exactly once by the algorithm. $\qquad\square$

Lemma 6.5. *Each reverse spanning tree induces* $2\binom{2\mu-1}{\mu-1}^{2^{\ell-1}}$ *distinct sequences of* $\beta(\ell, \mu)$ *that are generated by the algorithm.*

Proof. At each vertex $v \neq 0^{\ell-1}$ there are two types of out-edges $v \to u_1$ and $v \to u_2$, μ un-starred edges of one type and $\mu - 1$ un-starred edges of the second type. These $2\mu - 1$ edges are chosen arbitrarily on the first $2\mu - 1$ times in which v is visited, for $\binom{2\mu-1}{\mu-1}$ distinct choices. Similarly, at the vertex $0^{\ell-1}$ there are $\binom{2\mu}{\mu} = 2\binom{2\mu-1}{\mu-1}$ distinct choices. There are a total of $2^{\ell-1}$ distinct vertices in $\Gamma_{\ell,\mu}$ and the claim of the lemma follows. $\qquad\square$

Corollary 6.3. *The total number of distinct* (ℓ, μ)-*BdB words generated by the algorithm is*

$$2^{2^{\ell-1}-\ell+1} \left(\frac{2\mu - 1}{\mu - 1}\right)^{2^{\ell-1}}.$$

Proof. This follows from Lemma 6.5 and the fact that by Corollary 4.4 there are $2^{2^{\ell-1}-\ell}$ reverse spanning trees in $\Gamma_{\ell,\mu}$. $\qquad\square$

Theorem 6.8. *For each positive integers* $\ell > 1$ *and* $\mu > 0$ *we have*

$$|\beta(\ell, \mu)| = 2^{2^{\ell-1}} \left(\frac{2\mu - 1}{\mu - 1}\right)^{2^{\ell-1}}.$$

Proof. Let $S \in \beta(\ell, \mu)$ be a word that starts with $\ell - 1$ *zeros* and consider the period of S.

Since the length of S is $\mu 2^{\ell}$ and each ℓ-tuple appears μ times in S, it follows by Corollary 3.1 that the period of S is $\lambda 2^{\ell}$, where λ divides μ. This implies that the sequence S (when considered as a sequence of length $\lambda 2^{\ell}$) is associated with exactly 2λ positions with $\ell - 1$ *zeros*. Hence, there are exactly 2λ shifts of S that yield distinct sequences in $\beta(\ell, \mu)$, each one starting with $\ell - 1$ *zeros* but in another point of the prefix of length $\lambda 2^{\ell}$ of S. These 2λ distinct sequences, generated by the algorithm, are associated with exactly $\lambda 2^{\ell}$ distinct words of

$\beta(\ell, \mu)$. Since, by Corollary 6.3, the algorithm generates $2^{2^{\ell-1}-\ell+1}\binom{2\mu-1}{\mu-1}^{2^{\ell-1}}$ sequences starting with $\ell - 1$ *zeros*, it follows that their number in $\beta(\ell, \mu)$ (compared to the number of distinct (ℓ, μ)-BdB sequences generated by the algorithm as was computed in Corollary 6.3) is multiplied by $\frac{\lambda 2^\ell}{2\lambda} = 2^{\ell-1}$. Thus

$$|\beta(\ell, \mu)| = 2^{2^{\ell-1}}\left(\frac{2\mu - 1}{\mu - 1}\right)^{2^{\ell-1}}.$$ $\square$

Theorem 6.9. *For each positive integers $\ell > 1$ and $\mu > 0$ we have*

$$|\mathcal{B}(\ell, \mu)| = \frac{2^{2^{\ell-1}-\ell}}{\mu} \sum_{\substack{i=1 \\ \lambda=\text{g.c.d.}(\mu,i)}}^{\mu} \left(\frac{2\lambda - 1}{\lambda - 1}\right)^{2^{\ell-1}}.$$

Proof. The number of words in $\beta(\ell, \mu)$ was computed in Theorem 6.8. Burnside's lemma (see Theorem 1.2) is applied on these $2^{2^{\ell-1}}\binom{2\mu-1}{\mu-1}^{2^{\ell-1}}$ sequences of $\beta(\ell, \mu)$, where the group G consists of the $\mu 2^\ell$ cyclic permutations on the sequences of $\beta(\ell, \mu)$. Sequences can be left "fixed" only by shifts that are multiples of 2^ℓ. The period of a sequence is $\lambda 2^\ell$, where λ is a divisor of μ. Hence, by Theorem 6.8 and Theorem 1.2 we have that

$$|\mathcal{B}(\ell, \mu)| = \frac{1}{\mu 2^\ell} \sum_{\substack{i=1 \\ \lambda=\text{g.c.d.}(\mu,i)}}^{\mu} |\beta(\ell, \lambda)| = \frac{1}{\mu 2^\ell} \sum_{\substack{i=1 \\ \lambda=\text{g.c.d.}(\mu,i)}}^{\mu} 2^{2^{\ell-1}}\left(\frac{2\lambda - 1}{\lambda - 1}\right)^{2^{\ell-1}},$$

which implies the claim of the theorem. $\square$

Let $\mathcal{P}(\ell, \mu)$ be the set of (ℓ, μ)-BdB sequences whose period is exactly $\mu \cdot 2^\ell$.

Theorem 6.10. *For each positive integers $\ell > 1$ and $\mu > 0$ we have*

$$|\mathcal{P}(\ell, \mu)| = \frac{1}{\mu \cdot 2^\ell} \sum_{d|\mu} \mu(d) \cdot 2^{2^{\ell-1}}\left(\frac{\frac{2\mu}{d} - 1}{\frac{\mu}{d} - 1}\right)^{2^{\ell-1}}.$$

Proof. The size of $\beta(\ell, \mu)$, $|\beta(\ell, \mu)|$, can be computed in two different ways. By Theorem 6.8 we have that

$$|\beta(\ell, \mu)| = 2^{2^{\ell-1}}\left(\frac{2\mu - 1}{\mu - 1}\right)^{2^{\ell-1}}.$$

When λ divides μ, a sequence whose least period is $\lambda 2^\ell$ can be extended to length $\mu 2^\ell$ and least period $\lambda 2^\ell$. Such a sequence contributes with its cyclic shifts $\lambda 2^\ell$ sequences to $\beta(\ell, \mu)$. Hence, we have

$$|\beta(\ell, \mu)| = \sum_{\lambda|\mu} \lambda 2^\ell |\mathcal{P}(\ell, \lambda)|.$$

Now, let $g(\mu) = 2^{2^{\ell-1}} \binom{2\mu-1}{\mu-1}^{2^{\ell-1}}$ and let $f(\mu) = \mu 2^\ell |\mathcal{P}(\ell, \mu)|$. By the Möbius inversion formula (see Theorem 1.10) we have

$$\mu 2^\ell |\mathcal{P}(\ell, \mu)| = \sum_{d|\mu} \mu(d) \cdot 2^{2^{\ell-1}} \binom{\frac{2\mu}{d}-1}{\frac{\mu}{d}-1}^{2^{\ell-1}},$$

which proves the claim of the theorem. $\qquad\square$

Example 6.3. Let $\ell = 2$ and $\mu = 2$. There are five $(2, 2)$-BdB sequences as follows:

$$[00011101]$$
$$[00011011]$$
$$[00010111]$$
$$[00100111]$$
$$[00110011]$$

and hence $|\mathcal{B}(2, 2)| = 5$. The first four sequence are in $\mathcal{P}(2, 2)$ and hence $|\mathcal{P}(2, 2)| = 4$. The eight shifts of these four sequences are distinct, while only four shifts of the last sequence are distinct since its period is 4. Thus $|\beta(2, 2)| = 4 \cdot 8 + 4 = 36$. $\qquad\blacksquare$

We have generated now a hierarchy between all the binary sequences of even length whose weight is half of their length. This hierarchy is based on the number of appearances of each window in the sequence. For $n = \mu \cdot 2^\ell$, where ℓ is odd we have that

$$\mathcal{B}(\ell, \mu) \subset \mathcal{B}(\ell-1, 2\mu) \subset \mathcal{B}(\ell-2, 4\mu) \subset \cdots \subset \mathcal{B}(2, 2^{\ell-2}\mu) \subset \mathcal{B}(1, 2^{\ell-1}\mu)$$

and the classification by this hierarchy is based on the containment between these sets.

Next, we present another classification for balanced sequences that is based on derivatives of balanced sequences.

The ith derivative of S, $S^{(i)}$, is the derivative of $S^{(i-1)}$ for $1 \leq i \leq k-1$, where $S^{(0)} = S$.

A sequence S is called k-*order balanced* for $k \geq 0$, if $S^{(i)}$ is a balanced sequences for each $1 \leq i \leq k-1$.

Lemma 6.6. *If S is an (ℓ, μ)-BdB sequence, then S is ℓ-order balanced.*

Proof. For every two positive integers ℓ' and μ' we have that an (ℓ', μ')-BdB sequence is a balanced sequence. As in the proof of Theorem 5.2, it is easy to verify that the derivative of an (ℓ, μ)-BdB sequence is an $(\ell - 1, 2\mu)$-BdB sequence. Thus S is an ℓ-order balanced sequence. $\qquad\square$

We can define a hierarchy for balanced sequences of even length that is based on its balanced order. Clearly, by definition, if a sequence is k-order balanced for some $k > 0$, then it is also r-order balanced for each $0 \le r < k$. By Lemma 6.1 the two hierarchies coincide for balanced sequences and also coincide for 2-order balanced sequences that by Lemma 6.1 are $(2, \frac{n}{4})$-BdB sequences. Lemma 6.6 implies that an (ℓ, μ)-BdB sequence is also an ℓ-order balanced sequence. Is the other direction also true? In other words, given an ℓ-order balanced sequence S is it also an (ℓ, μ)-BdB sequence? The answer is no for many sequences, as can be observed from the following example.

Example 6.4. Let $\ell = 3$ and $\mu = 3$, which implies that $n = 24$. The sequence

$$S = [000101111101001110010010]$$

is a balanced sequence. Similarly

$$\mathbf{D}S = [001110000111010010110110]$$

and

$$\mathbf{D}^2 S = [010010001001110111011010]$$

are balanced sequences. Therefore S is a 3-order balanced sequence. However, the 3-tuple 111 appears four times as a window of S, and hence S is not a $(3, 3)$-BdB sequence. $\qquad\blacksquare$

6.4 The depth of a word

The linear complexity of a sequence, which was defined and discussed in Section 5.1, was defined only for cyclic sequences, but the definition can be easily adapted for acyclic sequences (which will be referred to as words). For a word of length n, the recursion should be defined in a way that given the first k symbols as the initial state of length k, all the other $n - k$ bits can be computed, each one by the same linear function from the previous k bits. It should be noted that not always can such a linear function be found (for example when the word is $(0^{n-1}1)$). When there is no such linear function, the linear complexity will be defined to be n. However, we can adopt another concept to replace the linear complexity in the case of words. This concept that will be called the depth will use the operator $\mathbf{D}$. For binary words whose length is a power of 2, it coincides with the linear complexity. Some of the results that follow can be compared with results on the linear complexity, where the proofs are very similar, but there are some differences between the corresponding results. Moreover, we will also use this measure to classify words and also for the characterization of linear codes, as will be demonstrated in this section. In the following, we will use the letter c for a word as these words will be considered as codewords later in this section. We start to consider words and codes over $\mathbb{F}_q$ and after that, we will consider only binary words and codes.

Definition 6.2. The *depth* of a word c of length n, over $\mathbb{F}_q$, depth(c), is the smallest integer i such that $\mathbf{D}^i c = (0^{n-i})$, where $\mathbf{D}^i c = \mathbf{D}(\mathbf{D}^{i-1}c)$ for $i > 1$. If no such i exists, then the depth of c is defined to be n.

As an immediate consequence of the definitions, we have the following lemma that can be compared with Corollary 5.2, and Lemmas 5.4 and 5.19.

Lemma 6.7. *The depth of a word c of length n is i, $n > i$, if and only if $\mathbf{D}^{i-1}c = (\alpha^{n-i+1})$, for some $\alpha \in \mathbb{F}_q^*$. The depth of a word of length n is at most n.*

Proof. If $x = (x_1x_2\cdots x_k)$ is a nonzero word in $\mathbb{F}_q^n$ and $\mathbf{D}x = (0^{k-1})$, then $\mathbf{D}x = (x_2 - x_1, x_3 - x_2, \ldots, x_k - x_{k-1}) = (0^{k-1})$, which implies that $x_i = x_{i+1}$ for $1 \le i \le k-1$, i.e., $x_i = \alpha$, $1 \le i \le k$, for some nonzero $\alpha \in \mathbb{F}_q$. On the other hand, if $x = (\alpha^k)$, then $\mathbf{D}x = (0^{k-1})$. This implies that the depth of a word c of length n is i, $n > i$, if and only if $\mathbf{D}^{i-1}c = (\alpha^{n-i+1})$ for some $\alpha \in \mathbb{F}_q^*$.

Hence, if for some j and $\ell \ge 1$, $\mathbf{D}^j c = (0^\ell)$, then the depth of c is less than n. If no such j and ℓ exist, then by Definition 6.2 the depth of c is n. Thus the depth of a word of length n is at most n. $\square$

The next lemma is the analog of the second part in Lemma 5.5.

Lemma 6.8. *If c_1 is a word of length n and depth i, and c_2 is a word of length n and depth j, $j < i$, then $c = c_1 + c_2$ is a word with depth i.*

Proof. Since c_1 is of depth i, it follows by Lemma 6.7 that $\mathbf{D}^{i-1}c_1 = (\alpha^{n-i+1})$ for some nonzero $\alpha \in \mathbb{F}_q$. Since c_2 is of depth j, $j < i$, it follows by Definition 6.2 that $\mathbf{D}^{i-1}c_2 = (0^{n-i+1})$. Thus $\mathbf{D}^{i-1}(c_1 + c_2) = (\alpha^{n-i+1})$, and hence by Lemma 6.7 we have that $c = c_1 + c_2$ has depth i. $\square$

Lemma 6.9. *If c is a word of length n over $\mathbb{F}_q$ and α is a nonzero element of $\mathbb{F}_q$, then αc and c have the same depth.*

Proof. This is an immediate observation from the fact that by Definition 6.2 we have that $\mathbf{D}(\alpha c) = \alpha \mathbf{D}c$. $\square$

The immediate consequence of Lemmas 6.8 and 6.9 is the following corollary.

Corollary 6.4. *If $c_1, c_2, \ldots, c_k$ are nonzero words of length n and distinct depths then $c_1, c_2, \ldots, c_k$ are linearly independent.*

Proof. Assume, on the contrary, that a subset of r words of length n and different depths from $\{c_1, c_2, \ldots, c_k\}$ are linearly dependent. W.l.o.g. assume that $c_r = \sum_{i=1}^{r-1} a_i c_i$, where $a_i \in \mathbb{F}_q^*$, $1 \leq i \leq r-1$, and

$$\text{depth}(c_1) < \text{depth}(c_2) < \cdots < \text{depth}(c_{r-1}) < \text{depth}(c_r),$$

for some $3 \leq r \leq k$.

By Lemmas 6.8 and 6.9, we have that $\text{depth}(a_1 c_1 + a_2 c_2) = \text{depth}(c_2)$. Hence, $\text{depth}(a_1 c_1 + a_2 c_2) < \text{depth}(c_3)$, which implies by Lemmas 6.8 and 6.9 that $\text{depth}(a_1 c_1 + a_2 c_2 + a_3 c_3) = \text{depth}(c_3)$. Therefore when we continue by induction and have that

$$\text{depth}(c_r) = \text{depth}\left(\sum_{i=1}^{r-1} a_i c_i\right) = \text{depth}(c_{r-1}),$$

a contradiction. Thus $c_1, c_2, \ldots, c_k$ are linearly independent. $\qquad\square$

It is easily verified that the converse of Corollary 6.4 is not correct. For example, any two nonzero words c and $\bar{c}$ are linearly independent but have the same depth.

Lemma 6.10. *Let c_1 and c_2 be two words of length n and depth i over $\mathbb{F}_q$. If α is a primitive element in $\mathbb{F}_q$, then there exists an integer j, $0 \leq j \leq q-2$, such that $c_1 + \alpha^j c_2$ is of depth m, $m < i$ (note that in this lemma and its proof only the superscripts for α are powers of the element in the field, while other superscripts refer to the multiplicity of a symbol in a string).*

Proof. By Lemma 6.7 we have that $\mathbf{D}^{i-1} c_1 = (\beta_1^{n-i+1})$ for some nonzero $\beta_1 \in \mathbb{F}_q$ and $\mathbf{D}^{i-1} c_2 = (\beta_2^{n-i+1})$ for some nonzero $\beta_2 \in \mathbb{F}_q$. Let j_1 and j_2 be two integers such that $0 \leq j_1, j_2 \leq q-2$, $\beta_1 = \alpha^{j_1}$, and $-\beta_2 = \alpha^{j_2}$. Let j_3 be an integer such that $0 \leq j_3 \leq q-2$ and $j_1 \equiv j_2 + j_3 \pmod{q-1}$, which implies that $\alpha^{j_2} \alpha^{j_3} = \alpha^{j_1} = \beta_1$. Since $\alpha^{j_3} \alpha^{j_2} = \alpha^{j_1}$, it follows that

$$\mathbf{D}^{i-1}(c_1 + \alpha^{j_3} c_2) = \mathbf{D}^{i-1} c_1 + \mathbf{D}^{i-1}(\alpha^{j_3} c_2) = (\beta_1^{n-i+1}) + \alpha^{j_3}(\beta_2^{n-i+1})$$
$$= (\beta_1^{n-i+1}) - ((\alpha^{j_3} \alpha^{j_2})^{n-i+1}) = (\beta_1^{n-i+1}) - (\beta_1^{n-i+1}) = (0^{n-i+1})$$

and hence $c_1 + \alpha^{j_3} c_2$ has depth less than i. $\qquad\square$

Definition 6.3. Given a code (a set of words) C of length n, let D_i be the number of codewords in C of depth i. The integers $D_0, D_1, \ldots, D_n$ are called the **depth distribution** of C.

Theorem 6.11. *The depth distribution of the nonzero codewords of an $[n, k]_q$ code consists of exactly k nonzero values.*

Proof. By Corollary 6.4, the depth distribution of the nonzero codewords of an $[n, k]_q$ code consists of at most k nonzero values. Assume, on the contrary, that the depth distribution of the nonzero codewords of an $[n, k]_q$ code C consists of m, $m < k$ nonzero values. Let C_1 be the subcode of C that consists of the $q^m - 1$ nontrivial linear combinations of m nonzero codewords $c_1, c_2, \ldots, c_m$, where

$$\text{depth}(c_m) > \text{depth}(c_{m-1}) > \cdots > \text{depth}(c_2) > \text{depth}(c_1).$$

Let c be a nonzero codeword in $C \setminus C_1$ with the smallest depth. W.l.o.g. assume that $\text{depth}(c) = \text{depth}(c_i)$, for some $1 \leq i \leq m$. If α is a primitive element in $\mathbb{F}_q$, then clearly, $\alpha^j c_i + c$ is a codeword in $C \setminus C_1$ for all $0 \leq j \leq q - 2$. By Lemma 6.10 there exists an integer r, $0 \leq r \leq q - 2$, such that $\text{depth}(\alpha^r c_i + c) < \text{depth}(c)$, a contradiction to the assumption that c is a codeword with the smallest depth in $C \setminus C_1$. Thus the depth distribution of the nonzero codewords of C consists of exactly k nonzero values. $\square$

Corollary 6.5. *Let* $c_1, c_2, \ldots, c_k$ *be nonzero codewords with distinct depths in an* $[n, k]_q$ *code, where*

$$\text{depth}(c_k) > \text{depth}(c_{k-1}) > \cdots > \text{depth}(c_2) > \text{depth}(c_1).$$

If $i_j = \text{depth}(c_j)$*, where* $1 \leq j \leq k$*, then* $D_{i_m} = q^m - q^{m-1}$ *for* $1 \leq m \leq k$*.*

Some immediate consequences from Theorem 6.11 and Corollary 6.4 are the following results.

Corollary 6.6. *Any* k *codewords of an* $[n, k]_q$ *code with distinct nonzero depths can form a generator matrix of the code.*

We will concentrate now more on the binary case, although most of the results can be extended to $\mathbb{F}_q$. The following lemma can be readily verified, but its strength is beyond its simplicity.

Lemma 6.11. *If* $x = (x_1, x_2, \ldots, x_n)$ *is a binary word whose depth is* δ*, then the depth of the word* $(x_1, x_2, \ldots, x_n, b)$*,* $b \in \{0, 1\}$ *is either* δ *or* $n + 1$*. The depth of* $(x_1, x_2, \ldots, x_n, b)$ *is* δ *if and only if the depth of* $(x_1, x_2, \ldots, x_n, \bar{b})$ *is* $n + 1$*.*

Proof. If the depth of $x = (x_1, x_2, \ldots, x_n)$ is δ, then $\mathbf{D}^{\delta-1} x = (1^{n-\delta+1})$. This implies that for some $b \in \{0, 1\}$ we have $\mathbf{D}^{\delta-1}(xb) = (1^{n-\delta+2})$ and $\mathbf{D}^{\delta-1}(x\bar{b}) = (1^{n-\delta+1}0)$. Therefore $\mathbf{D}^{\delta}(xb) = (0^{n-\delta+1})$, i.e., the depth of $(x_1, x_2, \ldots, x_n, b)$ is δ. On the other hand, it implies that $\mathbf{D}^n(x\bar{b}) = (1)$, i.e., the depth of $(x_1, x_2, \ldots, x_n, \bar{b})$ is $n + 1$. $\square$

Similarly to Lemma 6.11 we have the related result for $\mathbb{F}_q$.

Lemma 6.12. *If* $x = (x_1, x_2, \ldots, x_n)$ *is a word over* $\mathbb{F}_q$ *whose depth is* δ*, then the depth of the word* $(x_1, x_2, \ldots, x_n, \beta)$*,* $\beta \in \mathbb{F}_q$ *is either* δ *or* $n + 1$*. For exactly one* $\beta \in \mathbb{F}_q$ *the depth of* $(x_1, x_2, \ldots, x_n, \beta)$ *is* δ *and for* $\gamma \in \mathbb{F}_q \setminus \{\beta\}$ *the depth of* $(x_1, x_2, \ldots, x_n, \gamma)$ *is* $n + 1$*.*

Example 6.5. Consider the word (00110) whose depth is 3. The depth of the word (001100) is 3, while the depth of the word (001101) is 6. The depth of the word (0011010) is 7. The depth of the word (00110101) is also 7, while the depth of the word (00110100) is 8, the same as its linear complexity. The linear complexity of the word (00110) is 4 (it is generated by the irreducible polynomial $x^4 + x^3 + x^2 + x + 1$), while its depth is 3. The linear complexity of the word (0011101), which is one of the shifts of a span 3 M-sequence, is 3, while its depth is 7. Another shift (1101001) has linear complexity 3, while its depth is 5. ∎

An important tool in the understanding of the properties of words of certain depths, and for using the depth, is an algorithm for computing the depth of a word. We will give the algorithm only for binary words. This algorithm is a generalization of the Games–Chan algorithm (see Section 5.1) for computing the linear complexity of a cyclic word of length 2^n. A generalization for $\mathbb{F}_q$, $q > 2$, is quite simple and will follow the lines of a generalization of the Games–Chan algorithm for $\mathbb{F}_q$ (the non-binary linear complexity algorithm). The algorithm that follows is presented recursively.

Algorithm depth:

Let $v = (v_1, v_2, \ldots, v_n)$ be a binary word of length n and let r be the largest integer such that $2^r < n$. Let

$$v' = (v_1, v_2, \ldots, v_{2^r})$$

and

$$u = (v_1 + v_{2^r+1}, v_2 + v_{2^r+2}, \ldots, v_{n-2^r} + v_n).$$

Compute a function $\delta(v)$ recursively as follows:

- if $v = (0^n)$, then $\delta(v) = 0$.
- if $v = (1^n)$, then $\delta(v) = 1$.
- if $u \neq (0^{n-2^r})$, then $\delta(v) = 2^r + \delta(u)$.
- if $u = (0^{n-2^r})$, then $\delta(v) = \delta(v')$.

The output of the algorithm is the depth $\delta(v)$. ∎

For the proof of the next theorem we will use the operators **R** and **L** (see Section 1.2).

Theorem 6.12. *If $v = (v_1, v_2, \ldots, v_n)$ is a binary word, then after algorithm depth is applied recursively we have that $\delta(v) = \mathrm{depth}(v)$.*

Proof. Let $v = (v_1, v_2, \ldots, v_n)$ be a binary word of length n.

If $v = (0^n)$, then obviously $\mathrm{depth}(v) = 0$ and if $v = (1^n)$, then obviously $\mathrm{depth}(v) = 1$.

Let r, u, and v' be defined as in the algorithm. We recall that by Definition 6.2, we have that $\mathrm{depth}(v) \leq n$, and by Lemma 6.7, we have that depth$(v) = \delta$ if and only if $(\mathbf{R} - \mathbf{L})^{\delta-1} v = (1^{n-\delta+1})$. Also, note that over $\mathbb{F}_2$ we have $(\mathbf{R} - \mathbf{L})^{2^m} = \mathbf{R}^{2^m} - \mathbf{L}^{2^m}$ since $\binom{2^m}{k}$ is even for $1 \le k \le 2^m - 1$ and the operators $\mathbf{L}$ and $\mathbf{R}$ commute. Moreover, $(\mathbf{R} - \mathbf{L})^{2^r} v = \mathbf{R}^{2^r} v - \mathbf{L}^{2^r} v = u$. Therefore clearly $u \ne (0^{n-2^r})$ if and only if depth$(v) > 2^r$ and hence we have that depth$(v) = 2^r + \text{depth}(u)$.

The last case is when $u = (0^{n-2^r})$. Clearly, $u = (\mathbf{R} - \mathbf{L})^{2^r} v = (0^{n-2^r})$ if and only if depth$(v) \le 2^r$ since v is obtained from v' by attaching $n - 2^r$ bits to the end of v'. By Lemma 6.11 it implies that the depth of v is equal to the depth of v'.

Thus by the recursive definition of the function $\delta(v)$ in algorithm depth we have that $\delta(v) = \text{depth}(v)$. $\qquad\qquad\Box$

If n is a power of 2, then the algorithm depth for computing the depth coincides with the Games–Chan algorithm for computing the linear complexity of a cyclic sequence whose length is n. Hence, we have

Corollary 6.7. *If v is a binary word of length 2^n, then its depth as an acyclic word equals its linear complexity as a cyclic word.*

Note that if the length of the word is not a power of two, then usually the depth and the linear complexity are not related.

We turn now to consider the depth distribution of linear codes that can be used as a way to classify linear codes. In all the following lemmas we consider only binary words and codes. The first lemma characterizes some of the properties of words with length 2^n (cyclic or acyclic) and certain depths (linear complexities). Some of these properties are well known (see Section 5.1) and all of them can be easily derived from algorithm depth for computing the depth of a word or the Games–Chan algorithm, although the definitions of depth and linear complexity seem different for acyclic words and cyclic words, respectively.

Lemma 6.13. *Let v be a word of length 2^n.*

(1) *v has depth 2^n if and only if v has odd weight.*
(2) *v has depth $2^i + 1$ if and only if v has the form $(X\bar{X}X\bar{X} \cdots X\bar{X})$, where X is a word of period 2^i.*
(3) *If the period of v equals its length, then v has weight two only if v has depth $\sum_{i=m}^{n-1} 2^i = 2^n - 2^m$, for some m, $0 \le m \le n - 1$.*

Proof.

(1) This follows from Lemma 5.6.
(2) This follows from Lemma 5.7.
(3) Let u be a word of length 2^k and weight two. If the period of u is smaller than it length, then its period is 2^{k-1}, in which case depth$(u) = 2^{k-1}$ by (1). If the period of u is 2^k, then by Corollary 6.7 and the Games–Chan algorithm we have that depth$(u) = 2^{k-1} + \text{depth}(L(u) + R(u))$, where $L(u)$ is the prefix of u of length 2^{k-1} and $R(u)$ is the suffix of u of length 2^{k-1}.

Moreover, $L(u) + R(u)$ has length 2^{k-1} and weight two. Thus we can continue by induction to prove that $\text{depth}(v) = \sum_{i=m}^{n-1} 2^i = 2^n - 2^m$. $\qquad\square$

Two binary words $u = (u_1, u_2, \ldots, u_n)$ and $v = (v_1, v_2, \ldots, v_n)$ are called *orthogonal* if $\sum_{i=1}^{n} u_i v_i = 0$.

Lemma 6.14. *If v is a nonzero word of length 2^n and depth i, $1 \le i \le 2^{n-1}$, and u is a word of length 2^n and depth $2^n + 1 - i$, then u and v are not orthogonal.*

Proof. We will prove by induction that for each i, $1 \le i \le 2^{n-1}$, any word of length 2^n and depth i is not orthogonal to any word of length 2^n and depth $2^n + 1 - i$. The basis of the induction is $i = 1$, where the only word of depth 1, is (1^{2^n}), and by Lemma 6.13(1), a word of length 2^n has depth 2^n if and only if it has odd weight and hence the claim follows.

Assume now that the claim is true for some i, $1 \le i \le 2^{n-1} - 1$, i.e., each word of length 2^n and depth i is not orthogonal to any word of length 2^n and depth $2^n + 1 - i$.

Let $v = (v_1, v_2, \ldots, v_{2^n})$ be a word of length 2^n and depth $i + 1$, and $u = (u_1, u_2, \ldots, u_{2^n})$ be a word of length 2^n and depth $2^n - i$. By definition

$$(E + 1)v = (v_1 + v_2, v_2 + v_3, \ldots, v_{2^n - 1} + v_{2^n}, v_{2^n} + v_1),$$

whose depth is i, and by Lemma 6.13(1) we have that the weight of u, i.e., $\sum_{j=1}^{2^n} u_i$ is even. Hence, there exists a word y such that $u = Dy = D\bar{y}$, where

$$y = \left(0 = \sum_{j=1}^{2^n} u_j, u_1, u_1 + u_2, u_1 + u_2 + u_3, \ldots, \sum_{j=1}^{2^n - 1} u_j\right)$$

has depth $n - i + 1$. We also have that

$$Ey = \left(u_1, u_1 + u_2, u_1 + u_2 + u_3, \ldots, \sum_{j=1}^{2^n - 1} u_j, \sum_{j=1}^{2^n} u_j\right).$$

By carefully multiplying $(E + 1)v$ by Ey we have that

$$((E + 1)v) \cdot (Ey) = ((E + 1)v) \cdot (E\bar{y}) = \sum_{j=1}^{2^n} v_j u_j.$$

Since u is of length 2^n, it follows by Corollary 6.7 that the depth of u is equal to its linear complexity. Moreover, a cyclic shift does not affect the linear complexity of a word, and hence $\text{depth}(u) = \text{depth}(Eu)$. Since $\text{depth}(u) = \text{depth}(Eu)$, it follows that $(E + 1)v$ and y are orthogonal if and only if v and u are orthogonal. However, by the induction hypothesis, we have that $(E + 1)v$ (whose depth is i) and y (whose depth is $2^n + 1 - i$) are not orthogonal and hence v and u are not orthogonal. $\qquad\square$

Recall that by Lemma 1.10, for an $[n, k]_q$ self-dual code we have that $n = 2k$ and hence by Theorem 6.11 if the length of a self-dual code is 2^n, then it has exactly $2^{n-1} + 1$ nonzero values in its depth distribution, for which one is $D_0 = 1$.

Corollary 6.8. *If* $\{D_{i_0}, D_{i_1}, \ldots, D_{i_{2^n-1}}\}$ *is the set of nonzero values of the depth distribution of a self-dual binary code of length* 2^n, *then for any two integers* j *and* m, $i_j + i_m \neq 2^n + 1$.

Corollary 6.9. *In a self-dual code of length* 2^n *we have* $D_0 = 1$ *and for each* i, $1 \leq i \leq 2^{n-1}$, *either* $D_i = 0$ *and* $D_{2^n+1-i} \neq 0$, *or* $D_i \neq 0$ *and* $D_{2^n+1-i} = 0$.

It is important to note in Corollaries 6.8 and 6.9 that the claim is true for any coordinate permutation of the code, since a coordinate permutation on the code will preserve its self-duality, although the depths of the codewords can be changed (in fact usually their depths will be changed).

The first-order Reed–Muller code is an $[2^n, n + 1, 2^{n-1}]$ code. This code is unique, i.e., all linear codes with the same parameters are equivalent to the first-order Reed–Muller code. Recall that the simplex code is a $[2^n - 1, n, 2^{n-1}]$ code that can be constructed by the $2^n - 1$ cyclic shifts of a span n M-sequence with the addition of the all-zeros word. The first-order Reed–Muller code is obtained from the simplex code by adding a parity bit to all the codewords and adding all their complements to the code. The Reed–Muller code is one of the more useful codes in practice.

Lemma 6.15. *For any given* n, *any generator matrix with* $n + 1$ *rows, where row* i, $1 \leq i \leq n$, *is any word of length* 2^n *and depth* $2^{i-1} + 1$, *and row* $n + 1$ *is the only word of length* 2^n *and depth* 1, *is a generator matrix of the* $[2^n, n + 1, 2^{n-1}]$ *first-order Reed–Muller code.*

Proof. By Corollary 6.4, all the $n + 1$ defined rows are linearly independent. By Theorem 6.11, the depths of the nonzero codewords are 1 and $2^j + 1$, $0 \leq j \leq n - 1$. Therefore by Lemma 6.13(2), the weights of each codeword that is neither the all-zeros codeword nor the all-ones codeword is 2^{n-1} and the lemma follows. $\square$

The Hamming code is the unique $[2^n - 1, 2^n - n - 1, 3]$ code whose parity-check matrix contains all the $2^n - 1$ nonzero column vectors of length n. This parity-check matrix can be generated also from n linearly independent shifts of a span n M-sequence. Such shifts exist by Theorem 2.22. It is the dual to the $[2^n - 1, n, 2^{n-1}]$ simplex code. This can be observed as follows. Let G be a generator matrix for the $[2^n - 1, n, 2^{n-1}]$ simplex code obtained from n linearly consecutive shifts of span n M-sequence. All the columns of G are distinct since each one is associated with a different window of length n in the M-sequence. Therefore the minimum number of columns from G that sum to the all-zeros vector is three, and hence G is a parity-check matrix of a $[2^n - 1, 2^n - n - 1, 3]$ code that is the Hamming code. Since G is the generator matrix or the simplex code and the parity-check matrix of the Hamming code, it follows that these two codes are dual. The extended Hamming code obtained by adding a parity bit to the codewords of the $[2^n - 1, 2^n - n - 1, 3]$ Hamming code is the unique $[2^n, 2^n - n - 1, 4]$ code. It is the dual code of the $[2^n, n + 1, 2^{n-1}]$ first-order Reed–Muller code.

Example 6.6. The following generator matrix in standard form

$$\begin{bmatrix} 1 & 0 & 0 & 0 & 1 & 1 & 0 \\ 0 & 1 & 0 & 0 & 1 & 0 & 1 \\ 0 & 0 & 1 & 0 & 0 & 1 & 1 \\ 0 & 0 & 0 & 1 & 1 & 1 & 1 \end{bmatrix}$$

is a basis for the $[7, 4, 3]$ Hamming code C. However, the four codewords that form this matrix are not of four distinct depths. The following generator matrix G is a basis for the same code C:

$$G = \begin{bmatrix} 1 & 0 & 0 & 0 & 1 & 1 & 0 \\ 0 & 1 & 0 & 0 & 1 & 0 & 1 \\ 0 & 1 & 0 & 1 & 0 & 1 & 0 \\ 1 & 1 & 1 & 1 & 1 & 1 & 1 \end{bmatrix}.$$

The four rows of this matrix have the following depths, depth$(1000110) = 6$, depth$(0100101) = 5$, depth$(0101010) = 2$, and depth$(1111111) = 1$. The depth distribution of C is $D_0 = 1$, $D_1 = 1$, $D_2 = 2$, $D_5 = 4$, and $D_6 = 8$. The dual code of C, $C^\perp$, is the $[7, 3, 4]$ simplex code and its generator matrix can be taken as

$$\begin{bmatrix} 1 & 1 & 0 & 1 & 1 & 0 & 0 \\ 1 & 0 & 1 & 1 & 0 & 1 & 0 \\ 1 & 0 & 1 & 0 & 1 & 0 & 1 \end{bmatrix}.$$

The three rows of this matrix have the following depths, depth$(1101100) = 6$, depth$(1011010) = 5$, and depth$(1010101) = 2$. The depth distribution of $C^\perp$ is $D_0 = 1$, $D_2 = 1$, $D_5 = 2$, and $D_6 = 4$.

By an appropriate permutation on the columns, a generator matrix for the simplex code can be taken as shifts of the span 3 M-sequence $[0010111]$ and it is written as follows:

$$\begin{bmatrix} 0 & 0 & 1 & 0 & 1 & 1 & 1 \\ 1 & 0 & 0 & 1 & 0 & 1 & 1 \\ 1 & 1 & 1 & 0 & 0 & 1 & 0 \end{bmatrix}.$$

The three rows of this matrix have the following depths, depth$(0010111) = 7$, depth$(1001011) = 5$, and depth$(1110010) = 6$. The depth distribution of this code is $D_0 = 1$, $D_5 = 1$, $D_6 = 2$, and $D_7 = 4$.

The generator matrix of the $[8, 4, 4]$ extended Hamming code, which is also a first-order Reed–Muller code, can be constructed by adding parity for the rows of G as follows:

$$G' = \begin{bmatrix} 1 & 0 & 0 & 0 & 1 & 1 & 0 & 1 \\ 0 & 1 & 0 & 0 & 1 & 0 & 1 & 1 \\ 0 & 1 & 0 & 1 & 0 & 1 & 0 & 1 \\ 1 & 1 & 1 & 1 & 1 & 1 & 1 & 1 \end{bmatrix}.$$

The code is self-dual and the four codewords of this matrix have the following depths, $\mathrm{depth}(10001101) = 6$, $\mathrm{depth}(01001011) = 5$, $\mathrm{depth}(01010101) = 2$, and $\mathrm{depth}(11111111) = 1$. The depth distribution of this code is $D_0 = 1$, $D_1 = 1$, $D_2 = 2$, $D_5 = 4$, and $D_6 = 8$. ∎

Lemma 6.16. *For any given n, any generator matrix with $2^n - n - 1$ rows that contains any word of length 2^n and depth i for each i, $1 \le i \le 2^n - 1$, $i \ne 2^n - 2^j$, for each j, $0 \le j \le n - 1$, as a row, is a generator matrix for the $[2^n, 2^n - n - 1, 4]$ extended Hamming code.*

Proof. By Corollary 6.4 we have that the $2^n - n - 1$ rows selected are linearly independent and hence by Lemma 6.8 and Theorem 6.11, these are exactly the depths of the codewords in the generated code. By Lemma 6.13(1) all these chosen rows have even weight and hence the code cannot have any codeword of weight 1 or 3. By Lemma 6.13(3), the chosen depths do not contain words of weight 2. Thus the code has minimum distance 4 and hence it is a $[2^n, 2^n - n - 1, 4]$ code. □

Similarly to Lemma 6.16, we can obtain the following lemma.

Lemma 6.17. *For any given n, any generator matrix with $2^n - n - 1$ rows that contains any word of length $2^n - 1$ and depth i, for each i, $1 \le i \le 2^n - 1$, $i \ne 2^n - 2^j$, for each j, $0 \le j \le n - 1$, as a row, is a generator matrix for the $[2^n - 1, 2^n - n - 1, 3]$ Hamming code.*

Proof. One can verify from the algorithm for computing the depth of a word, that a word of length $2^n - 1$ and weight either one or two has depth $2^n - 2^j$ for some j, $0 \le j \le n - 1$. The lemma follows now from Lemmas 6.8 and 6.13(3) since the linear span of the given generator cannot contain nonzero codewords with weights one or two. □

Example 6.7. The following generator matrix of the $[8, 4, 4]$ extended Hamming code has words with depths 1, 2, 3, and 5:

$$\begin{bmatrix} 0 & 1 & 0 & 1 & 0 & 1 & 0 & 1 \\ 0 & 0 & 1 & 1 & 0 & 0 & 1 & 1 \\ 0 & 0 & 0 & 0 & 1 & 1 & 1 & 1 \\ 1 & 1 & 1 & 1 & 1 & 1 & 1 & 1 \end{bmatrix}.$$

∎

6.5 Notes

Since the number of binary sequences of length n is 2^n and those of weight w and length n is $\binom{n}{w}$ it is very important to classify them for diverse applications. The classifications presented in this chapter represent only a small fraction of these classifications. We have concentrated mainly on classifications for de Bruijn sequences, shortened de Bruijn sequences, and balanced sequences, which are the center of interest in this book. For example, some classifications are associated with constrained coding. Balanced sequences are important in many applications such as optical recording, magnetic recording, etc. They have been subject to extensive research. In Section 6.3 these sequences were classified based on the multiplicity of their ℓ-tuples and their balanced derivatives. Both classifications are based on different constraints that the sequences should satisfy. Indeed, the property of exactly one occurrence of each n-tuple in a span n de Bruijn sequence is a type of constraint. However, there are other constraints of balanced sequences. These constraints can form a hierarchy between all the balanced sequences or subsets of these sequences. One of the most investigated constraints is the longest run of a symbol is called the **run-length** constraint. A second one is the **maximum accumulated charge** constraint, which is the absolute value of the difference between the number of *ones* and the number of *zeros* in any prefix of a word. For some of the techniques and applications of such codes, the reader is referred to the book of Immink [21]. Balanced codes that have the run-length constraint and the accumulated charge constraint were extensively studied and some examples are given in the work of Calderbank, Herro, and Telang [7], Etzion [13], and van Tilborg and Blaum [44]. Other interesting constrained balanced codes are the spectral-null codes of various orders that were extensively studied in Karabed and Siegel [24] and Roth, Siegel, and Vardy [38] and further studied in Skachek, Etzion, and Roth [39], and Tallini and Bose [41]. The window property required by de Bruijn sequences is also a type of constraint. A relaxed window property in local segments of the sequence was discussed and analyzed in Chee, Etzion, Kiah, Marcovich, Vardy, Vu, and Yaakobi [9].

Section 6.1. Classification of sequences of length $2^n - 1$ with 2^{n-1} *ones* and $2^{n-1} - 1$ *zeros* was carried out by Golomb [17] and most of the material in the section is based on this paper. The counterexample of length 127 of a sequence in $R \cap C$ that is not an M-sequence was found by Cheng and Golomb [10]. The classification depicted in Fig. 6.2 was further analyzed by Golomb [17], where more subclasses are demonstrated. More analysis, related sequences, and families of sequences that can be added to this hierarchy can be found in Bromfield and Piper [5], Chung and No [11], No, Golomb, Gong, Lee, and Gaal [36], and No, Yang, Chung, and Song [37].

Section 6.2. The linear complexities of shortened de Bruijn sequences were considered by Mayhew and Golomb [32], and later by Kyureghyan [27]. The material in this section for these linear complexities is based on the work of

Mayhew and Golomb [32], where also the complexities distribution of shortened de Bruijn sequences are taken from their paper. The minimum number of products in the sum of products form function of the FSR_n that generates the shortened de Bruijn sequence was considered in Mayhew and Golomb [33]. Classification of de Bruijn sequences by their weight functions was considered first by Mayhew [30] from which the table given in this section was taken. The weight function was extensively studied and analyzed in the papers of Hauge and Mykkeltveit [18,19]. More related tables were given in Mayhew [31]. Finally, another classification method was suggested by Chan and Games [8] who examined the quadratic span of de Bruijn sequences.

Problem 6.6. Continue to provide a comprehensive analysis of the number of sequences in the weight functions (especially for the minimum and maximum ones) of de Bruijn sequences.

Golomb [16] proved that all irreducible factors of the trinomial $x^{2^n+1} + x + 1$ have degrees dividing $3n$, and therefore, periods dividing $2^{3n} - 1$. He proved that all irreducible factors of $x^{2^n} + x + 1$ have degrees dividing $2n$, and therefore, periods dividing $2^{2n} - 1$. He also conjectured that all irreducible factors of $x^{2^n+1} + x^{2^{n-1}-1} + 1$ have degrees dividing $6(n - 1)$ and periods dividing $2^{6(n-1)} - 1$. Mills and Zierler [34] proved that the degree of every irreducible polynomial that divides $x^{2^n+1} + x^{2^{n-1}-1} + 1$ divides either $2(n - 1)$ or $3(n - 1)$, but not $n - 1$. Information on irreducible trinomials with degrees up to 1000 was given by Zierler and Brillhart [49,50]. Finally, if the degree of an irreducible trinomial is the exponent of a Mersenne prime, then the polynomial is a primitive trinomial since the order of a root of an irreducible polynomial of degree n must divide $2^n - 1$. Such trinomials were considered by Zierler [48]. More work on primitive trinomials can be found in Goldstein and Zierler [15].

Section 6.3. Balanced codes were extensively studied over the years, e.g., Al-Bassam and Bose [1], Alon, Bergmann, Coppersmith, and Odlyzko [2], Hollmann and Immink [20], Immink and Weber [22], Knuth [25], Roth, Siegel, and Vardy [38], Skachek, Etzion, and Roth [39], Tallini and Bose [40,41], Tallini, Capocelli, and Bose [42], and Weber and Immink [45]. The family of balanced codes is the most important subclass of constant-weight codes, see Agrell, Vardy, and Zeger [3] and Brouwer, Shearer, Sloane, and Smith [6]. Knuth's algorithm was presented in [25]. The algorithm was later improved and modified by Al-Bassam and Bose [1], Immink and Weber [22], Tallini and Bose [40], Tallini, Capocelli, and Bose [42], and Immink and Weber [45]. The enumerative coding method was presented by Cover [12].

The hierarchy that is based on the multiplicity of ℓ-tuples in a sequence and the classification based on the number of balanced derivatives was carried out by Marcovich, Etzion, and Yaakobi [29], where also the generalized de Bruijn graph with multiple edges was presented. The enumerations of the associated types of sequences were also carried out in that paper. Sequences with a multiplicity of ℓ-tuples were also extensively studied by Tesler [43], where other enumeration results were provided. Some of the results are for the same objects, but the equivalent formula is different.

Section 6.4. The depth distribution was defined and considered first by Etzion [14]. The definition was considered for infinite sequences of finite depth by Mitchell [35], where it was proved that these sequences correspond to a set of equivalence classes of rational polynomials. This work also characterizes infinite sequences of finite depth in terms of their periodicity. Finally, the depth distribution of all cyclic codes is given. Further results were obtained in Kai, Wang, and Zhu [23], Kong, Zheng, and Ma [26], Luo, Fu, and Wei [28], Yuan, Zhu, and Kai [46], and Zeng, Luo, and Gong [47]. The depth was generalized and analyzed for sequences of length $n = p^\beta$ with entries taken from $\mathbb{Z}_{p^\alpha}$ by Bar Yehuda, Etzion, and Moran [4] and the computation is performed in the ring $\mathbb{Z}_{p^\alpha}$.

In the same way that the linear complexity could be defined by the operators $\mathbf{D}$ and $\mathbf{E}$ and also by a polynomial representation, a similar definition can be given to the depth. Let α be an element in $\mathbb{F}_q$, let $c = (c_0, c_1, \ldots, c_{n-1})$ be a word of length n, and let $c(x) = \sum_{j=0}^{n-1} c_j x^j$ be the characteristic polynomial of c. We say that c has depth i, if i is the least integer such that

$$(x - \alpha)^i c(x) \equiv 0 \ (\text{mod} \ (x - \alpha)^n).$$

References

[1] S. Al-Bassam, B. Bose, On balanced codes, IEEE Trans. Inf. Theory 36 (1990) 406–408.

[2] N. Alon, E.E. Bergmann, D. Coppersmith, A.M. Odlyzko, Balancing sets of vectors, IEEE Trans. Inf. Theory 34 (1988) 128–130.

[3] E. Agrell, A. Vardy, K. Zeger, Upper bounds for constant-weight codes, IEEE Trans. Inf. Theory 46 (2000) 2373–2395.

[4] R. Bar Yehuda, T. Etzion, S. Moran, Rotating-table games and derivatives of words, Theor. Comput. Sci. 108 (1993) 311–329.

[5] A.J. Bromfield, F.C. Piper, Linear recursion properties of uncorrelated binary sequences, Discrete Appl. Math. 27 (1990) 187–193.

[6] A.E. Brouwer, J.B. Shearer, N.J.A. Sloane, W.D. Smith, A new table of constant weight codes, IEEE Trans. Inf. Theory 36 (1990) 1334–1380.

[7] A.R. Calderbank, M.A. Herro, V. Telang, A multilevel approach to the design of dc-line codes, IEEE Trans. Inf. Theory 35 (1989) 579–583.

[8] A.H. Chan, R.A. Games, On the quadratic spans of de Bruijn sequences, IEEE Trans. Inf. Theory 36 (1990) 822–829.

[9] Y.M. Chee, T. Etzion, H.M. Kiah, S. Marcovich, A. Vardy, V.K. Vu, E. Yaakobi, Locally-constrained de Bruijn codes: properties, enumeration, code constructions, and applications, IEEE Trans. Inf. Theory 67 (2021) 7857–7875.

[10] U. Cheng, S.W. Golomb, On the characterization of PN sequences, IEEE Trans. Inf. Theory 29 (1983) 600.

[11] H. Chung, J.-S. No, Linear span of extended sequences and cascaded GMW sequences, IEEE Trans. Inf. Theory 45 (1999) 2060–2065.

[12] T. Cover, Enumerative source encoding, IEEE Trans. Inf. Theory 19 (1973) 73–77.

[13] T. Etzion, Constructions of error-correcting DC-free block codes, IEEE Trans. Inf. Theory 36 (1990) 899–905.

[14] T. Etzion, The depth distribution – a new characterization for linear codes, IEEE Trans. Inf. Theory 43 (1997) 1361–1363.

[15] R.M. Goldstein, N. Zierler, On trinomial recurrences, IEEE Trans. Inf. Theory 14 (1968) 150–151.

[16] S.W. Golomb, Shift Register Sequences, Holden Day, San Francisco, 1967, Aegean Park, Laguna Hills, CA, 1980, World Scientific, Singapore, 2017.

[17] S.W. Golomb, On the classification of balanced sequences of period $2^n - 1$, IEEE Trans. Inf. Theory 26 (1980) 730–732.

[18] E.R. Hauge, J. Mykkeltveit, On the classification of de Bruijn sequences, Discrete Math. 148 (1996) 65–83.

[19] E.R. Hauge, J. Mykkeltveit, The analysis of de Bruijn sequences of non-extremal weight, Discrete Math. 189 (1998) 133–147.

[20] H.D.L. Hollmann, K.A.S. Immink, Performance of efficient balanced codes, IEEE Trans. Inf. Theory 37 (1991) 913–918.

[21] K.A.S. Immink, Codes for Mass Data Storage Systems, Shannon Foundation Publisher, Eindhoven, The Netherlands, 2004.

[22] K.A.S. Immink, J.H. Weber, Very efficient balanced codes, IEEE J. Sel. Areas Commun. 28 (2010) 188–192.

[23] X. Kai, L. Wang, S. Zhu, The depth spectrum of negacyclic codes over $\mathbb{Z}_4$, Discrete Math. 340 (2017) 345–350.

[24] R. Karabed, P.H. Siegel, Matched spectral-null codes for partial-response channels, IEEE Trans. Inf. Theory 37 (1991) 818–855.

[25] D. Knuth, Efficient balanced codes, IEEE Trans. Inf. Theory 32 (1986) 51–53.

[26] B. Kong, X. Zheng, H. Ma, The depth spectrum of constacyclic codes over finite chain rings, Discrete Math. 338 (2015) 256–261.

[27] G.M. Kyureghyan, Minimal polynomials of the modified de Bruijn sequences, Discrete Appl. Math. 156 (2008) 1549–1553.

[28] Y. Luo, F.-W. Fu, V.K.-W. Wei, On the depth distribution of linear codes, IEEE Trans. Inf. Theory 46 (2000) 2197–2203.

[29] S. Marcovich, T. Etzion, E. Yaakobi, On hierarchies of balanced sequences, IEEE Trans. Inf. Theory 69 (2023) 2923–2939.

[30] G.L. Mayhew, Weight class distribution of de Bruijn sequences, Discrete Math. 126 (1994) 425–429.

[31] G.L. Mayhew, Further results on de Bruijn weight classes, Discrete Math. 232 (2001) 171–173.

[32] G.L. Mayhew, S.W. Golomb, Linear spans of modified de Bruijn sequences, IEEE Trans. Inf. Theory 36 (1990) 1166–1167.

[33] G.L. Mayhew, S.W. Golomb, Characterizations of generators for modified de Bruijn sequences, Adv. Appl. Math. 13 (1992) 454–461.

[34] W.H. Mills, N. Zierler, On a conjecture of Golomb, Pac. J. Math. 28 (1969) 635–640.

[35] C.J. Mitchell, On integer-valued rational polynomials and depth distributions of binary codes, IEEE Trans. Inf. Theory 44 (1998) 3146–3150.

[36] J.-S. No, S.W. Golomb, G. Gong, H.-K. Lee, P. Gaal, Binary pseudorandom sequences of period $2^n - 1$ with ideal autocorrelation, IEEE Trans. Inf. Theory 44 (1998) 814–817.

[37] J.-S. No, K. Yang, H. Chung, H.-Y. Song, New construction for families of binary sequences with optimal correlation properties, IEEE Trans. Inf. Theory 43 (1997) 1596–1602.

[38] R.M. Roth, P.H. Siegel, A. Vardy, High-order spectral-null codes – constructions and bounds, IEEE Trans. Inf. Theory 40 (1994) 1826–1840.

[39] V. Skachek, T. Etzion, R.M. Roth, Efficient encoding algorithm for third-order spectral-null codes, IEEE Trans. Inf. Theory 44 (1998) 846–851.

[40] L.G. Tallini, B. Bose, Balanced codes with parallel encoding and decoding, IEEE Trans. Comput. 48 (1999) 794–814.

[41] T.L. Tallini, B. Bose, On efficient high-order spectral-null codes, IEEE Trans. Inf. Theory 45 (1999) 2594–2601.

[42] L.G. Tallini, R.M. Capocelli, B. Bose, Design of some new efficient balanced codes, IEEE Trans. Inf. Theory 42 (1996) 790–802.

[43] G. Tesler, Multi de Bruijn sequences, J. Comb. 8 (2017) 439–474.

[44] H. van Tilborg, M. Blaum, On error-correcting balanced codes, IEEE Trans. Inf. Theory 35 (1989) 1091–1095.

[45] J.H. Weber, K.A.S. Immink, Knuth's balanced codes revisited, IEEE Trans. Inf. Theory 56 (2010) 1673–1679.

[46] J. Yuan, S. Zhu, X. Kai, On the depth spectrum of repeated-root constacyclic codes over finite chain rings, Discrete Math. 343 (2020) 111647.

[47] M. Zeng, Y. Luo, G. Gong, Sequences with good correlation property based on depth and interleaving techniques, Des. Codes Cryptogr. 77 (2015) 255–275.

[48] N. Zierler, Primitive trinomials whose degree is a Mersenne prime, Inf. Control 15 (1969) 67–69.

[49] N. Zierler, J. Brillhart, On primitive trinomial (mod 2), Inf. Control 13 (1968) 541–554.

[50] N. Zierler, J. Brillhart, On primitive trinomial (mod 2), II, Inf. Control 14 (1969) 566–569.

Chapter 7

One-dimensional applications

Cryptography, verification, Gray codes, and games

In this chapter, we consider applications of the theory that was developed so far. These applications are for important technologies on the one hand and for theoretical problems and games on the other hand.

The first application, discussed in Section 7.1, was already known in the ancient world. A (long) message has to be sent from one party to a second party. Since the message can be obtained by a third malicious party, it follows that the message has to be encrypted so that when falling into the hands of a third party, he will not be able to find the source message. The idea is to encrypt the message by adding to it a long (de Bruijn) sequence that was agreed upon by the two parties. Is this encryption method secure? This will be discussed in this section.

Section 7.2 considers a completely different problem. Given a very large number of binary inputs, for an electronic chip (device), and from many relatively small subsets of them Boolean functions are defined for a chip to be designed for the device. We want to verify whether each such function is well functioning, i.e., it provides the required answer for each possible input. If each function is based on t inputs, then 2^t values have to be tested to check each possible input. This makes such a verification very expensive since the number of Boolean functions is very large. Can we decrease dramatically the number of tests required to check all the possible inputs for each Boolean function? Sequences generated by irreducible polynomials will help us to achieve this goal in this section.

Section 7.3 is devoted to one type of Gray code. Gray codes have many applications and for each application, they should satisfy certain properties. The codewords are organized in a matrix for which each codeword is a row (or a column) in the matrix. In this section, the requirement for the Gray code is that all the columns of the code (when the codewords are written as rows) will be cyclic shifts of the other columns. Surprisingly, their construction resembles constructions of full cycles. One of the constructions that has some similarities to the construction of de Bruijn sequences with minimal complexity seems to be very close to optimality. A proof of the nonexistence of a better code than this code is done using the linear complexity of sequences.

In Section 7.4 we will see how the concept of linear complexity can help to find a winning strategy for a rotating-table game with two players.

Sequences and the de Bruijn Graph. https://doi.org/10.1016/B978-0-44-313517-0.00013-5

7.1 Stream ciphers

In cryptography, one of the main questions is how to submit messages between two parties in a channel that is open to the public, in a way that any third party who sees the information sent on the channel, will not be able to reveal the original content of the message. There are many modern ways to do this task, but in this section we will concentrate on an old, simple, and efficient method.

Alice and Bob want to communicate in a secure channel. Alice wants to send Bob a very long message, using a key, which is agreed upon between her and Bob, from which the whole message is constructed. She wants to encode her message by adding to it bit by bit a very long text, constructed from the key, in a way that if Mallory, the malicious, who knows a small fraction of the text (or a small fraction of the key) will not be able to generate the whole text (and the original message). It is assumed that Mallory knows the method in which the message is encoded, but he is missing the key that is shared by Alice and Bob. If Mallory will know this key, he will be able to find the original message. The key, from which the whole sequence is constructed, is applied to generate the long text that is added to the original message. It is also assumed that there is a set with a large number of keys, all of them are known to Mallory, but he does not know which key Alice and Bob choose to use each communication. His target is to generate this key. The protocol starts when Alice generates the long text with the agreed key from which the long text will be generated.

Before they start their communication, the first step that Alice and Bob should take is to generate a large set of keys. They want the generated long text to be as random as possible so that Mallory cannot find the text. The number of keys should be also very large so that Mallory will not be able to try all of them to find the right one that Alice and Bob are using.

M-sequences are long sequences with interesting properties, where some properties (such as **R-1**, **R-2**, and **R-3**) imply that M-sequences behave like random sequences. Assume that the long text $s_0 s_1 \cdots$ was generated by an M-sequence whose shift-register feedback function (which is the given short key) is

$$x_{n+1} = f(x_1, x_2, \ldots, x_n) = \sum_{i=1}^{n} c_i x_{n+1-i}, \quad i \in \mathbb{F}_q.$$

Hence, the sequence $\{s_k\}$ satisfies the linear recursion

$$s_k = \sum_{i=1}^{n} c_i s_{k-i}, \quad k = n, n+1, \ldots.$$

We assume now that Mallory can find a short subsequence of the long text (i.e., the M-sequence), which was added to the message. If the given bits found by Mallory are $s_m, s_{m+1}, s_{m+2}, \ldots, s_{m+2n-1}$, for some m, he forms the following n

equations that are satisfied by any section of length $2n$ in the sequence

$$s_{m+j} = \sum_{i=1}^{n} c_i s_{m+j-i}, \quad n \le j \le 2n - 1. \tag{7.1}$$

It is easily verified that in Eq. (7.1) there are n linearly independent equations. The n equations have n variables (the c_is). Hence, this set of n equations has a unique solution for the c_is and once Mallory solves this set of equations and finds the c_is, the whole M-sequence can be revealed to Mallory. Using the revealed M-sequence, Mallory can decode the message that was sent on the channel and reveals the original message. This makes this system not secure from a cryptographic point of view and Alice and Bob will be advised not to use it. Therefore we would like to use another similar sequence that cannot be revealed from a small fraction of the sequence.

The next method is to use any span n de Bruijn sequence as the long text added to the source message. Generally, we cannot find the sequence from a small set of linear equations, since it was proved in Theorem 5.2 that the linear complexity of a span n de Bruijn sequence is larger than half of its length. A similar property on the linear complexities is shared by most of the shortened de Bruijn sequences (see Section 6.2). Moreover, these sequences have the desired randomness properties such as **R-1** and **R-2**. We have to guarantee not to use a sequence from the small set that has a low linear complexity. Such a large set of de Bruijn sequences can be any set generated by any algorithm to merge cycles as presented in Section 4.3. There should be a tradeoff between the parameters so that the set will be large enough. The method used in this case can be based on either the set of bridging states V in the algorithm to merge PCR_n cycles or on the set of bridging states U in the algorithm to merge CSR_n cycles. To generate the whole sequence, it seems that one must know the whole stored set V of bridging states (or the set U of bridging states, respectively) and this set can be made as large as we like to make this method relatively secured. Each set V, respectively U, of size k can be used to form about 2^k different keys, so we can control the length of the key and the number of keys associated with the number of generated sequences that will be added to the source message. Furthermore, the algorithms to generate the whole sequence from the key are simple and efficient, as was demonstrated and proved in Section 4.3.

7.2 VLSI testing

In a very large-scale integration (VLSI) an electronic chip (device) can have a few hundred inputs and a similar number of outputs, i.e., logic Boolean functions, each depending on a smaller number of input variables. Due to the large number of inputs involved and a large number of associated outputs, testing these chips for proper functioning is a formidable task. One approach is to configure the input register (which contains all the s inputs of the device) as a shift register during testing and to shift through it a sequence that exhausts all the possible inputs for each function on the chip. The results of the output sequences are then compared with the stored correct ones. This procedure is very expensive as s is very large and we have to shift through the register a sequence of length 2^s something that is completely not practical. Another method is to consider each function separately and if a function has t inputs, then a shift register of length t will be enough to test it. The cost of this procedure is proportional to the lengths of the test sequences and it tends to grow exponentially with the largest order of a function on the chip, where the order of a function is the number of variables on which it depends. For example, if each function depends on t inputs and there are η functions, then $\eta \cdot 2^t$ tests are required to check such a chip. This number can be very large (since η can be huge) and hence the verification process can be very costly. Finally, another method might be to order the inputs on the chip in a way that each function is based on variables that occupy no more than t consecutive positions. This time a span t de Bruijn sequence (or a span t M-sequence) can be shifted through the register, but this method requires to order the variables in a way that t is small as possible, something that might not be feasible. For these reasons, the design of the shortest possible test sequence is of great importance. The idea will be to use an FSR_n to generate a sequence S, which will be shifted through all the inputs and the positions associated with each function will contain each nonzero word during one of the shifts of S. The all-zeros sequence will be also shifted through all the inputs to examine the results associated with the all-zeros input. This will complete the process. If, for example, all the functions were confined to t consecutive inputs, then again a span t M-sequence S will do the work. Moreover, also any span t de Bruijn sequence will do this task. However, in practice, the positions of each function required for the chip are not consecutive on the chip and we consider the largest gap between the first input and the last input of a function. If this gap is of size $t - 1$, then the associated function is confined to t consecutive inputs of the chip and a span t sequence will be required. However, again, the length 2^t of such a sequence can be so huge as to make this approach impractical.

We continue to prove that in many cases a set of binary sequences obtained from irreducible polynomials and in particular M-sequences, not necessarily of the same length, can be used to verify the functionality the Boolean functions.

Definition 7.1. A sequence *exercises* a set of t register positions if and only if all nonzero t-tuples appear in those positions.

Definition 7.2. For $t < s$, a binary sequence is called (s, t)-*universal* if when shifted through a register of length s (where s is the total number of input variables), it exercises every subset of t register positions.

Definition 7.3. For a set $R = \{r_0, r_1, \ldots, r_{t-1}\}$ of t register positions, the *set polynomial* $g_R(x)$ is defined by

$$g_R(x) \triangleq \prod_{Q \subseteq R} \sum_{r_i \in Q} x^{r_i}.$$

We are given an irreducible polynomial $f(x) = x^n + \sum_{i=1}^{n} c_i x^{n-i}$, where $n \geq t$, and its associated sequence $A = a_0 a_1 a_2 a_3 \cdots$ (an M-sequence if the polynomial is primitive and a few sequences, of the same length, if the polynomial is irreducible and not primitive) that satisfies the recurrence

$$a_k = \sum_{i=1}^{n} c_i a_{k-i}, \qquad (7.2)$$

with the initial nonzero n-tuple $(a_{-n} a_{-n+1} \cdots a_{-1})$.

Consider all the possible shifts of the nonzero sequences generated by $f(x)$ as rows in a matrix B, which has $L = 2^n - 1$ rows, and let T be the $L \times n$ matrix that is formed by a projection of any n columns of B. Note that by Corollary 2.2 all the sequences have the same period and hence there is no ambiguity.

Lemma 7.1. *Every nonzero n-tuple appears as a row of the matrix T if and only if the columns of T are linearly independent.*

Proof. Assume first that each n-tuple appears as a row in T. This immediately implies that the n columns of T are linearly independent.

Assume now that the columns of T are linearly independent. Since each n-tuple appears as a window exactly once in one of the nonzero sequences generated by $f(x)$, it follows that every n consecutive columns of B contain each one of the $2^n - 1$ nonzero n-tuples as a row. Hence, the first n columns of B contain each nonzero n-tuple exactly once. These column vectors can be used as rows for the generator matrix of the $[2^n - 1, n, 2^{n-1}]$ simplex code. Each other column of B can be represented as a linear combination of the first n columns of B. This linear combination is defined by the recursion induced by $f(x)$ given in Eq. (7.2). Hence, all these linear combinations coincide with the codewords of the simplex code. This implies that every n linearly independent column contains each nonzero n-tuple as a row in T. $\qquad \square$

Lemma 7.2. *If Q is a nonempty subset of R and $q(x) = \sum_{r_i \in Q} x^{r_i}$, then $f(x)$ divides $q(x)$ if and only if the columns of B that are associated with the subset Q sum to* zero.

Proof. If the columns in B that are associated with the subset Q sum to *zero*, then one of the columns is a sum of the other columns, i.e., this column is a linear combination of the other columns. This linear combination is induced by the polynomial $f(x)$ and hence $q(\beta) = 0$, where β is a root of $f(x)$. Since we also have $f(\beta) = 0$ and $f(x)$ is an irreducible polynomial, it follows that $f(x)$ divides $q(x)$.

If $f(x)$ divides $q(x)$, then $f(\alpha) = 0$ implies that $q(\alpha) = 0$ and hence since the columns of B are defined by the recursion induced by $f(x)$, it follows that associated columns of B defined by Q sum to *zero*. $\qquad \square$

Theorem 7.1. *Given an irreducible polynomial $f(x)$ of degree n and a set R of t register positions, where $n \geq t$, the nonzero sequences generated by $f(x)$ exercises the set R if and only if $g_R(x)$ is not divisible by $f(x)$.*

Proof. The nonzero sequences generated by the polynomial $f(x)$ are now shifted through the input register. The patterns appearing across the given set $R = \{r_0, r_1, \ldots, r_{t-1}\}$ of register positions are given by

$$A_i = (a_{i+r_0}, a_{i+r_1}, \ldots, a_{i+r_{t-1}}), \quad i \geq 0,$$

where A is taken over all the nonzero sequences generated by $f(x)$. For simplicity, assume the A is an M-sequence (the same arguments will work for the sequences generated by any other irreducible polynomial). Since the period of the M-sequence is equal to $L = 2^n - 1$, it suffices to examine the first L such patterns A_i, $0 \leq i \leq L - 1$.

Consider now a new $L \times t$ matrix T, where $T_{ij} = a_{i+r_j}$ (this matrix is associated with the t positions that are checked). By Lemma 7.1 and since $n \geq t$, every nonzero t-tuple appears as a row in A if and only if the columns of T are linearly independent. The columns of T are linearly dependent if and only if a nonempty subset of the columns in T sums to *zero*.

By Lemma 7.2 we have that $f(x)$ divides the polynomial $\sum_{r_i \in Q} x^{r_i}$, where Q is a nonempty subset of R, if and only if the associated subset of columns of T sums to *zero*.

Since the polynomial $f(x)$ is irreducible, it follows that $f(x)$ divides the set polynomial $g_R(x)$ if and only if there exists a subset $Q \subseteq R$ such that $f(x)$ divides the factor $\sum_{r_i \in Q} x^{r_i}$ of $g_R(x)$. Hence, by Lemmas 7.1 and 7.2 the proof is completed. □

Corollary 7.1. *Let $\{f^{(i)}(x)\}_{i=1}^k$ be a set of k distinct irreducible polynomials (not necessarily of the same degree), let S be the concatenation of the associated nonzero sequences generated by these polynomials, and let R be a given set of positions in the register of length s. If $g_R(x)$ is not divisible by*

$$\prod_{i=1}^k f^{(i)}(x),$$

then S exercise R.

Proof. The set polynomial $g_R(x)$ is not divisible by the product of the k irreducible polynomials if and only if it is not divisible by at least one of them. By Theorem 7.1 the segment of S contributed by the sequences, for which the characteristic polynomial does not divide $g_R(x)$, exercises the given set R. □

Corollary 7.1 provides a sufficient condition that the sequence obtained by a concatenation of the distinct sequences, is an (s, t)-universal sequence. This condition implies the existence of a test sequence for verification of the functionality of the chip in practice. Note that for one M-sequence, to be an (s, t)-universal sequence, the condition of Theorem 7.1 is necessary and sufficient.

7.3 Single-track Gray codes

Originally, Gray codes were defined as Hamiltonian cycles in the well-known binary n-cube (hypercube). The binary n-cube is an undirected graph with 2^n vertices represented by the 2^n binary n-tuples. Two vertices $(x_1, x_2, \ldots, x_n)$ and $(y_1, y_2, \ldots, y_n)$ are connected by an edge if and only if they differ in exactly one position, i.e., their Hamming distance is one. The simplest Gray code is defined recursively as follows. Let $\mathcal{M}$ be a $2^n \times n$ binary matrix whose rows represent the codewords of a Gray code in the n-cube. Let $\mathcal{M}_0$ the $2^n \times (n + 1)$ be the matrix defined from $\mathcal{M}$ by appending a column of *zeros* before the first column of $\mathcal{M}$. Let $\mathcal{M}_1$ the $2^n \times (n + 1)$ be the matrix defined from $\mathcal{M}$ by appending a column of *ones* before the first column of $\mathcal{M}$. It is easy to verify that the $2^{n+1} \times (n+1)$ matrix obtained from the rows of $\mathcal{M}_0$ followed by the reverse order rows of $\mathcal{M}_1$ is a Gray code for the $(n + 1)$-cube. This Gray code is called the *reflected Gray code*.

Example 7.1. For $n = 2$ an order of the words of length 2 in a Gray code is by 00, 01, 11, 10. The 4×3 matrices $\mathcal{M}_0$ and $\mathcal{M}_1$ are given by

$$
\begin{array}{ll}
000 & 100 \\
001 & 101 \\
011 & 111 \\
010 & 110
\end{array}
$$

The associated reflected Gray code of order 3 is given by the 8×3 matrix formed by column concatenation of the first matrix with the reverse of the second matrix. The associated 8×3 reflected Gray code of order 3 can be also represented by the following 3×8 transposed matrix, where each column represents a codeword

$$
\begin{array}{l}
00001111 \\
00111100 \\
01100110
\end{array}
$$

∎

We will present now the formal definition of a Gray code and the necessary definitions for a single-track Gray code. A length n **Gray code** is an order of distinct binary n-tuples, called the codewords,

$$W_0, W_1, \ldots, W_{\pi-1}$$

226 Sequences and the de Bruijn Graph

having the property that any two adjacent codewords W_i and W_{i+1} differ in exactly one component. If this property holds for $W_{\pi-1}$ and W_0 as well, then we say that the Gray code is *cyclic* with *period* π. Otherwise, we say that the Gray code is *acyclic*. Originally, a Gray code was defined to contain all the 2^n binary n-tuple and was cyclic. However, over time, applications for such codes that do not require all the binary n-tuples, but require other properties based on the application, were introduced. Such a code can be introduced in two ways by a matrix. The first way is by a $\pi \times n$ matrix whose rows represent the codewords of the code. This is the representation that will be used more frequently in this section. The second way is by an $n \times \pi$ matrix whose columns represent the codewords.

As a typical example of an application, a length n, period π Gray code can be used to record the absolute angular positions of a rotating wheel by encoding (e.g., optically) the codewords on n concentrically arranged tracks. The rotating wheel has n reading heads, mounted in parallel across the tracks sufficient to recover the codewords. When the heads are nearly aligned with the division between two codewords, any components that change between those words will be in doubt, and a spurious position value may result. Such quantization errors are minimized by using a Gray encoding, for then exactly one component can be in doubt, and the two codewords that could result identify the positions bordering the division, resulting in a small angular error.

When a high resolution is required, the need for a large number of concentric tracks results in encoders with large physical dimensions. This poses a problem in the design of small-scale or high-speed devices. Single-track Gray codes were proposed as a way of overcoming these problems. Let C be a length n cyclic Gray code with codewords $W_0, W_1, \ldots, W_{\pi-1}$ and write $W_i = [w_i^0, w_i^1, \ldots, w_i^{n-1}]$, so that w_i^j denotes component j of codeword i. We call the sequence

$$t_j(C) \triangleq [w_0^j, w_1^j, \ldots, w_{\pi-1}^j]$$

of period π, the jth *track* of C.

Definition 7.4. Let C be a length n Gray code with codewords $W_0, W_1, \ldots, W_{\pi-1}$ and let s_i, $0 \le i < \pi - 1$, denote the unique component in which W_i and W_{i+1} differ, which implies that $0 \le s_i \le n - 1$. Then, the sequence

$$S = s_0, s_1, \ldots, s_{\pi-2}$$

is called the *coordinate sequence* of C. If C is cyclic, then the sequence

$$S = s_0, s_1, \ldots, s_{\pi-2}, s_{\pi-1},$$

where $s_{\pi-1}$ denotes the unique component in which $W_{\pi-1}$ and W_0 differ, is called the *cyclic coordinate sequence* of C.

Theorem 7.2. *Let $S = s_0 s_1 \cdots s_{\pi-1}$ be a sequence with terms from the integers $0, 1, \ldots, n - 1$. S is the coordinate sequence of a length n Gray code with $\pi + 1$ codewords if and only if, in every subsequence $s_i, s_{i+1}, \ldots, s_h$ of S with $0 \le i < h \le \pi - 1$, some symbol occurs an odd number of times. The sequence S is a cyclic coordinate sequence of a length n, period π Gray code if and only if, in every subsequence $s_i, s_{i+1}, \ldots, s_h$ of S with $0 \le i < h \le \pi - 2$, some symbol occurs an odd number of times, while in S itself, every symbol occurs an even number of times.*

Proof. First, consider the case of an acyclic coordinate sequence. Suppose that S is a coordinate sequence of a length n Gray code and that in some subsequence $s_i, s_{i+1}, \ldots, s_h$ of S, every symbol occurs an even number of times. Then, in obtaining codeword W_{h+1} from codeword W_i, we make an even number of changes to every component of W_i. Thus $W_{h+1} = W_i$, contradicting the distinctness of the words in a Gray code. On the other hand, suppose that S on the symbols $0, 1, \ldots, n - 1$ has the property that in every subsequence $s_i, s_{i+1}, \ldots, s_h$, some symbol occurs an odd number of times. Then, if we choose an arbitrary n-tuple W_0 and generate a list of n-tuples by interpreting s_i as the unique component in which consecutive n-tuples W_i and W_{i+1} differ, it is clear that $W_{h+1} \ne W_i$ for every choice of i and h. Thus the list of $\pi + 1$ n-tuples obtained from S and W_0 is a Gray code whose coordinate sequence is S.

A similar argument also applies to cyclic coordinate sequences, the only difference to note being that the property that every symbol occurs an even number of times in S guarantees that for any W_0, a cyclic Gray code is obtained. The reason is that a symbol appears in $s_i, s_{i+1}, \ldots, s_h$ an odd number of times if and only if it appears in $s_{h+1}, \ldots, s_{\pi-1}, s_0, \ldots, s_{i-1}$ an odd number of times. $\square$

Definition 7.5. A *single-track Gray code* is a list of π distinct binary words of length n, such that every two consecutive words, including the last and the first, differ in exactly one position and when looking at the list as a $\pi \times n$ array, each column of the array is a cyclic shift of the first column. In other words, there exist integers

$$k_0, k_1, \ldots, k_{n-1}$$

called the *head positions*, where $k_0 = 0$, such that

$$t_i(\mathcal{C}) = \mathbf{E}^{k_i} t_0(\mathcal{C})$$

for each $0 \le i \le n - 1$. For each i, $0 \le i \le n - 1$, k_i is called the *position of the ith head*.

Lemma 7.3. *If $S = s_0 s_1 \cdots s_{\pi-1}$ is the cyclic coordinate sequence of a length n, period π, single-track Gray code, then for each symbol j with $1 \le j < n$, the positions where symbol j occurs in S are cyclic shifts of the positions where* *symbol 0 occurs in S. Conversely, if S is a sequence with this property and the properties of a cyclic coordinate sequence given in Theorem 7.2, then there exists a choice of W_0 such that S is the cyclic coordinate sequence of a single-track Gray code with the first codeword W_0.*

Proof. In a single-track Gray code $\mathcal{C}$, the jth track $t_j(\mathcal{C})$, $1 \le j \le n - 1$, is a cyclic shift by some k_j of $t_0(\mathcal{C})$. Symbol j occurs in position i of the coordinate sequence S if and only if term i and term $i + 1$ of $t_j(\mathcal{C})$ differ, or equivalently, term $i + k_j$, and term $i + k_j + 1$ of $t_0(\mathcal{C})$ differ. In turn, this holds precisely when symbol 0 occurs in position $i + k_j$ of S. Thus we see that the positions where symbol j occurs in S are a shift, by k_j, of the positions where symbol 0 occurs in S.

Conversely, suppose S has this property and the properties of a cyclic coordinate sequence given in Theorem 7.2. Then, for any choice of W_0, S is a cyclic coordinate sequence of length n, period π, Gray code $\mathcal{C}$ whose first codeword is W_0. Also, the positions where changes occur in track $t_j(\mathcal{C})$ of this code are cyclic shifts by k_j of the positions where changes occur in $t_0(\mathcal{C})$. Thus $t_j(\mathcal{C})$ is equal to the cyclic shift by k_j of either $t_0(\mathcal{C})$ or the complement of $t_0(\mathcal{C})$. Whether or not the complement occurs for a specific j depends only on component j of W_0, i.e., on w_0^j. Therefore by an appropriate choice of W_0, we can ensure that $t_j(\mathcal{C})$ is equal to a cyclic shift of $t_0(\mathcal{C})$ for every j, where $1 \le j < n$. For this choice, the sequence S is a coordinate sequence of a single-track Gray code whose first codeword is W_0. $\square$

Lemma 7.4. *If there exists a length n, period π single-track Gray code $\mathcal{C}$, then π is an even multiple of n and $2n \le \pi \le 2^n$.*

Proof. Let S be the cyclic coordinate sequence of $\mathcal{C}$. Suppose that symbol 0 occurs ℓ times in S, where $\ell \ge 2$. By Theorem 7.2, ℓ is even, and by Lemma 7.3, every symbol $0, 1, \ldots, n - 1$, occurs ℓ times in S. Hence, $\pi = n\ell$ and π is an even multiple of n. On the other hand, $\mathcal{C}$ is a list of π distinct n-tuples so $\pi \le 2^n$. $\square$

Let $\mathcal{C}$ be a single-track Gray code of length n and period π. By Lemma 7.4, there is a theoretical possibility that $\pi = 2^n$, but then, necessarily, n is a power of 2. We are going to show that there is no such code whose period is larger than 4.

Theorem 7.3. *There is no ordering of all the 2^n binary words of length $n = 2^m$, $m > 2$, in a list that satisfies all the following requirements:*

1. *Each two adjacent words have a different parity.*
2. *The list has the single-track property.*
3. *Each word appears exactly once.*

Proof. Let us assume the contrary, i.e., let s be the track of a single-track code in which each n-tuple appears exactly once and each two adjacent words have different parity. Let $s(x)$ be the generating function of s and θ_1 the largest integer for which there exists a polynomial $p_1(x)$ that satisfies

$$s(x) \equiv (x-1)^{\theta_1} p_1(x) \pmod{x^{2^n} - 1}. \tag{7.3}$$

Let $k_0, k_1, \ldots, k_{n-1}$ be the head positions in the list, let

$$h(x) \triangleq \sum_{i=0}^{n-1} x^{k_i}$$

be the **head locator** polynomial in the list, and let h be the word of length 2^n associated with $h(x)$ considered as a generating function. Let θ_2 be the largest integer for which there exists a polynomial $p_2(x)$ that satisfies

$$h(x) \equiv (x-1)^{\theta_2} p_2(x) \pmod{x^{2^n} - 1}. \tag{7.4}$$

Since $x^{2^n} - 1 = (x-1)^{2^n}$, it follows that $0 \le \theta_1, \theta_2 \le 2^n - 1$. Since every two adjacent words have different parity, it follows that

$$(x-1)h(x)s(x) \equiv 1 + x + x^2 + \cdots + x^{2^n-1} \pmod{x^{2^n} - 1}. \tag{7.5}$$

Since $(x-1)^{2^n} = x^{2^n} - 1$ and

$$(x-1)^{2^n-1} \equiv 1 + x + x^2 + \cdots + x^{2^n-1} \pmod{x^{2^n} - 1},$$

it follows from Eqs. (5.1), (7.3), (7.4), and (7.5), that

$$\theta_1 + \theta_2 = 2^n - 2. \tag{7.6}$$

Eqs. (5.1), (7.3), (7.4), and (7.5) also imply that $\theta_1 + 2$ is the linear complexity of h, and $\theta_2 + 2$ is the linear complexity of s. These two claims are verified from the following two equations,

$$(x-1)h(x)(x-1)^{\theta_1} p_1(x) \equiv 1 + x + x^2 + \cdots + x^{2^n-1} \pmod{x^{2^n} - 1},$$
$$(x-1)s(x)(x-1)^{\theta_2} p_2(x) \equiv 1 + x + x^2 + \cdots + x^{2^n-1} \pmod{x^{2^n} - 1}.$$

Since each word appears in the list exactly once, it follows that s is not periodic, and hence by Corollary 5.3 we have that

$$\theta_2 \ge 2^{n-1} - 1. \tag{7.7}$$

If we assume that h is periodic, then

$$\{k_i\}_{i=0}^{n-1} = \{2^{n-1} + k_i\}_{i=0}^{n-1}$$

and therefore the ith word and the $(i + 2^{n-1})$th word have the same bit values (a permutation of each other). This implies that the all-zeros word, which appears somewhere in the list, appears twice in the list, a contradiction. Thus h is not periodic, and therefore by Corollary 5.3 we have that

$$\theta_1 \geq 2^{n-1} - 1.$$

Self-dual sequences of length 2^n have weight 2^{n-1} and since h has weight n, it follows that h is not self-dual when $2^n \geq 4$, and hence by Lemma 5.7 the linear complexity of h is not $2^{n-1} + 1$. Therefore

$$\theta_1 \geq 2^{n-1}. \tag{7.8}$$

Summing Eqs. (7.7) and (7.8) implies that

$$\theta_1 + \theta_2 \geq 2^n - 1,$$

in contradiction to Eq. (7.6). Thus no such single-track code with track s exists. $\square$

Corollary 7.2. *There does not exist a single-track Gray code of length n and period 2^n.*

Theorem 7.4. *Let $S_0, S_1, \ldots, S_{r-1}$ be r length n binary pairwise inequivalent full-order necklaces, such that for each i, $0 \leq i < r - 1$, S_i and S_{i+1} differ in exactly one coordinate, and there also exists an integer ℓ, where g.c.d.$(\ell, n) = 1$, such that S_{r-1} and $\mathbf{E}^\ell S_0$ differ in exactly one coordinate, then the following words (read row by row) form a length n, period nr single-track Gray code:*

S_0	S_1	$\cdots$	S_{r-1}
$\mathbf{E}^\ell S_0$	$\mathbf{E}^\ell S_1$	$\cdots$	$\mathbf{E}^\ell S_{r-1}$
$\mathbf{E}^{2\ell} S_0$	$\mathbf{E}^{2\ell} S_1$	$\cdots$	$\mathbf{E}^{2\ell} S_{r-1}$
$\vdots$	$\vdots$	$\vdots$	$\vdots$
$\mathbf{E}^{(n-1)\ell} S_0$	$\mathbf{E}^{(n-1)\ell} S_1$	$\cdots$	$\mathbf{E}^{(n-1)\ell} S_{r-1}$

Proof. Since ℓ is relatively prime to n, the integers $0, \ell, 2\ell, \ldots, (n-1)\ell$ are the distinct residues modulo n. It is then clear from the properties of the words $S_0, S_1, \ldots, S_{r-1}$ that the list of words in the statement of the theorem does form a cyclic Gray code. We need only to show that this code has the single-track property. Suppose that the words $S_0, S_1, \ldots, S_{r-1}$ are written in a vertical list to form an $r \times n$ binary array. Let $C_0, C_1, \ldots, C_{n-1}$ be the columns of this array. Then, it is easy to see that jth track of the code is $C_j, C_{j+\ell}, C_{j+2\ell}, \ldots, C_{j+(n-1)\ell}$ (with subscripts taken modulo n), formed by the concatenation of the columns in this order. In particular, track 0 is just $C_0, C_\ell, C_{2\ell}, \ldots, C_{(n-1)\ell}$ and contains all the columns in some order. Now, since ℓ and n are relatively primes, it follows that for every j we have that $j = m_j \ell$ modulo n for some m_j. Then, the jth track is the sequence

$$C_{m_j\ell}, C_{m_j\ell+\ell}, C_{m_j\ell+2\ell}, \ \ldots, C_{m_j\ell+(n-1)\ell},$$

which is simply a cyclic shift of track 0. Hence, the code is a single-track Gray code.   □

Example 7.2. For $n = 5$, the list of full-order necklaces [00001], [00011], [10011], [11011], [11010], [10010], satisfies the conditions of Theorem 7.4 and leads to a length 5, period 30, single-track Gray code. The code is written in a 5×30 array, where each column represents one codeword, as follows:

$$001111\ 000110\ 000000\ 011111\ 111100$$
$$000110\ 000000\ 011111\ 111100\ 001111$$
$$000000\ 011111\ 111100\ 001111\ 000110\ .$$
$$011111\ 111100\ 001111\ 000110\ 000000$$
$$111100\ 001111\ 000110\ 000000\ 011111$$

Track 0 equals

$$0, 0, 1, 1, 1, 1, 0, 0, 0, 1, 1, 0, 0, 0, 0, 0, 0, 0, 0, 1, 1, 1, 1, 1, 1, 1, 1, 1, 0, 0$$

and the coordinate sequence (counting components from the top by 0,1,2,3,4) is

$$3, 0, 1, 4, 1, 0, 2, 4, 0, 3, 0, 4, 1, 3, 4, 2, 4, 3, 0, 2, 3, 1, 3, 2, 4, 1, 2, 0, 2, 1. \ ■$$

It is believed that whenever n is a prime, there exists an arrangement of the $\frac{2^n-2}{n}$ full-order necklaces of length n into a single-track Gray code of length n and period $2^n - 2$. Such an arrangement can be found relatively easily by computer search up to $n = 19$.

Example 7.3. For $n = 7$, there are 18 full-order necklaces. These necklaces can be ordered as follows to satisfy the conditions of Theorem 7.4 and to lead to a length 7, period 126, single-track Gray code

S_0	=	[0000001]	S_9	=	[0110101]
S_1	=	[0000101]	S_{10}	=	[0110111]
S_2	=	[0001101]	S_{11}	=	[0100111]
S_3	=	[0001001]	S_{12}	=	[0100101]
S_4	=	[1001001]	S_{13}	=	[1100101] .
S_5	=	[1011001]	S_{14}	=	[1000101]
S_6	=	[1111001]	S_{15}	=	[1000111]
S_7	=	[1111101]	S_{16}	=	[0000111]
S_8	=	[0111101]	S_{17}	=	[0000011]

For a self-dual word $S = [X \ \bar{X}]$ we have that for each i, $\mathbf{E}^i S = [Y \ \bar{Y}]$ for some Y. Hence, we have the following lemma, which is an immediate consequence of the discussion on the CCR$_n$ in Chapter 3.

Lemma 7.5. *If S_1 and S_2 are two inequivalent full-order self-dual necklaces of length $2n$, then $2n$ distinct n-tuples appear as subsequences of consecutive bits in each of S_1 and S_2, while none of the n-tuples appearing in S_1 appear in S_2.*

Proof. The claim follows immediately from the observation that one n-tuple in the self-dual sequence of period $2n$ determines the whole self-dual sequence. $\square$

Lemma 7.5 leads to the following idea for constructing single-track Gray codes. The idea is described in the following theorem whose proof is analogous to the proof of Theorem 7.4.

Theorem 7.5. *Let $S_0, S_1, \ldots, S_{r-1}$ be r binary self-dual pairwise inequivalent full-order sequences of length $2n$. For each $i = 0, 1, \ldots, r - 1$, let $S_i = [s_i^0, s_i^1, \ldots, s_i^{2n-1}]$ and define*

$$\mathbf{F}^j S_i = [s_i^j, s_i^{j+1}, \ldots, s_i^{j+n-1}],$$

where superscripts are taken modulo $2n$.

If for each $0 \le i < r - 1$, S_i and S_{i+1} differ in exactly two coordinates, and there also exists an integer ℓ, where g.c.d.$(\ell, 2n) = 1$, such that S_{r-1} and $\mathbf{E}^\ell S_0$ differ in exactly two coordinates, then the following words form a length n, period $2nr$ single-track Gray code.

$\mathbf{F}^0 S_0$	$\mathbf{F}^0 S_1$	$\cdots$	$\mathbf{F}^0 S_{r-1}$
$\mathbf{F}^\ell S_0$	$\mathbf{F}^\ell S_1$	$\cdots$	$\mathbf{F}^\ell S_{r-1}$
$\mathbf{F}^{2\ell} S_0$	$\mathbf{F}^{2\ell} S_1$	$\cdots$	$\mathbf{F}^{2\ell} S_{r-1}$
$\vdots$	$\vdots$	$\vdots$ $\vdots$	
$\mathbf{F}^{(2n-1)\ell} S_0$	$\mathbf{F}^{(2n-1)\ell} S_1$	$\cdots$	$\mathbf{F}^{(2n-1)\ell} S_{r-1}$

Example 7.4. For $n = 5$, the list of full-order self-dual inequivalent necklaces [0000011111], [0100010111], [0100110110], satisfies the conditions of Theorem 7.5 and leads to a length 5, period 30, single-track Gray code. The code is written in a 10×30 array, where the first five entries of each column represent one codeword, as follows (but also any other 5 consecutive rows can represent

such a Gray code).

$$
\begin{array}{l}
000011\ 000000\ 001111\ 100111\ 111110 \\
011000\ 000001\ 111100\ 111111\ 110000 \\
000000\ 001111\ 100111\ 111110\ 000011 \\
000001\ 111100\ 111111\ 110000\ 011000 \\
\underline{001111\ 100111\ 111110\ 000011\ 000000} \\
111100\ 111111\ 110000\ 011000\ 000001 \\
100111\ 111110\ 000011\ 000000\ 001111 \\
111111\ 110000\ 011000\ 000001\ 111100 \\
111110\ 000011\ 000000\ 001111\ 100111 \\
110000\ 011000\ 000001\ 111100\ 111111
\end{array}
$$

Track 0 equals

$$0,0,0,0,1,1,0,0,0,0,0,0,0,0,1,1,1,1,1,0,0,1,1,1,1,1,1,1,1,0$$

and the coordinate sequence (counting components from the top by 0,1,2,3,4) is

$$1,4,1,0,3,0,4,2,4,3,1,3,2,0,2,1,4,1,0,3,0,4,2,4,3,1,3,2,0,2.$$

■

Problem 7.1. Provide a construction for a single-track Gray code of length p, p prime, and period $2^p - 2$, based on full-order necklaces or self-dual full-order necklaces of period $2p$, or show some p for which such a code does not exist. Similarly, we have the same problem for period $2^{2p} - 4$ with self-dual full-order necklaces of period $4p$.

We will now present a recursive construction based on Theorem 7.5. This construction can be compared with the construction of binary de Bruijn sequences of minimal complexity presented in Section 5.2. Let $S_0, S_1, \ldots, S_{r-1}$ be the set of all inequivalent full-order self-dual necklaces of length $2n$ and let $\mathcal{Y}(n)$ denote the set of 2^{n-1} elements consisting of the $2^{n-1} - 1$ words of the form $(1, y_1, \ldots, y_{n-1})$, where at least one of the y_is is a *zero*, together with the word (0^n). For each $S = [X, \bar{X}]$ of length $2n$ and for $Y \in \mathcal{Y}(n)$, let

$$S_Y = [Y\ \ X+Y\ \ \bar{Y}\ \ X+\bar{Y}].$$

The proof of the following lemma is along the same lines as the proof of Lemma 5.13.

Lemma 7.6. *The set of sequences*

$$\bigcup_{i=0}^{r-1} S_i(n),$$

where

$$S_i(n) = \bigcup_{Y \in \mathcal{Y}(n)} (S_i)_Y,$$

contains $2^{n-1}r$ inequivalent self-dual necklaces of length $4n$.

We will continue the construction, but restrict ourselves to n, which is a power of 2. When n is a power of 2, there are $r = \frac{2^n}{2n}$ inequivalent full-order self-dual necklaces of length $2n$ and these contain all the n-tuples as subsequences. Assume that $S_0, S_1, \ldots, S_{r-1}$, the set of all inequivalent self-dual necklaces of length $2n$, are arranged so that the following three properties hold:

(C1) For each i, S_i and S_{i+1}, where subscripts taken modulo r, differ in exactly two positions k and $k + n$.

(C2) Let $\text{diff}^*(S_i, S_{i+1})$ denote the first position on which S_i and S_{i+1} differ and let

$$\mathcal{D}_n \triangleq \{\text{diff}^*(S_i, S_{i+1}) : 0 \leq i < r - 2\}.$$

Then,

$$\mathcal{D}_n = \{0, 1, \ldots, n - 1\}.$$

(C3) $E(S_{r-2})$ differs in exactly two positions from S_0. More precisely, we require that

$$
\begin{aligned}
S_{r-2} &= [0^{n-4}10001^{n-4}0111] \\
S_{r-1} &= [0^{n-4}10011^{n-4}0110] \; \cdot \\
S_0 &= [0^{n-4}00011^{n-4}1110]
\end{aligned}
$$

Example 7.5. For $n = 8$, the 16 self-dual necklaces of length 16 are ordered below so that Properties **(C1)**, **(C2)**, and **(C3)** hold

S_0	=	[0000000111111110]	S_8	=	[1111000100001110]
S_1	=	[1000000101111110]	S_9	=	[1101000100101110]
S_2	=	[1000001101111100]	S_{10}	=	[1101100100100110]
S_3	=	[1100001100111100]	S_{11}	=	[0101100110100110]
S_4	=	[1100011100111000]	S_{12}	=	[0101100010100111]
S_5	=	[1101011100101000]	S_{13}	=	[0100100010110111]
S_6	=	[1101010100101010]	S_{14}	=	[0000100011110111]
S_7	=	[1111010100001010]	S_{15}	=	[0000100111110110]

■

Lemma 7.7. *For any $Y \in \mathcal{Y}(n)$ the list of words*

$$S(Y) = (S_0)_Y, (S_1)_Y, \ldots, (S_{r-1})_Y,$$

satisfies Property **(C1)**.

Proof. If X_i and X_{i+1} differ in exactly one position, then so do $[Y \ X_i + Y]$ and $[Y \ X_{i+1} + Y]$. This implies that $[Y \ X_i + Y \ \bar{Y} \ X_i + \bar{Y}]$ and $[Y \ X_{i+1} + Y \ \bar{Y} \ X_{i+1} + \bar{Y}]$ differ in exactly two positions that are $2n$ positions apart. Since this is true for each i, $0 \le i \le r - 1$, it follows that the list $S(Y)$ satisfied Property **(C1)**. □

Lemma 7.8. *If Y and Z differ in exactly one position d and $diff^*(S_i, S_{i+1}) = d$, then the list of words*

$$(S_i)_Y, (S_{i+1})_Z, (S_{i+2})_Z, \ldots, (S_{r-1})_Z, (S_0)_Z, \ldots, (S_i)_Z, (S_{i+1})_Y,$$

*satisfies Property **(C1)** above. The first and the last pair of words differ only in positions d and $d + 2n$, while for every $n \le d' < 2n$, some pair of consecutive words in the list differ only in positions d' and $d' + 2n$.*

Proof. If $S_i = [X_i \ \bar{X}_i]$, where X_i and X_{i+1} differ in exactly position d and Y and Z also differ exactly in position d, then $X_i + Y = X_{i+1} + Z$ and hence $[Y \ X_i + Y]$ and $[Z \ X_{i+1} + Z]$ differ exactly in position d. Similarly, $[Z \ X_i + Z]$ and $[Y \ X_{i+1} + Y]$ differ exactly in position d. The statement about position n up to $2n - 1$ follows from the construction of the words $(S_j)_Z$ and Property **(C2)** of $S_0, S_1, \ldots, S_{r-1}$. □

Lemma 7.9. *If the set of self-dual necklaces of length $2n = 2^{m+1}$ can be arranged to satisfy Properties **(C1)**, **(C2)**, and **(C3)**, then so can the set of self-dual sequences of length $4n$.*

Proof. We start by forming the list of words $S(Y)$ for $Y \in \mathcal{Y}(n)$, as in Lemma 7.7. Next, we merge these lists in the set $\{S(Y)\}$ using Lemma 7.8. We order the words of $\mathcal{Y}(n)$ as follows: we take $Y_0 = (0^n)$, $Y_j = (1^j 0^{n-j})$, $1 \le j \le n - 1$, and $Y_n = (10^{n-2}1)$. Then, we order the remaining words of $\mathcal{Y}(n)$ so that each Y_i, $i \ge n$, differs in exactly one position from some Y_j, $j < i$. Note that for $1 \le j < n$, Y_j differs from Y_{j-1} in position $j - 1$, while Y_n differs from Y_1 in position $n - 1$.

Given an initial list $S(Y_0)$, assume that the lists $S(Y_1), S(Y_2), \ldots, S(Y_{\ell-1})$ have been successfully inserted into the main list. We will show now that $S(Y_\ell)$ can also be introduced.

Now, there exists a word Y_j with $j < \ell$ such that Y_j and Y_ℓ differ in exactly one position, say d, and, for some $0 \le i < r - 2$, there exist a pair of words $S_i = [X_i, \bar{X}_i]$ and $S_{i+1} = [X_{i+1}, \bar{X}_{i+1}]$ such that X_i and X_{i+1} also differ in exactly position d. We claim that the words

$$S_{iY_j} = [Y_j \ X_i + Y_j \ \bar{Y}_j \ X_i + \bar{Y}_j]$$

and

$$S_{(i+1)Y_j} = [Y_j \ X_{i+1} + Y_j \ \bar{Y}_j \ X_{i+1} + \bar{Y}_j]$$

still lie adjacent in the main list. For if not, then some list $S(Y_m)$, $m < \ell$, must have been inserted between them. This only occurs if Y_j and Y_m differ in exactly position d. This, in turn, implies that $Y_\ell = Y_m$, a contradiction, since these words are distinct. Therefore we can insert a cyclic shift of the code $\mathcal{S}(Y_\ell)$ between the words S_{iY_j} and $S_{(i+1)Y_j}$ using Lemma 7.8, extending the main list.

Applying this process, beginning with $\mathcal{S}(Y_1)$ and ending with $\mathcal{S}(Y_{2^{n-1}-1})$, we obtain a list of all $2^{n-1}r$ inequivalent self-dual necklaces that obviously satisfy Property (C1).

Observe that in the above procedure, we never insert any words in positions between the last two words and the first word of the initial list $\mathcal{S}(Y_0)$. These three words are

$$[0^{2n-4}10001^{2n-4}0111]$$
$$[0^{2n-4}10011^{2n-4}0110] \ .$$
$$[0^{2n-4}00011^{2n-4}1110]$$

Thus these words remain the last two words and the first word of the final list, so that the final list satisfy Property (C3).

Examining the last list inserted, we see that Lemma 7.8 guarantees that there are pairs of consecutive words in the list that differ in positions n up to $2n - 1$. Moreover, from the choice of words $Y_0, \ldots, Y_n$ and Lemma 7.8, there are pairs of consecutive words in the list that differ in positions 0 up to $n - 1$. Hence, Property (C2) is satisfied. □

An immediate consequence from Example 7.5 and Lemma 7.9 is the following theorem.

Theorem 7.6. *For every $m \geq 3$, there exists an arrangement of the self-dual sequences of length $2n = 2^{m+1}$ satisfying Properties (C1), (C2), and (C3).*

For $m \geq 3$ and $n = 2^m$, let the list of words in Theorem 7.6 be $S_0, S_1, \ldots, S_{r-1}$, where $r = \frac{2^n}{2n}$. Consider the list $S_0, S_1, \ldots, S_{r-2}$. Now, for each $0 \leq i < r - 2$, S_i and S_{i+1} differ in exactly two positions, while $\mathbf{E}S_{r-2}$ differs in exactly two positions from S_0. Thus Theorem 7.5 applies, with $j = 2n - 1$, to show:

Theorem 7.7. *If n is a power of 2, $n \geq 8$, then there exists a single-track Gray code of length n and period $2^n - 2n$.*

7.4 Rotating-table games

The next application that we consider shows that linear complexity can be used to have a winning strategy in games. Consider the following game of two players Alice and Bob seated by a rotating round table. The game starts when Alice (the *adversary*) puts n drinking glasses evenly spaced on the edge of the table, such that some of the glasses are in the upright position, and others are upside down. The goal of Bob is to set all the glasses in the upright position, while Alice tries to prevent him from doing so. The first round of the game starts when Bob points to some of the glasses, and asks Alice to invert them. Next, Alice rotates the table, and then she inverts the glasses at the locations pointed out by Bob (which might be different glasses, from those that were pointed out by Bob). This completes the first round. The second round starts similarly, by having Bob select a subset of the glasses to invert, and so forth.

We can view the n glasses on the table as a cyclic binary sequence of length n of *zeros* (for upright glasses) and *ones* (for upside down). We can view the instructions of Bob also as a binary sequence, where *one* means to invert the glass in the associated position and *zero* means to leave the glass as is. It can be viewed as a cyclic sequence since the table is rotated that has the same effect as having a cyclic shift in the positions of the glasses (or a cyclic shift on the sequence chosen by Bob). The game continues in rounds as before, where at the ith round ($i \geq 1$), the position of the glasses on the table is represented by a sequence W_{i-1}, and the new positions of the glasses are generated as a binary sequence W_i, as follows:

1. Bob gives Alice a binary word called key_i. This word denotes which glasses should be inverted after the table is rotated.
2. Alice selects an integer α_i in the range between 0 and $n - 1$. This integer α_i indicates that the table should be rotated by $\frac{360\alpha_i}{n}$ degrees. This implies that $W_i = W_{i-1} + \mathbf{E}^{\alpha_i} key_i$.

Bob wins the game if he can force Alice to generate the sequence $[0^n]$. For which values of n does Bob has a winning strategy, and when he has a winning strategy, what is the number of rounds required for his win? The game has two versions: an open game and a blind game. Up to now, the open game has been described. The blind game is the same as the open game, with one important exception: Bob is blindfolded from the very beginning of the game. He should point to the glasses that he wants to invert by sending a sequence of length n.

The blind game can be also described in a different way, which is more convenient to handle. Since Bob receives no information during the game, the sequence $(key_1, key_2, \ldots, key_m)$ that he generates during the game depends only on the number of glasses n. Therefore we can describe the blind game as a *one* player game, in which Alice plays against the sequence KEY $= (key_1, key_2, \ldots, key_m)$ as follows:

- Initially, the sequence KEY is given to Alice.
- Using this sequence, Alice generates the sequence $S = W_0, W_1, \ldots, W_m$ as follows:
  1. Choose an arbitrary vector as W_0.
  2. Given W_{i-1}, then select an integer α_i in the range $[0, 1, \ldots, n-1]$, and set $W_i = W_{i-1} + \mathbf{E}^{\alpha_i} key_i$.
- Alice loses the game if one of the W_is is the sequence $[0^n]$.

A sequence KEY is a ***universal sequence*** if Alice must lose the game when playing against this sequence.

Lemma 7.10. *If Bob has a winning strategy, then the number of glasses n is a power of 2.*

Proof. Assume first that n is odd. Alice can start with the sequence $[10^{n-1}]$ (the initial status of the glasses on the table) and W_{i-1} has the subsequences 01 and 10. Since n is odd, it follows that key_i has the subsequence 00 or the subsequence 11. This implies that Alice can have a 10 subsequence and a 01 subsequence in W_i, if W_{i-1} has such subsequence. Thus Bob cannot have a winning strategy.

Assume now that $n = 2^m k$, where $k > 1$ is an odd integer. Consider now 2^m words of length k obtained from W_{i-1}, where the entries in the jth word, $0 \leq j \leq 2^m - 1$, are those that were in the same position j modulo 2^m in W_{i-1}. Consider similarly 2^m words of length k obtained from key_i, where the entries in the jth word, $0 \leq j \leq 2^m - 1$, are those that were in the same place j modulo 2^m in key_i. Alice starts again with the sequence $[10^{n-1}]$, where the *one* is in position 0. Consider now the words of length k obtained from W_{i-1} and key_i. A word obtained from the positions that are congruent to 0 modulo 2^m should have the subsequences 01 and 10. W_0', obtained from those positions of W_0, is $[10^{k-1}]$ and hence it has these subsequences. Assume now that W_{i-1}', obtained from those positions of $W_{i=1}$, has these subsequences. In the associated words of key_i, we have a subsequence 00 or a subsequence 11. Hence, with the right cyclic shift Alice can make sure that W_i', will have the required subsequences. We continue with the arguments as in the case where n is odd. Thus Bob cannot have a winning strategy.

Therefore Bob can have a winning strategy only if n is a power of 2. $\qquad\square$

When n is a power of 2 we will show now that the necessary condition of Lemma 7.10 are also sufficient for both the open game and the blind game. For the open game, the strategy that Bob should use is very simple and it is based on the linear complexity of the sequence S that represents the glasses on the round table. When Bob sees the sequence S associated with the status of the glasses on the table, he chooses the same sequence (or another sequence with the same linear complexity) to change the status of the glasses on the table. Assume that T is the sequence chosen by Bob. Alice makes a shift by i to the sequence T and changes the status of the glasses in the positions with a *one* in $\mathbf{E}^i T$. By Lemma 5.5 we have that the linear complexity of the sequence associated with the new status of the glasses on the table, $\mathbf{E}^i T + S$, has linear complexity less than the linear complexity of S. This implies that the complexity of the sequence associated with the glasses on the table is reducing from one round to another. When this complexity is zero all the glasses will be in their upright position and Bob will win the game. This will be done using at most $c(S)$ rounds, where $c(S)$ is the linear complexity of the initial sequence of glasses on the table.

For the blind game, we will first show that a universal sequence must contain all the nonzero words of length n.

Lemma 7.11. *If* $\text{KEY} = (key_1, key_2, \ldots, key_m)$ *is a universal sequence, then in every play of the game all the nonzero binary words must be generated and hence* $m \geq 2^n - 1$.

Proof. Assume, on the contrary, that KEY is a universal sequence, but a nonzero binary word Y is not generated in some play of the game. We will show that in this case, Alice can win the game. Assume further that the first word generated by Alice in this play, which will be called the original play, was $W = W_0$.

Consider now another play of the game, in which Alice makes the exact moves as in the original play, with one exception. The first word that Alice generates is not W, but it is $W - Y$. It is easy to verify that a binary word $X - Y$ is generated in the current play if and only if the word X is generated in the original play. In particular, $Y - Y = 0^n$ is not generated in the current game. This means that KEY is not a universal sequence, a contradiction. Thus every nonzero binary word is generated by a universal sequence and hence $m \geq 2^n - 1$. $\quad\square$

Now, we will describe a universal sequence of length $2^n - 1$ that is an optimal sequence by Lemma 7.11. The construction is defined recursively and the proof that it is indeed a universal sequence will be done inductively in $n + 1$ steps, where at step i, $0 \leq i \leq n$, we construct a sequence KEY_i of length $2^i - 1$, having the following properties:

P1$_i$: All the words in KEY_i are of linear complexity at most i.
P2$_i$: Let W_0 be the first word generated by Alice. If $c(W_0) \leq i$, then Alice must lose the game when playing against KEY_i.

By property **P2$_n$** we have that the sequence $\text{KEY} = \text{KEY}_n$ is universal.

KEY_0 is the empty sequence of length 0 and hence it satisfies properties **P1$_0$** and **P2$_0$**. Assume now that we are given a sequence KEY_i of length $\ell_i = 2^i - 1$ that satisfies **P1$_i$** and **P2$_i$** for some $0 \leq i \leq n - 1$. A sequence KEY_{i+1} of length $2^{i+1} - 1$ that satisfies properties **P1$_{i+1}$** and **P2$_{i+1}$** is constructed as follows.

If X is an arbitrary word such that $c(X) = i + 1$, then set

$$\text{KEY}_{i+1} := \text{KEY}_i \circ X \circ \text{KEY}_i,$$

where $A \circ B$ is the sequence of keys obtained by taking the keys in A followed by the keys in B.

It is easy to verify that ℓ_{i+1}, the length of KEY_{i+1}, is $2\ell_i + 1 = 2^{i+1} - 1$. It remains to show that properties **P1$_{i+1}$** and **P2$_{i+1}$** are satisfied.

Property **P1$_{i+1}$** holds since by property **P1$_i$** all the words in KEY_i have complexity at most i and the complexity of X is $i + 1$. To see that property **P2$_{i+1}$** is satisfied, assume first that $c(W_0) \leq i$. Then, by the induction hypothesis, Alice loses the game during the first application of KEY_i on W_0. Thus we are left with the case where $c(W_0) = i + 1$. Now, after Alice plays only against KEY_i, by Lemma 5.5 we have that the word Y that is generated at the end of the process is of complexity $i + 1$. By Lemma 5.5 we also have that $c(X + Y) < i + 1$ since $c(Y) = c(X) = i + 1$. Hence, after Alice plays against $\text{KEY}_i \circ X$, the word W

that is generated has complexity at most i. Now, again we use the induction hypothesis, and after Alice takes W and continues to play against KEY_i, by property $\mathbf{P2}_i$ Alice loses the game. Thus $KEY_i \circ X \circ KEY_i$ is a universal sequence against a word with complexity at most $i + 1$ and $\mathbf{P2}_{i+1}$ is satisfied.

7.5 Notes

Linear shift registers and in particular M-sequences have many applications. We have presented in this chapter only a small number of examples for these applications. The same, on a somewhat smaller scale, is true for nonlinear shift-register sequences. The same is also true for the de Bruijn graph, de Bruijn cycles, linear complexities of sequences, and balanced sequences. For example Hsish [14,15] suggested that de Bruijn sequences can be used for structured light patterns that can acquire the range data of an object with the use of single camera for three-dimensional imaging systems. It is intriguing to know what new applications for all these concepts the future will bring.

Section 7.1. Stream ciphers is the most ancient method in encoding a message for security. Although today many advanced technologies are known, stream ciphers still have an important role in this area. For some work that was done on stream ciphers, the reader can be referred to the work of Klein [20], Lempel [23], and Rueppel [31].

Section 7.2. VLSI testing was considered to be an important problem that received the attention of several companies during the 1980s. The discussion in this section is mainly due to the direction and the results suggested by Lempel and Cohen [24] who considered only primitive polynomials. The generalization for irreducible polynomials taken in this section demanded some different proofs. A slightly different approach, using M-sequences, was suggested by Barzilai, Coppersmith, and Rosenberg [3] and Tang and Chen [38]. The approach was generalized to other $LFSR_n$s by Hollmann [17]. Other techniques for such VLSI testing were suggested by Kagaris Makedon, and Tragoudas [19], Rajski and Tyszer [30], Seroussi and Bshouty [35], and Wang and McCluskey [39].

Section 7.3. Single-track codes are used in high-technology companies using for example sensitive sensing instruments, instruments for saving energy, and even instruments for drilling oil.

Single-track Gray codes were introduced first in Hiltgen, Paterson, and Brandestini [16] and were further studied by Etzion and Paterson [8] and Schwartz and Etzion [34]. The basic definitions and Theorem 7.2 were given by Hiltgen, Paterson, and Brandestini [16]. The construction of single-track Gray codes of length $m = 2^n$, period $2^m - 2m$ is due to Etzion and Paterson [8] who also gave several constructions based on full-order sequences. The proof for the nonexistence of single-track Gray codes of length $m = 2^n$, period 2^m appears in Schwartz and Etzion [34] who also proved some properties and gave several constructions for single-track Gray codes. The material in this section is taken from these three papers.

In Schwartz and Etzion [34] there is another recursive construction based on the existence of two single-track Gray codes, one of length n and period $2^n - c_n$ and a second one of length k and period $2^k - c_k$, obtained by full-order necklaces ordering, with certain properties (which are usually not difficult to satisfy). The construction is again obtained by an appropriate ordering of necklaces has length nk and period $2^{nk} - c_{nk}$, where

$$c_{nk} = 2^{nk}(c_k 2^{-k} + c_n 2^{-n} - c_k c_n 2^{-(n+k)}).$$

If we further assume that we have sequences of single-track Gray codes such that

$$\lim_{n \to \infty} \frac{c_n}{2^n} = 0 \qquad \lim_{k \to \infty} \frac{c_k}{2^k} = 0,$$

then we have

$$\lim_{n,k \to \infty} \frac{c_{nk}}{2^{nk}} = 0.$$

Consider now $n = 2k + 1$ and the vertices of the n-cube whose weights are k or $k + 1$. Is there a Hamiltonian cycle in the n-cube passing only these vertices? The **middle-levels problem** is to order these binary words in a way that every two consecutive words differ in exactly one coordinate. This implies that in any two consecutive words, there is one word of weight k and one word of weight $k + 1$.

The middle-levels conjecture was presented by Buck and Wiedemann [5], Havel [13] and extensive work was done on the problem over the years, e.g., see the papers of Alpar-Vajk [1], Johnson [18], Savage and Winkler [33], and Shields, Shields and Savage [36]. The middle-levels conjecture was eventually solved by Mütze [27].

All the necklaces that contain the vertices of the middle levels are full-order necklaces since g.c.d.$(k, 2k + 1) = $ g.c.d.$(k + 1, 2k + 1) = 1$. An interesting question is whether there exists a solution for the middle-levels problem that yields a single-track Gray code? The idea of constructing a single-track Gray code based on full-order necklaces (see Theorem 7.4) might also work here.

Example 7.6. Consider $n = 5$ and the following list with the 4 necklaces of length 5 and weight 2 or 3.

1) [00011]
2) [00111]
3) [00101]
4) [10101]

Clearly, E[10101] = [01011] and [00011] differ exactly in the second position and hence this ordering yields a single-track middle-levels Gray code of length 5 and period 20 given by the following 5×20 array.

$$
\begin{array}{l}
0001\ 1111\ 1100\ 0111\ 0000 \\
0000\ 0001\ 1111\ 1100\ 0111 \\
0111\ 0000\ 0001\ 1111\ 1100\ . \\
1100\ 0111\ 0000\ 0001\ 1111 \\
1111\ 1100\ 0111\ 0000\ 0001
\end{array}
$$

∎

Example 7.7. Another example is the following list with the 10 necklaces of length 7 and weights 3 or 4.

1) [0000111]
2) [0001111]
3) [0001101]
4) [0011101]
5) [0011001]
6) [1011001] .
7) [1011000]
8) [1011010]
9) [1001010]
10) [1001011]

Clearly, $E[1001011]$ and $[0000111]$ differ exactly in the third position and hence this ordering yields a single-track middle levels Gray code of length 7. ∎

Contrary to the unsolved Problem 7.1, there is a construction for the middle-levels problem based on all the full-order necklaces. This solution was presented by Merino, Mička, and Mütze [26].

We have proved that there is no length n single-track Gray code with period 2^n. If n is a sum of two powers of 2, it was proved in Gregor, Merino, and Mütze [12] that there exists a length n Gray code with period 2^n with exactly two different tracks.

In general, Gray codes were extensively studied in the literature. They were found by Gray [11] and introduced later in Gilbert [10] as a listing of all the binary n-tuples in a list such that any two successive tuples in the list differ in exactly one position.

Finally, for an excellent survey on Gray codes, the interested reader is referred to the work of Savage [32] that was extended and updated later by Mütze [28].

Different generalizations for the concept of Gray codes have been given over the years. Such generalizations include the arrangements of other combinatorial objects in a such way that any two consecutive elements in the list differ in some pre-specified, usually small way, as was considered by Chung, Diaconis, and Graham [6].

Section 7.4. Combinatorial games have been very popular in mathematics over the years with an excellent survey given by Fraenkel [9] and attracted computer scientists when computer science started as a discipline. An excellent series of four books were written by Berlkamp, Conway, and Guy [4]. Martin Gardner was known for his puzzles given in the column "Recreational Mathematics" published in Scientific American. Indeed, a rotating-table game was suggested by Lasser and Ramshaw [22] in a book dedicated to him, where several versions and the history of the game were presented. The original version was given by Lewis and Willard [25]. The version presented in this section is due to Bar Yehuda, Etzion, and Moran [2]. A sequence of follow-up works on this problem can be found in Ehrenborg and Skinner [7], Korsky [21], Rabinovich [29], and Sidana and Sharma [37].

The rotating-table game described in this section was generalized by Bar Yehuda, Etzion, and Moran [2] for words over alphabets of arbitrary size $\sigma > 2$, as follows. Instead of n drinking glasses, we now have on the rotating table n roulettes of σ sides each. Denote the sides of the roulettes by $0, 1, \ldots, \sigma - 1$. Each round starts when Bob selects some of the roulettes, and for each selected roulette, Bob also selects an angle between 0 and $\frac{\sigma-1}{\sigma}360$ degrees, by which it should be rotated. After receiving these instructions, Alice first rotates the table, and then she follows instructions of Bob regarding the roulettes, which after the rotation are at the locations originally selected by Bob. Bob wins the game if he can force Alice to set all the roulettes in a way that the side of each roulette that is closest to the center of the table is the one marked by a *zero*. Describing this in the notation of words over the alphabet $\{0, 1, \ldots, \sigma - 1\}$, we obtain a description similar to the one for binary words, where the addition is performed modulo σ. The necessary condition for a winning strategy for Bob is generalized as follows.

Theorem 7.8. *If Bob has a winning strategy, then there exists a prime p such that $\sigma = p^{\alpha}$ and $n = p^{\beta}$ for some integers $\alpha > 0$ and $\beta > 0$.*

A sequence KEY is a (σ, n) **universal sequence** if Alice must lose the game when playing against this sequence. Lemma 7.11 can be generalized easily using the same proof as follows.

Lemma 7.12. *If $\text{KEY} = (key_1, key_2, \ldots, key_m)$ is a (σ, n) universal sequence, then in every play of the game all the nonzero words, over $\{0, 1, \ldots, \sigma - 1\}$, must be generated and hence $m \geq \sigma^n - 1$*

Theorem 7.9. *Bob can win the rotating table game (open or blind) if and only if $n = p^{\beta}$ and $\sigma = p^{\alpha}$, where $\alpha, \beta \geq 1$.*

It is interesting to note that the proofs of these claims are based on a generalization of the depth of sequences over the ring $\mathbb{Z}_{p^{\alpha}}$ rather than over a field.

References

[1] K. Alpar-Vajk, A particular Hamiltonian cycle on middle levels in the de Bruijn digraph, Discrete Math. 312 (2012) 608–613.

[2] R. Bar-Yehuda, T. Etzion, S. Moran, Rotating-table games and derivatives of words, Theor. Comput. Sci. 108 (1993) 311–329.

[3] Z. Barzilai, D. Coppersmith, A.L. Rosenberg, Exhaustive generation of bit patterns with applications to VLSI self-testing, IEEE Trans. Comput. 32 (1983) 190–194.

[4] E.R. Berlekamp, J.H. Conway, R.K. Guy, Winning Ways for Your Mathematical Plays, 4 volumes, CRC Press, New York, 2004.

[5] M. Buck, D. Wiedemann, Gray codes with restricted density, Discrete Math. 48 (1984) 163–171.

[6] F. Chung, P. Diaconis, R. Graham, Universal cycles for combinatorial structures, Discrete Math. 110 (1992) 43–59.

[7] R. Ehrenborg, C.M. Skinner, The Blind Bartender's problem, J. Comb. Theory, Ser. A 70 (1995) 249–266.

[8] T. Etzion, K.G. Paterson, Near optimal single-track Gray codes, IEEE Trans. Inf. Theory 42 (1996) 779–789.

[9] A. Fraenkel, Combinatorial games: selected bibliography with a succinct gourmet introduction, Electron. J. Comb. (2012) DS2.

[10] E.N. Gilbert, Gray codes and paths on the n-cube, Bell Syst. Tech. J. 37 (1958) 815–826.

[11] F. Gray, Pulse code communications, US Patent (1953) 2632058.

[12] P. Gregor, A. Merino, T. Mütze, The Hamilton compression of highly symmetric graphs, https://arxiv.org/abs/2205.08126, 2022.

[13] I. Havel, Semipaths in directed cubes, in: Graphs and Other Combinatorial Topics, Prague, 1982, in: Teubner-Texte zur Mathematik, vol. 59, Teubner, Leipzig, Germany, 1983, pp. 101–108.

[14] Y.-C. Hsieh, A note on the structured light of three-dimensional imaging systems, Pattern Recognit. Lett. 19 (1998) 315–318.

[15] Y.-C. Hsieh, Decoding structured light patterns for three-dimensional imaging systems, Pattern Recognit. 34 (2001) 343–349.

[16] A.P. Hiltgen, K.G. Paterson, M. Brandestini, Single-track Gray codes, IEEE Trans. Inf. Theory 42 (1996) 1555–1561.

[17] H. Hollmann, Design of test sequences for VLSI self-testing using LFSR, IEEE Trans. Inf. Theory 36 (1990) 386–392.

[18] J.R. Johnson, Long cycles in the middle two layers of the discrete cube, J. Comb. Theory, Ser. A 105 (2004) 255–271.

[19] D. Kagaris, F. Makedon, S. Tragoudas, A method for pseudo-exhaustive test pattern generation, IEEE Trans. Comput.-Aided Des. Integr. Circuits Syst. 13 (1994) 1170–1178.

[20] A. Klein, Stream Ciphers, Springer, London, 2013.

[21] S. Korsky, Permutations of counters on a table, https://arxiv.org/abs/2112.04965, 2021.

[22] W.T. Lasser, L. Ramshaw, Probing the rotating table, in: D.A. Klarner (Ed.), The Mathematical Gardner, Prindle, Weber and Schmidt, Boston, MA, 1981, pp. 285–307.

[23] A. Lempel, Cryptology in transition, Comput. Surv. 11 (1979) 285–303.

[24] A. Lempel, M. Cohn, Design of universal test sequences for VLSI, IEEE Trans. Inf. Theory 31 (1985) 10–17.

[25] T. Lewis, S. Willard, The rotating table, Math. Mag. 53 (1980) 174–179.

[26] A. Merino, O. Mička, T. Mütze, On a combinatorial generation problem of Knuth, SIAM J. Comput. 51 (2022) 379–423.

[27] T. Mütze, Proof of the middle levels conjecture, Proc. Lond. Math. Soc. 112 (2016) 677–713.

[28] T. Mütze, Combinatorial Gray codes – an updated survey, https://arxiv.org/abs/2202.01280, 2022.

[29] Y. Rabinovich, A generalization of the blind rotating table games, Inf. Process. Lett. 176 (2022) 106233.

[30] J. Rajski, J. Tyszer, Recursive pseudoexhaustive test pattern generation, IEEE Trans. Comput. 42 (1993) 1517–1521.

[31] R.A. Rueppel, Analysis and Design of Stream Ciphers, Springer-Verlag, Berlin, 1986.

[32] C. Savage, A survey of combinatorial Gray codes, SIAM Rev. 39 (1997) 605–629.

[33] C.D. Savage, P. Winkler, Monotone Gray codes and the middle levels problem, J. Comb. Theory, Ser. A 70 (1995) 230–248.

[34] M. Schwartz, T. Etzion, The structure of single-track Gray codes, IEEE Trans. Inf. Theory 45 (1999) 2383–2396.

[35] G. Seroussi, N.H. Bshouty, Vector sets for exhaustive testing of logic circuits, IEEE Trans. Inf. Theory 34 (1988) 513–522.

[36] I. Shields, B.J. Shields, C.D. Savage, An update on the middle levels problem, Discrete Math. 309 (2009) 5271–5277.

[37] T. Sidana, A. Sharma, Roulette games and depths of words over commutative ring, Des. Codes Cryptogr. 89 (2021) 641–678.

[38] D.T. Tang, C. Chen, Logic test-pattern generation using linear codes, IEEE Trans. Comput. 33 (1984) 845–850.

[39] L.-T. Wang, E.J. McCluskey, Circuits for pseudoexhaustive test pattern generation, IEEE Trans. Comput.-Aided Des. 7 (1988) 1068–1080.

Chapter 8

DNA sequences and DNA codes

The genome assembly and DNA storage

In this chapter, we will consider applications of sequences and the de Bruijn graph for concepts associated with DNA. The applications are mainly in two directions, the Human Genome Project and DNA storage. We will describe these two applications and how the de Bruijn graph and its sequences can be used for these two applications. Three combinatorial structures associated with these two applications and other concepts of DNA and RNA will be also discussed: constant-weight de Bruijn sequences, reconstruction of a sequence from its subsequences, and non-overlapping codes.

Section 8.1 is devoted to the human genome project, a project that was considered towards the end of the 1980s. A large number of subsequences from a genome sequence have to be merged into the original genome. This process, which is called DNA sequencing in the genome assembly, was the main target of the project. Several methods were developed for this purpose. One of the most successful methods is based on paths and cycles in the de Bruijn graph. The basic elements of the project and the de Bruijn graph method will be discussed in this section.

A somewhat dual problem is to reconstruct a codeword of a given code from a subset of its subsequences. Such a problem has an important application in one of the most fascinating storage media of the 21st century, namely DNA storage. This storage media is the topic of Section 8.2. We will give a short introduction to this research area. The next three sections will be devoted to graphs and codes associated with sequences related to biology, e.g., DNA and RNA.

Section 8.3 is devoted to cycles that cover a certain subset of edges in G_n, where the goal is to have a path that represents all the words of length n and weight w. Such a path based on the vertices of G_n does not exist. However, the words of length n and weight w can be represented by the words of length $n - 1$ and weights $w - 1$ or w since there is a unique way to complete them to words of length n and weight w by appending one more bit. Similarly, removing the last bit of all the words of length n and weight w yields all the words of length $n - 1$ and weights $w - 1$ or w. A graph, whose edges represent all the words of length $n + 1$ and weights between w_1 and w_2 and their associated vertices in G_n, is constructed. In this graph, there exist Eulerian cycles, as will be proved in this section. The associated sequences will be called constant-weight de Bruijn sequences.

Sequences and the de Bruijn Graph. https://doi.org/10.1016/B978-0-44-313517-0.00014-7

In Section 8.4, a reconstruction problem of a sequence from its subsequences will be discussed. In both DNA sequencing and DNA storage we are given a set of subsequences of a long sequence and we have to reconstruct the long sequence. This is the reconstruction problem that is important mainly for the genome assembly.

Overlapping and non-overlapping of codewords is also an important concept in the reconstruction of a sequence from its subsequences. Section 8.5 will consider codes where no two words are overlapping.

8.1 The genome assembly

The **Human Genome Project (HGP)** was a very ambitious project in the 20th century. The target of the project was to have a complete mapping of all the genes of a human being. The genome is the set of all our genes. The HGP has revealed that there are more than 20 000 human genes. The impact of the HGP was incredible, but we will concentrate on the application of the de Bruijn graph to the project.

We start with some basic concepts used to understand the foundations of the problem. We have no intention of obtaining a deep knowledge and understanding of the biology behind all the concepts. A *nucleotide* is the basic building block of nucleic acid. DNA and RNA are polymers made of long chains of nucleotides. **Deoxyribonucleic Acid (DNA)** contains four letters (amino acids):

- *A* - Adenine;
- *T* - Thymine;
- *C* - Cytosine;
- *G* - Guanine.

These letters come in pairs $\{A, T\}$ and $\{C, G\}$, as depicted in Fig. 8.1, where a schematic structure of DNA is depicted. Each one of the components has its chemical structure.

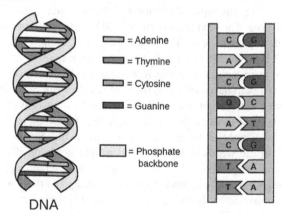

DNA

FIGURE 8.1 DNA with its amino acids.

Genome assembly (and its DNA sequencing) is a computational process that receives as an input a large number of short subsequences from a long DNA sequence. The computational process deciphers these short subsequences and produces as an output the original representation of the chromosome of the long DNA sequence. We can distinguish between three types of solutions, i.e., techniques for the computational process.

1. Greedy.
2. Overlap layout consensus (OLC), which is based on a Hamiltonian path.
3. A de Bruijn graph (DBG), which is based on an Eulerian path in a subgraph of the de Bruijn graph.

The *greedy* computational process receives the short subsequences (called *strands*) and tries to find overlaps of the prefix from one strand with a suffix from another strand to form the long DNA chromosome.

Example 8.1. Assume we have the following four strands: CAACCGTAG, AGTTTCCA, CAATTCAGT, and AGTGTACACAA. These four strands induce the following four overlaps:

- the suffix of length 2 of the strand CAACCGTAG with the prefix of length 2 of the strand AGTTTCCA.
- the suffix of length 2 of the strand AGTTTCCA with the prefix of length 2 of the strand CAATTCAGT.
- the suffix of length 3 of the strand CAATTCAGT with the prefix of length 3 of the strand AGTGTACACAA.
- the suffix of length 3 of the strand AGTGTACACAA with the prefix of length 3 of the strand CAACCGTAG

We combine these four overlaps (in this given order) to obtain the circular genome

$$CAACCGTAGTTTCCAATTCAGTGTACA.$$

It is readily verified that each strand is a subsequence in this circular genome of length 27 that can be written in any of its cyclic shifts, e.g.,

$$AGTTTCCAATTCAGTGTACACAACCGT$$

or

$$CAATTCAGTGTACACAACCGTAGTTTC.$$

However, there are also other overlaps between these four strands. Consider now the following four overlaps:

- the suffix of length 2 of the strand CAACCGTAG with the prefix of length 2 of the strand AGTGTACACAA.
- the suffix of length 3 of the strand AGTGTACACAA with the prefix of length 3 of the strand CAATTCAGT.

- the suffix of length 3 of the strand CAATTCAGT with the prefix of length 3 of the strand AGTTTCCA.
- the suffix of length 2 of the strand AGTTTCCA with the prefix of length 2 of the strand CAACCGTAG.

We combine these four overlaps (in this given order) to obtain another circular genome whose length is also 27 as follows:

CAACCGTAGTGTACACAATTCAGTTTC.

∎

The greedy algorithm might be sometimes efficient and can produce the correct genome, but as we saw in Example 8.1 it might produce different genomes, so the goal is to have a better systematic algorithm.

Example 8.2. Consider, for example, the circular genome depicted in Fig. 8.2, where the amino acids (DNA letters) are read from left to right (clockwise)

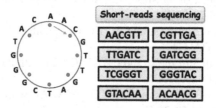

FIGURE 8.2 Short-reads sequencing from a circular genome.

The circular genome can be read as AACGTTGATCGGGTAC. Consider now the eight short reads of length 6 for sequencing: AACGTT, CGTTGA, TTGATC, GATCGG, TCGGGT, GGGTAC, GTACAA, and ACAACG. Consider now the following large overlaps:

- the suffix of length 4 of GATCGG with the prefix of length 4 of TCGGGT.
- the suffix of length 4 of TCGGGT with the prefix of length 4 of GGGTAC.
- the suffix of length 4 of GGGTAC with the prefix of length 4 of GTACAA.
- the suffix of length 4 of GTACAA with the prefix of length 4 of ACAACG.
- the suffix of length 4 of ACAACG with the prefix of length 4 of AACGTT.
- the suffix of length 4 of AACGTT with the prefix of length 4 of CGTTGA.
- the suffix of length 4 of CGTTGA with the prefix of length 4 of TTGATC.
- the suffix of length 4 of TTGATC with the prefix of length 4 of GATCGG.

From these eight overlaps we construct an **overlap graph** depicted in Fig. 8.3 and the circular genome sequence GATCGGGTACAACGTT by the order of the edges in the graph. This genome is a cyclic sequence of the genome in Fig. 8.2.

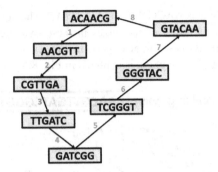

FIGURE 8.3 The overlap graph.

■

Although we have constructed the correct circular genome in Example 8.2, we might have some practical problems that can cause the production of an incorrect genome. These problems can be:

1. Not all the short subsequences are read. In this case, the number of overlaps might not be enough to reconstruct the whole sequence of the genome.
2. The strands that are read might lead to a few different genomes.
3. There might be errors in the reads, which can produce an incorrect genome, especially if the reads are relatively long, although they are short compared to the long genome.
4. Some of the reads might repeat and this can cause some confusion.
5. The genome might not be circular as in our current examples.

One partial solution can be obtained by using even shorter reads that will be called *k-mers*. The two computational methods, the overlap layout consensus, and the de Bruijn graph are based on this idea of the k-mers.

In the OLC method we construct a graph whose vertices are the k-mers and the direct edges represent $(k-1)$-mers. There is a directed edge $u \to v$ in the graph if and only if the suffix of length $k-1$ of the k-mer in u is the prefix of length $k-1$ of the k-mer in v. The goal now is to find a Hamiltonian cycle in the graph. From such a Hamiltonian cycle the circular genome is derived. If the genome is not circular, then instead of a Hamiltonian cycle, it will be required to find a Hamiltonian path in the graph. We note that the graph is a subgraph of the de Bruijn graph, but the labels on the edges are different in these two graphs.

Example 8.3. Consider the circular genome AACGTTGATCGGGTAC of Example 8.2 and partition the reads into 3-mers that will represent the vertices of the graph shown in Fig. 8.4. Every two vertices for which the suffix of length two of the first equals the prefix of length two of the second are connected by an edge from the first vertex to the second vertex. Now, the target is to find a directed Hamiltonian cycle in the directed graph. If the genome was acyclic, then only a Hamiltonian path had to be found.

In this example, each vertex is a 3-mer. There are two Hamiltonian cycles in the graph leading to two possible circular genomes. The first one is the original one and the second one is AACGGGTTGATCGTAC. The numbers on the edges in Fig. 8.4 indicate the order of the edges in these two cycles. To avoid this ambiguity it is required to use larger k-mers. In our example, we use 4-mers for the vertices. The appropriate graph is depicted in Fig. 8.5.

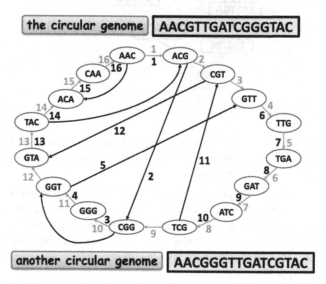

FIGURE 8.4 Two circular genomes found by the overlap layout consensus method.

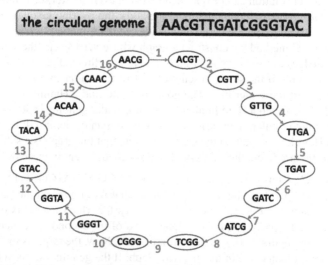

FIGURE 8.5 A circular genome found by the overlap layout consensus method.

We see from Example 8.3 that as the length of the genome is larger, it will be required to use k-mers with larger k. However, the main problem in the process of the OLC is that finding a Hamiltonian cycle (path) in a graph is an NP-complete problem. It might not be difficult in some graphs (like in the de Bruijn graph), but we have no indication when in the constructed graph (which is a subgraph of the de Bruijn graph $G_{4,n}$) there will be an efficient method to find the Hamiltonian cycle. This is the motivation for the next computational method of sequencing.

In the DBG method for the genome assembly, the target will be to find an Eulerian cycle or an Eulerian path. The problem of finding such Eulerian cycles and paths is solved in polynomial time using an algorithm, as was done in the proof of Theorem 1.15. Again, we construct a graph whose edges are k-mers. A directed edge $e = u \rightarrow v$ connects the vertex u that represents the prefix of e, whose length is $k - 1$, to the vertex v that represents the suffix of e, whose length is $k - 1$. The goal now is to find an Eulerian cycle in the graph. From such an Eulerian cycle the circular genome is derived. If the genome is not circular, then instead of an Eulerian cycle, it will be required to find an Eulerian path in the graph.

Example 8.4. Consider the same circular genome AACGTTGATCGGGTAC as in Example 8.3. Using 3-mers for the edges and 2-mers for the vertices, we generate the graph of Fig. 8.6. In this graph, it is required to find an Eulerian cycle. Exactly as in Example 8.3 there will be two solutions based on two Eulerian cycles. The numbers on the edges indicate the order of the edges in these two cycles.

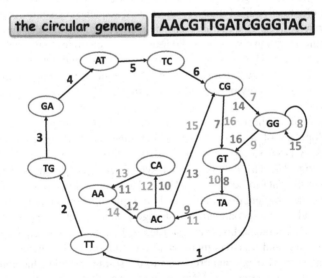

FIGURE 8.6 A circular genome found by the de Bruijn graph method.

We note again that contrary to Example 8.4, taking larger k-mers will make it more probable that the outcome will be exactly one possible genome sequence.

8.2   DNA storage codes

The storage demands for all computerized systems are increasing from year to year. Writing some approximate figures for the storage demands will not be productive. The reason is that any given figures will be dramatically increased in just a few years, as is expected today. Hence, there is an important need for a new storage medium that is small in size, reliable, and with a long lifespan. The concept of DNA storage seems to be the solution to this problem for many years, as its size is dramatically small and DNA is not damaged for tens of thousands of years. There are of course the physical problems – how to store information in such a medium, how to retrieve it, and how to correct possible errors while reading from and writing into the medium? These are important problems, but there is a strong belief that these problems will be solved with time.

Although the requirement for such storage is a problem that occurred during the beginning of the 21st century, the concept of DNA storage was first suggested in 1959 by Richard Feynman. The interest in storage solutions based on DNA molecules was increased as a result of the HGP in which there was remarkable progress in sequencing and assembly methods, as was discussed in Section 8.1. Some techniques and computations have involved variations of the de Bruijn graph.

As described in Section 8.1, DNA consists of four types of nucleotides: adenine (A), cytosine (C), guanine (G), and thymine (T). A single DNA strand, also called an oligonucleotide (oligo), is an ordered sequence of some combination of these nucleotides. DNA strands can be synthesized chemically and modern DNA synthesizers can concatenate the four DNA nucleotides to form almost any possible sequence. This process enables us to store digital data in the strands. The data can be read back with common DNA sequencers, while the most popular ones use DNA polymerase enzymes and are referred to as sequencing by synthesis.

Progress in synthesis and sequencing technologies has paved the way for the development of a non-volatile data-storage technology based on DNA molecules. A DNA storage system consists of three important entities, as depicted in Fig. 8.7. The first is a DNA synthesizer that produces the strands that encode the data to be stored in DNA. To produce strands with an acceptable error rate, the length of the strands is typically limited to no more than 250 nucleotides. The second part is a storage container with compartments that stores the DNA strands, however, unordered. Lastly, a DNA sequencer reads back the strands and transfers them back to digital data. The encoding and decoding stages are two external processes to the storage systems that convert the binary user data into strands of DNA in such a way that even in the presence of errors (the 4 nucleotides colored in underlined red in Fig. 8.7), it will be possible to revert to the original binary data of the user. DNA as a storage system

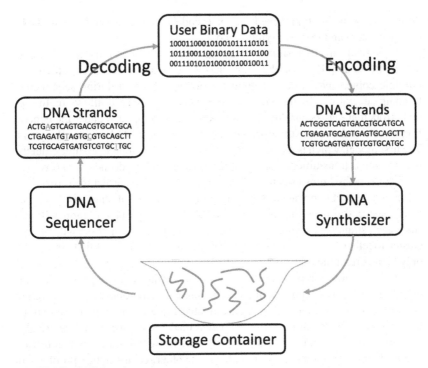

FIGURE 8.7 DNA-based storage system.

has several attributes that distinguish it from any other storage system. The most outstanding one is that the strands are not ordered in the memory and thus it is not possible to know the order in which they were stored. Usually, this constraint can be overcome by using block addresses, also called indices, that are stored as part of the strand. Note that this limitation already imposes the capacity of DNA storage to be strictly less than 2 bits per nucleotide. This structure also prevents random access to the stored data since it is not possible to read a given strand in the pool and most of the proposed systems have to read the entire pool to retrieve even a single strand.

This described model is quite complicated and deserves much thought about the encoding and decoding to DNA strands and from DNA strands, respectively. To store and retrieve information in DNA storage, one starts with the desired information (sequences) encoded into a set of sequences over the alphabet $\{A, T, G, C\}$ as those are the letters used for DNA. The DNA sequencer part is somehow related to the DNA sequencing that was described in Section 8.1, but there is at least one major difference between the two processes. While in DNA sequencing, the DNA strands, which were called k-mers are produced from a genome of an organ body, the DNA strands for DNA storage are produced via the technology based on some codebook that is used for the DNA storage system and they are synthesized.

Sequencing of DNA strands is usually preceded by polymerase chain reaction (PCR) amplification that replicates the strands. Hence, every strand has multiple copies and several of them are going to be read during sequencing. Reading several copies for each strand is beneficial since it allows us to correct errors that may occur during this process. This setup falls under the general framework of the reconstruction problem that has been studied over the years. The first model, which is the most appropriate for DNA storage, assumes that the information is transmitted over multiple channels, and the decoder that observes all channel estimations uses this inherited redundancy to correct the errors. The main studied problem was finding the minimum number of channels that guarantees successful decoding in the worst case. Here, our task is more difficult since each strand has several copies. Hence, it follows that first we will need to cluster together the read strands that belong to the same amplified strand and only then correct these errors.

The reconstruction problem has many other variants that depend on the input received for the reconstruction. The reconstruction problems, which are not associated with multiple channels, are related more to the genome assembly, but they were also mentioned in the context of DNA storage, and they are also interesting from a combinatorial point of view. We will discuss one reconstruction problem in Section 8.4. We mention some other reconstruction problems in Section 8.6 and concentrate now on some of the appropriate definitions and how the de Bruijn graph can be used to solve the reconstruction problem. Although our alphabet consists of the four letters A, C, G, and T, we will continue to make most of our discussion with $\mathbb{F}_2$, but $\mathbb{F}_q$ will not be ignored.

For the ℓ-mers (they are also called k-grams for DNA storage) of an acyclic sequence S, let $p(S; q, \ell)$ denote the **profile vector** of length q^ℓ, indexed by all the words of $\mathbb{Z}_q^\ell$ ordered lexicographically. The jth entry in $p(S; q, \ell)$ denotes the number of occurrences of the jth word of this lexicographic order in S. The sum of entries in $p(S; q, \ell)$ is $n - \ell + 1$ since there are $n - \ell + 1$ distinct windows of length ℓ in S. For DNA storage, there is some balancing between the different symbols and hence we would like to have some kind of a de Bruijn graph with "almost" balanced words. We continue to consider the binary case but keep in mind that the generalization for a larger alphabet is required. This generalization is usually straightforward. Let the **de Bruijn graph** $G_n(w_1, w_2)$ be the subgraph of G_n that contains all the edges whose weight is between w_1 and w_2 and the vertices that are connected by these edges.

Example 8.5. Consider the graph $G_3(2, 3)$ presented in Fig. 8.8 and its sequence $S = 01100110101$ of length 11.

The profile vector of length $2^4 = 16$ is

$$p(S; 2, 4) = (0, 0, 0, 1, 0, 1, 2, 0, 0, 1, 1, 0, 1, 1, 0, 0).$$

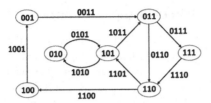

FIGURE 8.8 The graph $G_3(2,3)$.

The sequence S is associated with a path of length 8 in $G_3(2,3)$ as follows:

$$011 \xrightarrow{0110} 110 \xrightarrow{1100} 100 \xrightarrow{1001} 001 \xrightarrow{0011} 011 \xrightarrow{0110} 110 \xrightarrow{1101} 101 \xrightarrow{1010} 010 \xrightarrow{0101} 101.$$

■

Many questions can be asked just on the brief exposition that was given. Given w_1 and w_2 such that $1 \le w_1 < w_2 \le n$, is there an Eulerian cycle in $G_n(w_1, w_2)$? Given a sequence S and its profile vector $p(S; q, \ell)$, is the profile vector sufficient to find the sequence S? The answer is no in many cases, but there are many cases in which the profile vector is sufficient to reconstruct the sequence S, as was discussed in Section 8.1. We should remember that while in the DNA sequencing discussed in Section 8.1, there is exactly one unknown sequence that was picked from some organ body, in DNA storage the sequence required to be reconstructed was taken from a codebook that contains many sequences known in advance, but still it is not an easy task to reconstruct the sequence (codeword) taken from the codebook.

Example 8.6. Assume first that the codebook contains the $2^{2^{n-1}-n}$ span n acyclic de Bruijn sequences (each of length $2^n + n - 1$ starting with n *zeros*) and we consider ℓ-mers with $\ell = n + 1$. For each sequence S in the codebook, the profile vector $p(S; 2, n + 1)$ of length 2^{n+1} has exactly 2^n nonzero entries, all equal to *one*. If S is a sequence from the codebook, then we can form a unique truth table that consists of 2^n rows (the first n bits of the edge is the state and the last bit is the result of the feedback function for these n bits). Each $(n + 1)$-mer defines exactly one row in the truth table. This immediately implies that we can reconstruct any de Bruijn sequence taken from the codebook.

Now, if the codebook contains again the $2^{2^{n-1}-n}$ span n acyclic de Bruijn sequences, starting with n *zeros*, and we consider ℓ-mers with $\ell = n$. The profile vector $p(S; 2, n)$ of length 2^n and all its 2^n entries are nonzero, all equal to *one*. Therefore we will not be able to distinguish between any two sequences of the codebook as they all have the same profile vector. ■

Example 8.6 shows that we have to be careful in choosing the codebook and the value of ℓ for the ℓ-mers. However, more parameters force the chosen codebook and other parameters of the DNA storage system. This is beyond our exposition, but it will be also briefly discussed in Section 8.6.

8.3 Constant-weight de Bruijn sequences

The first goal of this section is to form a sequence from the de Bruijn graph G_n that contains all the words of length n and weight w. This is possible only when $w \in \{0, 1, n-1, n\}$. Therefore we will achieve this goal with a simple correspondence between the words in the path and the words of length n and weight w. By simple combinatorial arguments, we can see that the words of length $n - 1$ and weights $w - 1$ or w can represent all the words of length n and weight w. This can be also easily derived from the combinatorial identity

$$\binom{n}{w} = \binom{n-1}{w} + \binom{n-1}{w-1}.$$

A word of length $n - 1$ and weight w is mapped to a word of length n and weight w by appending a *zero* to its end. A word of length $n - 1$ and weight $w - 1$ is mapped to a word of length n and weight w by appending a *one* to its end. A word of length n and weight w is mapped to a word of length $n - 1$ and weight either $w - 1$ or w by removing its last entry.

The associated problem is whether $G_n(w_1, w_2)$, where $w_1 < w_2$ has an Eulerian cycle? An edge of $G_n(w_1, w_2)$ represented by the $(n + 1)$-tuple $(x_0, x_1, \ldots, x_{n-1}, x_n)$ is directed from the vertex $(x_0, x_1, \ldots, x_{n-1})$ to the vertex $(x_1, \ldots, x_{n-1}, x_n)$. Since the edge $(x_0, x_1, \ldots, x_{n-1}, x_n)$ has any weight between w_1 and w_2, it follows that the vertices can have any weight between $w_1 - 1$ and w_2. A vertex $(z_1, z_2, \ldots, z_n)$ with weight w_2 can have only one in-edge (which starts with a *zero*) and only one out-edge (which ends with a *zero*) to avoid an edge whose weight is larger than w_2. A vertex $(z_1, z_2, \ldots, z_n)$ with weight $w_1 - 1$ can have only one in-edge (which starts with a *one*) and only one out-edge (which ends with a *one*) to avoid an edge whose weight is smaller than w_1. A vertex $(z_1, z_2, \ldots, z_n)$ with weight between w_1 and $w_2 - 1$ has no constraints on its in-edges and out-edges since its incident edges in any case will be with weights between w_1 and w_2. Hence, each such vertex has in-degree two and out-degree two. Therefore we have

Lemma 8.1. *The vertices of $G_n(w_1, w_2)$ can have any weight between $w_1 - 1$ and w_2. The in-degree of each vertex of $G_n(w_1, w_2)$ is equal to its out-degree.*

Given Lemma 8.1 and Theorem 1.15, to prove that there exists an Eulerian cycle in $G_n(w_1, w_2)$ we have only to prove that the graph $G_n(w_1, w_2)$ is a strongly connected graph. Consider a possible path from a vertex $x = (x_1, x_2, \ldots, x_n)$ to a vertex $z = (z_1, z_2, \ldots, z_n)$. Assume that the first bit in which x and z differ is bit i, i.e., $x_i \neq z_i$ and $x_j = z_j$ for $1 \leq j < i$. First, we will show a path from $x = (x_1, x_2, \ldots, x_n)$ to $(z_1, z_2, \ldots, z_i, y_1, \ldots, y_{n-i})$ for some $y_1, \ldots, y_{n-i}$. The proof will continue iteratively when in the next step we consider a path from $(z_1, z_2, \ldots, z_i, y_1, \ldots, y_{n-i})$ to $z = (z_1, z_2, \ldots, z_n)$. In the last step, the path will end at the vertex $z = (z_1, z_2, \ldots, z_n)$.

For the first step, we distinguish between three cases depending on whether the weight of x is $w_1 - 1$, between w_1 and $w_2 - 1$, or w_2.

Case 1: If $w_1 \leq \text{wt}(x_1, x_2, \ldots, x_n) \leq w_2 - 1$, then consider the path defined by the sequence $x_1, x_2, \ldots, x_n, z_1, \ldots, z_{i-1}, z_i$ (where each of the $n + 1$ consecutive bits define an edge and each n consecutive bits define a vertex) and the claim is proved since the weight of each vertex, with the possible exception of the last one, is the same as the weight of x. The weight of the last vertex is between $\text{wt}(x) - 1 \geq w_1 - 1$ and $\text{wt}(x) + 1 \leq w_2$, which is in the required range. The path can be continued with any $n - i$ edges, from $G_n(w_1, w_2)$, to obtain the required result.

Case 2: If $\text{wt}(x_1, x_2, \ldots, x_n) = w_1 - 1$, then we distinguish again between two cases depending on whether $x_i = 0$ or $x_i = 1$.

- If $x_i = 0$ and $z_i = 1$, then consider the path defined by the sequence $x_1, x_2, \ldots, x_n, 1, z_1, \ldots, z_{i-1}, z_i$ and the claim is proved since the weight of each vertex on this path is either $\text{wt}(x) = w_1 - 1$ or $\text{wt}(x) + 1 = w_1$. The path can be continued with any $n - i$ edges, from $G_n(w_1, w_2)$, to obtain the required result.

- If $x_i = 1$ and $z_i = 0$, then we claim that there exists a j, $i < j \leq n$, with $x_j = 0$. If no such j exists, then $\text{wt}(x) = \text{wt}(x_1, x_2, \ldots, x_i, 1, \ldots, 1) = w_1 - 1$ and hence $\text{wt}(z_1, z_2, \ldots, z_i, \ldots, z_n) < w_1 - 1$, a contradiction. Consider now a path defined by the sequence $x_1, x_2, \ldots, x_n, x_1, \ldots, x_{j-1}, 1, x_{j+1}, \ldots, x_n$. This path exists in the graph since the weight of each vertex on this path is either $\text{wt}(x) = w_1 - 1$ or $\text{wt}(x) + 1 = w_1$. The last vertex on this path is $(x_1, \ldots, x_{j-1}, 1, x_{j+1}, \ldots, x_n)$ and its weight is w_1. Now, we can use Case 1 to prove the existence of a path from $(x_1, \ldots, x_{j-1}, 1, x_{j+1}, \ldots, x_n)$ to a vertex $(z_1, z_2, \ldots, z_i, y_1, \ldots, y_{n-i})$ as required.

Case 3: If $\text{wt}(x_1, x_2, \ldots, x_n) = w_2$, then we distinguish again between two cases depending on whether $x_i = 0$ or $x_i = 1$.

- If $x_i = 1$ and $z_i = 0$, then consider the path defined by the sequence $x_1, x_2, \ldots, x_n, 0, z_1, \ldots, z_{i-1}, z_i$ and the claim is proved since the weight of each vertex on this path is either $\text{wt}(x) = w_2$ or $\text{wt}(x) - 1 = w_2 - 1$. The path can be continued with any $n - i$ edges, from $G_n(w_1, w_2)$, to obtain the required result.

- If $x_i = 0$ and $z_i = 1$, then we claim that there exists a j, $i < j \leq n$, with $x_j = 1$. If no such j exists, then $\text{wt}(x) = \text{wt}(x_1, x_2, \ldots, x_i, 0, \ldots, 0) = w_2$ and hence $\text{wt}(z_1, z_2, \ldots, z_i, \ldots, z_n) > w_2$, a contradiction. Consider now a path defined by the sequence $x_1, x_2, \ldots, x_n, x_1, \ldots, x_{j-1}, 0, x_{j+1}, \ldots, x_n$. This path exists in the graph since the weight of each vertex on this path is either $\text{wt}(x) = w_2$ or $\text{wt}(x) - 1 = w_2 - 1$. The last vertex on this path is $(x_1, \ldots, x_{j-1}, 0, x_{j+1}, \ldots, x_n)$ and its weight is $w_2 - 1$. Now, we can use Case 1 to prove the existence of a path from $(x_1, \ldots, x_{j-1}, 0, x_{j+1}, \ldots, x_n)$ to $(z_1, z_2, \ldots, z_i, y_1, \ldots, y_{n-i})$.

Once the first step is completed we have shown the existence of a path from vertex $x = (x_1, x_2, \ldots, x_n)$ to a vertex $z = (z_1, z_2, \ldots, z_i, y_1, \ldots, y_{n-i})$. We now have to show that there exists a path in $G_n(w_1, w_2)$ starting in the vertex $z = (z_1, z_2, \ldots, z_i, y_1, \ldots, y_{n-i})$ and ending in the vertex $z = (z_1, z_2, \ldots, z_n)$. This is accomplished by at most $n - i$ iterations of the first step.

Thus we have proved the following lemma.

Lemma 8.2. *The graph $G_n(w_1, w_2)$ is a strongly connected graph.*

Lemmas 8.1 and 8.2 lead to the following consequence from Theorem 1.15.

Corollary 8.1. *There exists an Eulerian cycle in $G_n(w_1, w_2)$.*

The description we gave for a path from a vertex $x = (x_1, x_2, \ldots, x_n)$ to a vertex $z = (z_1, z_2, \ldots, z_n)$ in $G_n(w_1, w_2)$ does not seem to be efficient. However, it should be noted that the purpose of this description was only to show that the graph $G_n(w_1, w_2)$ is strongly connected. To have an efficient algorithm to generate an Eulerian cycle in this graph we can use the Eulerian cycle algorithm presented in Section 1.2. To obtain the sequence of words of length n and weight w, we have to consider the Eulerian cycle in the graph $G_{n-1}(w - 1, w)$ and append a *one* or a *zero*, respectively, to each word along the path, depending on whether its weight is $w - 1$ or w, respectively.

The same approach can be used for a larger alphabet of size σ and the graph $G_{\sigma,n}$, by defining a graph $G_{\sigma,n}(w_1, w_2)$, whose edges contain all the words whose length is $n + 1$ and weight is between w_1 and w_2.

Lemma 8.3. *The in-degree of each vertex of $G_{\sigma,n}(w_1, w_2)$ is equal to its out-degree.*

Proof. As in $G_n(w_1, w_2)$ we have to consider the same three cases for the weight of a vertex x in $G_{\sigma,n}(w_1, w_2)$.

1. If $\mathrm{wt}(x) = w_1 - 1$, then the symbol preceding x on an edge must be a nonzero symbol and the same is for the symbol succeeding x on an edge. Hence, the in-degree of x is $\sigma - 1$ and the out-degree of x is also $\sigma - 1$.
2. If $\mathrm{wt}(x) = w_2$, then the symbol preceding x on an edge must be *zero* and the same is for the symbol succeeding x on an edge. Hence, the in-degree of x is *one* and the out-degree of x is also *one*.
3. If $w_1 \leq \mathrm{wt}(x) \leq w_2 - 1$, then each one of the σ symbols can be appended to x to start an in-edge and each one of the σ symbols can be appended to x to end an out-edge. Hence, the in-degree and the out-degree of x is σ.   □

Again, in view of Lemma 8.3 and Theorem 1.15, there exists an Eulerian cycle in $G_{\sigma,n}(w_1, w_2)$ if and only if $G_{\sigma,n}(w_1, w_2)$ is a strongly connected graph.

Lemma 8.4. *The graph $G_{\sigma,n}(w_1, w_2)$ is a strongly connected graph.*

Proof. The proof is exactly the same as the proof of Lemma 8.2, where the *one* in $G_n(w_1, w_2)$ is replaced by any of the $\sigma - 1$ nonzero symbols as required by the associated vertices.   □

Corollary 8.2. *There exists an Eulerian cycle in* $G_{\sigma,n}(w_1, w_2)$.

8.4 Reconstruction of a sequence from subsequences

This is one basic combinatorial problem that is associated with DNA sequencing as part of the genome assembly and for reconstructing a sequence for DNA storage. Each one requires a different type of input. Without elaborating more on the distinction between the two, it should be clear that there are many reconstruction problems, some of which are more theoretical than practical.

If there is some information about the sequence S it might be of help to reconstruct it, as was demonstrated in Example 8.6. We elaborate now on one possible reconstruction problem. We will assume that all the sequences are binary unless mentioned otherwise. It should be also mentioned that the reconstruction problem that will be considered in this section is more theoretical than practical, and it is given here also for its combinatorial interest. Other reconstruction problems will be mentioned in Section 8.6.

The reconstruction problem, which will be considered, is known as the k-deck problem, where the ordered information of all the projections of k positions in the sequence S is given as an input. Let $X = (x_1, x_2, \ldots, x_n)$ be a binary word of length n. For $A \subseteq \{1, 2, \ldots, n\}$, we use $X(A)$ to denote the subsequence with indices in A. In other words, $X(A) = (x_{a_1}, x_{a_2}, \ldots, x_{a_k})$, where $A = \{a_1, a_2, \ldots, a_k\}$ and $a_1 < a_2 < \cdots < a_k$. For $k < n$, the k-**deck** of X, denoted by $D_k(X)$, refers to the multiset of all the $\binom{n}{k}$ subsequences of X of length k. We represent the k-deck of a word X by an integer-valued vector of length 2^k. Specifically,

$$D_k(X) \triangleq (X_\alpha)_{\alpha \in \{0,1\}^k},$$

where X_α denotes the number of occurrences of α as a subsequence of X and the indices in $\{0, 1\}^k$ are presented in increasing lexicographic order.

Let $D_k(n)$ be the number of distinct k-decks of all the words of length n.

Example 8.7. Consider the sequence $X = (101000110)$. The subsequence 010 is contained 12 times in X as follows:

$$X(\{2,3,4\}) = X(\{2,3,5\}) = X(\{2,3,6\}) = X(\{2,3,9\}) = X(\{2,7,9\})$$
$$= X(\{2,8,9\}) = X(\{4,7,9\}) = X(\{4,8,9\}) = X(\{5,7,9\}) = X(\{5,8,9\})$$
$$= X(\{6,7,9\}) = X(\{6,8,9\}) = 010$$

and hence $X_{010} = 12$.

With the same computations for all subsequences of lengths 1, 2, and 3, we have the following distribution

$D_1(X) = (X_0, X_1) = (5, 4)$,

$D_2(X) = (X_{00}, X_{01}, X_{10}, X_{11}) = (10, 9, 11, 6)$,

$D_3(X) = (X_{000}, X_{001}, X_{010}, X_{011}, X_{100}, X_{101}, X_{110}, X_{111}) = (10, 12, 12, 6, 16, 15, 9, 4)$.

We will present now a sequence of results concerning the parameters defined for the k-deck. The first lemma is a trivial observation.

Lemma 8.5. *If X is a binary sequence of length n and weight w, then*
1. $D_1(X) = (n - w, w)$.
2. $D_1(n) = n + 1$.

Lemma 8.6. *If X is a binary sequence of length n and weight w, then*
1. $D_2(X) = \left(\binom{n-w}{2}, t, w(n-w) - t, \binom{w}{2}\right)$, *where t can be any integer between 0 and $w(n - w)$.*
2. $D_2(n) = (n^3 + 5n + 6)/6$.

Proof. Let X be a binary sequence of length n and weight w.

1. For a sequence with w *ones* we have $X_{11} = \binom{w}{2}$ and $X_{11} = \binom{n-w}{2}$. For the sequence $(1^w 0^{n-w})$ we have $X_{01} = 0$ and for the sequence $(0^{n-w} 1^w)$ we have $X_{01} = w(n-w)$. Two sequences of weight w that differ only in two consecutive positions, where one has 01 and the other has 10, in these two positions, differ by exactly one in the value of their X_{01} and by exactly one in the value of their X_{10}. Hence, all the values between 0 and $w(n-w)$ are attained for X_{01} with sequences of length n and weight w. Therefore for a sequence X of length n and weight w we have

$$D_2(X) = \left(\binom{n-w}{2}, t, w(n-w) - t, \binom{w}{2}\right),$$

where t can be any integer between 0 and $w(n - w)$, depending on X.
2. Once t between 0 and $w(n - w)$ is determined, the whole 2-deck of the sequence is determined. Therefore the total number of 2-decks is

$$\sum_{w=0}^{n} (w(n-w) + 1) = \sum_{w=0}^{n} wn - \sum_{w=0}^{n} w^2 + n + 1$$

and since it is well known that $\sum_{w=0}^{n} w^2 = n(n + 1)(2n + 1)/6$, it follows that

$$\sum_{w=0}^{n} (w(n-w) + 1) = \frac{n^2 + n^3}{2} - \frac{2n^3 + 3n^2 + n}{6} + n + 1 = \frac{n^3 + 5n + 6}{6}$$

and thus, $D_2(n) = (n^3 + 5n + 6)/6$. □

Lemma 8.7. *The k-deck of a sequence X of length n induces the $(k - 1)$-deck of the sequence.*

Proof. Each projection of k positions in the sequence X induces k projections of $k - 1$ positions. Each $k - 1$ projection in the sequence X is contained in

$n - k + 1$ projections of k positions in X. Hence, summing the $(k - 1)$-decks of the sequences of length k in the k-deck of X and dividing by $n - k + 1$ yields the $(k - 1)$-deck of X. □

Corollary 8.3. *If two sequences X and Y have the same k-deck, $k > 1$, then they have the same $(k - i)$-deck for each $1 \leq i \leq k - 1$.*

Corollary 8.4. *If a sequence X of length n has a unique k-deck for some $k > 0$, then X has unique $(k + i)$-deck for each $1 \leq i \leq n - i$.*

Proof. If two sequences X and Y have the same $(k + i)$-deck, then by Corollary 8.3 they have the same k-deck. Therefore if X does not have a unique $(k + i)$-deck, then it does not have a unique k-deck. □

Lemma 8.8. *All the sequences of length n over an alphabet of size $\sigma > 2$ have a unique k-deck if and only if all the binary sequences of length n have a unique k-deck.*

Proof. Assume first that all the sequences of length n over an alphabet of size $\sigma > 2$ have a unique k-deck. Since this set of sequences contains also all the binary sequences, it follows that all the binary sequences of length n have a unique k-deck.

Assume now that all the binary sequences of length n have a unique k-deck. Let X be a sequence over an alphabet of size $\sigma > 2$ and let α be a nonzero symbol in the alphabet. Let X' be a sequence obtained from X by replacing all symbols different from α by a *zero*. Clearly, X' is a binary sequence and hence it has a unique k-deck, which implies that we can determine the positions of α in X. Repeating the same procedure with the other symbols yields the positions of each symbol of the alphabet in X. Thus X has a unique k-deck. □

Let $S(k)$ be the smallest value of n such that there exist two distinct sequences of length n with the same k-deck.

Let $T(n)$ be the smallest value of k such that all the sequences of length n have unique k-decks.

The following lemma is an immediate consequence of these two definitions.

Lemma 8.9.

- *If $S(k) \leq n$, then $T(n) > k$.*
- *If $T(n) \leq k$, then $S(k) > n$.*

Proof.

- If $S(k_0) \leq n_0$, then the smallest n for which there exists two distinct sequences of length n with the same k_0-deck is at most n_0. Hence, there are sequences of length n_0 that have the same k_0-deck. Therefore the smallest k such that all the sequences of length n_0 have a unique k-deck is larger than k_0. Thus $T(n_0) > k_0$, i.e., if $S(k) \leq n$, then $T(n) > k$.

- If $T(n_0) \leq k_0$, then the smallest k such that all sequences of length n_0 have a unique k-deck is at most k_0. Hence, all the sequences of length n_0 have a unique k_0-deck. Therefore the smallest n for which there exist two distinct sequences with the same k_0-deck is larger than n_0. Thus $S(k_0) > n_0$, i.e., if $T(n) \leq k$, then $S(k) > n$. □

We will now derive two simple lower and upper bounds on $T(n)$, which is the smallest value of k for which all sequences of length n have a unique k-deck. The lower bound will be proved by a simple counting argument.

Lemma 8.10. *For each ϵ, $0 < \epsilon < 1$, there exists an N such that*

$$T(n) > \epsilon \cdot \log_2 n$$

for all $n > N$.

Proof. The proof is by a counting argument. We first compute the number of possible k-decks (including those that might be impossible to obtain). We have to choose $\binom{n}{k}$ elements (for the k-subsets of a sequence of length n) from the 2^k possible ordered k-tuples. This number is a combination with repetitions that is equal to

$$\binom{\binom{n}{k} + 2^k - 1}{\binom{n}{k}} = \binom{\binom{n}{k} + 2^k - 1}{2^k - 1}.$$

For $3 \leq k < n$ we have that

$$\binom{\binom{n}{k} + 2^k - 1}{2^k - 1} < \binom{n}{k}^{2^k - 1} < n^{k(2^k - 1)}.$$

For a fixed k and sufficiently large n we have that $n^{k(2^k - 1)} < 2^n$ and hence the number of possible k-decks is smaller than the number of sequences of length n. This implies that there are sequences of length n whose k-decks are not unique.

Let $k = \epsilon \cdot \log_2 n$ for some $0 < \epsilon < 1$ and again consider the inequality

$$n^{k(2^k - 1)} < 2^n,$$

which by substitution of $k = \epsilon \cdot \log_2 n$ becomes

$$n^{\epsilon \cdot (2^{\epsilon \cdot \log_2 n} - 1) \cdot \log_2 n} < 2^n,$$

which is equivalent to

$$\epsilon \cdot \left(\log_2 n\right)^2 \left(n^{\epsilon} - 1\right) < n,$$

which is again true for sufficiently large n. Thus for any ϵ, $0 < \epsilon < 1$ and sufficiently large n we have that

$$T(n) > \epsilon \cdot \log_2 n.$$

□

For an upper bound on $T(n)$ we will use a different presentation for binary sequences. For a binary sequence X of length n and weight w we define the *zero vector* to be a vector of length $w + 1$, $(i_0, i_1, \ldots, i_w)$ whose jth entry i_j, $1 \le j \le w - 1$, is the number of *zeros* between the jth *one* and the $(j + 1)$th *one* in the sequence, i_0 is the number of *zeros* before the first *one* in the sequence and i_w is the number of *zeros* after the last *one* in the sequence. By definition, the sum of entries of the zero vector of X is $n - w$. Similarly, we define the *one vector* of X to be the vector that indicates the number of *ones* between the *zeros* of X. The one vector of X is equal to the zero vector of $\bar{X}$. The sum of the entries of the one vector of X is w.

Example 8.8. For the sequence $X = (101000110)$, the zero vector is $(0, 1, 3, 0, 1)$ and the one vector is $(1, 1, 0, 0, 2, 0)$. ∎

The following lemma is a simple observation.

Lemma 8.11. *A sequence is uniquely defined by its zero (one) vector.*

Lemma 8.12. *A sequence of length n and weight $k - 1$ has a unique k-deck.*

Proof. Let X be a sequence of length n and weight $k - 1$ with zero vector $(i_0, i_1, \ldots, i_{k-1})$ and Y be a subsequence of X with one vector $(j, k - j - 1)$, where $0 \le j \le k - 1$, i.e., Y is a subsequence of length k and weight $k - 1$ with a unique *zero* in its $(j + 1)$th position. Clearly, Y has weight $k - 1$ as the weight of X, and hence the number of appearances of Y as a projection of k positions of X is exactly i_j. This implies that the k-deck determines the exact zero vector of X. Therefore by Lemma 8.11 the k-deck of X is unique. □

Corollary 8.5.

$$T(n) \le \frac{n}{2} + 1 .$$

Proof. A sequence X of length n has at most $\frac{n}{2}$ *zeros* or at most $\frac{n}{2}$ *ones*. Therefore by Lemma 8.12, X has a unique k-deck for some $k \le \frac{n}{2} + 1$. This implies by Corollary 8.4 that X has a unique $\left(\left\lfloor \frac{n}{2} \right\rfloor + 1\right)$-deck. Thus

$$T(n) \le \frac{n}{2} + 1.$$ □

8.5 Synchronization codes

Various types of synchronization codes are generated for a unique deciphering of a text message. There are a few families of codes, which are considered in this section. These families of codes have found applications associated with DNA.

A code $\mathcal{C}$ is called a ***comma-free code*** if it contains words of length n over an alphabet of size σ and for any two codewords (not necessarily distinct)

$(a_1, a_2, \ldots, a_n)$ and $(b_1, b_2, \ldots, b_n)$, any subsequence of

$$a_1 a_2 \cdots a_n b_1 b_2 \cdots b_n$$

is not a codeword in $\mathcal{C}$, except for the prefix of length n and the suffix of length n. Let $CF(n, \sigma)$ be the maximum number of codewords in a comma-free code of length n over an alphabet of size σ.

Theorem 8.1. *For each positive integers $n \geq 1$ and $\sigma \geq 2$,*

$$CF(n, \sigma) \leq \frac{1}{n} \sum_{d \mid n} \mu(d) \cdot \sigma^{n/d}.$$

Proof. Let $\mathcal{C}$ be a comma-free code of length n over an alphabet of size σ. For any given word $(a_1, a_2, \ldots, a_n)$ the sequence

$$a_1, a_2, \ldots, a_n, a_1, a_2, \ldots, a_n$$

contains all the cyclic shifts of $(a_1, a_2, \ldots, a_n)$ as subsequences and hence from each full-order necklace of length n over σ beads at most one word can be contained in $\mathcal{C}$. If the word $(a_1, a_2, \ldots, a_n)$ has period d that is less than n, then the sequence

$$a_1, a_2, \ldots, a_n, a_1, a_2, \ldots, a_n$$

has a subsequence

$$a_{d+1}, a_{d+2}, \ldots, a_n, a_1, a_2, \ldots, a_d = a_1, a_2, \ldots, a_n$$

and hence it cannot be contained in $\mathcal{C}$. Therefore $\mathcal{C}$ contains at most one word from each full-order necklace of length n over σ beads. By Theorem 3.6 the number of such full-order necklaces is

$$\frac{1}{n} \sum_{d \mid n} \mu(d) \cdot \sigma^{n/d},$$

which completes the proof. $\qquad\qquad\qquad\qquad\qquad\qquad\qquad\qquad\qquad\square$

A much stronger condition than the one for comma-free codes is required in the following type of codes called non-overlapping codes.

A code $\mathcal{C}$ is called a ***non-overlapping code*** if all the codewords have the same length and for any two codewords $c_1, c_2 \in \mathcal{C}$ (not necessarily distinct) any nonempty prefix of c_1 is not a suffix of c_2. Let $NO(n, \sigma)$ be the maximum number of codewords in a non-overlapping code of length n over an alphabet of size σ.

We have that $NO(1, \sigma) = \sigma$, and hence we will assume from now that $n \geq 2$. It is also easy to verify by the definition that a non-overlapping code is also a comma-free code, but not the converse.

Theorem 8.2. *If $n \geq 2$ and $\sigma \geq 2$, then*

$$\text{NO}(n, \sigma) < \frac{\sigma^n}{2n - 1}.$$

Proof. Let C be a non-overlapping code of length n over an alphabet Σ with σ letters. Define the following set A.

$$A \triangleq \{(x, i) : x = (x_0, \ldots, x_{2n-2}) \in \Sigma^{2n-1}, x_i \cdots x_{i+n-1} \in C, 0 \leq i \leq 2n-2\},$$

where the subscripts in the subsequence $x_i \cdots x_{i+n-1}$ are taken modulo $2n - 1$.

There are $|C|$ codewords in C and each one can start in any of the $2n - 1$ positions of a word x in $(x, i) \in A$. The codeword is of length n and hence there are σ^{n-1} distinct ways to complete the word x in (x, i). Since C is a non-overlapping code, it follows that two codewords (not necessarily distinct) cannot appear as distinct sub-words of the same word of length $2n - 1$. Therefore

$$|A| = (2n - 1) |C| \sigma^{n-1}.$$

This also implies that for each word x of length $2n - 1$ over Σ, there exists at most one choice of an integer i such that $(x, i) \in A$ and hence A contains at most σ^{2n-1} codewords. Moreover, the words of length $2n - 1$ that contain $2n - 1$ repetitions of the same symbol cannot have a sub-word from a non-overlapping code, and hence

$$|A| \leq \sigma^{2n-1} - \sigma < \sigma^{2n-1}.$$

Thus we have

$$(2n - 1) |C| \sigma^{n-1} = |A| < \sigma^{2n-1},$$

which implies the claim of the theorem. □

Non-overlapping code construction:

Let k be an integer such that $1 \leq k \leq n - 1$. Let C be the set of all words in Σ^n, where $(x_1, x_2, \ldots, x_n) \in C$ if the following three conditions are satisfied:

- $x_i = 0$ for all $1 \leq i \leq k$;
- $x_{k+1} \neq 0$ and $x_n \neq 0$;
- the sequence $x_{k+2}, x_{k+3}, \ldots, x_{n-1}$ does not contain k consecutive *zeros*.

Theorem 8.3. *The code C constructed in the non-overlapping code construction is a non-overlapping code.*

Proof. Since a codeword in C starts with k consecutive *zeros* followed by a nonzero symbol and ends with a nonzero symbol, it follows that to have an overlap of a prefix of a codeword x with a suffix of a codeword y, y must have a run of k *zeros* except for the initial one. However, the only run of k *zeros*, is in the first k positions. Therefore C is a non-overlapping code. □

Lemma 8.13. *If σ is a fixed integer greater than 1, then the codes of the non-overlapping code construction imply that*

$$\liminf_{n \to \infty} \frac{\text{NO}(n, \sigma)}{\sigma^n / n} \geq \frac{(\sigma - 1)^2(2\sigma - 1)}{4\sigma^4}.$$

Proof. For a given k, any sequence that starts with k *zeros* followed by a nonzero symbol, ends with a nonzero symbol, and is not generated by the non-overlapping code construction, contains a run of k *zeros* in the other $n - k - 2$ positions. This run of *zeros* can start in $n - k - 2 - (k - 1) = n - 2k - 1$ possible positions (as it cannot overlap the last symbol). For the other $n - 2k - 2$ positions (outside these two runs of k *zeros*) there are σ^{n-2k-2} possibilities (some words will be counted more than once) and therefore there are at most $(n - 2k - 1)\sigma^{n-2k-2}$ such sequences, on these $n - k - 2$ positions, not fixed by the non-overlapping code construction. Therefore there are at least

$$\sigma^{n-k-2} - (n - 2k - 1)\sigma^{n-2k-2} > \sigma^{n-k-2} - n \cdot \sigma^{n-2k-2}$$

such sequences, on these $n - k - 2$ positions, which are generated by the construction, if $n \geq 2k + 2$. There are also $(\sigma - 1)^2$ distinct ways to choose the nonzero symbols for the $(k + 1)$th position and the last position. This implies that for the given k, the code $\mathcal{C}$ generated by the construction has at least

$$(\sigma - 1)^2 \left(\sigma^{n-k-2} - n\sigma^{n-2k-2} \right) = \left(\frac{\sigma - 1}{\sigma} \right)^2 \sigma^n \left(\sigma^{-k} - n\sigma^{-2k} \right)$$

codewords. The expression $\sigma^{-k} - n\sigma^{-2k}$ is maximized when $k = \log_\sigma(2n) + \delta$, where δ is chosen so that $|\delta| < 1$ and k is an integer. In this case, the value of $\sigma^{-k} - n\sigma^{-2k}$ is bounded below by $\frac{2\sigma-1}{4n\sigma^2}$ if δ is nonnegative. Thus

$$|\mathcal{C}| \geq \frac{(\sigma - 1)^2(2\sigma - 1)}{4n\sigma^4}\sigma^n,$$

which implies the claim of the lemma. $\qquad \square$

The family of constant-weight codes is always interesting and we turn now to binary constant-weight non-overlapping codes.

Constant-weight non-overlapping code construction:

Let w be a positive integer and $n = 2w - 1$. Consider the following binary code of length n and words of weight w. The code $\mathcal{C}$ contains all the codewords such that $x = (x_1, x_2, \ldots, x_n) \in \mathcal{C}$ if the following three conditions are satisfied:

- $x_1 = 1$;
- $\sum_{i=1}^{n} x_i = w$;
- In each nonempty prefix of x, the number of *ones* is strictly larger than the number of *zeros*.

Theorem 8.4. *The code C obtained in the constant-weight non-overlapping code construction is a non-overlapping code of length $n = 2w - 1$, and weight w.*

Proof. By definition all the words in C have length $2w - 1$, w *ones*, and $w - 1$ *zeros*. For any codeword x in C, each nonempty prefix of x has more *ones* than *zeros*. Hence, since x has w *ones* and $w - 1$ *zeros*, it follows that in any suffix of x, the number of *zeros* is at least the same as the number of *ones*. Therefore a prefix of the word x cannot be the suffix of a word $y \in C$. Thus C obtained in the constant-weight non-overlapping code construction is a non-overlapping code of length $n = 2w - 1$, and weight w. $\qquad\square$

8.6 Notes

Sequences and the de Bruijn graph are heavily used in biology and bioinformatics. This is no surprise since concepts in biology like DNA and RNA are presented as sequences. The de Bruijn graph was also used in antibody sequencing, see Bandeira, Pham, Pevzner, Arnott, and Lill [6], in synteny block reconstruction, see Pham and Pevzner [72], and in RNA assembly, see Grabherr et al. [40].

Section 8.1. Sequencing of the human genome was discussed first in Lander et al. [52] and in Venter et al. [85]. The number of references on the application of the de Bruijn graph for DNA sequencing is quite large. An excellent introduction to the problem was given in Compeau, Pevzner, and Tesler [23]. The direction discussed in this part of the chapter is taken from their paper.

Before the OLC method and DBG method there were other methods for sequencing such as sequencing by hybridization, see Drmanac, Labat, Brukner, and Crkvenjakov [26]. Another algorithm was suggested, for example, by Idury and Waterman [43]. The OLC method is usually not efficient since finding a Hamiltonian path is an NP-problem, see Garey and Johnson [35]. The method was developed by Kececioglu and Myers [48] and further, by Adams et al. [3]. Nevertheless, it was successfully applied by Fleischmann et al. [30] to obtain the first microbial genome. The de Bruijn graph method was first suggested by Pevzner [69]. The performances of the OLC algorithm and the DBG algorithm were compared by Schatz, Delcher, and Salzberg [76].

Errors in reads are unavoidable and handling errors in the reads is discussed, for example, in Butler, MacCallum, Kleber, Shlyakhter, Belmonte, Lander, Nusbaum, and Jaffe [12], Chaisson and Pevzner [14], Li et al. [60], Miller, Koren, and Sutton [64], Paszkiewicz and Studholme [68], Pevzner, Tang, and Tesler [70], Pevzner, Tang, and Waterman [71], Simpson, Wong, Jackman, Schein, Jones, and Birol [80], and Zerbino and Birney [89].

Section 8.2. DNA-based storage has attracted significant attention due to recent demonstrations of the viability of storing information in macromolecules. Unlike classical optical and magnetic storage technologies, DNA-based storage does not require an electrical supply to maintain data integrity, and given the trends in cost decreases of DNA synthesis and sequencing, it is now acknowledged that shortly DNA storage may become a highly competitive archiving technology.

The potential for using macromolecules for ultra-dense storage was recognized as early as in the 1960s by Feynman [29]. DNA molecules, which may be abstracted as strings over the four symbol alphabet $\{A, C, G, T\}$, stand out due to several unique properties:

1. Self-assembly potential (DNA has been successfully used as a building block of a number of small-scale self-assembly-based computers, see Nadrian [65]).

2. Stability (DNA can be recovered from 30 000 year-old Neanderthal and 700000 year-old horse bones, see Saey [75]), and capacity (a single human cell, with a tiny mass, hosts DNA strands encoding 6.4 GB of information).

Furthermore, the technologies for synthesizing (writing) artificial DNA and for massive sequencing (reading) have reached unprecedented levels of efficiency and accuracy, see Shendure and Aiden [79]. As a result, DNA storage systems may be the most plausible DNA-based platform to materialize shortly.

Storing data in DNA is not a new idea. One of the first experiments was conducted by Clelland, Risca, and Bancrof [21], where they recovered a message consisting of 23 characters. Shorty after, Leier, Richter, Banzhaf, and Rauhe [53] managed to successfully store three sequences of nine bits each. A more significant accomplishment, concerning the amount of data stored successfully, was reported by Gibson et al. [36]. They stored 1280 characters in a bacterial genome, that is, in vivo storage. The first large-scale experiments that demonstrated the potential of in vitro DNA storage were reported by Church, Gao, and Kosuri [19] who recovered 643 kB of data, and by Goldman, Bertone, Chen, Dessimoz, LeProust, Sipos, and Birney [38] who accomplished the same task for a 739 kB message. However, both of these groups did not recover the entire message successfully due to the lack of using the appropriate coding solutions to correct errors. Church, Gao, and Kosuri [19] had 10-bit errors, and Goldman, Bertone, Chen, Dessimoz, LeProust, Sipos, and Birney [38] lost two strands of 25 nucleotides. Later, Grass, Heckel, Puddu, Paunescu, and Stark [41], stored and recovered successfully a 81 kB message and Bornholt, Lopez, Carmean, Ceze, Seelig, and Strauss [11] similarly succeeded while storing a 42 kB message. Another progress in the amount of stored data was reported in Blawat, Gaedke, Hütter, Chen, Turczyk, Inverso, Pruitt, and Church [10] who successfully stored 22 MB of data. More recently, 2.11 MB of data were stored with a high storage rate, as shown by Erlich and Zielinski [28]. Organick et al. [67] succeeded to store 200 MB of data, thereby storing an order of magnitude more data than the previous experiment reported by Blawat, Gaedke, Hütter, Chen, Turczyk, Inverso, Pruitt, and Church [10]. A method that offers both random access and rewritable storage was developed in Yazdi, Yuan, Ma, Zhao, and

Milenkovic [87]. The work on sequences, coding, and storing data in DNA storage is rapidly developing as of 2023. To complete this part we will also mention the work by Song, Cai, and Immink [81].

The microscopic world in which the DNA molecules reside induces error patterns that are fundamentally different from their digital counterparts. This distinction results from the specific error behavior in DNA and the method by which DNA strands are stored together. Hence, to maintain reliability in reading and writing, new coding schemes must be developed, and first attempts for such solutions were already implemented in proof-of-concept storage systems, e.g., see Erlich and Zielinski [28] and Yazdi, Yuan, Ma, Zhao, and Milenkovic [87]. The definitions and the analysis of the DNA storage channel were done by Kiah, Puleo, and Milenkovic [49] who also found its application to DNA storage and made the connection of this analysis to the de Bruijn graph. More work on these topics can be found in newly developed works that cite these references.

Beyond all these, there are other research studies carried out on DNA storage codes that involve the de Bruijn graph and de Bruijn sequences. We have the feeling that such research studies will continue for a long time.

Section 8.3. The graph $G_n(w_1, w_2)$ was considered first by Ruskey, Sawada, and Williams [74]. They also described an efficient algorithm to generate an Eulerian cycle in this graph. Its generalization for $G_{\sigma,n}(w_1, w_2)$ was carried out by Kiah, Puleo, and Milenkovic [49]. The description and the proofs that are given in this section are slightly different from those given in these papers.

Section 8.4. The k-deck problem was presented first by Kalashnik [47]. The lower and upper bounds on $T(n)$ proved in Lemma 8.10 and Corollary 8.5, respectively, were presented in the paper of Manvel, Meyerowitz, Schwenk, Smith and Stockmeyer [61], which started the relatively large amount of research on this topic. Dudik and Schulman [25] presented the following bounds of $S(k)$:

Theorem 8.5.

- *If* $7 \le k \le 28$, *then* $S(k) \le 1.75 \cdot 1.62^k$.
- *If* $29 \le k \le 84$, *then* $S(k) \le 0.25 \cdot 1.17^k k^3 \log_2 k$.
- *If* $85 \le k$, *then* $S(k) \le 3^{(1.5+o(1))(\log_3 k)^2}$.

Rigo and Salimov [73] suggested Lemmas 8.5 and 8.6. They proved that $D_k(n) \le \prod_{i=1}^{k} \left(\binom{n}{i} + 1 \right)^{(2^i - 1)}$ and for a fixed k they proved that we have $D_k(n) = O\left(n^{2((k-1)2^k + 1)} \right)$. Chrisnata, Kiah, Karingula, Vardy, Yaakobi, and Yao [17] proved that $D_k(n) = O\left(n^{(k-1)2^{k-1} + 1} \right)$. They proved an asymptotic lower bound $D_k(n) = \Omega(n^k)$ and improved it for $k = 3$ to $D_3(n) = \Omega(n^6)$. Other papers that considered the k-deck problem are by Choffrut and Karhumäki [22], Krasikov and Roditty [51] and Scott [78].

There are many other reconstruction problems, but we will concentrate now only on one of them. Motivated by protein sequencing the following reconstruction problem was suggested by Acharya, Das, Milenkovic, Orlitsky, and

Pan [2]. Assume that we are given a multiset that contains all the contiguous subsequences of a sequence S, where the symbols of the subsequences are unordered, i.e., only the number of appearances of each symbol in the subsequence is given. Can the sequence S be reconstructed from all its subsequences?

Example 8.9. The sequence CTCAG is decomposed to a set with $\binom{5}{2} = 10$ subsequences as follows:

$$\{A, C, C, G, T, AC, AG, CT, CT, ACG, ACT, C^2T, AC^2T, ACGT, AC^2GT\}.$$

∎

We start by showing that in a similar way to the k-deck problem, it is sufficient to consider only binary sequences, by using similar arguments.

Lemma 8.14. *If all binary sequences of a certain length are reconstructible from their contiguous unordered subsequences, then sequences of the same length over any finite alphabet are reconstructible from their contiguous unordered subsequences.*

Proof. Assume that all the binary sequences of length n can be reconstructed from their contiguous unordered subsequences. Let X be a sequence of length n over an alphabet of size $\sigma > 2$ and let α be a nonzero symbol in the alphabet. Let X' be a sequence obtained from X by replacing all symbols different from α by *zeros*. Clearly, X' is a binary sequence of length n and hence it is reconstructible from its contiguous unordered symbols. Hence, we can determine the positions of α in X. Repeating the same procedure with the other symbols yields the positions of each symbol of the alphabet in X. Thus X is reconstructible from its contiguous unordered symbols. □

Another trivial observation is that the sequences S and S^R have the same decomposition into contiguous unordered subsequences. Hence, these two sequences will not be distinguished as different sequences. The following results were proved by Acharya, Das, Milenkovic, Orlitsky, and Pan [2].

Lemma 8.15. *All sequences of length at most 7 are reconstructible from their contiguous unordered subsequences.*

Lemma 8.16. *If $n + 1$ is a product of two integers greater than 2, then there exists a pair of sequences of length $n + 1$ that have the same decomposition.*

Theorem 8.6.

1. *All sequences whose length is one less than a prime are reconstructible from their contiguous unordered subsequences.*
2. *All sequences whose length is one less than twice a prime are reconstructible from their contiguous unordered subsequences.*

3. *For each other length there are pairs of sequences that have the same decomposition into contiguous unordered subsequences.*

Reconstruction of sequences from their subsequences is one of the main problems in DNA storage. This has attracted much attention. The first important papers on reconstructions associated with information theory were written by Levenshtein [56,57]. This work, on reconstruction with multiple channels, which was followed by others, is the main direction required for DNA storage. The deletion channel is one of the most important ones for DNA storage and the reconstruction model of Levenshtein was considered for this channel by Gabrys and Yaakobi [33]. Further studies on reconstruction with a different direction were carried out by Cheraghchi, Gabrys, Milenkovic, and Ribeiro [16], and Gabrys and Milenkovic [32].

The information stored in the DNA is subject to errors and several types of errors can occur in codewords stored in DNA. Although errors occurred in the received strands, it is still required to overcome these errors and reconstruct the codewords. This is one of the main targets of research on DNA storage. Many types of errors are typical for DNA storage. One type of error is a duplication of a small subsequence that is attached exactly in the place in which it was duplicated. Such an error called tandem-duplication and reconstruction of codewords in the presence of tandem-duplication errors, was considered by Jain, Farnoud, Schwartz, and Bruck [45] and by Yehezkeally and Schwartz [88]. Another type of error is a deletion of a symbol, an insertion of a symbol, or a substitution error. An error that can be any one of these three types of errors is called an edit error. Reconstruction in the presence of an edit error was considered first by Abu-Sini and Yaakobi [1] and later by Cai, Kiah, Nguyen, and Yaakobi [13] and by Chrisnata, Kiah, and Yaakobi [18]. A combination of tandem-duplication errors and edit errors was considered by Tang and Farnoud [83]. Edit errors combined with adjacent transposition errors are called Damerau errors and these errors were considered by Gabrys, Yaakobi, and Milenkovic [34]. A similar study was carried out by Gabrys, Kiah, and Milenkovic [31] for codes in the asymmetric Lee distance for DNA storage.

Section 8.5. Comma-free codes were defined first by Crick, Griffith, and Orgel [24] in connection with protein synthesis. They considered a reconstruction of a sequence of length 20 or amino acids with a comma-free code. The mathematical analysis of this type of code was carried out a year later by Golomb, Gordon, and Welch [39]. An excellent survey of the results in this area until 1987 was given by Levenshtein [58]. Further work on comma-free codes can be found in Churchill [20], Eastman [27], Jiggs [46], King and Gaborit [50], Scholtz [77], and Tang, Golomb, and Graham [84].

Non-overlapping codes were reintroduced several times over the years. Gilbert [37] was the first to consider non-overlapping codes, but they were more directly approached first by Levenshtein [55]. They were called later cross-bifix-free codes by Bajić and Stojanović [4], Bajić, Stojanović, and Lindner [5], and Stefanovic and Bajić [82]. Yazdi, Kiah, Gabrys, and Milenkovic [86] have shown how these codes can be used for reconstruction in DNA storage.

Blackburn [9] gave a review of non-overlapping code with new results and simpler proofs for older results. Theorems 8.2 and 8.3, and Lemma 8.13 are based on his ideas.

Theorem 8.2 can be improved with the following result obtained by Levenshtein [55] whose proof requires more sophisticated analysis.

Theorem 8.7. *If $n \geq 2$ and $\sigma \geq 2$, then*

$$\mathrm{NO}(n, \sigma) \leq \frac{1}{n} \left(\frac{n-1}{n} \right)^{n-1} \sigma^n,$$

where e is the base of the natural logarithm and when $n \to \infty$

$$\mathrm{NO}(n, \sigma) \leq \frac{1}{e} \cdot \frac{\sigma^n}{n-1}.$$

Gilbert [37] and Levenshtein [54] proved that when σ is fixed and k (of the non-overlapping code construction) is chosen appropriately as a function of n, we have the following bound on the code $\mathcal{C}$ obtained from the non-overlapping code construction.

$$|\mathcal{C}| \gtrsim \frac{\sigma-1}{\sigma \cdot e} \cdot \frac{\sigma^n}{n},$$

where e is the base of the natural logarithm, and $n \to \infty$ over the subsequence $\left\{ n = \frac{\sigma^i - 1}{\sigma - 1} \right\}_{i=0}^{\infty}$. More results and improvements can be found in Levy and Yaakobi [59].

The non-overlapping code construction requires sequences with no runs of k consecutive *zeros*. Such sequences are known as run-length limited codes (RLL codes) and their number was extensively studied. A survey on these codes was given by Immink [44] and a later excellent survey that covers also other constrained codes was written by Marcus, Roth, and Siegel [62].

For large σ compared to n, Blackburn [9] has proved that the bound of Theorem 8.7 can be attained if n divides σ. When σ is large and n does not divide σ, the bound of Theorem 8.7 is attained asymptotically for $\sigma \to \infty$, as was shown by Blackburn [9]. For $\sigma \in \{2, 3\}$, Blackburn [9] managed to compute the exact value of $\mathrm{NO}(n, \sigma)$ as follows.

Theorem 8.8.

1. *For any $\sigma \geq 2$ we have that $\mathrm{NO}(2, \sigma) = \left\lceil \frac{\sigma}{2} \right\rceil \cdot \left\lfloor \frac{\sigma}{2} \right\rfloor$;*
2. *for any $\sigma \geq 2$ we have that $\mathrm{NO}(3, \sigma) = [2\sigma/3]^2 (\sigma - [2\sigma/3])$;*

where $[x]$ denotes the nearest integer to the real number x.

Constant-weight non-overlapping codes were constructed by Markov and Noskov [63] and the constant-weight non-overlapping code construction given in this section was introduced by Bilotta, Pergola, and Pinzani [8]. Further work on non-overlapping codes can be found in Bilotta, Grazzini, Pergola, and Pinzani [7], Chee, Kiah, Purkayastha, and Wang [15], Levy and Yaakobi [59], and Morita, van Wijngaarden, and Han Vinck [66].

Other codes for synchronization based on other types of non-overlapping prefixes and suffixes were produced by Guibas and Odlyzko [42].

References

[1] M. Abu-Sini, E. Yaakobi, On Levenshtein's reconstruction problem under insertions, deletions, and substitutions, IEEE Trans. Inf. Theory 67 (2021) 7132–7158.

[2] J. Acharya, H. Das, O. Milenkovic, A. Orlitsky, S. Pan, String reconstruction from substring compositions, SIAM J. Discrete Math. 29 (2015) 1340–1371.

[3] M.D. Adams, et al., The genome sequence of Drosophila melanogaster, Science 287 (2000) 2185–2195.

[4] D. Bajic, J. Stojanovic, Distributed sequences and search process, in: Proc. IEEE Int. Conf. Commun., Paris, 2004, pp. 514–518.

[5] D. Bajic, J. Stojanovic, J. Lindner, Multiple window-sliding search, in: Proc. IEEE Int. Symp. on Infor. Theory (ISIT), Yokohoma, Japan, 2003, p. 249.

[6] N. Bandeira, V. Pham, P. Pevzner, D. Arnott, J.R. Lill, Automated de novo protein sequencing of monoclonal antibodies, Nat. Biotechnol. 26 (2008) 1336–1338.

[7] S. Bilotta, E. Grazzini, E. Pergola, R. Pinzani, Avoiding cross-bifix-free binary words, Acta Inform. 50 (2013) 157–173.

[8] S. Bilotta, E. Pergola, R. Pinzani, A new approach to cross-bifix-free sets, IEEE Trans. Inf. Theory 58 (2012) 4058–4063.

[9] S.R. Blackburn, Non-overlapping code, IEEE Trans. Inf. Theory 61 (2015) 4890–4894.

[10] M. Blawat, K. Gaedke, I. Hütter, X.-M. Chen, B. Turczyk, S. Inverso, B.W. Pruitt, G.M. Church, Forward error correction for DNA data storage, Int. Conf. Comput. Sci. 80 (2016) 1011–1022.

[11] J. Bornholt, R. Lopez, D.M. Carmean, L. Ceze, G. Seelig, K. Strauss, A DNA-based archival storage system, in: Proc. of the Twenty-First Int. Conf. on Architectural Support for Programming Languages and Operating Systems (ASPLOS), Atalnta, GA, 2016, pp. 637–649.

[12] J. Butler, I. MacCallum, M. Kleber, I.A. Shlyakhter, M.K. Belmonte, E.S. Lander, C. Nusbaum, D.B. Jaffe, ALLPATHS: de novo assembly of whole-genome shotgun microread, Genome Res. 18 (2008) 810–820.

[13] K. Cai, H.M. Kiah, T.T. Nguyen, E. Yaakobi, Coding for sequence reconstruction for single edits, IEEE Trans. Inf. Theory 68 (2022) 66–79.

[14] M.J. Chaisson, P.A. Pevzner, Short read fragment assembly of bacterial genomes, Genome Res. 18 (2008) 324–330.

[15] Y.M. Chee, H.M. Kiah, P. Purkayastha, C. Wang, Cross-bifix-free codes within a constant factor of oprimality, IEEE Trans. Inf. Theory 59 (2013) 4668–4674.

[16] M. Cheraghchi, R. Gabrys, O. Milenkovic, J. Ribeiro, Coded trace reconstruction, IEEE Trans. Inf. Theory 66 (2020) 6084–6103.

[17] J. Chrisnata, H.M. Kiah, S.R. Karingula, A. Vardy, E. Yaakobi, H. Yao, On the number of distinct k-decks: enumeration and bounds, Adv. Math. Commun. 17 (2023) 960–978.

[18] J. Chrisnata, H.M. Kiah, E. Yaakobi, Correcting deletions with multiple reads, IEEE Trans. Inf. Theory 68 (2022) 7141–7158.

[19] G.M. Church, Y. Gao, S. Kosuri, Next-generation digital information storage in DNA, Science 337 (2012) 1628.

[20] A.L. Churchill, Restrictions and generalizations on comma-free codes, Electron. J. Comb. 16 (2009), #R25.

[21] C.T. Clelland, V. Risca, C. Bancroft, Hiding messages in DNA microdots, Nature 399 (1999) 533–534.

[22] C. Choffrut, J. Karhumäki, Combinatorics of words, in: G. Rozenberg, A. Salomaa (Eds.), Handbook of Formal Languages, vol. I, 1997, pp. 329–438.

[23] P.E.C. Compeau, P.A. Pevzner, G. Tesler, How to apply de Bruijn graph to genome assembly, Nat. Biotechnol. 29 (2011) 987–991.

[24] F.H.C. Crick, J.S. Griffith, L.E. Orgel, Codes without commas, Proc. Natl. Acad. Sci. 43 (1957) 416–421.

[25] M. Dudik, L.J. Schulman, Reconstruction from subsequences, J. Comb. Theory, Ser. A 103 (2003) 337–348.

[26] R. Drmanac, I. Labat, I. Brukner, R. Crkvenjakov, Sequencing of megabase plus DNA by hybridization: theory of the method, Genomics 4 (1989) 114–128.

[27] W.L. Eastman, On the construction of comma-free codes, IEEE Trans. Inf. Theory 11 (1965) 263–267.

[28] Y. Erlich, D. Zielinski, DNA fountain enables a robust and efficient storage architecture, Science 355 (2017) 950–954.

[29] R. Feynman, There's plenty of room at the bottom, Eng. Sci., California Inst. Technol. 23 (1960) 22–36.

[30] R.D. Fleischmann, et al., Whole-genome random sequencing and assembly of Haemophilus influenzae Rd, Science 269 (1995) 496–512.

[31] R. Gabrys, H.M. Kiah, O. Milenkovic, Asymmetric Lee distance codes for DNA-based storage, IEEE Trans. Inf. Theory 63 (2017) 4982–4995.

[32] R. Gabrys, O. Milenkovic, Unique reconstruction of coded strings from multiset substring spectra, IEEE Trans. Inf. Theory 65 (2019) 7682–7696.

[33] R. Gabrys, E. Yaakobi, Sequence reconstruction over the deletion channel, IEEE Trans. Inf. Theory 64 (2018) 2924–2931.

[34] R. Gabrys, E. Yaakobi, O. Milenkovic, Codes in the Damerau distance for deletion and adjacent transposition correction, IEEE Trans. Inf. Theory 64 (2018) 2550–2570.

[35] M.R. Garey, D.S. Johnson, Computers and Intractability: A Guide to the Theory of NP-Completeness, W. H. Freeman and Company, New York, 1979.

[36] D.G. Gibson, et al., Creation of a bacterial cell controlled by a chemically synthesized genome, Science 329 (2010) 52–56.

[37] E.N. Gilbert, Synchronization of binary messages, IRE Trans. Inf. Theory 6 (1960) 470–477.

[38] N. Goldman, P. Bertone, S. Chen, C. Dessimoz, E.M. LeProust, B. Sipos, E. Birney, Towards practical, high-capacity, low-maintenance information storage in synthesized DNA, Nature 494 (2013) 77–80.

[39] S.W. Golomb, B. Gordon, L.R. Welch, Comma-free codes, Can. J. Math. 10 (1958) 202–209.

[40] M.G. Grabherr, et al., Full-length transcriptome assembly from RNA-Seq data without a reference genome, Nat. Biotechnol. 29 (2011) 644–652.

[41] R.N. Grass, R. Heckel, M. Puddu, D. Paunescu, W.J. Stark, Robust chemical preservation of digital information on DNA in silica with error-correcting codes, Angew. Chem., Int. Ed. 54 (2015) 2552–2555.

[42] L.J. Guibas, A.M. Odlyzko, Maximal prefix-synchronized code, SIAM J. Appl. Math. 35 (1978) 401–418.

[43] R.M. Idury, M.S. Waterman, A new algorithm for DNA sequence assembly, J. Comput. Biol. 2 (1995) 291–306.

[44] K.A.S. Immink, Runlength-limited sequences, Proc. IEEE 78 (1990) 1745–1759.

[45] S. Jain, F. Farnoud, M. Schwartz, J. Bruck, Duplication-correcting codes for data storage in the DNA of living organisms, IEEE Trans. Inf. Theory 63 (2017) 4996–5010.

[46] B.H. Jiggs, Recent results in comma-free codes, Can. J. Math. 15 (1963) 178–187.

[47] L.O. Kalashnik, The reconstruction of a word from fragments, in: Numerical Mathematics and Computer Technology, in: Akad. Nauk. Ukrain. SSR Inst. Mat., vol. IV, 1973, pp. 56–57.

[48] J.D. Kececioglu, E.W. Myers, Combinatorial algorithms for DNA sequence assembly, Algoritmica 13 (1995) 7–51.

[49] H.M. Kiah, G.J. Puleo, O. Milenkovic, Codes for DNA sequence profiles, IEEE Trans. Inf. Theory 62 (2016) 3125–3146.

[50] O.D. King, P. Gaborit, Binary templates for comma-free DNA codes, Discrete Appl. Math. 155 (2007) 831–839.

[51] I. Krasikov, Y. Roditty, On a reconstruction problem for sequences, J. Comb. Theory, Ser. A 77 (1997) 344–348.

[52] E.S. Lander, et al., Initial sequencing and analysis of the human genome, Nature 409 (2001) 860–921.

[53] A. Leier, C. Richter, W. Banzhaf, H. Rauhe, Cryptography with DNA binary strands, Biosystems 57 (2000) 13–22.

[54] V.I. Levenshtein, Decoding automata which are invariant with respect to their initial state, Probl. Kibern. 12 (1964) 125–136 (in Russian).

[55] V.I. Levenshtein, Maximum number of words in codes without overlaps, Probl. Pereda. Inf. 6 (1970) 88–90 (in Russian), Probl. Inf. Transm. 6 (1970) 355–357 (in English).

[56] V.I. Levenshtein, Efficient reconstruction of sequences, IEEE Trans. Inf. Theory 47 (2001) 2–22.

[57] V.I. Levenshtein, Efficient reconstruction of sequences from their subsequences or supersequences, J. Comb. Theory, Ser. A 93 (2001) 310–332.

[58] V.I. Levenshtein, Combinatorial problems motivated by comma-free codes, J. Comb. Des. 12 (2004) 184–196.

[59] M. Levy, E. Yaakobi, Mutually uncorrelated codes for DNA storage, IEEE Trans. Inf. Theory 65 (2019) 3671–3691.

[60] R. Li, et al., De novo assembly of human genomes with massively parallel short read sequencing, Genome Res. 20 (2010) 265–272.

[61] B. Manvel, A. Meyerowitz, A. Schwenk, K. Smith, P. Stockmeyer, Reconstruction of sequences, Discrete Math. 94 (1991) 209–219.

[62] B.H. Marcus, R.M. Roth, P.H. Siegel, Constrained systems and coding for recording channels, in: V. Pless, W. Huffman (Eds.), Handbook of Coding Theory, Elsevier, Amsterdam, The Netherlands, 1998, pp. 1635–1764.

[63] A.A. Markov, V.V. Noskov, Construction and properties of binary constant-weight codes without overlaps, Discrete Anal. 18 (1971) 49–65 (in Russian).

[64] J.R. Miller, S. Koren, G. Sutton, Assembly algorithms for next-generation sequencing data, Genomics 95 (2010) 315–327.

[65] N.C. Seemen, An overview of structural DNA nanotechnology, Mol. Biotechnol. 37 (2007) 246–257.

[66] H. Morita, A.J. van Wijngaarden, A.J. Han Vinck, On the construction of maximal prefix-synchronized codes, IEEE Trans. Inf. Theory 42 (1996) 2158–2166.

[67] L. Organick, et al., Random access in large-scale DNA data storage, Nat. Biotechnol. 36 (2018) 242–248.

[68] K. Paszkiewicz, D.J. Studholme, De novo assembly of short sequence reads, Brief. Bioinform. 11 (2010) 457–472.

[69] P.A. Pevzner, ℓ-tuple DNA sequencing: computer analysis, J. Biomol. Struct. Dyn. 7 (1989) 63–73.

[70] P.A. Pevzner, H. Tang, G. Tesler, De novo repeat classification and fragment assembly, Genome Res. 14 (2004) 1786–1796.

[71] P.A. Pevzner, H. Tang, M.S. Waterman, An Eulerian path approach to DNA fragment assembly, Proc. Natl. Acad. Sci. 98 (2001) 9748–9753.

[72] S.K. Pham, P.A. Pevzner, DRIMM-Synteny: decomposing genomes into evolutionary conserved segments, Bioinformatics 26 (2010) 2509–2516.

[73] M. Rigo, P. Salimov, Another generalization of abelian equivalence: binomial complexity of infinite words, Theor. Comput. Sci. 601 (2015) 47–57.

[74] F. Ruskey, J. Sawada, A. Williams, de Bruijn sequences for fixed-weight binary strings, SIAM J. Discrete Math. 26 (2012) 605–617.

[75] T.H. Saey, Story one: ancient horse's DNA fills in picture of equine evolution: a 700,000-year-old fossil proves astoundingly well preserved, Sci. News 184 (2013) 5–6.

[76] M.C. Schatz, A.L. Delcher, S.L. Salzberg, Assembly of large genomes using second-generation sequencing, Genome Res. 20 (2010) 1165–1173.

[77] R.A. Scholtz, Maximal and variable word-length comma-free codes, IEEE Trans. Inf. Theory 15 (1969) 300–306.

[78] A.D. Scott, Reconstructing sequences, Discrete Math. 175 (1997) 231–238.

[79] J. Shendure, E.L. Aiden, The expanding scope of DNA sequencing, Nat. Biotechnol. 30 (2012) 1084–1094.

[80] J.T. Simpson, K. Wong, S.D. Jackman, J.E. Schein, S.J.M. Jones, I. Birol, ABySS: a parallel assembler for short read sequence data, Genome Res. 19 (2009) 1117–1123.

[81] W. Song, K. Cai, K.A.S. Immink, Sequence-subset distance and coding for error control in DNA-based data storage, IEEE Trans. Inf. Theory 66 (2020) 6048–6065.

[82] C. Stefanovic, D. Bajic, On the search for a sequence from a predefined set of sequences in random and framed data streams, IEEE Trans. Commun. 60 (2012) 189–197.

[83] Y. Tang, F. Farnoud, Error-correcting codes for short tandem duplication and edit errors, IEEE Trans. Inf. Theory 68 (2022) 871–880.

[84] B. Tang, S.W. Golomb, R.L. Graham, A new result on comma-free codes of even word-length, Can. J. Math. 39 (1987) 513–526.

[85] J.C. Venter, et al., The sequence of the human genome, Science 291 (2001) 1304–1351.

[86] S.M.H.T. Yazdi, H.M. Kiah, R. Gabrys, O. Milenkovic, Mutually uncorrelated primers for DNA-based data storage, IEEE Trans. Inf. Theory 64 (2018) 6283–6296.

[87] S.M.H.T. Yazdi, Y. Yuan, J. Ma, H. Zhao, O. Milenkovic, A rewritable, random-access DNA-based storage system, Nat. Sci. Rep. 5 (2015) 14138.

[88] Y. Yehezkeally, M. Schwartz, Reconstruction codes for DNA storage with uniform tandem-duplication errors, IEEE Trans. Inf. Theory 66 (2020) 2658–2668.

[89] D.R. Zerbino, E. Birney, Velvet: algorithms for de novo short read assembly using de Bruijn graphs, Genome Res. 18 (2008) 821–829.

Chapter 9

Two-dimensional arrays

Perfect maps and distinct differences arrays

After having considered one-dimensional sequences, our goal now is to examine whether the definitions and results can be generalized into two-dimensional arrays. In particular, we want to generalize de Bruijn sequences, shortened de Bruijn sequences, and M-sequences. In this chapter, we restrict ourselves only to binary arrays, although the definitions and the results can be straightforwardly generalized for non-binary arrays. The definitions for a generalization for de Bruijn sequences and shortened de Bruijn sequences are straightforward. Perfect maps (or de Bruijn arrays) are $r \times t$ binary two-dimensional cyclic (doubly periodic) arrays in which all the $n \times m$ binary matrices appear exactly once as windows in one period of the array. A shortened de Bruijn array (or shortened perfect map) is an $r \times t$ binary two-dimensional cyclic array in which each one of the nonzero $n \times m$ binary matrices appears exactly once as a window in one period of the array.

As for M-sequences, we would like to have a definition for which an associated array has properties that generalize **R-1**, **R-2**, and **R-3** (sec Section 2.2). Generalizing property **R-1** is straightforward. Instead of **R-2** we will settle with the window property for such arrays. Instead of **R-3** we will demand a stronger property namely, the shift-and-add property. A pseudo-random array is an $r \times t$ binary two-dimensional cyclic array in which each one of the nonzero $n \times m$ binary matrices appears exactly once as a window in one period of the array and also if the array A is added to a nontrivial shift of A, then the outcome is another shift of A. As for M-sequences, such arrays have various applications.

We start in Section 9.1 by representing the problem of finding de Bruijn arrays and shortened de Bruijn arrays as a graph problem in a similar way as was done for de Bruijn sequences and shortened de Bruijn sequences. Two different representations will be presented. These representations will be of help in the constructions that follow in the other sections.

In Section 9.2 we discuss one type of construction for perfect maps. The construction in this section will be based on the graph whose vertices represent matrices, where each $n \times m$ binary matrix is contained exactly once in a matrix represented by one of the vertices. A Hamiltonian cycle in this graph is equivalent to a perfect map. The construction of the matrices for the vertices will require a structure called a perfect factor in G_n and these factors will also

Sequences and the de Bruijn Graph. https://doi.org/10.1016/B978-0-44-313517-0.00015-9

be a subject for discussion in this section. A shortened de Bruijn array will be constructed with a similar method, where another type of factor, namely, a zero factor, will replace the perfect factor.

Pseudo-random arrays are shortened perfect maps with the exception that they have the shift-and-add property as M-sequences. These arrays are the topic of Section 9.3. A construction that is based on folding an M-sequence will yield arrays with similar properties to those of M-sequences.

Section 9.4 is devoted to a recursive construction and also a generalization of the operator **D** into two dimensions.

Section 9.5 will consider a generalization of one-dimensional difference patterns into two-dimensional patterns with distinct differences.

9.1   Graph representations of perfect maps

The goal of this section is to represent de Bruijn arrays and shortened de Bruijn arrays by graphs, similarly as was done for de Bruijn sequences and shortened de Bruijn sequences with the de Bruijn graph. First, the formal definitions for these arrays should be given.

Definition 9.1. A *perfect map* (or a *de Bruijn array*) is an $r \times t$ binary (doubly) cyclic array, such that each binary $n \times m$ matrix appears exactly once as a window in the array. Such an array will be called an $(r, t; n, m)$-PM.

The definition of a perfect map immediately implies the following lemma.

Lemma 9.1. *If A is an $(r, t; n, m)$-PM, then*

1. $r > n$ or $r = n = 1$;
2. $t > m$ or $t = m = 1$; and
3. $rt = 2^{nm}$.

Lemma 9.1 immediately implies that both r and t must be powers of 2, restricting the possible parameters of a perfect map.

Definition 9.2. A *shortened perfect map* (or a *shortened de Bruijn array*) is an $r \times t$ binary (doubly) cyclic array, such that each nonzero binary $n \times m$ matrix appears exactly once as a window in the array. Such an array will be called an $(r, t; n, m)$-SPM.

The definition of a shortened perfect map immediately implies the following lemma.

Lemma 9.2. *If A is an $(r, t; n, m)$-SPM, then*

1. $r > n$ or $r = n = 1$;

2. $t > m$ or $t = m = 1$; and

3. $rt = 2^{nm} - 1$.

Lemma 9.2 immediately implies that both r and s divide $2^{nm} - 1$, restricting the possible parameters of shortened perfect maps. Moreover, a shortened perfect map cannot be constructed from a perfect map in the same way that a shortened de Bruijn sequence is constructed from a de Bruijn sequence.

The graph that represents span n de Bruijn sequences and shortened de Bruijn sequences is unique for each n. In contrast, the graph for de Bruijn arrays will not be unique for each pair (n, m). The graphs for de Bruijn arrays will be also different from the graphs for shortened de Bruijn arrays. Moreover, for a given pair (n, m), the graph for the associated de Bruijn arrays or the associated shortened de Bruijn arrays might not be unique. We will show that for most sets of parameters, the graph for de Bruijn arrays is not unique. These graphs are associated with some specific factors in G_n. We will consider first the graphs for perfect maps.

A *perfect factor* PF(n, k) in G_n is a factor with 2^{n-k} cycles of length 2^k. On each cycle in the perfect factor, one state will be chosen (arbitrarily) as a *zero state*. We will choose the zero state to be the state on the cycle with minimum value when considered as a binary number, but any other choice can be also used. This choice will not affect the arrays that will be constructed by the graph. The cycles of the factor will be ordered in ascending order of their zero states. The *location* of a cycle c_i in this order is denoted by $L(c_i)$, $1 \leq i \leq 2^{n-k}$, $0 \leq L(c_i) \leq 2^{n-k} - 1$.

Example 9.1. For $n = 6$ and $r = 2^k = 8$, the following eight cycles form a perfect factor in G_6, presented in their zero shift and ordered in ascending order of their zero states:

$$
\begin{aligned}
c_0 &= [00000011] \\
c_1 &= [00001001] \\
c_2 &= [00010111] \\
c_3 &= [00011101] \\
c_4 &= [00101011] \\
c_5 &= [00110101] \\
c_6 &= [00111111] \\
c_7 &= [01101111]
\end{aligned}
$$

For the cycle $c_3 = [00011101]$ we have that $L(c_3) = 3$, its zero state is (000111), and its 8 shifts are as follows:

zero shift	[00011101]
shift one	[00111010]
shift two	[01110100]
shift three	[11101000]
shift four	[11010001]
shift five	[10100011]
shift six	[01000111]
shift seven	[10001110]

■

Define a graph $G_{PF(n,k),m}$ whose vertices are $2^k \times m$ matrices. Each matrix is an m-***state*** $X = (x_0, (x_1, s_1), \ldots, (x_{m-1}, s_{m-1}))$, where x_i is a cycle in $PF(n, k)$, represented as a column vector and s_i is a shift of x_i compared to its zero state. The first cycle x_1 is taken with its zero shift and the other cycles are taken in all possible shifts. An immediate consequence of this definition is the following lemma.

Lemma 9.3. *The number of vertices in the graph* $G_{PF(n,k),m}$ *is* 2^{nm-k}.

Example 9.2. For $n = 6$, $m = 5$, and $r = 8$ consider the factor $PF(6, 3)$ of Example 9.1. The following three 8×5 arrays are vertices, i.e., m-states in $G_{PF(6,3),5}$: $(c_0, (c_3, 3), (c_3, 5), (c_5, 1), (c_4, 0)), (c_5, (c_2, 5), (c_1, 4), (c_7, 0), (c_7, 1))$, and $(c_7, (c_6, 0), (c_6, 0), (c_2, 3), (c_5, 0))$. These three m-states form the following 8×5 matrices that are vertically periodic:

$$
\begin{bmatrix}
0 & 1 & 1 & 0 & 0 \\
0 & 1 & 0 & 1 & 0 \\
0 & 1 & 1 & 1 & 1 \\
0 & 0 & 0 & 0 & 0 \\
0 & 1 & 0 & 1 & 1 \\
0 & 0 & 0 & 0 & 0 \\
1 & 0 & 1 & 1 & 1 \\
1 & 0 & 1 & 0 & 1
\end{bmatrix},
\begin{bmatrix}
0 & 1 & 1 & 0 & 1 \\
0 & 1 & 0 & 1 & 1 \\
1 & 1 & 0 & 1 & 0 \\
1 & 0 & 1 & 0 & 1 \\
0 & 0 & 0 & 1 & 1 \\
1 & 0 & 0 & 1 & 1 \\
0 & 1 & 0 & 1 & 1 \\
1 & 0 & 0 & 1 & 0
\end{bmatrix},
\text{ and }
\begin{bmatrix}
0 & 0 & 0 & 1 & 0 \\
1 & 0 & 0 & 0 & 0 \\
1 & 1 & 1 & 1 & 1 \\
0 & 1 & 1 & 1 & 1 \\
1 & 1 & 1 & 1 & 0 \\
1 & 1 & 1 & 0 & 1 \\
1 & 1 & 1 & 0 & 0 \\
1 & 1 & 1 & 0 & 1
\end{bmatrix},
$$

respectively. ■

Lemma 9.4. *Each* $n \times m$ *binary matrix is contained exactly once as a window in one of the vertices of the graph* $G_{PF(n,k),m}$.

Proof. The proof is done by induction on m.
Basis: If $m = 1$, then each $n \times 1$ binary matrix is contained exactly once in one vertex since the vertices represent the cycles of the perfect factor $PF(n, k)$ in G_n.
Induction hypothesis: Assume the claim holds for $n \times (m - 1)$ windows in $G_{PF(n,k),m-1}$.

Induction step: Let T_m be an $n \times m$ binary matrix and let T_{m-1} be the matrix that consists of the first $m - 1$ column of T_m. By the induction hypothesis, T_{m-1} is contained exactly once in one vertex of $G_{\text{PF}(n,k),m-1}$. The vertices of $G_{\text{PF}(n,k),m}$ can be generated by taking the $2^k \times (m - 1)$ matrix of each vertex of $G_{\text{PF}(n,k),m-1}$ and attaching to its end all the cycles of $\text{PF}(n, k)$ in all their 2^k possible shifts. Since each binary n-tuple is contained exactly once as a window in one of the cycles of $\text{PF}(n, k)$, it follows that exactly one of those shifts attaches the last column of T_m to T_{m-1} and form T_m. $\qquad\square$

The graph $G_{\text{PF}(n,k),m}$ has $2^{n(m+1)-k}$ directed edges as follows. From the vertex $X = (x_0, (x_1, s_1), \ldots, (x_{m-1}, s_{m-1}))$ there exists a directed edge to the vertex $Y = (y_0, (y_1, t_1), \ldots, (y_{m-1}, t_{m-1}))$ if and only if for each j, $1 \leq j \leq m - 2$, $(y_j, t_j) = (x_{j+1}, s_{j+1} - s_1)$, where $s_{j+1} - s_1$ is taken modulo 2^k, $y_0 = x_1$, $y_{m-1} \in \text{PF}(n, k)$, and $t_{m-1} \in \{0, 1, \ldots, 2^k - 1\}$. Such an edge can be represented by a $2^k \times (m + 1)$ matrix $(x_0, (x_1, s_1), \ldots, (x_{m-1}, s_{m-1}), (x_m, s_m))$, where $x_m = y_{m-1}$, and $s_m = s_1 + t_{m-1} \pmod{2^k}$. Thus we have the following lemma.

Lemma 9.5. *The out-degree of each vertex in $G_{\text{PF}(n,k),m}$ is 2^n and this is also the in-degree of each vertex.*

Similarly to the proof of Lemma 9.4 we can prove the following lemma.

Lemma 9.6. *Each $n \times (m + 1)$ binary matrix is contained exactly once as a window in one of the edges of $G_{\text{PF}(n,k),m}$.*

Finally, similarly to the proof of Lemma 1.15, we have the following lemma.

Lemma 9.7. *Given two vertices $u, v \in G_{\text{PF}(n,k),m}$, there are exactly 2^k paths of length m from u to v and hence $G_{\text{PF}(n,k),m}$ is a connected graph.*

The reason that there is no unique path from a vertex u to a vertex v in $G_{\text{PF}(n,k),m}$ compared to the unique path in $G_{\sigma,n}$ is that the vertex v can be taken in each one of its 2^k cyclic shifts compared to vertex u, and each such cyclic shift will lead to a different path from u to v.

Corollary 9.1. *There exists an Eulerian cycle in $G_{\text{PF}(n,k),m}$ for each $m \geq 1$.*

Again, similarly to Theorem 1.16, we can prove the following theorem.

Theorem 9.1. *The line graph of $G_{\text{PF}(n,k),m}$, is $G_{\text{PF}(n,k),m+1}$.*

Theorem 9.1 implies that exactly as in the de Bruijn graph, where an Eulerian cycle in $G_{\sigma,n-1}$ implies the existence of a Hamiltonian cycle in $G_{\sigma,n}$, we have the following corollary.

Corollary 9.2. *There exists a Hamiltonian cycle in $G_{\text{PF}(n,k),m}$ for each $m \geq 1$.*

It is tempting to say that an Eulerian cycle obtained based on Corollary 9.1 is associated with an $(r, t; n, m + 1)$-PM. This is usually the case, but we have to prove that the cycle ends in the same cyclic shifts with which it began. We will delay the proof of the related claim to Theorem 9.5 that will show that there exists one exception to this claim. An Eulerian cycle will usually yield an associated de Bruijn array, but to generate such an array or many such arrays in the next section we will prefer a technique based on construction for Hamiltonian cycles, as was done in Section 4.3 in constructions of de Bruijn sequences.

The existence of the graph $G_{PF(n,k),m}$ depends on the existence of a perfect factor $PF(n, k)$. By the definition of a perfect factor $PF(n, k)$ there are simple necessary conditions for the existence of such a factor.

Lemma 9.8. *If there exists a perfect factor* $PF(n, k)$*, then* $k \leq n < 2^k$*.*

Proof. A perfect factor $PF(n, k)$ contains 2^{n-k} cycles of length 2^k in G_n. This immediately implies that $n - k \geq 0$, i.e., $n \geq k$. Since each n-tuple is contained in one of the cycles, including the all-ones and the all-zeros n-tuples, it follows that the length of a cycle must be larger than n, i.e., $n < 2^k$. Thus the claim of the lemma is proved. □

Theorem 9.2. *If n and k are integers such that $k \leq n < 2^k$, then there exists a perfect factor in G_n with 2^{n-k} cycles of length 2^k.*

Proof. We distinguish between two cases.
Case 1: $2^{k-1} \leq n < 2^k$. An associated perfect factor, namely $\Omega(n)$, exists by Corollary 5.7.
Case 2: $k \leq n \leq 2^{k-1}$. Let S be a de Bruijn sequence of length 2^k and minimum complexity $2^{k-1} + k$. Such a sequence exists by Theorem 5.4. By Corollaries 4.5, 4.6, and 5.5, $\mathbf{D}^{-(n-k)}S$ contains disjoint sequences (cycles) in $G_{k+n-k} = G_n$ with complexity $2^{k-1} + k + (n - k) = 2^{k-1} + n$. Moreover, since $n \leq 2^{k-1}$, it follows that $2^{k-1} + n \leq 2^k$ and hence the sequences in $\mathbf{D}^{-(n-k)}S$ are of length 2^k. Also, by Corollary 5.5 there are 2^{n-k} disjoint sequences in $\mathbf{D}^{-(n-k)}S$. Thus $\mathbf{D}^{-(n-k)}S$ is a factor is G_n that contains 2^{n-k} cycles of length 2^k. □

Corollary 9.3. *A perfect factor* $PF(n, k)$ *exists if and only if* $k \leq n < 2^k$*.*

By Corollary 9.3 there exists a perfect factor $PF(n, k)$ in G_n for each set of parameters (n, k), where $k \leq n < 2^k$. For most such sets of parameters, there exist many such factors. Therefore the graph $G_{PF(n,k),m}$ is usually not unique. If $k \leq n \leq 2^{k-1}$, then using Theorems 5.4 and 9.2 we have that there exist many such factors based on de Bruijn sequences of length 2^k and minimum complexity $2^{k-1} + k$. Many such sequences were constructed in Section 5.2. Moreover, for most parameters, other de Bruijn sequences with specified complexities can be used. As for $2^{k-1} \leq n < 2^k$ the factor that we showed is $\Omega(n)$. This is just one such factor, but many others can be constructed for some of these parameters. It is interesting to note that the two sets of parameters intersect in $n = 2^{k-1}$.

We continue and define a related graph to form shortened de Bruijn arrays. Instead of a perfect factor $PF(n, k)$ with 2^{n-k} cycles of length 2^k in G_n, it will be required to define a factor in G_n in which one of the cycles is the all-zeros self-loop vertex and all the other cycles have the same length that divides $2^n - 1$. A *zero factor* $ZF(n, k)$ in G_n, where $2^n - 1 = d \cdot k$, is a factor with the cycle $[0]$ and d simple cycles of length k. We will also say that the *exponent* of the factor is k. Note that the length of a cycle in a zero factor $ZF(n, k)$ is k, while the length of a cycle in a perfect factor $PF(n, k)$ is 2^k.

With a zero factor $ZF(n, k)$ we define a graph $G_{ZF(n,k),m}$ in the same way that we defined a graph $G_{PF(n,k),m}$. For each nonzero cycle in $ZF(n, k)$ we define a zero state. The all-zeros cycle of the zero factor is considered exactly as all the other d cycles of length k with one exception, it does not have distinct shifts as all its shifts are equal. Similarly to the cycles in a perfect factor, we define a zero state as one of the states of the cycles. The other states of a cycle will be considered with their shifts related to the zero state. The cycles will be ordered in some order and each cycle will be given a location in this ordering. The vertices in $G_{PF(n,k),m}$ are $k \times m$ matrices, whose columns are cycles of the zero factor $ZF(n, k)$, where the first nonzero cycle in a matrix is taken in its zero shift and each other cycle is taken in any one of its k shifts compared to its zero state, except for the cycle $[0]$, which is taken as is. Each vertex contains at least one nonzero cycle of $ZF(n, k)$. The edges in the graph are $k \times (m + 1)$ matrices whose columns are cycles of the zero factor $ZF(n, k)$, where the first m columns are associated with the starting vertex of the edge and the last m columns are associated with the end vertex of the edge. The shifts of the starting vertex are the same as the shifts of the first m columns of the edge, and the related shifts of the end vertex are the shifts of the last m columns of the edge. The following results on $G_{ZF(n,k),m}$ can be proved similarly to those in the lemmas associated with $G_{PF(n,k),m}$.

Lemma 9.9. *The number of vertices in the graph $G_{ZF(n,k),m}$ is $d\frac{2^{nm}-1}{2^n-1}$.*

Proof. Each $k \times m$ matrix has at least one column that contains a nonzero cycle. The first such column can be in each one of the m columns. Once the first such column is determined, there are d possible cycles that can be used for this first nonzero column. Each one of the next columns can start with any of the 2^n states (the all-zeros state is associated with the all-zeros cycle). Therefore the number of distinct $k \times m$ matrices is

$$\sum_{i=1}^{m} d \cdot 2^{n(m-i)} = d \sum_{i=0}^{m-1} 2^{n \cdot i} = d\frac{2^{nm} - 1}{2^n - 1},$$

which is also the number of vertices in $G_{ZF(n,k),m}$. □

The construction of the matrices associated with the vertices of $G_{ZF(n,k),m}$ immediately implies the following lemma.

Lemma 9.10. *Each $n \times m$ binary matrix is contained exactly once as a window in one of the vertices of $G_{ZF(n,k),m}$.*

Unfortunately, we cannot generalize Lemma 9.5 on the in-degree and the out-degree of each vertex in $G_{PF(n,k),m}$. The degrees in $G_{ZF(n,k),m}$ are given in the following lemma.

Lemma 9.11. *The in-degree of a vertex in $G_{ZF(n,k),m}$ whose last $m - 1$ columns are all-zeros is 2^n and the out-degree of such a vertex is d. The in-degree of a vertex in $G_{ZF(n,k),m}$ whose first $m - 1$ columns are all-zeros is d and the out-degree of such a vertex is 2^n. For the other vertices of $G_{ZF(n,k),m}$ the in-degree of each vertex is 2^n and this is also its out-degree.*

Proof. To each vertex v of $G_{ZF(n,k),m}$ whose last $m - 1$ columns contain only *zeros* we can add any one of the d nonzero cycles of $ZF(n, k)$ as the new last column. The shift in which such a cycle is added can be ignored since all these shifts are the same as related to the previous $m - 1$ consecutive $m - 1$ all-zeros columns. Therefore the out-degree of such vertex v is d. The first column of v must be nonzero and hence before the first column, we can add any cycle with all possible shifts and hence its in-degree is 2^n. The same arguments hold for the in-degree and the out-degree of a vertex whose first $m - 1$ columns contain only *zeros*. Hence, the in-degree of a such vertex is d and its out-degree is 2^n. For any other vertex v the same arguments hold and since neither all its first $m - 1$ columns nor all its last $m - 1$ columns are *zeros*, it follows that the in-degree of v is 2^n and the same holds for its out-degree. $\square$

Similarly to Lemma 9.7, we have the following lemma.

Lemma 9.12. *Given two vertices $u, v \in G_{ZF(n,k),m}$, there are exactly k paths of length m from u to v and hence $G_{ZF(n,k),m}$ is a connected graph.*

In view of Lemma 9.11 there is no Eulerian cycle in $G_{ZF(n,k),m}$. Moreover, Lemma 9.6 cannot be generalized for $G_{ZF(n,k),m}$ since all the $n \times (m + 1)$ matrices whose either first m columns are all-zeros or last m columns are all-zeros do not appear as $n \times (m + 1)$ windows of some edges in $G_{ZF(n,k),m}$. On the other hand, in view of Lemma 9.10 a Hamiltonian cycle in $G_{ZF(n,k),m}$ will be associated with a shortened de Bruijn array and its construction will be the same as the construction of de Bruijn arrays. For both graphs defined (for perfect factors and zero factors), a necklaces factor will be defined for the construction. The construction of a necklaces factor for $G_{PF(n,k),m}$ and for $G_{ZF(n,k),m}$ will be described in the next section.

The next step is to construct zero factors in G_n in the same way that perfect factors were constructed. Unfortunately, unlike perfect factors that exist for all possible parameters, there is still a considerable gap in the knowledge on zero factors in G_n. The theory of $LFSR_n$ presented in Section 2.1 implies some types of zero factors.

Theorem 9.3. *If the characteristic polynomial $f(x)$ of a shift register is irreducible, then the shift register produces a zero factor with exponent k, where k is the smallest integer such that $f(x)$ divides $x^k - 1$.*

Proof. The claim follows immediately from Definition 2.1, Theorem 2.3, and Corollary 2.2. □

The following Theorem is an immediate observation from Corollary 2.6.

Theorem 9.4. *Let $f_i(x)$, $1 \le i \le r$, be r different irreducible polynomials of degree n, and let their corresponding shift registers have zero factors with exponent e. Then, the feedback shift register that has the characteristic polynomial $\prod_{i=1}^{r} f_i(x)$ produces a zero factor with exponent e.*

Example 9.3. To obtain a zero factor with 9 cycles of length 7 in G_6 we consider the two primitive polynomials of degree 3 and their associated M-sequences of length seven. These two M-sequences are $S_1 = [0011101]$ and $S_2 = [0010111]$. We multiply these two polynomials and obtain the characteristic polynomial

$$(x^3 + x^2 + 1)(x^3 + x + 1) = x^6 + x^5 + x^4 + x^3 + x^2 + x + 1,$$

whose LFSR$_6$, which is the PSR$_6$, forms the nine cycles

$$
\begin{array}{rl}
1) & [0001010] \\
2) & [0101101] \\
3) & [1100011] \\
4) & [1111110] \\
5) & [1000100] \ . \\
6) & [0110000] \\
7) & [1011001] \\
8) & [0011101] \\
9) & [0010111]
\end{array}
$$

These cycles can be also obtained by applying Lemma 2.5 and Theorem 2.5 on the two M-sequences of length 7, as follows:

$$S_1 + S_2 = [0011101] + [0010111] = [0001010]$$
$$\mathbf{E}S_1 + S_2 = [0111010] + [0010111] = [0101101]$$
$$\mathbf{E}^2 S_1 + S_2 = [1110100] + [0010111] = [1100011]$$
$$\mathbf{E}^3 S_1 + S_2 = [1101001] + [0010111] = [1111110]$$
$$\mathbf{E}^4 S_1 + S_2 = [1010011] + [0010111] = [1000100]$$
$$\mathbf{E}^5 S_1 + S_2 = [0100111] + [0010111] = [0110000]$$
$$\mathbf{E}^6 S_1 + S_2 = [1001110] + [0010111] = [1011001]$$

$$S_1 + [0] = [0011101] + [0000000] = [0011101]$$
$$[0] + S_2 = [0000000] + [0010111] = [0010111].$$

∎

We continue with another graph representation for de Bruijn arrays and shortened de Bruijn arrays.

We define a graph $G_{n,m}(V, E_1, E_2)$, where V is the set of all $n \times m$ binary arrays, E_1 and E_2 are sets of edges defined as follows. There is an edge in E_1 from the $n \times m$ binary array A_1 to the $n \times m$ binary array A_2 if and only if the $(n-1) \times m$ array obtained from the last $n-1$ rows of A_1 is equal to the $(n-1) \times m$ array obtained from the first $n-1$ rows of A_2. There is an edge in E_2 from the $n \times m$ binary array B_1 to the $n \times m$ binary array B_2 if and only if the $n \times (m-1)$ array obtained from the last $m-1$ columns of B_1 is equal to the $n \times (m-1)$ array obtained from the first $m-1$ columns of B_2. We have that the edges of E_1 are vertical edges and the edges of E_2 are horizontal edges related to the vertices of size $n \times m$.

An $(r, t; n, m)$-PM is represented by a subgraph of $G_{n,m}(V, E_1, E_2)$ that contains all the vertices of V, a factor in $G_{n,m}(V, E_1, E_2)$ that contains t cycles of length r with edges only from E_1, and a factor in $G_{n,m}(V, E_1, E_2)$ that contains r cycles of length t with edges only from E_2. The edges from E_1 of the factor are determined by the order of the $n \times m$ windows projected by any m consecutive columns of the array. The edges from E_2 of the factor are determined by the order of the $n \times m$ windows projected by any n consecutive rows of the array. However, not every two such factors can form an $(r, t; n, m)$-PM.

An $(r, t; n, m)$-SPM is represented by a subgraph of $G_{n,m}(V \setminus \{0\}, E_1, E_2)$ that contains all the vertices of V, except of the all-zeros vertex, a factor in $G_{n,m}(V \setminus \{0\}, E_1, E_2)$ that contains t cycles of length r with edges only from E_1, and a factor in $G_{n,m}(V \setminus \{0\}, E_1, E_2)$ that contains r cycles of length t with edges only from E_2. The edges from E_1 of the factor are determined by the order of the $n \times m$ windows projected by any m consecutive columns of the array. The edges from E_2 of the factor are determined by the order of the $n \times m$ windows projected by any n consecutive rows of the array.

The degree of a vertex in $G_{n,m}(V, E_1, E_2)$ is easily computed. We distinguish between the degree associated with the set E_1 and the degree associated with the set E_2.

Lemma 9.13. *The in-degree of a vertex $v \in V$ of $G_{n,m}(V, E_1, E_2)$ with edges from E_1 is 2^m. The out-degree of a vertex $v \in V$ of $G_{n,m}(V, E_1, E_2)$ with edges from E_1 is 2^m. The in-degree of a vertex $v \in V$ of $G_{n,m}(V, E_1, E_2)$ with edges from E_2 is 2^n. The out-degree of a vertex $v \in V$ of $G_{n,m}(V, E_1, E_2)$ with edges from E_2 is 2^n.*

The two distinct graph representations defined for de Bruijn arrays and shortened de Bruijn arrays will be used to construct such arrays in the following sections.

9.2 Constructions by merging cycles

We will generalize a construction from Section 4.3 for generating de Bruijn sequences to a construction for perfect maps and shortened perfect maps. For this purpose, we will use a graph $G_{\mathrm{PF}(n,k),m}$ and find a Hamiltonian cycle in it by merging cycles of a factor in the graph using the merge-or-split method.

For an m-state X, in $G_{\mathrm{PF}(n,k),m}$, where

$$X = (x_0, (x_1, s_1), \ldots, (x_{m-1}, s_{m-1})),$$

the *set of companions* $\mathbb{X}'$ is a set of m-states, where

$$(y_0, (y_1, t_1), \ldots, (y_{m-1}, t_{m-1})) \in \mathbb{X}'$$

if and only if $y_i = x_i$ and $t_i = s_i$, $1 \leq i \leq m - 2$, $y_{m-1} \in \mathrm{PF}(n, k)$, $y_0 = x_0$, and one of the following holds:

- $y_{m-1} \neq x_{m-1}$ and $t_{m-1} \in \{0, 1, \ldots, 2^k - 1\}$;
- $y_{m-1} = x_{m-1}$, $t_{m-1} \in \{0, 1, \ldots, 2^k - 1\}$, and $t_{m-1} \neq s_{m-1}$.

A cycle $\mathcal{C} = [T^{(1)}, T^{(2)}, \ldots, T^{(\ell)}]$ in $G_{\mathrm{PF}(n),m}$ contains ℓ ordered vertices, where $T^{(i)}$, $1 \leq i \leq \ell$, are the consecutive vertices of the cycle and there is a directed edge between any two consecutive vertices cyclically. The cycle $\mathcal{C}$ will be represented also by an $2^k \times (\ell + m - 1)$ matrix

$$R(\mathcal{C}) = \left(T_0^{(1)}, T_0^{(2)} \cdots T_0^{(\ell-1)}, T_0^{(\ell)}, T_1^{(\ell)} \cdots T_{m-1}^{(\ell)} \right),$$

where $T_i^{(j)}$ corresponds to the ith column of $T^{(j)}$, and each m consecutive columns are associated with a vertex in $G_{\mathrm{PF}(n),m}$. Note that we can erase the last $m - 1$ columns if they are in the same shift as the first $m - 1$ columns and consider the $2^k \times \ell$ matrix R as a doubly periodic matrix. To avoid any confusion, however, we will not erase those columns unless the matrix corresponds to a perfect map.

We continue to explore the connection between an $(r, t; n, m)$-PM and a Hamiltonian cycle in $G_{\mathrm{PF}(n,k),m}$.

Theorem 9.5. *A sufficient condition for the existence of a $(2^k, 2^{nm-k}; n, m)$-PM, where $n < 2^k \leq 2^n$ and $m \neq 2$ if $k = n$, is the existence of a Hamiltonian cycle in $G_{\mathrm{PF}(n,k),m}$.*

Proof. By the definition of the matrix representation $R(\mathcal{C})$ of a cycle $\mathcal{C}$ and by Lemma 9.4, each $n \times m$ window appears in the matrix representation $R(\mathcal{C})$ of the Hamiltonian cycle in $G_{\mathrm{PF}(n,k),m}$. Hence, we only have to prove that the last $m - 1$ columns of $R(\mathcal{C})$, are in the same shift as the first $m - 1$ columns of $R(\mathcal{C})$; this would imply that we can erase the last $m - 1$ columns of $R(\mathcal{C})$ to obtain a $(2^k, 2^{nm-k}; n, m)$-PM. For this, we have to sum the shifts of the columns in $R(\mathcal{C})$, where the shift of a column is relative to the shift of the previous column.

In $R(C)$ there are $2^{nm-k} + m - 1$ columns. Each of the first 2^{nm-k} columns is associated with a different vertex in $G_{PF(n,k),m}$. There are 2^k possible shifts. Each shift appears $\frac{2^{nm-k}}{2^k} = 2^{nm-2k}$ times. Hence, the sum of all the shifts is equal to

$$2^{nm-2k}(0 + 1 + 2 + 3 + \cdots + (2^k - 2) + (2^k - 1)) = (2^k - 1)2^{nm-k-1}.$$

Since the vertical size of the array is 2^k, it follows that the sum of the shifts should be taken modulo 2^k. To obtain the last $m - 1$ columns in the same shift as the first $m - 1$ columns, we must have that the sum of the shifts is 0 modulo 2^k, i.e., $(2^k - 1)2^{nm-k-1} \equiv 0 \pmod{2^k}$, which is equivalent to $nm - k - 1 \geq k$. Since $n > 1$ and $m > 1$, it implies that a $(2^k, 2^{nm-k}; n, m)$-PM will be generated whenever $n < 2^k \leq 2^n$, unless both $k = n$ and $m = 2$. $\qquad\square$

To generate a Hamiltonian cycle, we take a factor in $G_{PF(n,k),m}$ and join its cycles into a single Hamiltonian cycle. This is done using the following theorem, which is analogous to Lemma 1.19.

Theorem 9.6. *Two cycles C_1 and C_2 in a factor of $G_{PF(n,k),m}$, with an m-state X on C_1 and an m-state Y on C_2 such that $Y \in \mathbb{X}'$, form a single cycle when the predecessors of X and Y are interchanged.*

The **necklaces factor** (NF) is a factor of $G_{PF(n,k),m}$ that is defined by the following property: The two m-states $X = (x_0, (x_1, s_1), \ldots, (x_{m-1}, s_{m-1}))$ and $Y = (y_0, (y_1, t_1), \ldots, (y_{m-1}, t_{m-1}))$ are on the same NF-cycle if and only if X is a cyclic shift of Y, i.e., there exists an i such that $y_0 = x_i$ and for each $j, 1 \leq j \leq m - 1$, $(y_j, t_j) = (x_{i+j}, s_{i+j} - s_i)$, where subscripts are taken modulo m and $s_{i+j} - s_i$ is taken modulo 2^k.

For an m-state $X = (x_0, (x_1, s_1), \ldots, (x_{m-1}, s_{m-1}))$, the σ-**weight**, $\mathrm{wt}_\sigma(X)$, of X, where for each x_i, $1 \leq i \leq m - 1$, $L(x_i) \leq \sigma$, is the number of entries in X for which $L(x_j) = \sigma$. The σ-weight $\mathrm{wt}_\sigma(C)$ of a cycle C from the NF is the σ-weight of each of its m-states. Clearly, for a given $\sigma, 0 \leq \sigma \leq 2^{n-k} - 2$, the σ-weight is defined only for some of the m-states of $G_{PF(n,k),m}$. For $\sigma = 2^{n-k} - 1$ the σ-weight is defined for all the m-states.

Lemma 9.14. *Let C_1 be a cycle from the NF of σ-weight ω, where $\sigma > 0$ and $\omega > 0$. Then, there exists an m-state X on C_1 with an m-state $Y \in \mathbb{X}'$ such that Y is an m-state on a different cycle C_2 whose σ-weight is $\omega - 1$.*

Proof. Since $\mathrm{wt}_\sigma(C_1) = \omega > 0$, it follows that there exists on C_1 an m-state of the form $X = (x_0, (x_1, s_1), \ldots, (x_{m-2}, s_{m-2}), (x_{m-1}, s_{m-1}))$, where $L(x_{m-1}) = \sigma$. Hence, each m-state of the form

$$Y = (x_0, (x_1, s_1), \ldots, (x_{m-2}, s_{m-2}), (y_{m-1}, t_{m-1})),$$

where $L(y_{m-1}) < \sigma$, has σ-weight $\omega - 1$. Therefore Y is an m-state on an NF-cycle C_2 with $\mathrm{wt}_\sigma(C_2) = \omega - 1$ and $Y \in \mathbb{X}'$. $\qquad\square$

The *shift weight*, swt(X), of an m-state $X = (x, (x, s_1), \ldots, (x, s_{m-1}))$, where $L(x) = 0$, is the number of values of i for which $s_i \neq 0$. The shift weight swt($\mathcal{C}$) of a cycle from the NF, where $\text{wt}_1(\mathcal{C}) = 0$ (i.e., $\text{wt}_0(\mathcal{C}) = m$), is the minimum shift weight among its m-states. For any other m-state, the shift weight is not defined.

Lemma 9.15. *If $\mathcal{C}_1$ is a cycle from the NF, whose shift weight is $\omega > 0$, then there exists an m-state X on $\mathcal{C}_1$ with an m-state $Y \in \mathbb{X}'$ such that Y is on a different cycle $\mathcal{C}_2$ whose shift weight is $\omega - 1$.*

Proof. Since swt($\mathcal{C}_1$) $= \omega > 0$, it follows that there exists an m-state of the form $X = (x, (x, s_1), \ldots, (x, s_{m-2}), (x, s_{m-1}))$ on $\mathcal{C}_1$, where $L(x) = 0$, swt(X) $= \omega$, and $s_{m-1} > 0$ (if $s_{m-1} = 0$, then we can take on $\mathcal{C}_1$ another m-state, e.g., $X = (x, (x, 0), (x, s_1), \ldots, (x, s_{m-2}))$), where swt($X$) $= \omega$ (and this process continues if $s_{m-2} = 0$). Hence, the m-state $Y = (x, (x, s_1), \ldots, (x, s_{m-2}), (x, 0))$ is an m-state on another NF-cycle $\mathcal{C}_2$ for which swt($\mathcal{C}_2$) $= \omega - 1$ and $Y \in \mathbb{X}'$. $\square$

Construction merge NF:

Lemmas 9.14 and 9.15, along with Theorem 9.6, suggest a simple method of joining all the NF-cycles to construct a Hamiltonian cycle in $G_{\text{PF}(n),m}$. At each step, we have a main cycle obtained in the previous steps by joining a subset of the NF-cycles. Initially, the main cycle contains all the m-states of 1-weight *zero*. This cycle is constructed in $m + 1$ initial steps. In initial step 0, the main cycle is the unique cycle of 1-weight *zero* and shift weight *zero*. Before initial step ℓ, $1 \le \ell \le m$, the main cycle contains all the m-states with 1-weight *zero* and shift weight less than or equal to $\ell - 1$. In step ℓ we extend the main cycle by joining to it all the NF-cycles of 1-weight *zero* and shift weight ℓ in an arbitrary order. This is always possible because the current main cycle contains all the states whose shift weight is less than ℓ and since each NF-cycles of shift weight $\ell \ge 1$ contains an m-state of the form $X = (x, (x, s_1), \ldots, (x, s_{m-2}), (x, s_{m-1}))$, where $L(x) = 0$, $s_{m-1} > 0$, and swt(X) $= \ell$. It can be joined (see Theorem 9.6 and Lemma 9.15) to the current main cycle. After all the NF-cycles with 1-weight *zero* have been joined to the main cycle, the main cycle is extended by adjoining all the cycles of 1-weight *one*. In general step $jm + i$, $0 \le j \le 2^{n-k} - 2$, $1 \le i \le m$, we extend the main cycle by adjoining all the NF-cycles of $(j + 1)$-weight i in arbitrary order. This is always possible because the current main cycle contains all the states whose $(j + 1)$-weight is less than i and since each of $(j + 1)$-weight $i \ge 1$ has an m-state of the form $X = (x, (x_1, s_1), \ldots, (x_{m-2}, s_{m-2}), (x_{m-1}, s_{m-1}))$, where $L(x_{m-1}) = j + 1$, it can be joined (see Theorem 9.6 and Lemma 9.14) to the current main cycle. This procedure ends when all the NF-cycles have been joined together. ∎

An immediate consequence is the following theorem.

Theorem 9.7. *Construction merge NF yields a Hamiltonian cycle in $G_{\text{PF}(n,k),m}$.*

A specific procedure for generating the bits of the perfect maps can be given in a way similar to the procedure for generating t-ary de Bruijn cycles, as explained in Section 4.3. There are many ways to choose the bridging states. Thus many perfect maps can be efficiently constructed in this way.

We continue to discuss shortened de Bruijn arrays. The difference between a de Bruijn sequence and a shortened de Bruijn sequence is the all-zeros n-tuple that is not contained in a shortened de Bruijn sequence. The same difference is between de Bruijn arrays and shortened de Bruijn arrays, where the all-zeros $n \times m$ binary matrix is not contained in a shortened de Bruijn array.

It is quite easy to verify that in a sequence of length $2^n - 1$ in which all the windows of length n are distinct, the only missing n-tuple is either the all-zeros n-tuple or the all-ones n-tuple. Just consider a graph obtained from G_{n-1}, where one edge was removed. Unless the removed edge is one of the self-loops, the obtained graph does not have an Eulerian cycle. The analogous proof for the two-dimensional arrays is slightly more complicated.

Lemma 9.16. *If in an $r \times t$ array, $r > n$ and $t > m$, $rt = 2^{nm} - 1$, all the $n \times m$ windows are distinct, then the only $n \times m$ window that does not appear in the array is either the all-zeros matrix or the all-ones matrix.*

Proof. Consider the graph $G_{n,m}(V, E_1, E_2)$ and assume, on the contrary, that there exists such an array A without the $n \times m$ window associated with a vertex v that is not the all-zeros matrix or the all-ones matrix.

Let $v \to u$ be an edge in E_1. By Lemma 9.13, there are 2^m out-edges from v in E_1. Let $\mathcal{U} \triangleq \{u_1 = u, u_2, \ldots, u_{2^m}\}$ be the set of vertices such that $v \to u_i$, $1 \leq i \leq 2^m$ (the $n \times m$ matrices associated with these vertices have the same projection of their first $n - 1$ rows). By Lemma 9.13 there are 2^m in-edges to u in E_1. Let $\mathcal{V} \triangleq \{v_1 = v, v_2, \ldots, v_{2^m}\}$ be the set of vertices such that $v_i \to u$, $1 \leq i \leq 2^m$ (the $n \times m$ matrices associated with these vertices have the same projection of their last $n - 1$ rows). By the definition of $G_{n,m}(V, E_1, E_2)$ we have that $v_i \to u_j$ for all $1 \leq i, j \leq 2^m$ and there are no other out-edges in E_1 for any of the v_is and no other in-edges in E_1 for any of the u_js.

If such an $r \times t$ array A exists, then there exists a factor in $G_{n,m}(V \setminus \{v\}, E_1, E_2)$ with t cycles of length r and edges in E_1. Since all the out-edges from $\mathcal{V}$ are to vertices in $\mathcal{U}$ and all the vertices in $\mathcal{U}$ are contained in A (unless $v \in \mathcal{U}$), it follows that to have such a factor we must have that $v \in \mathcal{U}$ (as otherwise there will be no edge $v' \to u'$ for some $u' \in \mathcal{U}$). However, a vertex v is in $\mathcal{V}$ and also in $\mathcal{U}$ if and only if it is either the all-zeros vertex or the all-ones vertex. $\square$

The construction for a shortened de Bruijn array from a graph $G_{\text{ZF}(n,k),m}$ is identical to the construction of a de Bruijn array from a graph $G_{\text{PF}(n,k),m}$. A necklaces factor (NF) in $G_{\text{ZF}(n,k),m}$ is defined in the same way that it is defined in $G_{\text{PF}(n,k),m}$, where two m-states $X = (x_0, (x_1, s_1), \ldots, (x_{m-1}, s_{m-1}))$ and $Y = (y_0, (y_1, t_1), \ldots, (y_{m-1}, t_{m-1}))$ are on the same NF-cycle if and only if X is a cyclic shift of Y, i.e., there exists an i such that $y_0 = x_i$ and for each j,

$1 \le j \le m - 1$, $(y_j, t_j) = (x_{i+j}, s_{i+j} - s_i)$, where subscripts are taken modulo m and $s_{i+j} - s_i$ is taken modulo k.

Merging all the cycles of the NF in $G_{ZF(n,k),m}$ is done by the merge-or-split method, exactly in the same way it was done in the NF of the graph $G_{PF(n,k),m}$. We have to prove an analog to Theorem 9.5 that is even easier for $G_{ZF(n,k),m}$ since the last $m - 1$ columns are always in the same shift as the first $m - 1$ column. This is because each shift appears the same number of times and for each shift of size ℓ_1, there exists a shift of size ℓ_2, such that $\ell_1 + \ell_2 = k$. Therefore we have the following theorem.

Theorem 9.8. *If there exists a zero factor* ZF(n, k), *where* $n < k < 2^n$, *and a Hamiltonian cycle in* $G_{ZF(n,k),m}$, *then there exists a* $(k, (2^{nm} - 1)/k; n, m)$-*SPM.*

9.3 Pseudo-random arrays

A *pseudo-random array* is a shortened perfect map that has the shift-and-add property. In other words, A is a pseudo-random array if it is a shortened de Bruijn array, and for each cyclic horizontal and cyclic vertical shift A_1 of A (which together do not form a trivial shift), we have that $A_2 = A + A_1$ is another such shift of A. Such an array will be called an $(r, t; n, m)$-PRA.

A shortened perfect map is related to a perfect map exactly like a shortened full cycle is related to a full cycle. However, only a small fraction of these shortened full cycles are M-sequences. Similarly, only a small fraction of the shortened perfect maps are pseudo-random arrays. One construction for these arrays will be based on the folding of M-sequences into rectangular arrays.

Assume that $\eta = 2^{k_1 k_2} - 1$, $r = 2^{k_1} - 1$, and $t = \frac{\eta}{r}$, where g.c.d.$(r, t) = 1$. Let $S = s_0 s_1 s_2 \cdots$ be a span $k_1 k_2$ M-sequence obtained from a primitive polynomial $p(z)$ of degree $k_1 k_2$. Write S down the right diagonals of an $r \times t$ array $B = \{b_{ij}\}$, $0 \le i \le r - 1, 0 \le j \le t - 1$, starting at b_{00}, b_{11}, b_{22} and so on, where the last position is $b_{r-1,t-1}$. After b_{ij} we continue to write $b_{i+1,j+1}$, where $i + 1$ is taken modulo r and $j + 1$ is taken modulo t.

Example 9.4. For $k_1 = k_2 = 2$, $r = 3$, and $t = 5$, consider the span 4 M-sequence $S = [000111101011001]$, with positions numbered from $0, 1$, up to 14. Consider now the 3×5 array B with the entries b_{ij}, $0 \le i \le 2, 0 \le j \le 4$, where the positions 0 through 14, of the sequence, are folded into B as follows:

$$B = \begin{bmatrix} b_{00} & b_{01} & b_{02} & b_{03} & b_{04} \\ b_{10} & b_{11} & b_{12} & b_{13} & b_{14} \\ b_{20} & b_{21} & b_{22} & b_{23} & b_{24} \end{bmatrix}, \quad \begin{bmatrix} 0 & 6 & 12 & 3 & 9 \\ 10 & 1 & 7 & 13 & 4 \\ 5 & 11 & 2 & 8 & 14 \end{bmatrix}.$$

The M-sequence S is folded into the array B keeping the order of the entries in S according to the order defined by B. The outcome is the array

$$\begin{bmatrix} 0 & 1 & 0 & 1 & 0 \\ 1 & 0 & 0 & 0 & 1 \\ 1 & 1 & 0 & 1 & 1 \end{bmatrix},$$

which forms a $(3, 5; 2, 2)$-PRA. ∎

A horizontal shift and/or a vertical shift of the array will be equivalent to an array obtained by folding from another point of the M-sequence. Since the M-sequence has the shift-and-add property, it follows that if we make any such shift, the two arrays will sum to another shift of the array. This is the shift-and-add property of the array.

Example 9.5. Consider the array and the M-sequence S of Example 9.4. We shift the array horizontally by 2 and vertically by 1 and add them as follows, where the first bit of S is in bold:

$$\begin{bmatrix} \mathbf{0} & 1 & 0 & 1 & 0 \\ 1 & 0 & 0 & 0 & 1 \\ 1 & 1 & 0 & 1 & 1 \end{bmatrix} + \begin{bmatrix} 1 & 1 & 1 & 1 & 0 \\ 1 & 0 & \mathbf{0} & 1 & 0 \\ 0 & 1 & 1 & 0 & 0 \end{bmatrix} = \begin{bmatrix} 1 & 0 & 1 & 0 & \mathbf{0} \\ 0 & 0 & 0 & 1 & 1 \\ 1 & 0 & 1 & 1 & 1 \end{bmatrix}.$$

The M-sequence S starts in the leftmost array in b_{00}, in the middle array at b_{12}, and in the rightmost array at b_{04}. ∎

To verify that in general, the constructed array has the $k_1 \times k_2$ window property we have to use the Chinese reminder theorem (see Theorem 1.9 and Corollary 1.2). Note that for our purpose we need Corollary 1.2 only for $s = 2$, but it can be used for any s, i.e., the results can be generalized for s-dimensional arrays. We write the M-sequence S in an $r \times t$ array B, where s_ℓ, $0 \le \ell < \eta$, is written in b_{ij}, where

$$\ell \equiv i \pmod{r}$$
$$\ell \equiv j \pmod{t}$$

The array B can be represented as a polynomial (generating function), $b(x, y)$, where $x^r = 1$ and $y^t = 1$.

$$b(x, y) = \sum_{i=0}^{r-1} \sum_{j=0}^{t-1} b_{ij} x^i y^j.$$

If $s(z)$ is the polynomial (generating function) representing the M-sequence $S = s_0 s_1 s_2 \cdots$ of length $\eta = 2^{k_1 k_2} - 1$ and $b(x, y)$ is the polynomial that represents the $r \times t$ array B, then $b(x, y) = s(z)$, where $z = xy$. The term $s_\ell z^\ell$ is replaced by the term $b_{ij} x^i y^j$, where $\ell \equiv i \pmod{r}$, $0 \le i < r$ and $\ell \equiv j \pmod{t}$, $0 \le j < t$. Thus s_ℓ will be the value of b_{ij} as required.

Example 9.6. Consider the $(3, 5; 2, 2)$-PRA

$$\begin{bmatrix} 0 & 1 & 0 & 1 & 0 \\ 1 & 0 & 0 & 0 & 1 \\ 1 & 1 & 0 & 1 & 1 \end{bmatrix}$$

of Example 9.4, where $x^3 = 1$ and $y^5 = 1$. Clearly,

$$b(x, y) = (y + y^3) + x(1 + y^4) + x^2(1 + y + y^3 + y^4).$$

For the associated M-sequence $S = 000111101011001$ we have

$$s(z) = z^3 + z^4 + z^5 + z^6 + z^8 + z^{10} + z^{11} + z^{14}$$

and

$$s(xy) = y^3 + xy^4 + x^2 + y + x^2y^3 + x + x^2y + x^2y^4 = b(x, y),$$

as required. ∎

Theorem 9.9. *Each $k_1 \times k_2$ binary nonzero matrix is contained as a window in the $r \times t$ array B, where $r = 2^{k_1} - 1$ and $t = (2^{k_1k_2} - 1)/r$, exactly once.*

Proof. Let α be a primitive element in GF($2^{k_1k_2}$) and let S be its associated M-sequence. Furthermore, let G be a $(k_1k_2) \times (2^{k_1k_2} - 1)$ generator matrix for a $[\eta = 2^{k_1k_2} - 1, k_1k_2, 2^{k_1k_2-1}]$ simplex code $\mathcal{C}$ (see Definition 2.9). This generator matrix can be taken as a $(k_1k_2) \times (2^{k_1k_2} - 1)$ matrix defined by k_1k_2 consecutive shifts of the M-sequence S. The columns of G can be rearranged in a way that they are considered as the representation of $\alpha^0, \alpha^1, \alpha^2, \ldots, \alpha^{\eta-1}$ in this order. Let $\ell(i, j) \equiv i \pmod{r}$, $\ell(i, j) \equiv j \pmod{t}$, where $0 \le \ell(i, j) < 2^{k_1k_2} - 1$.

We claim that the k_1k_2 columns of G associated with the elements $\alpha^{\ell(i,j)}$, where $\ell(i, j) \equiv i \pmod{r}$, $\ell(i, j) \equiv j \pmod{t}$, $0 \le i \le k_1 - 1$ and $0 \le j \le k_2 - 1$, are linearly independent. These columns are associated with the $k_1 \times k_2$ subarray in the upper left corner of B.

Assume, on the contrary, that

$$\sum_{i=0}^{k_1-1} \sum_{j=0}^{k_2-1} c_{ij}\alpha^{\ell(i,j)} = 0,$$

where not all the c_{ij} are *zeros*. Since g.c.d.$(r, t) = 1$, it follows by Theorem 1.5 that there exist two integers μ and ν such that $\mu r + \nu t = 1$. Let $\beta = \alpha^{\nu t}$, $\gamma = \alpha^{\mu r}$, which implies that β is an element of order r and γ is an element of order t since $rt = 2^{k_1k_2} - 1$ and hence $\alpha^{rt} = 1$. This also implies that $\alpha = \alpha^{\mu r + \nu t} = \beta\gamma$ and, hence, $\alpha^{\ell(i,j)} = \beta^{\ell(i,j)}\gamma^{\ell(i,j)} = \beta^i\gamma^j$. Moreover,

$rt = 2^{k_1 k_2} - 1$ implies that $\alpha^{rt} = 1$ and, hence, $\beta^r = \alpha^{vtr} = 1$ and $\gamma^t = \alpha^{\mu rt} = 1$. Therefore we have

$$0 = \sum_{i=0}^{k_1-1} \sum_{j=0}^{k_2-1} c_{ij} \alpha^{\ell(i,j)} = \sum_{i=0}^{k_1-1} \sum_{j=0}^{k_2-1} c_{ij} (\beta\gamma)^{\ell(i,j)} = \sum_{j=0}^{k_2-1} \left(\sum_{i=0}^{k_1-1} c_{ij} \beta^i \right) \gamma^j. \quad (9.1)$$

The order of β is $r = 2^{k_1} - 1$, and, hence, β is a primitive element in $\mathbb{F}_{2^{k_1}}$. Therefore the coefficient of γ^j in Eq. (9.1) is an element of $\mathbb{F}_{2^{k_1}}$. Let m be the smallest integer such that $\gamma^{2^m-1} = 1$.

Clearly, $\alpha^{\mu r (2^m-1)} = \gamma^{2^m-1} = 1$ and since the order of α is rt, it follows that rt divides $\mu r (2^m - 1)$ and t divides $2^m - 1$ (since by $\mu r + vt = 1$ we have that g.c.d.$(\mu, t) = 1$). The binary representation of $t = \frac{\eta}{r} = \frac{2^{k_1 k_2}-1}{2^{k_1}-1}$ is $10^{k_1-1}10^{k_1-1}1 \cdots 0^{k_1-1}1$, where the number of *ones* in this representation is k_2. Hence, this binary representation contains $k_1(k_2 - 1) + 1$ digits. The binary representation of $2^m - 1$ is $11 \cdots 1$, where the number of *ones* in this representation is m. Hence, by considering binary multiplication, the smallest m for which t divides $2^m - 1$ is $k_1 k_2$. Thus $m = k_1 k_2$ is the smallest positive integer such that $\gamma^{2^m} = \gamma$ and the k_2 elements

$$\gamma, \gamma^{2^{k_1}}, \gamma^{2^{2k_1}}, \ldots, \gamma^{2^{(k_2-1)k_1}}$$

are distinct. Therefore

$$\prod_{i=0}^{k_2-1} (x - \gamma^{2^{ik_1}})$$

is the minimal polynomial of γ in $\mathbb{F}_{2^{k_1}}$. This polynomial has degree k_2 in $\mathbb{F}_{2^{k_1}}$. Now, the polynomial in Eq. (9.1) is a polynomial in γ with a smaller degree $k_2 - 1$ that equals *zero*. This implies that all the coefficients of γ^j in Eq. (9.1) are equal to *zero*. Therefore for each $0 \le j \le k_2 - 1$ we have

$$\sum_{i=0}^{k_1-1} c_{ij} \beta^i = 0,$$

a contradiction since this is a polynomial in β of degree less than k_1 and the minimal zero polynomial of β has degree k_1. This completes the proof of the claim that the $k_1 k_2$ columns of G associated with the elements $\alpha^{\ell(i,j)}$, where $\ell(i, j) \equiv i \pmod{r}$, $\ell(i, j) \equiv j \pmod{t}$, $0 \le i \le k_1 - 1$ and $0 \le j \le k_2 - 1$, are linearly independent.

This claim implies that the $k_1 \times k_2$ array in the upper leftmost corner of B is nonzero. This $k_1 \times k_2$ window can be chosen arbitrarily since the M-sequence can start at any nonzero initial $(k_1 k_2)$-tuple. This $k_1 \times k_2$ window determines the rest of the codeword of the simplex code $\mathcal{C}$. Hence, by the shift-and-add property, there are no two equal $k_1 \times k_2$ windows, as otherwise we can have an all-zeros $k_1 \times k_2$ window by adding the associated to such shifts with two equal $k_1 \times k_2$ windows. Thus we have the window property. □

9.4 Recursive constructions for perfect maps

This section is devoted to several recursive constructions of perfect maps with various parameters.

The first method is to construct $(2^n, 2^{n(m-1)}; n, m)$-PM, where $m \geq 3$. The method will be based on two de Bruijn sequences. Let $K = [k_1 k_2 \cdots k_{2^{n(m-1)}}]$ be a span $m - 1$ de Bruijn sequence over $\mathbb{Z}_{2^n}$ and let S be a span n binary de Bruijn sequence defined as a column vector. Let $\mathbf{E}^i S$ denote the shift of S (as a column vector) by i positions and define the following $2^n \times 2^{n(m-1)}$ array

$$A = \left[\mathbf{E}^{k_1} S \; \mathbf{E}^{k_1+k_2} S \; \mathbf{E}^{k_1+k_2+k_3} S \; \cdots \; \mathbf{E}^{\sum_{i=1}^{2^{n(m-1)}} k_i} S \right]. \tag{9.2}$$

Theorem 9.10. *The array A defined in Eq. (9.2) is a $(2^n, 2^{n(m-1)}; n, m)$-PM.*

Proof. As in the proof of Theorem 9.5 we show that the array A is cyclic, i.e., the first $m - 1$ columns are in the appropriate shift related to the last $m - 1$ columns, i.e., $k_1 + \sum_{i=1}^{2^{n(m-1)}} k_i \equiv k_1 \pmod{2^n}$, i.e., $\sum_{i=1}^{2^{n(m-1)}} k_i \equiv 0 \pmod{2^n}$. This is an immediate consequence as each symbol of $\mathbb{Z}_{2^n}$ is contained exactly $\frac{2^{n(m-1)}}{2^n} = 2^{n(m-2)}$ times in K. Therefore summing all the shifts we have (recall that $m \geq 3$)

$$\sum_{i=1}^{2^{n(m-1)}} k_i = \sum_{i=0}^{2^n-1} (2^{n(m-2)} \cdot i) \equiv 0 \pmod{2^n},$$

which implies that the array A is cyclic.

The number of entries in the array A is $2^n \cdot 2^{n(m-1)} = 2^{nm}$ and hence to complete the proof it is sufficient to show that each $n \times m$ binary matrix appears as a window in A. Let $X = (x_1, x_2, \ldots, x_m)$ be such an $n \times m$ matrix, where x_j, $1 \leq j \leq m$, is a column vector of length n. Let i_j, $1 \leq j \leq m$, be the position of x_j in the span n binary de Bruijn sequence S. Let

$$Y = \mathbf{D}(i_1, i_2, \ldots, i_m) = (i_2 - i_1, i_3 - i_2, \ldots, i_m - i_{m-1}),$$

where the computations are performed modulo 2^n. The m-tuple $(i_1, i_2, \ldots, i_m)$ represents the consecutive shifts in which the sequence S is taken in A. The $(m - 1)$-tuple Y also represents these shifts, where $i_j - i_{j-1}$ is the shift of the jth column of A related to the $(j - 1)$th column of A, associated with the $n \times m$ array X in A. The $(m - 1)$-tuple Y is contained in a window $(k_{\delta+1}, k_{\delta+2}, \ldots, k_{\delta+m-1})$ of the span $m - 1$ de Bruijn sequence K, where each k_j is considered as an integer in $\mathbb{Z}_{2^n}$. Consider the sub-matrix of A,

$B = (B_\delta, B_{\delta+1}, \ldots, B_{\delta+m-1})$, which is a projection of m columns of A that starts at column δ and ends at column $\delta + m - 1$. By the definition of A we have that $B_\ell = \mathbf{E}^{k_\ell} B_{\ell-1}$, for $\delta + 1 \leq \ell \leq \delta + m - 1$. Therefore the matrix X is contained in the sub-matrix B as an $n \times m$ window. □

Example 9.7. Let $n = 2$, $m = 3$,

$$S = [0011]$$

and

$$K = [0032112331022013].$$

The array A with the 2×3 window property, obtained from the construction in Eq. (9.2), is

$$
A = \begin{bmatrix}
0 & 0 & 1 & 0 & 1 & 1 & 0 & 0 & 1 & 0 & 0 & 1 & 0 & 0 & 0 & 0 \\
0 & 0 & 0 & 1 & 1 & 0 & 1 & 0 & 0 & 0 & 0 & 1 & 0 & 0 & 1 & 0 \\
1 & 1 & 0 & 1 & 0 & 0 & 1 & 1 & 0 & 1 & 1 & 0 & 1 & 1 & 1 & 1 \\
1 & 1 & 1 & 0 & 0 & 1 & 0 & 1 & 1 & 1 & 1 & 0 & 1 & 1 & 0 & 1
\end{bmatrix}.
$$

■

Theorem 9.11. *If there exists an* $(r, t; n, m)$*-PM whose columns are sequences with period r and even weight, then there exists an* $(r, 2^m t; n + 1, m)$*-PM.*

Proof. Let A be an $(r, t; n, m)$-PM whose columns are sequences with period r and even weight and define

$$\mathcal{A} \triangleq [\overbrace{A\, A\, \cdots\, A}^{2^m \text{ times of } A}].$$

Clearly, $\mathcal{A}$ is an $r \times (2^m t)$ array. Theorem 4.4 and Corollary 4.5 imply that if we apply the operator $\mathbf{D}^{-1}$ on a column from A, the outcome is two complementary sequences of the same length r. These sequences will be used to construct a new array $\mathcal{B}$. By Theorem 4.4, all these sequences have period r. Let

$$S = [s_0, s_1, \ldots, s_{2^m - 1}]$$

be a span m binary de Bruijn sequence whose first m bits are *zeros* and hence the sequence $[s_1, s_2, \ldots, s_{2^m-1}]$ is the shortened de Bruijn sequence obtained from S. Define the sequence

$$\mathcal{B} \triangleq (\overbrace{0, 0, \ldots, 0}^{t \text{ times}}, \overbrace{s_1, s_2, \ldots, s_{2^m-1}, s_1, s_2, \ldots, s_{2^m-1}, \ldots, s_1, s_2, \ldots, s_{2^m-1}}^{t \text{ times of } s_1, s_2, \ldots, s_{2^m-1}})$$

of length $t + (2^m - 1)t = 2^m t$ and denote $\mathcal{B} = (b_1, b_2, \ldots, b_{2^m t})$.

We define a new array $\mathcal{B}$. Let the ith column of $\mathcal{B}$, $\mathcal{B}_i = \mathbf{D}_{b_i}^{-1} \mathcal{A}_i$, where $\mathcal{A}_i$ is the ith column of $\mathcal{A}$ and $1 \leq i \leq 2^m t$. Clearly, $\mathcal{B}$ is an $r \times (2^m t)$ array and hence it has $r2^m t$ windows of size $(n+1) \times m$. Since $r2^m t = 2^m 2^{nm} = 2^{(n+1)m}$, it follows that to complete the proof it is sufficient to show that either each $(n+1) \times m$ binary array is contained at least once as an $(n+1) \times m$ window in $\mathcal{B}$ or that all the $(n+1) \times m$ windows of $\mathcal{B}$ are distinct.

Assume, on the contrary, that there exist two $(n+1) \times m$ windows X and Y in $\mathcal{B}$ for which $X = Y$. Let $\mathcal{X}$ and $\mathcal{Y}$ be the two $n \times m$ arrays obtained from X and Y, respectively, by applying $\mathbf{D}$ on the columns of X and Y, respectively (note that these columns are acyclic sequences). Clearly, if $X = Y$, then $\mathcal{X} = \mathcal{Y}$. Since A is an $(r, t; n, m)$-PM, it follows that all its $n \times m$ windows are distinct and hence $\mathcal{X} = \mathcal{Y}$ implies that they are obtained from the same position of A in different locations of $\mathcal{A}$. Moreover, $\mathcal{X} = \mathcal{Y}$ implies that they are in columns of $\mathcal{B}$ projected from two sets of m columns associated with the same m-tuple in B. Recall that t is a power of 2 and hence g.c.d.$(t, 2^m - 1) = 1$. This implies that the projection of two sets of m columns on $\mathcal{A}$ and B (where the two sets of columns from $\mathcal{A}$ yield the same $r \times m$ window) cannot yield the same pair of m-tuples from B and hence cannot yield the same pair of $(n+1) \times m$ submatrices in $\mathcal{B}$, a contradiction. Thus $\mathcal{B}$ is an $(r, 2^m t; n+1, m)$-PM. $\qquad\square$

In the conditions of Theorem 9.11 it is required that if A is the $(r, t; n, m)$-PM from which the new perfect map is constructed, the columns of length r will have even weight. This is indeed the case in the array defined in (9.2) and in most of the de Bruijn arrays constructed from perfect factors in Section 9.2. In all the perfect factors, constructed in the proof of Theorem 9.2, we have that all the sequences have the same period $r = 2^k$ and the same linear complexity. Only when the linear complexity is also 2^k we have that the weight of the sequences is odd.

Theorem 9.12. *If there exists an $(r, t; n, m)$-PM whose column weights are odd, then there exists a $(2r, 2^{m-1}t; n+1, m)$-PM.*

Proof. The proof goes along the same lines as the proof of Theorem 9.11, but there are some important delicate differences.

Let A be an $(r, t; n, m)$-PM whose columns are sequences with period r and odd weight and define

$$\mathcal{A} \triangleq [\ \overbrace{A\ A\ \cdots\ A}^{2^{m-1}\ \text{times of}\ A}\ ].$$

Clearly, $\mathcal{A}$ is an $r \times (2^{m-1}t)$ array. By Theorem 4.5, we have that applying $\mathbf{D}^{-1}$ on a column of A, the result is a self-dual sequence of double length $2r$. Each such sequence can start with a *zero* or a *one*, depending on whether we use $\mathbf{D}_0^{-1}$ or $\mathbf{D}_1^{-1}$, respectively. These sequences will be used to construct a new array $\mathcal{B}$. By Theorem 4.5, all these sequences have period $2r$. Let

$$S = [s_0, s_1, \ldots, s_{2^{m-1}-1}]$$

be a span m half de Bruijn sequence (see Definition 4.2) whose first m bits are *zeros* and hence the sequence $[s_1, s_2, \ldots, s_{2^{m-1}-1}]$ is the "shortened" half de Bruijn sequence obtained from S. Define the sequence

$$B \triangleq (\overbrace{0, 0, \ldots, 0}^{t \text{ times}}, \overbrace{s_1, s_2, \ldots, s_{2^{m-1}-1}, s_1, s_2, \ldots, s_{2^{m-1}-1}, \ldots, s_1, s_2, \ldots, s_{2^{m-1}-1}}^{t \text{ times of } s_1, s_2, \ldots, s_{2^{m-1}-1}})$$

of length $t + (2^{m-1} - 1)t = 2^{m-1}t$ and define $B = (b_1, b_2, \ldots, b_{2^{m-1}t})$.

We define a new array $\mathcal{B}$ as follows. Let the ith column of $\mathcal{B}$ be $\mathbf{D}_{b_i}^{-1}\mathcal{A}_i$, where $\mathcal{A}_i$ is the ith column of $\mathcal{A}$ and $1 \le i \le 2^{m-1}t$. Clearly, $\mathcal{B}$ is an $(2r) \times (2^{m-1}t)$ array and hence it has $2r2^{m-1}t$ windows of size $(n + 1) \times m$. Since $2r2^{m-1}t = 2^m2^{nm} = 2^{(n+1)m}$, it follows that to complete the proof it is sufficient to show that either each $(n + 1) \times m$ binary array is contained at least once as a $(n + 1) \times m$ window in $\mathcal{B}$ or that all the $(n + 1) \times m$ windows of $\mathcal{B}$ are distinct.

Assume, on the contrary, that there exist two $(n + 1) \times m$ windows X and Y in $\mathcal{B}$ for which $X = Y$. Let $\mathcal{X}$ and $\mathcal{Y}$ be the two $n \times m$ arrays obtained from X and Y, respectively, by applying $\mathbf{D}$ on the columns of X and Y, respectively (note that these columns are acyclic sequences). Clearly, if $X = Y$, then $\mathcal{X} = \mathcal{Y}$. Since A is an $(r, t; n, m)$-PM, it follows that all its $n \times m$ windows are distinct and hence $\mathcal{X} = \mathcal{Y}$ implies that they are obtained from the same position of A in different locations of $\mathcal{A}$. Moreover, $\mathcal{X} = \mathcal{Y}$ implies that they are in columns of $\mathcal{B}$ projected from two sets of m columns associated with the same m-tuple in B. Recall that t is a power of 2 and hence g.c.d.$(t, 2^{m-1} - 1) = 1$. This implies that the projection of two sets of m columns on $\mathcal{A}$ and B (where the two sets of columns from $\mathcal{A}$ yield the same $r \times m$ window) cannot yield the same pair of m-tuples from B and hence cannot yield the same pair of $(n + 1) \times m$ submatrices in $\mathcal{B}$, a contradiction. Thus $\mathcal{B}$ is a $(2r, 2^{m-1}t; n + 1, m)$-PM. $\square$

The various constructions for de Bruijn arrays that were presented can be iterated to obtain such arrays with various parameters. Moreover, the transpose of each such array can be also used for such purpose. Finally, there are many ad-hoc constructions for small perfect maps on which the constructions can be applied.

9.5 Two-dimensional arrays with distinct differences

There is no two-dimensional concept of a difference set, but there are some structures that consider two-dimensional arrays with distinct differences. The rulers that are derived from difference sets (see Example 2.10) can be generalized to arrays with distinct differences as follows.

Definition 9.3. An $n \times m$ *distinct difference configuration* (DDC) is an $n \times m$ array in $\mathbb{Z}^2$ containing dots, such that any two lines connecting two dots are distinct in their length or their slope.

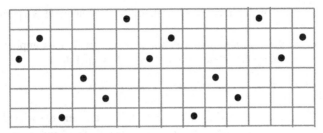

FIGURE 9.1   An array in which every 6 consecutive columns form a 6 × 6 Costas array obtained from the singly periodic Welch construction.

Definition 9.3 is the most general definition that generalizes the definition of a ruler. Examples of DDCs are each 6 × 6 array obtained from 6 consecutive columns in Fig. 9.1 and the arrays in Fig. 9.5. Later in this section and also in the next chapter we will generalize this definition for other shapes and also present an alternative definition for a set of dots without a specific shape (instead of the $n \times m$ rectangle). We can consider the dots in $\mathbb{Z}^2$ in two equivalent ways. The first one is to put the dot on the points of $\mathbb{Z}^2$. The second one, which will be used in this chapter and the next one, is as follows. Consider the unit square whose end-points are (i, j), $(i + 1, j)$, $(i, j + 1)$, and $(i + 1, j + 1)$. A dot at point (i, j) for the second option is at the center of this unit square. However, we start with a more specific definition of a structure that has several applications, has symmetry associated with rows and columns, and finally the structure can be described also by a one-dimensional sequence, which defines a permutation.

A ***Costas array*** of order n is an $n \times n$ permutation matrix with n **dots**, where the $\binom{n}{2}$ vectors connecting two dots in the matrix are all distinct as vectors (in their magnitude or their slope). The matrix will be also defined as a binary matrix with *ones* in the positions of the dots.

There are two basic constructions of Costas arrays, the Welch construction and the Golomb construction.

The Welch construction:

Let $\alpha \in \mathbb{F}_p$ be a primitive root modulo the prime p, and c a constant integer. Define the $(p - 1) \times (p - 1)$ binary matrix A by

$$A(j, i + c) = 1 \quad \text{if and only if} \quad \alpha^{i+c} \equiv j (\text{mod } p) \quad 1 \le i, j \le p - 1.$$

■

The *one* when $A(j, i) = 1$ (the same as $a_{j,i} = 1$) is in row j, column i, which means that the dot in the grid is in point (coordinate) (i, j). An example of an array derived from the Welch construction is depicted in Fig. 9.1.

Theorem 9.13. *The array A defined by the Welch construction is a Costas array of order $p - 1$.*

Proof. Since α is a primitive root modulo p, it follows that $p - 1$ successive powers of α form the multiplicative group modulo p, i.e., these $p - 1$ consecutive powers are distinct nonzero residues modulo p and hence A is a permutation matrix.

Assume, on the contrary, that A is not a Costas array. This implies that we can find two distinct pairs of entries with *ones* (dots) as follows:

$$\{(i + c, \alpha^{i+c}), (i + \delta + c, \alpha^{i+\delta+c})\} \quad \{(\ell + c, \alpha^{\ell+c}, (\ell + \delta + c, \alpha^{\ell+\delta+c})\},$$

where $1 \le i < \ell$, $1 \le \delta \le p - 3$, $\ell + \delta \le p - 1$, and the associated vectors $(\delta, \alpha^{i+\delta+c} - \alpha^{i+c})$ and $(\delta, \alpha^{\ell+\delta+c} - \alpha^{\ell+c})$ are equal. This implies that

$$\alpha^{i+c}(\alpha^{\delta} - 1) = \alpha^{\ell+c}(\alpha^{\delta} - 1).$$

Since $\alpha^{\delta} - 1 \not\equiv 0 \pmod{p}$, it follows that $\alpha^i = \alpha^{\ell}$ and hence $i = \ell$, a contradiction. Thus A is a $(p - 1) \times (p - 1)$ Costas array. $\qquad \square$

The Golomb construction:

Let α, β be two primitive elements in $\mathbb{F}_q$ (not necessarily distinct). Define the binary $(q - 2) \times (q - 2)$ matrix A by

$$A(j, i) = 1 \quad \text{if and only if} \quad \alpha^i + \beta^j = 1 \quad 1 \le i, j \le q - 2.$$

The dot is at coordinates $(i, \log_\beta(1 - \alpha^i))$. $\qquad \blacksquare$

Theorem 9.14. *The array A defined by the Golomb construction is a Costas array of order $q - 2$.*

Proof. Clearly, each one of the sets $\{\alpha^i : 1 \le i \le q-2\}$ and $\{\beta^j : 1 \le j \le q-2\}$ contains $\mathbb{F}_q \setminus \{0, 1\}$. This implies that for α^i, $1 \le i \le q - 2$, there exists exactly one j such that $\alpha^i + \beta^j = 1$ and therefore A is a permutation matrix.

For $\alpha^i + \beta^j = 1$ let $j = \log_\beta(1 - \alpha^i)$. Assume, on the contrary, that A is not a Costas array. Hence, there exist two pairs of points with vectors having the same magnitude and slope as follows:

$$(i, \log_\beta(1 - \alpha^i)), \quad (i + \delta, \log_\beta(1 - \alpha^{i+\delta})),$$
$$(\ell, \log_\beta(1 - \alpha^{\ell})), \quad (\ell + \delta, \log_\beta(1 - \alpha^{\ell+\delta})),$$

where $1 \le i < \ell$, $1 \le \delta \le q - 4$, $\ell + \delta \le q - 2$ and the corresponding vectors,

$$(\delta, \log_\beta(1 - \alpha^{i+\delta}) - \log_\beta(1 - \alpha^i))$$

and

$$(\delta, \log_\beta(1 - \alpha^{\ell+\delta}) - \log_\beta(1 - \alpha^{\ell}))$$

are equal. This implies the following equalities

$$\log_\beta \frac{1 - \alpha^{i+\delta}}{1 - \alpha^i} = \log_\beta \frac{1 - \alpha^{\ell+\delta}}{1 - \alpha^{\ell}}$$

$$\frac{1-\alpha^{i+\delta}}{1-\alpha^i} = \frac{1-\alpha^{\ell+\delta}}{1-\alpha^\ell}$$

$$\alpha^i(\alpha^\delta - 1) = \alpha^\ell(\alpha^\delta - 1).$$

Since $\alpha^\delta - 1 \neq 0$, it follows that $\alpha^i = \alpha^\ell$ and hence $i = \ell$, a contradiction. Thus A is a $(q-2) \times (q-2)$ Costas array. $\qquad\square$

One interesting question is the periodicity of a Costas array, i.e., can these arrays be made singly periodic or doubly periodic? The first natural related question to ask is whether it is possible to put dots in $\mathbb{Z}^2$ in such a way that each $n \times n$ window defines an $n \times n$ Costas array. Unless, $n = 1$ or $n = 2$, where the construction is trivial, we will prove that there is no such filling of the space. We can continue and ask whether there exists an $n \times \infty$ array such that each $n \times n$ array is a Costas array. We will prove that the Welch construction produces such an array. However, before that, we will give a weak definition for a doubly periodic array.

Definition 9.4. An $n \times n$ Costas array is *weakly doubly periodic* with period (n_1, n_2), where $n \leq n_1, n_2 \leq n + 1$, if copies of the Costas array can be placed in $\mathbb{Z}^2$ in such a way that each $n_2 \times n_1$ window is a DDC, it contains n dots, no row has more than one dot, and no column has more than one dot.

Definition 9.5. Let α be a primitive root modulo a prime p. We define the *Welch periodic array* to be the set

$$\mathcal{L}_p \triangleq \{(i, j) \in \mathbb{Z}^2 : \alpha^i \equiv j \pmod{p}\}.$$

The array $\mathcal{L}_p$ is weakly doubly periodic with period $(p-1, p)$, i.e., if $\mathcal{L}_p$ contains a dot at position (i, j) then it also contains dots at all positions of the form $(i + \lambda(p-1), j + \mu p)$, where $\lambda, \mu \in \mathbb{Z}$. It has a distinct difference property up to its periodicity with a proof that is almost the same as the proof of Theorem 9.13.

Theorem 9.15. *The array of dots $\mathcal{L}_p$ obtained by Definition 9.5, is a weakly doubly periodic Costas array with period $(p-1, p)$.*

Definition 9.6. We say that dots A and A' at positions (i, j) and (i', j'), of the Welch doubly periodic array, are *equivalent*, and we write $A \equiv A'$, if $i' = i + \lambda(p-1)$ and $j' = j + \mu p$ for some $\lambda, \mu \in \mathbb{Z}$.

Lemma 9.17. *Let d and e be positive integers such that $d \not\equiv 0 \pmod{p-1}$ and $e \not\equiv 0 \pmod{p}$. Suppose that $\mathcal{L}_p$ contains dots A and B at positions (i_1, j_1) and $(i_1 + d, j_1 + e)$, respectively, and dots A' and B' at positions (i_2, j_2) and $(i_2 + d, j_2 + e)$, respectively. Then, $A \equiv A'$ and $B \equiv B'$.*

Proof. By the definition of $\mathcal{L}_p$ we have

$$j_1 \equiv \alpha^{i_1} \pmod{p}$$

$$j_2 \equiv \alpha^{i_2} \pmod{p}$$

$$j_1 + e \equiv \alpha^{i_1+d} \pmod{p}$$

$$j_2 + e \equiv \alpha^{i_2+d} \pmod{p}.$$

By eliminating j_1 and j_2 from these equations we have that

$$\alpha^{i_1} + e \equiv \alpha^{i_1+d} \pmod{p}$$

$$\alpha^{i_2} + e \equiv \alpha^{i_2+d} \pmod{p},$$

which implies that

$$\alpha^{i_2} - \alpha^{i_1} \equiv (\alpha^{i_2} - \alpha^{i_1})\alpha^d \pmod{p}$$

and hence

$$(\alpha^d - 1)(\alpha^{i_1} - \alpha^{i_2}) \equiv 0 \pmod{p}.$$

Since $d \not\equiv 0 \pmod{p-1}$, it follows that $i_1 \equiv i_2 \pmod{p-1}$, which also implies that $j_1 \equiv j_2 \pmod{p}$.

Thus $A \equiv A'$ and $B \equiv B'$. $\qquad\qquad\qquad\qquad\qquad\qquad\qquad\qquad \Box$

Definition 9.7. The unordered pair $\{(i, j), (i + d, j + e)\}$ will be referred to in the following also as the vector (d, e).

By Definition 9.5, we have that if $\mathcal{L}_p$ contains dots at (i, j) and $(i + d, j)$, then $d \equiv 0 \pmod{p-1}$ and if it contains dots at (i, j) and $(i, j + e)$ then $e \equiv 0 \pmod{p}$. Therefore a vector (d, e) (a vector between a point (i, j) and a point $(i + d, j + e)$) can occur at most once as a difference between two of the dots of $\mathcal{L}_p$ that lie within any particular $p \times (p - 1)$ rectangle.

Definition 9.8. An $n \times n$ Costas array A is called a *singly periodic Costas array* if there exist an $n \times \infty$ array of dots (or an $\infty \times n$ array of dots) in which any $n \times n$ sub-array is a cyclic shift of A.

Corollary 9.4. *The $(p - 1) \times (p - 1)$ Costas array constructed by the Welch construction is singly periodic. It can be extended to a $(p - 1) \times \infty$ array of dots in which each $(p - 1) \times (p - 1)$ sub-array is a Costas array.*

Fig. 9.1 demonstrates the middle section of the $6 \times \infty$ singly periodic Costas array from the Welch construction.

Definition 9.9. Let α be a primitive element of $\mathbb{F}_q$, where q is a power of a prime. We define the *Golomb periodic array* to be the set

$$\mathcal{G}_q \triangleq \{(i, j) \in \mathbb{Z}^2 : \alpha^i + \alpha^j = 1\}.$$

Similarly to the proof of Theorem 9.14 we can prove the following theorem.

Theorem 9.16. *The Golomb periodic array* $\mathcal{G}_q$ *is weakly doubly periodic with period* $(q-1, q-1)$, *where columns and rows of* $\mathbb{Z}^2$, *whose indices are congruent 0 modulo* $q-1$ *are empty.*

Definition 9.10. An $n \times n$ Costas array A is called a ***doubly periodic Costas array*** if we can assign dots to $\mathbb{Z}^2$ such that any $n \times n$ sub-array is a cyclic shift (possibly horizontally and vertically) of A.

In other words, an $n \times n$ weakly doubly periodic Costas array with period (n_1, n_2), where $n = n_1 = n_2$, is a doubly periodic Costas array. Does there exist a doubly periodic $n \times n$ Costas array? It is easy to verify that there are such arrays for $n = 1$ and for $n = 2$. Unfortunately, there are no such arrays for larger n.

Theorem 9.17. *Unless* $n = 1$ *or* $n = 2$, *there is no array of dots in* $\mathbb{Z}^2$, *in which each* $n \times n$ *sub-array is a Costas array.*

Proof. The proof for $n = 3$ and $n = 4$ can be carried out in many ways and it is left as an exercise. Assume now that there exists a doubly periodic Costas array of order $n > 4$. By definition, each $n \times n$ window in such an array is an order n Costas array. We can color the dots in n colors, 1, 2, 3, and so on up to n such that each $n \times n$ window contains exactly one dot from each color.

The number of distinct possible difference vectors for which the two dots connected by a difference vector fit into a $k \times k$ window is $2(k-1)^2$. This can be verified by noting that each such difference vector can be fit into a $k \times k$ window, where one of the dots is either in the bottom left corner or in the bottom right corner.

Assume now that each such difference vector appears exactly once between two different colors, i.e., a repeat of such a vector does not occur within an $n \times n$ sub-array. We distinguish now between n odd and n even to show that our assumption is incorrect.

Case 1: Assume that $n = 2k - 1$.

We can count how many such difference vectors occur in a doubly periodic $(2k-1) \times (2k-1)$ Costas array. For each color, we can form a $(2k) \times (2k)$ array whose four corners have four dots colored by the same color. Inside the array, there are all the other $2k - 2$ dots, each one forms such a difference vector (fits into a $k \times k$ window) with one of the corner dots as depicted in Fig. 9.2 for a total of $2k - 2$ distinct difference vectors, confined to a $k \times k$ window. This can be done for each one of the $n = 2k - 1$ colors, taken as corner dots. Since each difference vector will be counted twice for each color on its end-point dots, it follows that the total number of such vectors is $\frac{(2k-1)(2k-2)}{2}$. However,

$$\frac{(2k-1)(2k-2)}{2} = 2(k-1)^2 + k - 1 > 2(k-1)^2,$$

a contradiction.

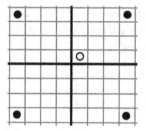

FIGURE 9.2   Each color such as the white in a black circle dot has to occur exactly once inside the $(2k) \times (2k)$ window having the four dots colored by black.

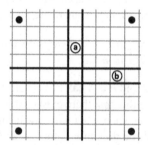

FIGURE 9.3   At most two colors like the **a** and **b** in the two dots fall outside every $k \times k$ window containing a dot colored by black in the $(2k) \times (2k)$ window having the four dots colored by black.

Case 2: Assume that $n = 2k$.

We can count how many such difference vectors occur in a doubly periodic $(2k) \times (2k)$ Costas array. For each color, we can form a $(2k + 1) \times (2k + 1)$ array whose four corners have four dots colored by the same color. Inside the array, there are all the other $2k - 1$ dots, each one, except one or two in the middle row and the middle column, forms such a difference vector with one of the corner dots, as depicted in Fig. 9.3 for a total of at least $2k - 3$ distinct difference vectors, confined to a $k \times k$ window. This can be done for each one of the $n = 2k$ colors. Since each such difference vector will be counted twice for each color on its end-point dots, it follows that the total number of such vectors is at least $\frac{2k(2k-3)}{2}$. However,

$$\frac{2k(2k - 3)}{2} = 2(k - 1)^2 + k - 2 > 2(k - 1)^2,$$

a contradiction.

Therefore by Case 1 and Case 2, there exists a repeat of a difference vector confined to a $k \times k$ window with vectors whose end-points dots have different colors.

We will prove now that there exists an $n \times n$ window in which there is such a repeated vector. This repeat cannot use only three colors, i.e., the two equal difference vectors share a point, as it will be confined to a $(2k - 1) \times (2k - 1)$ window, and hence also to an $n \times n$ window, which defines a Costas array.

Therefore the repeat consists of four dots colored with different colors. Since each $n \times n$ window contains all the n colors, it follows that we can choose such a window with one of these two difference vectors. W.l.o.g. we can assume that these two vectors are $\{(0, 0), (d_1, d_2)\}$, where $0 < d_1, d_2$, (the two dots are colored with **b** and **d**) and $\Upsilon = \{(i_1, i_2), (i_1 + d_1, i_2 + d_2)\}$ (the two dots are colored with **a** and **c**), where d_1 and d_2 are positive integers such that $0 < d_1, d_2 \le k - 1$. This scenario is depicted in Fig. 9.4. Note again that the array is doubly periodic with horizontal period $(0, n)$ and vertical period $(n, 0)$, and distinguish between two cases depending on whether n is odd or even.

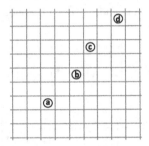

FIGURE 9.4 The two repeated vectors having four different colors.

Case 1: If $n = 2k - 1$, then the four points are within an $n \times n$ array if $-(k - 1) \le i_1, i_2 \le k - 1$. Since the two points $(-(k - 1), -(k - 1))$ and $(k - 1, k - 1)$ are the bottom left point and the upper right point, respectively, of an $n \times n$ array, it follows with the appropriate periodicity of the vector Υ, that w.l.o.g. $-(k - 1) \le i_1, i_2 \le k - 1$.

Case 2: If $n = 2k$, then the four points are within an $n \times n$ array if $-k \le i_1, i_2 \le k$. Since the two points $(-k, -k)$, (k, k) are the bottom left point and the upper right point, respectively, of an $(n + 1) \times (n + 1)$ array, it follows with the appropriate periodicity of the vector Υ, that w.l.o.g. $-k \le i_1, i_2 \le k$.

Both cases imply that the four points are within an $n \times n$ window and hence the repeated difference vector appears twice in the same $n \times n$ window, a contradiction.

Thus there is no $n \times n$ doubly periodic Costas array for $n > 2$. $\qquad\Box$

Now, we generalize the notion of Costas arrays and DDCs to other two-dimensional patterns with distinct differences.

Definition 9.11. A DDC $\mathrm{DD}(m, r)$ is a set of m dots placed on $\mathbb{Z}^2$ such that the following two properties are satisfied:

1. Any two of the dots in the configuration are at a (Euclidean) distance at most r apart.

2. All the $\binom{m}{2}$ differences between pairs of dots are distinct either in length or in slope.

Definition 9.11 can be slightly relaxed as follows. A **DDC DD**(m) is a set of m dots on $\mathbb{Z}^2$, with the property that the lines joining distinct pairs of dots are all different in length or slope. This DDC can be in any shape $\mathcal{S}$ and it is called an $\mathcal{S}$-DDC.

The motivation for generalizing the definition of a Costas array to DD(m, r) and DD(m) is for two reasons. One is just from a combinatorial point of view that these structures are interesting. The other is for an application that will be explained in Section 10.2. One question that can be asked is as follows. For a given integer r, what is the maximum integer m such that a DD(m, r) exists? The DDCs for the maximum m and $2 \leq r \leq 11$, are depicted in Fig. 9.5.

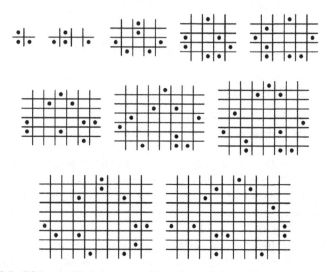

FIGURE 9.5 DD(m, r) with the largest possible m for $r = 2, 3, \ldots, 11$.

Let $\mathcal{A}$ be a (generally infinite) array of dots in $\mathbb{Z}^2$, and let η and κ be positive integers. We say that $\mathcal{A}$ is *doubly periodic* with period (η, κ) if

$$\mathcal{A}(i, j) = \mathcal{A}(i + \eta, j) \quad \text{and} \quad \mathcal{A}(i, j) = \mathcal{A}(i, j + \kappa)$$

for all integers i and j. Assume that each $\kappa \times \eta$ sub-array of $\mathcal{A}$ has exactly d dots. We define the *density* of $\mathcal{A}$ to be $d/(\eta\kappa)$. We write $(i, j) + \mathcal{S}$ for the shifted copy $\{(i + i', j + j') : (i', j') \in \mathcal{S}\}$ of $\mathcal{S}$. Let $\mathcal{A}$ be a doubly periodic array. We say that $\mathcal{A}$ is a *doubly periodic $\mathcal{S}$-DDC* if the dots contained in every shift $(i, j) + \mathcal{S}$ of $\mathcal{S}$ form a DDC.

The following theorem is an immediate consequence of the discussion on the Welch periodic arrays and the Golomb periodic arrays.

Theorem 9.18.

- *The Welch periodic array $\mathcal{L}_p$ is a doubly periodic DDC whose period is $(p-1, p)$ and its density is $1/p$.*
- *The Golomb periodic array $\mathcal{G}_q$ is a doubly periodic DDC whose period is $(q-1, q-1)$ and its density is $(q-2)/(q-1)^2$.*

Doubly periodic DDCs will be used to prove the existence of the DDCs with some given shape and a large number of dots. This will be done using the following theorem.

Theorem 9.19. *Let $\mathcal{S}$ be a shape, and let $\mathcal{A}$ be a doubly periodic $\mathcal{S}$-DDC of density δ. Then, there exists a set of at least $\lceil \delta|\mathcal{S}| \rceil$ dots contained in $\mathcal{S}$ that form a DDC.*

Proof. Let the period of $\mathcal{A}$ be (η, κ). Let $m_{i,j}$ be the number of dots of $\mathcal{A}$ contained in the shift $(i, j) + \mathcal{S}$ of $\mathcal{S}$. Since $\mathcal{A}$ is doubly periodic, it follows from the definition of the density of $\mathcal{A}$ that

$$\sum_{i=1}^{\eta}\sum_{j=1}^{\kappa} m_{i,j} = (\eta\kappa)\delta|\mathcal{S}|.$$

Hence, the average size of the integer $m_{i,j}$ is $\delta|\mathcal{S}|$. This implies that there exists an integer $m_{i',j'}$ such that $m_{i',j'} \geq \lceil \delta|\mathcal{S}| \rceil$. The $m_{i',j'}$ dots in $(i', j') + \mathcal{S}$ form a DDC, by our assumption on $\mathcal{A}$, and so the appropriate shift of these dots provides a DDC in $\mathcal{S}$ with at least $\lceil \delta|\mathcal{S}| \rceil$ dots, as required. $\qquad\square$

Next, we illustrate a general technique to construct a $DD(m, r)$ with m as large as possible.

Let $R = \lfloor r/2 \rfloor$, and let $\mathcal{S}$ be the set of points (a shape) in $\mathbb{Z}^2$ that are contained in a circle of radius R about the origin. We construct a DDC contained in $\mathcal{S}$ with many dots. Such a configuration is a $DD(m, r)$ for some value of m. The most straightforward approach is to find a large square contained in $\mathcal{S}$ (which will have sides of length $\sqrt{2}R$). Within this square, we define a Costas array of the largest order (whose order is smaller or equal to $\sqrt{2}R$). This will produce a $DD(m, r)$, where

$$m = \sqrt{2}R - o(R) = \tfrac{1}{\sqrt{2}}r - o(r) \approx 0.707r.$$

To motivate a better construction, we proceed as follows. We find a square of side n, where $n > \sqrt{2}R$, that partially overlaps the circle (see Fig. 9.6). The constructions of doubly periodic arrays based on Costas arrays show that there exist doubly periodic $n \times n$ DDCs that have density approximately $1/n$ (see Theorem 9.18). Hence, Theorem 9.19 shows that for any shape $\mathcal{S}'$ within the square, there exist DDCs in $\mathcal{S}'$ that have at least $|\mathcal{S}'|/n$ dots. Let $\mathcal{S}'$ be the intersection of the square with the circle $\mathcal{S}$ with radius R. Defining θ as in Fig. 9.6,

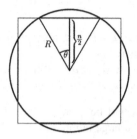

FIGURE 9.6   Square intersecting a circle.

some basic geometry shows that the area of $\mathcal{S}'$ is

$$|\mathcal{S}'| = \frac{\frac{\pi}{2} - 2\theta + \sin 2\theta}{2\cos^2\theta}|\mathcal{S}| = 2R^2\left(\frac{\pi}{2} - 2\theta + \sin 2\theta\right).$$

Since $n = 2R\cos\theta$, it follows from Theorem 9.19 that the density of dots within $\mathcal{S}'$ can be about $1/n = 1/(2R\cos\theta)$ when n is large. Hence, we can hope for at least μR dots, where μ is the maximum value of

$$(\pi/2 - 2\theta + \sin 2\theta)/\cos\theta$$

on the interval $0 \le \theta \le \pi/4$. A value of $\mu \approx 1.61589$, is achieved when $\theta \approx 0.41586$ (and for this value we have $n = r\cos\theta = cr$, where $c \approx 0.915$).

Theorem 9.20. *Let μ be defined as above. There exists a DD(m, r) in which $m = \frac{\mu r}{2} - o(r) \approx 0.808r$.*

Proof. Define $c \approx 0.915$ as above. Let q be the smallest prime power such that $q > cr$. By Theorem 1.26 on the gaps between primes we have that $cr < q < cr + (cr)^{5/8}$ and hence $q \sim cr$. By Theorem 9.18, there exists a doubly periodic $(q-1) \times (q-1)$ DDC $\mathcal{A}$ of density $(q-2)/(q-1)^2$. Let $\mathcal{S}'$ be the intersection between $\mathcal{S}$ and a circle of radius $\lfloor r/2 \rfloor$ about the origin. Therefore $\mathcal{A}$ is a doubly periodic $\mathcal{S}'$-DDC. By Theorem 9.19, there exists a DDC in $\mathcal{S}'$ with at least m dots, where $m = |\mathcal{S}'|(q-2)/(q-1)^2$. However, the geometric argument above shows that $|\mathcal{S}'|(q-2)/(q-1)^2 \sim \frac{\mu r}{2}$, and the theorem follows.   □

9.6   Notes

The necessity for solutions of two-dimensional and multi-dimensional coding problems that are generalizations of one-dimensional coding problems have been increasing from the last decades of the 20th century and into the 21st century. Such problems include error correction of burst errors, where the one-dimensional case was solved, for example, in Abdel-Ghaffar [1], Abdel-Ghaffar, McEliece, Odlyzko, and van Tilborg [2], and Reiger [50] and the two-dimensional case was considered, for example, in Breitbach, Bossert, Zyablov, and Sidorenko [5], Etzion and Yaakobi [14], Imai [30], and Roth and Seroussi [54]. Another problem concerns the capacity and coding for constrained channels, where the one-dimensional case is almost completely solved and has hundreds of research papers that are well documented in Marcus, Roth, and Siegel [36]. The two-dimensional case is far from being solved, but still has many papers, such as Etzion [11], Sharov and Roth [55], and Tal, Etzion, and Roth [60].

Randomness properties of two-dimensional arrays can be defined by window properties or by distinct differences. Combinatorics of such arrays were considered by Siu [56].

Section 9.1. Perfect factors, the graphs $G_{\text{PF}(n,k),m}$, zero factors, and the graph $G_{\text{ZF}(n,k),m}$ were defined by Etzion [10]. The graph $G_{n,m}(V, E_1, E_2)$ was defined by Fan, Fan, Ma, and Siu [15]. The existence of perfect factors for all admissible parameters was proved by Etzion [10]. Non-binary perfect factors were considered and constructed by Mitchell [37], Mitchell and Paterson [41], and Paterson [46]. For most parameters, binary perfect factors are not unique. It is not difficult to verify that $\text{PF}(2^k - 1, k)$ is unique and contains all the sequences of length 2^k with odd weight. However, generally, we have the following question.

Problem 9.1. For which n and k, is $\text{PF}(n, k)$ unique?

As for zero factors, except for the factors that are guaranteed by Theorems 9.3 and 9.4, which were mentioned in Etzion [10], the only other zero factors are obtained by the following theorem mentioned in Etzion [10], which is credited to Golomb [19].

Theorem 9.21. *Every factor e of $2^n - 1$ that is not a factor of any number $2^d - 1$ with $d < n$ occurs as the exponent of a zero factor that corresponds to an irreducible polynomial of degree n. There are $\frac{\phi(e)}{n}$ irreducible polynomials that correspond to zero factors with exponent e.*

The first set of parameters where a zero factor is not known to exist is $\text{ZF}(12, 15)$. This leads to the first specific open problem.

Problem 9.2. Does there exist a $\text{ZF}(12, 15)$?

Problem 9.3. Provide new constructions for zero factors. Analyze all parameters for which zero factors exist (or those parameters where their existence is unknown). Are there n, d, and k, such that $2^n - 1 = d \cdot k$, where $k > n$, and there is no zero factor with d cycles of length k in G_n?

A graph in which a Hamiltonian cycle is associated with an $(r, t; n, m)$-PM can be defined without perfect factors. The vertices of the graph are $r \times m$ matrices, where each $n \times m$ matrix is contained in exactly one vertex. There is an edge $u \rightarrow v$ if the last $m - 1$ columns of u are equal to the first $m - 1$ columns of v. A similar definition can be given to constructed shortened perfect maps. Perfect factors lead to a construction of such a graph.

Problem 9.4. Provide a general construction of such a graph that is not derived from a perfect factor and not from given perfect maps.

Section 9.2. The representation of the perfect map as a graph and a generalization of the **D**-morphism for two-dimensional arrays was introduced by Fan, Fan, Ma, and Siu [15]. The construction based on perfect factors, zero factors, and the necklaces factor was given in Etzion [10].

The first example of a perfect map was presented by Reed and Steward [49] who presented a $(4, 4; 2, 2)$-PM. They were later considered by Gordon [25] and by Clapham [6]. Non-binary perfect maps were constructed by Cock [7], Hurlbert and Isaak [28], Hurlbert, Mitchell, and Paterson [29], and Paterson [47,48]. Multi-dimensional perfect maps were considered in Hurlbert and Isaak [27]. All the perfect maps that are discussed in this section and all these papers are periodic. One-dimensional de Bruijn sequences are periodic and their aperiodic version is equivalent to the periodic one. This is not the case for perfect maps. Aperiodic perfect maps can be derived from periodic ones, but there can be aperiodic ones that are not derived from periodic arrays. Moreover, there are parameters of aperiodic perfect maps for which the aperiodic perfect maps cannot be obtained from periodic perfect maps. This is also true if the arrays are periodic only in one of the dimensions. This was considered by Mitchell [39] who proved that the necessary conditions for the existence of binary aperiodic perfect maps, in one or two dimensions, are also sufficient.

Choosing the appropriate perfect factors for the construction of associated perfect maps yields perfect maps that can be encoded and decoded more efficiently, as was described by Mitchell [38] and Mitchell and Paterson [40].

Section 9.3. The folding of an M-sequence into a rectangular array was carried out by MacWilliams and Sloane [35]. A generalization of the technique using irreducible polynomials to obtain a set of arrays that together have the window property can be given too. Folding M-sequences into arrays of different sizes is of interest. For example, folding an M-sequence of length 255 into a 5×51 rectangle yields some pseudo-random arrays, but the necessary theory was not developed. Some of these arrays, for example, can have the 4×2 window property, but not the 2×4 window property. Such a generalization can also use the lemmas of Section 7.2. The constructed arrays can be obtained also as what are called maximum-area matrices by Nomura, Miyakawa, Imai, and Fukuda [44]. Different definitions and constructions for pseudo-random arrays were given by Spann [59], by van Lint, MacWilliams, and Sloane [33], and by Soloveychik, Xiang, and Tarokh [57,58].

Problem 9.5. Provide new constructions for pseudo-random arrays.

Problem 9.6. Provide constructions for sets of arrays of the same size in which each $n \times m$ binary matrix is contained in exactly one array in a set. The same problem when only the nonzero $n \times m$ binary matrix is not contained in these arrays.

Section 9.4. The recursive construction based on two de Bruijn sequences is due to Ma [34]. The other recursive constructions were presented in Fan, Fan, Ma, and Siu [15]. Paterson [45] proved that the necessary conditions of Lemma 9.1 are sufficient. This was done by first constructing $(r, t; n, m)$-PM for some small values of n and m to use them as initial conditions for a recursive construction. After that, Paterson [45] applied the recursive constructions of Fan, Fan, Ma, and Siu [15] with appropriate arrays, or more precisely appropriate columns of the arrays, as was done in the constructions associated with Theorems 9.11 and 9.12. By using the transpose of these arrays, the conditions on the columns yield the same conditions on the rows. One of the keys in the construction is that all columns (rows) have the same appropriate linear complexities and this can be done by initially using the arrays obtained from the perfect factors in Sections 9.1 and 9.2.

Section 9.5. The Welch construction for Costas arrays was presented first by Golomb and Taylor [22], where also the construction due to Lempel was presented. The construction of Lempel considers $\beta = \alpha$ in the Golomb construction. The generalization for $\beta \neq \alpha$ was given by Golomb [20]. There are many variants for these constructions, some of which were presented in Golomb [20], but more detailed variants and an excellent early survey on these arrays, their structure and properties, were given by Golomb and Taylor [23].

The proof of Theorem 9.17 was carried out by Taylor [61]. Freedman and Levanon [16] proved that any two Costas arrays of order n have at least one vector in common. This property was also solved by Taylor [61].

There are a few types of arrays with distinct differences. A *sonar sequence* is an $n \times m$ DDC in which each column has exactly one dot. The main objective is for a given n to find the largest m such that an $n \times m$ sonar sequence exists. A bound

$$m < n + 3n^{2/3} + 2n^{1/3} + 9$$

was proved by Erdös, Graham, Ruzsa, and Taylor [8] using the method of Erdös and Turán [9] who presented bounds for arrays of dots with distinct slopes or lengths. Construction of sonar sequences from M-sequences was done by Games [17]. Moreno, Games, and Taylor [42] presented a table for constructions of $n \times m$ sonar sequences up to $m = 100$. Moreno, Golomb, and Corrada [43] consider a generalization of sonar sequences into arrays in which some columns with no dots are permitted.

Robinson [51] considered and analyzed two other types of rectangles. The first one was $n \times m$ binary arrays with exactly one *one* in each column and there were no repeated vectors when the array was shifted only horizontally. Such arrays are called **radar arrays**. The second type of array was $n \times m$ arrays with dots such that there are no repeated difference vectors in each such array. Such arrays are called **Golomb rectangles**. Construction of Golomb rectangles by the folding of Golomb rulers was presented by Robinson [52]. Robinson [53] considered efficient ways to find these types of arrays by computer search. Constructions and bounds for radar arrays were also given by Blokhuis and Tiersma [4], Ge, Ling, Miao [18], Hamkins and Zeger [26], and Zhang and Tu [62]. The difference triangle presented in Section 2.5 is a tool to check whether DDCs like sonar sequences have distinct differences. Some of the DDCs are periodic arrays and instead of the difference triangle, we should consider the difference cylinder.

Thus far, all the two-dimensional arrays that were discussed are defined on the square grid. However, for some applications, it is preferable to use other grids. In computer applications, pixels are circles that are the smallest addressable element in an image. It is desirable to pack them as densely as possible on a grid. It is easy to verify that they will have higher density if they are packed in a hexagonal grid rather than in a square grid, as depicted in Fig. 9.7 using associated DDCs.

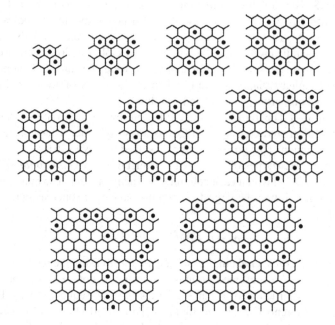

FIGURE 9.7 $DD^*(m, r)$ with the largest possible m for $r = 2, 3, \ldots, 10$.

Definition 9.12. A *hexagonal* DDC $DD^*(m, r)$ is a set of m dots placed on the hexagonal grid such that the following two properties are satisfied:

1. Any two dots in the configuration are at Euclidean distance at most r apart.

2. All the $\binom{m}{2}$ differences between pairs of dots are distinct either in length or in slope.

The DDCs in the hexagonal grid for the maximum m and $2 \leq r \leq 10$, are depicted in Fig. 9.7. This model and other related models were discussed and analyzed by Blackburn, Etzion, Martin, and Paterson [3].

Let $S_1, S_2, \ldots$ be an infinite sequence of similar shapes (same shape scaled in size), where $|S_{i+1}| > |S_i|$. Using the technique of Erdös and Turán [8,9] that was also used for a similar purpose by Erdös, Graham, Ruzsa, and Taylor [8], and for which a detailed proof is given in Blackburn, Etzion, Martin, and Paterson [3], it was proved by Etzion [12] that

Theorem 9.22. *An upper bound on the number of dots in* S_i, $i \to \infty$, *is* $\lim_{i \to \infty} (\sqrt{|S_i|} + o(\sqrt{|S_i|}))$.

Finally, we note that the problem of points with distinct differences is of interest also from a discrete geometry point of view. Some similar questions can be found in Lefmann and Thiele [32].

After it was proved in Theorem 9.17 that there is no doubly periodic Costas array, we want to find whether there exist singly periodic Costas arrays, except for those generated by the Welch construction (see Corollary 9.4). A step toward an answer to this question will be given using the concept of Tuscan squares introduced by Golomb and Taylor [24]. An $n \times n$ array is a ***Latin square*** if each one of the symbols $0, 1, \ldots, n - 1$ appears exactly once in each row and each column. An ***Italian square*** is an $n \times n$ array in which each one of the symbols $0, 1, \ldots, n - 1$ appears exactly once in each row. A ***Tuscan-k square*** is an Italian square in which for any two symbols a and b, and for each t from 1 to k, there is at most one row in which b is the tth symbol to the right of a.

A Tuscan-$(n - 1)$ square is called a ***Florentine square***. If the square is also Latin, then it will be called a ***Vatican square***.

A ***circular Tuscan-k array*** is an $n \times (n + 1)$ array in which each of the $n + 1$ symbols $0, 1, \ldots, n - 1, *$ appears exactly once in each one of the n rows, and in which the Tuscan-k property holds when the rows are taken to be circular.

A ***polygonal path*** $X_1, X_2, \ldots, X_n$ is a permutation of $0, 1, \ldots, n - 1$. A Latin square A defined by a polygonal path $X_1, X_2, \ldots, X_n$ is defined as follows:

$$A(i, j) \equiv X_j + i - 1 \pmod{n}, \quad 1 \leq i, j \leq n.$$

Polygonal path constructions and properties of the constructed arrays were analyzed by Golomb, Etzion, and Taylor [21].

Example 9.8. Let $n = 8$ and consider the permutation $0, 7, 1, 6, 2, 5, 3, 4$. The Latin square obtained from this permutation is

$$
\begin{array}{cccccccc}
0 & 7 & 1 & 6 & 2 & 5 & 3 & 4 \\
1 & 0 & 2 & 7 & 3 & 6 & 4 & 5 \\
2 & 1 & 3 & 0 & 4 & 7 & 5 & 6 \\
3 & 2 & 4 & 1 & 5 & 0 & 6 & 7 \\
4 & 3 & 5 & 2 & 6 & 1 & 7 & 0 \\
5 & 4 & 6 & 3 & 7 & 2 & 0 & 1 \\
6 & 5 & 7 & 4 & 0 & 3 & 1 & 2 \\
7 & 6 & 0 & 5 & 1 & 4 & 2 & 3
\end{array}
$$

This Latin square is a Tuscan-1 square as each ordered pair (i, j) appears as two consecutive elements in some row of the array exactly once.

The following lemma can be readily verified.

Lemma 9.18. *A necessary and sufficient condition that the polygonal path* $X_1, X_2, \ldots, X_n$ *forms a Tuscan-1 square is that*

$$\{X_{i+1} - X_i \ (\mathrm{mod}\ n) : 1 \leq i \leq n - 1\} \equiv \{i : 1 \leq i \leq n - 1\}.$$

In other words, Lemma 9.18 implies that to form a Tuscan-1 square by a polygonal $X_1, X_2, \ldots, X_n$ all the consecutive differences $X_{i+1} - X_i$ must be distinct modulo n. Similarly, we have a condition to obtain a circular Tuscan-k array and a circular Vatican array.

Lemma 9.19. *A necessary and sufficient condition that the polygonal path* $X_1, X_2, \ldots, X_n$ *forms a Tuscan-k square is that for each* j, $1 \leq j \leq k$

$$\{X_{i+j} - X_i \ (\mathrm{mod}\ n) : 1 \leq i \leq n - 1\} \equiv \{i : 1 \leq i \leq n - 1\},$$

where subscripts are taken modulo $n + 1$.

Corollary 9.5. *A necessary and sufficient condition that the polygonal path* $X_1, X_2, \ldots, X_n$ *forms a circular Vatican arrays is that for each* k, $1 \leq k \leq n - 1$

$$\{X_{i+k} - X_i \ (\mathrm{mod}\ n) : 1 \leq i \leq n - 1\} \equiv \{i : 1 \leq i \leq n - 1\},$$

where subscripts are taken modulo $n + 1$.

One can easily verify that the construction presented in the following theorem yields a circular Vatican array. This construction is akin to the construction of a singly periodic Costas array by the Welch construction.

Theorem 9.23. *If p is a prime and α is a primitive root modulo p, then the polygonal path* $X_1, X_2, \ldots, X_{p-1}$ *defined by*

$$X_i = j \ \text{if } \alpha^j \equiv i \ (\mathrm{mod}\ p), \quad 1 \leq i \leq p - 1, \ 0 \leq j \leq p - 2$$

forms a $(p - 1) \times p$ *circular Vatican array.*

Lemma 9.20. *If $X_1, X_2, \ldots, X_n$ is a polygonal path that forms a Tuscan-1 square, then n is even.*

Proof. By Lemma 9.18, each nonzero residue modulo n appears as a difference of the form $X_{i+1} - X_i$ (mod n), $1 \leq i \leq n - 1$, exactly once. This implies that

$$X_n \equiv X_1 + \sum_{i=1}^{n-1} i \equiv X_1 + \frac{n(n-1)}{2} \pmod{n}.$$

If n is odd, then $\frac{n(n-1)}{2} \equiv 0$ (mod n) and hence $X_n = X_1$, which contradicts the fact that $X_1, X_2, \ldots, X_n$ is a permutation of $0, 1, \ldots, n - 1$. Thus n is even.   □

Lemma 9.21. *If a polygonal path $X_1, X_2, \ldots, X_n$ produces an $n \times (n + 1)$ circular Tuscan-k array, then for each i, $1 \leq i \leq k$,*

$$X_{n+1-i} \equiv X_i + \frac{n}{2} \pmod{n}.$$

Proof. Letting the symbol $*$ be X_0 we will confine our attention to the sequence $X_0, X_1, X_2, \ldots, X_n$ that we may assume in the top row of the $n \times (n+1)$ circular Tuscan-k array. For a fixed i, with $1 \leq i \leq k$, look at all the distinct cycles (one or more depending on the divisors of $n + 1$) of the form $X_t, X_{t+i}, X_{t+2i}, \ldots$, for some $0 \leq t \leq n$, where the subscripts are taken modulo $n + 1$. Every symbol is hit exactly once by one of these cycles, and since the array is a circular Tuscan-i array, each one of the differences $1, 2, \ldots, n - 1$ modulo n must occur exactly once in only one cycle as $X_{t+(r+1)i} - X_{t+ri}$ (the differences with X_0 are ignored). Note that the sum of all these differences, in a cycle that does not contain X_0, is congruent to 0 modulo n. The one cycle that does hit X_0 will contain $\ldots, X_{n+1-i}, X_0, X_i, \ldots$ successively, which accounts for the two pairs of symbols, successive in a cycle, to which we assign no difference value modulo n. The remaining differences therefore sum to $\frac{n}{2}$ (mod n) since for even n

$$\sum_{i=1}^{n-1} i = \frac{n(n+1)}{2} \equiv \frac{n}{2} \pmod{n}. \tag{9.3}$$

Hence, we can calculate

$$(X_{n+1-i} - X_{n+1-2i}) + (X_{n+1-2i} - X_{n+1-3i}) + \cdots + (X_{3i} - X_{2i}) + (X_{2i} - X_i)$$
$$= X_{n+1-i} - X_i \equiv \frac{n}{2} \pmod{n},$$

since again we have the equality in Eq. (9.3) when n is even.   □

The following theorem makes use of the amazing Theorems 1.27 and 1.28 proved by Kelly [31] in 1954 to show a connection between the size of Florentine arrays and some primes. The theorem was proved by Etzion, Golomb, and Taylor [13].

Theorem 9.24. *If there exists an $n \times (n + 1)$ polygonal path for a circular Florentine array, then $n + 1$ is prime.*

Proof. Using Lemma 9.21 we aim to satisfy the conditions of Theorems 1.27 and 1.28.

Assume that the polygonal path $X_0, X_1, X_2, \ldots, X_n$ has $X_0 = *$, $X_1 = 0$, and that $X_2, \ldots, X_n$ is a permutation of the integers $1, 2, \ldots, n - 1$. Let

$$\mathcal{A} \triangleq \{i \; : \; X_i \text{ is even}\}, \quad \mathcal{B} \triangleq \{i \; : \; X_i \text{ is odd}\}.$$

With the subscripts taken modulo $n + 1$ it is clear that $\mathcal{A} \cap \mathcal{B} = \varnothing$, $\mathcal{A} \cup \mathcal{B} = \{1, 2, \ldots, n\}$, and $1 \in \mathcal{A}$. Since n is even, it follows that $|\mathcal{A}| = |\mathcal{B}| = \frac{n}{2}$.

If we take any subscript $t \in \mathcal{A} \cup \mathcal{B}$, it will partition $X_0, X_1, \ldots, X_n$ into cycles of the form

$$\ldots, X_{i-t}, X_i, X_{i+t}, X_{i+2t}, \ldots$$

Furthermore, since the polygonal path generates a circular Tuscan-n array, it follows that the successive differences in cycles, $X_{i+t} - X_i$, will take each of the values $1, 2, \ldots, n - 1$ exactly once, and take no values for $X_t - X_0$ and $X_0 - X_{-t}$ because $X_0 = *$. Thus in both cases analyzed below, there will be $\frac{n}{2}$ odd differences and $\frac{n}{2} - 1$ even differences.

One more general fact will be taken for granted. In any cycle of odd and even integers that does not contain $X_0 = *$, the number of times an odd integer is followed by an even integer equals the number of times even is followed by odd. We distinguish now between two cases depending on whether $n + 1$ is congruent to 1 modulo 4 or congruent to 3 modulo 4.

Case 1: $n + 1 = 4k + 1$. In this case $n/2$ is even, so we know by Lemma 9.21 that $X_t \equiv X_{-t} \pmod{2}$. This tells us that in the cycle that contains $X_0 = *$ (as well as in all the other cycles if there are any) it will occur the same number of times that X_i is odd and X_{i+t} is even as that X_i is even and X_{i+t} is odd. Consequently, there are k subscripts in the set $\{i \; : \; i \in \mathcal{A}, \; i + t \in \mathcal{B}\}$ as well as k subscripts in the set $\{i \; : \; i \in \mathcal{B}, \; i + t \in \mathcal{A}\}$, since the number of odd differences all together is $2k = n/2$. Considering the $2k - 1$ even differences depend on t as follows:

1. If $t \in \mathcal{A}$, then X_t is even, so X_{-t} is even, $X_{-t+t} = X_0 = *$ is not even. This guarantees that only $k - 1$ of the remaining subscripts for which $i \in \mathcal{A}$ are such that $i + t \in \mathcal{A}$. Thus we infer that there are exactly k subscripts for which $i \in \mathcal{B}$ and $i + t \in \mathcal{B}$. In other words, one condition of Theorem 1.27 is satisfied.

2. If $t \in \mathcal{B}$, then X_t is odd, so X_{-t} is odd, $X_{-t+t} = X_0 = *$ is not odd, and only $k - 1$ of the remaining subscripts for which $i \in \mathcal{B}$ are such that $i + t \in \mathcal{B}$. Thus we infer that there are k subscripts for which $i \in \mathcal{A}$ and $i + t \in \mathcal{A}$. In other words, the second condition of Theorem 1.27 is satisfied.

Thus $n + 1$ is prime in this case.

Case 2: $n + 1 = 4k - 1$. In this case $n/2$ is odd, hence we know by Lemma 9.21 that $X_t \not\equiv X_{-t}$ (mod 2).

1. We already know that $|\mathcal{A}| = |\mathcal{B}| = 2k - 1$.
2. If $t \in \mathcal{A}$, then X_t is even, so X_{-t} is odd, $X_{-t+t} = X_0$, $-t \in \mathcal{B}$, and $0 \in t + \mathcal{B}$. From this, we can also see that $k - 1$ is the number of times that X_i is odd and X_{i+t} is even, while k is the number of times that X_i is even and X_{i+t} is odd. The $2k - 2 = \frac{n}{2} - 1$ even differences can only be accounted for by having $k - 1$ times that X_i is even and X_{i+t} is even, beside $k - 1$ times that X_i is odd and X_{i+t} is odd. Altogether, there is one subscript $i \in \mathcal{B}$ such that $i + t = 0$, $k - 1$ subscripts for which $i \in \mathcal{B}$ and $i + t \in \mathcal{A}$, and $k - 1$ subscripts for which $i \in \mathcal{B}$ and $i + t \in \mathcal{B}$. In other words, the conditions of Theorem 1.28 are satisfied.

Thus $n + 1$ is prime in this case.

Therefore if an $n \times (n + 1)$ polygonal path circular Florentine array exists, then $n + 1$ is prime. $\square$

Theorem 9.25. *A Vatican square defined by the polygonal path* $X_1, X_2, \ldots, X_n$ *exists if and only if a singly periodic Costas array* $C_1, C_2, \ldots, C_n$ *exists, where* C_i *is the row in which the dot is located in the ith column.*

Proof. Assume first that $X_1, X_2, \ldots, X_n$ is a polygonal path pattern for a Vatican square. Define $C_j = i$ if and only if $X_i = j$, $1 \le i, j \le n$ and $C_\ell = C_{\ell+n}$ for each integer ℓ. Assume, on the contrary, that $C_1, C_2, \ldots, C_n$ does not define an $n \times \infty$ singly periodic Costas array. Then, we can find three integers i, j, and δ, where $1 \le i < j < j + \delta \le i + n - 1$, where

$$C_{i+\delta} - C_i = C_{j+\delta} - C_j .$$

If we denote $k_2 = C_{i+\delta}, k_1 = C_i, r_2 = C_{j+\delta}, r_1 = C_j$, then

$$X_{k_2} - X_{k_1} \equiv X_{r_2} - X_{r_1} \equiv \delta \pmod{n}$$

for

$$k_2 - k_1 = r_2 - r_1,$$

where the computation is performed modulo n since $i + \delta$, $j, j + \delta$ might be larger than n, but $C_\ell = C_{\ell+n}$ for any integer ℓ. Therefore $X_1, X_2, \ldots, X_n$ is not a polygonal path pattern for a Vatican square, a contradiction.

Assume now that $C_1, C_2, \ldots, C_n$ and $C_\ell = C_{\ell+n}$ for each integer ℓ, is an $n \times \infty$ singly periodic Costas array. Define $X_i = j$ if and only if $C_j = i$, $1 \le i, j \le n$. Assume, on the contrary, that $X_1, X_2, \ldots, X_n$ is not a polygonal path pattern for an $n \times n$ Vatican array. Then, we can find three integers i, j, δ such that $1 \le i \le j < j + \delta \le n$ such that

$$X_{i+\delta} - X_i \equiv X_{j+\delta} - X_j \pmod{n}.$$

If we denote $k_2 = X_{i+\delta}, k_1 = X_i, r_2 = X_{j+\delta}, r_1 = X_j$, then $1 \le k_1, k_2, r_1, r_2 \le n$ and

$$C_{k_2} - C_{k_1} = C_{r_2} - C_{r_1} = \delta$$

for

$$k_2 - k_1 \equiv r_2 - r_1 \pmod{n}.$$

We distinguish between two cases:

Case 1: If $r_2 - r_1 = k_2 - k_1$, then $C_{k_2} - C_{k_1} = C_{r_2} - C_{r_1} = \delta$ implies that $C_1, C_2, \ldots, C_n$ does not define an $n \times n$ Costas array, a contradiction.

Case 2: Either $r_2 - r_1 + n = k_2 - k_1$ or $r_2 - r_1 = k_2 - k_1 + n$ and w.l.o.g. we assume that $r_2 - r_1 = k_2 - k_1 + n$ that implies that $k_2 < k_1$.

If $k_2 < r_1$, then $C_{k_2+n} = C_{k_2}, r_1 < r_2, k_1, k_2 + n \le r_1 + n - 1$, and

$$C_{k_2+n} - C_{k_1} = C_{r_2} - C_{r_1}$$

with

$$(k_2 + n) - k_1 = r_2 - r_1$$

and hence $C_1, C_2, \ldots, C_n$ and $C_\ell = C_{\ell+n}$ for each integer ℓ, does not define an $n \times \infty$ singly periodic Costas array, a contradiction.

If $r_1 < k_2 < r_2$, then $C_{r_1+n} = C_{r_1}, k_2 < r_2, k_1, r_1 + n \le k_2 + n - 1$, and

$$C_{k_1} - C_{k_2} = C_{r_1+n} - C_{r_2}$$

with

$$k_1 - k_2 = (r_1 + n) - r_2$$

and hence $C_1, C_2, \ldots, C_n$ and $C_\ell = C_{\ell+n}$ for each integer ℓ, does not define an $n \times \infty$ singly periodic Costas array, a contradiction.

The two parts of the proof complete the proof of the theorem. $\qquad\square$

To examine whether an $n \times n$ Costas array is singly periodic we can use the difference cylinder. Assume we consider the array of dots as an $n \times \infty$ array. We consider any n consecutive columns of the array with dots at rows $C_0, C_1, \ldots, C_{n-1}$ in this order. In the jth row of the difference cylinder, $1 \le j \le n - 1$, the calculated differences are $C_{i+j} - C_i, 0 \le i \le 2n - j$, in this order, where subscripts are taken modulo n. For the array to be a singly periodic Costas array, it is required that any $n - j$ consecutive differences in the jth row must be distinct (this is true for any $n - j$ consecutive differences in a row). A proof that goes similarly to that of Theorem 9.25 yields the following slightly stronger result than the one of Theorem 9.25.

Theorem 9.26. *An $n \times (n + 1)$ circular Vatican array with the polygonal path $X_1, X_2, \ldots, X_n$ exists if and only if there exists a singly periodic Costas array*

$C_1, C_2, \ldots, C_n$ and $C_\ell = C_{\ell+n}$ for each integer ℓ. In such a Costas array for each j, $1 \leq j \leq n - 1$, every $n - j$ integers in the jth row of the difference cylinder are different modulo $n + 1$, i.e., it is a weakly doubly periodic array with period $(n, n + 1)$.

Corollary 9.6. If there exists a singly periodic Costas array $C_1, C_2, \ldots, C_n$ and $C_\ell = C_{\ell+n}$ for each integer ℓ, such that for each j, $1 \leq j \leq n - 1$, all the $n - j$ consecutive integers in the jth row of the difference cylinder are different modulo $n + 1$, then $n + 1$ is a prime.

Problem 9.7. Do there exist more singly periodic Costas arrays of order n except for those constructed from the Welch construction? Distinguish between the cases where $n + 1$ is a prime and $n + 1$ is not a prime.

References

[1] K.A.S. Abdel-Ghaffar, On the existence of optimum cyclic burst-correcting codes over GF(q), IEEE Trans. Inf. Theory 34 (1988) 329–332.

[2] K.A.S. Abdel-Ghaffar, R.J. McEliece, A.M. Odlyzko, H.C.A. van Tilborg, On the existence of optimum cyclic burst-correcting codes, IEEE Trans. Inf. Theory 32 (1986) 768–775.

[3] S.R. Blackburn, T. Etzion, K.M. Martin, M.B. Paterson, Two-dimensional patterns with distinct differences – constructions, bounds, and maximal anticodes, IEEE Trans. Inf. Theory 56 (2010) 1216–1229.

[4] A. Blokhuis, H.J. Tiersma, Bounds for the size of radar arrays, IEEE Trans. Inf. Theory 34 (1988) 164–167.

[5] M. Breitbach, M. Bossert, V. Zyablov, V. Sidorenko, Array codes correcting a two-dimensional cluster of errors, IEEE Trans. Inf. Theory 44 (1998) 2025–2031.

[6] C.R.J. Clapham, Universal tiling and universal (0, 1)-matrices, Discrete Math. 58 (1986) 87–92.

[7] J.C. Cock, Toroidal tilings from de Bruijn-Good cyclic sequences, Discrete Math. 70 (1988) 209–210.

[8] P. Erdős, R. Graham, I.Z. Ruzsa, H. Taylor, Bounds for arrays of dots with distinct slopes or lengths, Combinatorica 12 (1992) 39–44.

[9] P. Erdős, P. Turán, On a problem of Sidon in additive number theory and some related problems, J. Lond. Math. Soc. 16 (1941) 212–215.

[10] T. Etzion, Constructions for perfect maps and pseudorandom arrays, IEEE Trans. Inf. Theory 34 (1988) 1308–1316.

[11] T. Etzion, Cascading methods for runlength-limited arrays, IEEE Trans. Inf. Theory 43 (1997) 319–324.

[12] T. Etzion, Sequence folding, lattice tiling, and multidimensional coding, IEEE Trans. Inf. Theory 57 (2011) 4383–4400.

[13] T. Etzion, S.W. Golomb, H. Taylor, Tuscan-k squares, Adv. Appl. Math. 10 (1989) 164–174.

[14] T. Etzion, E. Yaakobi, Error-correction of multidimensional bursts, IEEE Trans. Inf. Theory 55 (2009) 961–976.

[15] C.T. Fan, S.M. Fan, S.L. Ma, M.K. Siu, On de Bruijn arrays, Ars Comb. 19A (1985) 205–213.

[16] A. Freedman, N. Levanon, Any two $N \times N$ Costas signals must have at least one common ambiguity sidelobe if $N > 3$ – a proof, Proc. IEEE 73 (1985) 1530–1531.

[17] R.A. Games, An algebraic construction of sonar sequences using M-sequences, SIAM J. Algebraic Discrete Methods 8 (1987) 753–761.

[18] G. Ge, A.C.H. Ling, Y. Miao, A systematic construction for radar arrays, IEEE Trans. Inf. Theory 54 (2008) 410–414.

[19] S.W. Golomb, Shift Register Sequences, Holden Day, San Francisco, 1967, Aegean Park, Laguna Hills, CA, 1980, World Scientific, Singapore, 2017.

[20] S.W. Golomb, Algebraic constructions for Costas arrays, J. Comb. Theory, Ser. A 37 (1984) 13–21.

[21] S.W. Golomb, T. Etzion, H. Taylor, Polygonal path constructions for Tuscan-k squares, Ars Comb. 30 (1990) 97–140.

[22] S.W. Golomb, H. Taylor, Two-dimensional synchronization patterns for minimum ambiguity, IEEE Trans. Inf. Theory 28 (1982) 600–604.

[23] S.W. Golomb, H. Taylor, Constructions and properties of Costas arrays, Proc. IEEE 72 (1984) 1143–1163.

[24] S.W. Golomb, H. Taylor, Tuscan squares – a new family of combinatorial designs, Ars Comb. 20B (1985) 115–132.

[25] B. Gordon, On the existence of perfect maps, IEEE Trans. Inf. Theory 12 (1966) 486–487.

[26] J. Hamkins, K. Zeger, Improved bounds on maximum size binary radar arrays, IEEE Trans. Inf. Theory 43 (1987) 997–1000.

[27] G. Hurlbert, G. Isaak, On the de Bruijn torus problem, J. Comb. Theory, Ser. A 64 (1993) 50–62.

[28] G. Hurlbert, G. Isaak, New constructions for de Bruijn tori, Des. Codes Cryptogr. 6 (1995) 47–56.

[29] G.H. Hurlbert, C.J. Mitchell, K.G. Paterson, On the existence of de Bruijn tori with two by two windows, J. Comb. Theory, Ser. A 76 (1996) 213–230.

[30] H. Imai, Two-dimensional Fire codes, IEEE Trans. Inf. Theory 19 (1973) 796–806.

[31] J.B. Kelly, A characteristic property of quadratic residues, Proc. Am. Math. Soc. 5 (1954) 38–46.

[32] H. Lefmann, T. Thiele, Point sets with distinct distances, Combinatorica 15 (1995) 379–408.

[33] J.H. van Lint, F.J. MacWilliams, N.J.A. Sloane, On pseudo-random arrays, SIAM J. Appl. Math. 36 (1979) 62–72.

[34] S.L. Ma, A note on binary arrays with a certain window property, IEEE Trans. Inf. Theory 30 (1984) 774–775.

[35] F.J. MacWilliams, N.J.A. Sloane, Pseudo-random sequences and arrays, Proc. IEEE 64 (1976) 1715–1729.

[36] B.H. Marcus, R.M. Roth, P.H. Siegel, Constrained systems and coding for recording channels, in: V. Pless, W. Huffman (Eds.), Handbook of Coding Theory, Elsevier, Amsterdam, The Netherlands, 1998, pp. 1635–1764.

[37] C.J. Mitchell, Constructing c-ary perfect factors, Des. Codes Cryptogr. 4 (1994) 341–368.

[38] C.J. Mitchell, de Bruijn sequences and perfect factors, SIAM J. Discrete Math. 10 (1997) 270–281.

[39] C.J. Mitchell, Aperiodic and semi-periodic perfect maps, IEEE Trans. Inf. Theory 41 (1995) 88–95.

[40] C.J. Mitchell, K.G. Paterson, Decoding perfect maps, Des. Codes Cryptogr. 4 (1994) 11–30.

[41] C.J. Mitchell, K.G. Paterson, Perfect factors from cyclic codes and interleaving, SIAM J. Discrete Math. 11 (1998) 241–264.

[42] O. Moreno, R.A. Games, H. Taylor, Sonar sequences from Costas arrays and the best known sonar sequences up to 100 symbols, IEEE Trans. Inf. Theory 39 (1993) 1985–1987.

[43] O. Moreno, S.W. Golomb, C.J. Corrada, Extended sonar sequences, IEEE Trans. Inf. Theory 43 (1997) 1999–2005.

[44] T. Nomura, H. Miyakawa, H. Imai, A. Fukuda, The theory of two-dimensional linear recurring arrays, IEEE Trans. Inf. Theory 18 (1972) 773–785.

[45] K.G. Paterson, Perfect maps, IEEE Trans. Inf. Theory 40 (1994) 743–753.

[46] K.G. Paterson, Perfect factors in the de Bruijn graph, Des. Codes Cryptogr. 5 (1995) 115–138.

[47] K.G. Paterson, New classes of perfect maps I, J. Comb. Theory, Ser. A 73 (1996) 302–334.

[48] K.G. Paterson, New classes of perfect maps II, J. Comb. Theory, Ser. A 73 (1996) 335–345.

[49] I.S. Reed, R.M. Stewart, Note on the existence of perfect maps, IRE Trans. Inf. Theory 8 (1962) 10–12.

[50] S.H. Reiger, Codes for correction of 'clustered' errors, IRE Trans. Inf. Theory 6 (1960) 16–21.

[51] J.P. Robinson, Golomb rectangles, IEEE Trans. Inf. Theory 31 (1985) 781–787.

[52] J.P. Robinson, Golomb rectangles as folded rulers, IEEE Trans. Inf. Theory 43 (1997) 290–293.

[53] J.P. Robinson, Generic search for Golomb arrays, IEEE Trans. Inf. Theory 46 (2000) 1170–1173.

[54] R.M. Roth, G. Seroussi, Reduced-redundancy product codes for burst error correction, IEEE Trans. Inf. Theory 44 (1998) 1395–1406.

[55] A. Sharov, R.M. Roth, Two-dimensional constrained coding based on tiling, IEEE Trans. Inf. Theory 56 (2010) 1800–1807.

[56] M.K. Siu, The combinatorics of binary arrays, J. Stat. Plan. Inference 62 (1997) 103–113.

[57] I. Soloveychik, Y. Xiang, V. Tarokh, Pseudo-Wigner matrices, IEEE Trans. Inf. Theory 64 (2018) 3170–3178.

[58] I. Soloveychik, Y. Xiang, V. Tarokh, Symmetric pseudo-random matrices, IEEE Trans. Inf. Theory 64 (2018) 3179–3196.

[59] R. Spann, A two-dimensional correlation property of pseudo-random maximal-length sequences, Proc. IEEE 53 (1963) 2137.

[60] I. Tal, T. Etzion, R.M. Roth, On row-by-row coding for 2-D constraints, IEEE Trans. Inf. Theory 55 (2009) 3565–3576.

[61] H. Taylor, Non-attacking rooks with distinct differences, Commun. Sci. Inst., Univ. Southern Calif., Tech. Rep. CSI-84-03-02, 1984.

[62] Z. Zhang, C. Tu, New bounds for the sizes of radar arrays, IEEE Trans. Inf. Theory 40 (1994) 1672–1678.

Chapter 10

Two-dimensional applications

Self-location, secure sensor networks, folding

In this chapter, we consider applications and techniques associated with two-dimensional arrays that have a two-dimensional window property or distinct differences property.

Section 10.1 considers the basic problem of how to find a position in a large area. The solution will be to use the uniqueness of windows in an array that maps the area. We will consider a simple scheme that can be encoded and decoded efficiently. The shape of the windows will be a type of cross and all possible windows with this shape will be considered. When we consider possible errors in the shape from which we want to recover the position, a rectangular window will be used and we show its optimality in error correction with rather a naive algorithm.

Section 10.2 considers an application of two-dimensional patterns with distinct differences in key-predistribution schemes for wireless sensor networks. We will show that Costas arrays are very efficient for this purpose, particularly those schemes obtained from the Welch construction.

Finally, Section 10.3 presents a method to obtain two-dimensional arrays having distinct differences with various shapes. This section will present unbounded two-dimensional arrays in which any window of the given shape has a pattern with distinct differences. The method is based on folding a one-dimensional sequence with either pattern with distinct differences or with a window property, into a two-dimensional array with the same property. The folding will be based on a tiling for the two-dimensional shape with a lattice. A generalization for multi-dimensional arrays will be also discussed. Also, different shapes with the window property can be obtained by this method, but this will not discussed in this chapter

10.1 Robust self-location two-dimensional patterns

Take a blindfolded man on a random one-hour walk around town and then remove his blindfold. How will he know where he is? He has several options, based on the information he can gather. The man could carefully count his steps and take note of every turn during the blindfolded walk to know his location relative to the beginning of his trip. Armed with a navigation tool such as a sextant or GPS unit, he could ask the stars or the GPS satellites where he is. Lastly,

Sequences and the de Bruijn Graph. https://doi.org/10.1016/B978-0-44-313517-0.00016-0
325

he could simply look around for a reference, such as a street sign, a landmark building, or even a city map with a little arrow saying *"You are here."*

There are numerous applications where a similar problem is encountered. We need to somehow measure the position of a mobile or movable device, using some sort of sensory input. Wheeled vehicles can count the turns of their wheels much like the man counting his steps. Similarly, many devices, from industrial machine stages to ball mice, employ sensors that are coupled with the mechanics and count small physical steps of a known length, in one or more dimensions. The small relative position differences can be accumulated to achieve relative *self-location* to a known starting point. Technologies such as those found in optical mice, use imaging sensors instead of mechanical encoders to estimate the relative motion by constantly inspecting the moving texture or pattern of the platform beneath them. We elaborate more on this in Section 10.4.

This section proposes a product construction to generate two-dimensional binary patterns for absolute self-location. We start by presenting the product construction based on two sequences with some one-dimensional window properties. A two-dimensional array with optimal self-location based on sensing a cross shape is obtained by this construction. After that, it will be proved that the same construction can be used for reasonably effective error correction of self-location with a rectangular shape.

The approach that we use for building two-dimensional arrays with self-location properties is based on a product of two sequences, one of which is a de Bruijn sequence and the other a half de Bruijn sequence.

For two binary sequences $\mathcal{T} = [t_1, \ldots, t_K]$ and $\mathcal{S} = [s_1, \ldots, s_N]$ the product $\mathcal{T} \otimes \mathcal{S}$ is a $K \times N$ array $\mathcal{R}$ in which $r_{i,j}$, $1 \leq i \leq K$, $1 \leq j \leq N$, contains the value $t_i + s_j$.

Take a span k half de Bruijn sequence $\mathcal{T} = [t_1, \ldots, t_K]$ and a span n de Bruijn sequence $\mathcal{S} = [s_1, \ldots, s_N]$, of lengths $K = 2^{k-1}$ and $N = 2^n$, respectively, and let $\mathcal{R} = \mathcal{T} \otimes \mathcal{S}$. Clearly, each row in $\mathcal{R}$ equals either $\mathcal{S}$ or $\bar{\mathcal{S}}$. Similarly, each column of $\mathcal{R}$ equals either $\mathcal{T}$ or $\bar{\mathcal{T}}$. Thus each row and each column retain their window property and can serve for self-location in each dimension.

Theorem 10.1. *Each cross-shaped pattern with k vertical entries and n horizontal entries appears exactly once as a pattern in the array $\mathcal{R}$.*

Proof. Let X be a column vector of length k and Y be a row vector of length n. By the definition of a half de Bruijn sequence, we have that either X or $\bar{X}$ is contained in the sequence $\mathcal{T}$. Let $\mathcal{X}$ be the pattern that appears. Both Y and $\bar{Y}$ appear in the sequence $\mathcal{S}$. Crosses with vertical vector X and horizontal vector Y appear in $\mathcal{R}$ only in the portions related to $Z_1 \triangleq \mathcal{X} \otimes Y$ and $Z_2 \triangleq \mathcal{X} \otimes \bar{Y}$. Moreover, the crosses in Z_1 are complements of the crosses in Z_2. For each cross inside Z_1 and Z_2, there are two possible assignments, depending on the mutual entry of the vertical and horizontal components $\mathcal{X}$ and Y, respectively. Each one of these assignments appears in either Z_1 or Z_2. $\square$

By Theorem 10.1, we can use a cross-sensor array to sample k vertical pixels and n horizontal pixels (with one mutual pixel) to obtain self-location.

Corollary 10.1. *The proposed method is optimal in terms of the number of sampled pixels required to achieve self-location with a cross of vertical length k and horizontal length n.*

Corollary 10.2. *In the array $\mathcal{R}$ each possible sampled $k \times n$ window has a unique location.*

Remark. In practice, the planar domain is generally not cyclic as we have considered it by using the cyclic sequences $\mathcal{T}$ and $\mathcal{S}$. To retain the ability to sense all $2^{k-1} \cdot 2^n$ possible locations with a sensor whose footprint is $k \times n$ pixels array, we extend $\mathcal{T}$ and $\mathcal{S}$ by appending their first $k-1$ bits and $n-1$ bits, respectively, to their ends. The result is now a $(2^{k-1} + k - 1) \times (2^n + n - 1)$ array.

Example 10.1. An example of the proposed two-dimensional grid pattern can be seen in Fig. 10.1. It was generated using a span 5 half de Bruijn sequence in the vertical axis and a span 4 de Bruijn sequence in the horizontal axis, resulting in a cyclic array of 16×16 pixels. The first column and the first row in the figure contain the location indexes. The second column and the second row contain $\mathcal{T}$ and $\mathcal{S}$, respectively. From the bit values inside the grid, we can decode our position. Examples of the readout for two such different sensors are marked in Fig. 10.1 with an underlined red color. The sensors are 5×4 crosses. In both, the vertical readout is 11101, and the horizontal readout is 1100 and its unique position can be easily decoded from $\mathcal{T}$ and $\mathcal{S}$.

		1	2	3	4	5	6	7	8	9	10	11	12	13	14	15	16
		0	0	0	0	1	1	1	1	0	1	1	0	0	1	0	1
1	1	1	1	1	1	0	0	0	0	1	0	0	1	1	0	1	0
2	1	1	1	1	1	0	0	0	0	1	0	0	1	1	0	1	0
3	1	1	1	1	1	0	0	0	0	1	0	0	1	1	0	1	0
4	1	1	1	1	1	0	0	0	0	1	0	0	1	1	0	1	0
5	1	1	1	1	1	0	0	0	0	1	0	0	1	1	0	1	0
6	0	0	0	0	0	1	1	1	1	0	1	1	0	0	1	0	1
7	1	1	1	1	1	0	0	0	0	1	0	0	1	1	0	1	0
8	0	0	0	0	0	1	1	1	1	0	1	1	0	0	1	0	1
9	1	1	1	1	1	0	0	0	0	1	0	0	1	1	0	1	0
10	1	1	1	1	1	0	0	0	0	1	0	0	1	1	0	1	0
11	0	0	0	0	0	1	1	1	1	0	1	1	0	0	1	0	1
12	1	1	1	1	1	0	0	0	0	1	0	0	1	1	0	1	0
13	1	1	1	1	1	0	0	0	0	1	0	0	1	1	0	1	0
14	1	1	1	1	1	0	0	0	0	1	0	0	1	1	0	1	0
15	0	0	0	0	0	1	1	1	1	0	1	1	0	0	1	0	1
16	0	0	0	0	0	1	1	1	1	0	1	1	0	0	1	0	1

FIGURE 10.1 The 16×16 product array of $\mathcal{R} = \mathcal{T} \otimes \mathcal{S}$. The marked cells illustrate a readout by two possible cross-shaped sensors.

The first step in the proposed method recovers the one-dimensional subsequences that correspond to the location in each dimension. Essentially, the two-dimensional problem is now reduced to two independent one-dimensional decoding problems. Decoding the location of a subsequence in a de Bruijn sequence is a well-known problem. Decoding of a half de Bruijn sequence is carried out similarly.

The cross-shaped sensor is rather 'spread out', so it might be a disadvantage in applications. This weakness becomes an advantage for robust self-location when a rectangular-shaped sensor is used. If we use a $k \times n$ pixel sensor (see Corollary 10.2), we can utilize the inherent redundancy within the kn bits to decode the location while overcoming a considerable number of faulty bit readings. This is also a very practical choice, considering that two-dimensional rectangular sensor grids are the most common variety and are the standard choice for most applications.

We assume that fewer than one-quarter of the bits in each row and fewer than one-half of the bits in each column of the $k \times n$ input array (which we want to locate) are in error. This is a fair assumption that can account for quite strong noise in practical terms. The following presented algorithm to obtain robust self-location is based on a simple majority decoding as follows.

Robust self-location algorithm:

The input for the algorithm is a rectangle

$$Z \triangleq \{z_{ij} : 1 \leq i \leq k, 1 \leq j \leq n\} = (X \otimes Y) + \mathcal{E},$$

where X is a vertical k-tuple of a given vertical span k half de Bruijn sequence $\mathcal{T}$; Y is a horizontal n-tuple of a given span n horizontal de Bruijn sequence $\mathcal{S}$; $\mathcal{E}$ is a $k \times n$ error pattern. We assume that fewer than $\frac{n}{4}$ of the bits in each row of $\mathcal{E}$ are *ones* and fewer than $\frac{k}{2}$ of the bits in each column of $\mathcal{E}$ are *ones*. The output is the original vertical and horizontal subsequences X and Y, respectively.

- Assume that the first bit of X is b (no value is assigned to b). Let D be the first row of Z.
- For each row A of Z (other than the first row) do
  - if more than half of the bits of $A + D$ are *zeroes* then the corresponding bit of X is b;
  - otherwise, the corresponding bit of X is $\bar{b}$.
- Assign 0 or 1 to b to obtain X that appears in $\mathcal{T}$ (it is known in advance which k-tuples appears in $\mathcal{T}$, the half de Bruijn sequence).
- For each column B of Z
  - if more than half of the bits of $B + X$ are *zeroes* then the corresponding bit in Y is a *zero*;
  - otherwise, the corresponding bit in Y is a *one*.
  - From the computed X and Y form $X \otimes Y$. ∎

Theorem 10.2. *Let $\mathcal{R}$ be a $2^{k-1} \times 2^n$ array and let Υ be a $k \times n$ pixel sensor. If fewer than one-quarter of the bits in each row of $\mathcal{R}$ and fewer than half of the bits in each column of $\mathcal{R}$ are in error, then the robust self-location algorithm accurately decodes the sensor location.*

Proof. Since the number of errors in a row of Υ is fewer than $\frac{n}{4}$ it follows that two rows that were originally the same will agree in more than half of their bits. One original row and one complement row will disagree in more than half of their bits. Therefore the related bits of the vertical k-tuple X will be the same or different, respectively, depending on whether the two rows agree in more than $\frac{n}{2}$ bits or disagree in at least $\frac{n}{2}$ bits, respectively. Having all the k bits of X in terms of the variable b, there is only one assignment of a legal k-tuple (from the two possible complement assignments associated with $b = 0$ and $b = 1$) since the vertical sequence is a half de Bruijn sequence and it is known in advance which k-tuples it contains.

Having the correct vertical subsequence X, since the number of errors in a column is fewer than $\frac{n}{2}$, it follows that if X agrees in more than $\frac{n}{2}$ bits with a column, then the corresponding bit of Y is a *zero*; if it disagrees in more than $\frac{n}{2}$ bits with a column then the corresponding bit of Y is a *one*. □

Remark. Decoding can be done also if more than one-quarter of the bits in some rows are in error. A slightly better condition would be to require that the number of distinct positions in error in any two rows is fewer than $\frac{n}{2}$. This requirement can be further improved.

A similar algorithm will also work if we will exchange between rows and columns, or equivalently if we will consider a transposed array. Therefore we can exchange our assumption on the number of wrong bits in a row or a column. However, having for example at least half of the bits wrong in a given column (or a given row) will cause a wrong identification of the original subsequences.

Lemma 10.1. *Let $\mathcal{R}$ be a $2^{k-1} \times 2^n$ array and let Υ be a $k \times n$ pixel sensor. If at least half of the bits in one of the columns of a pixel sensor are in error, then we cannot ensure accurate decoding of the original subsequences.*

Proof. Let Y and Y' be two n-tuples that differ only in the last bit. Both Y and Y' appear as windows of length n in the span n de Bruijn sequence $\mathcal{S}$. Let X be a k-tuple that appears as a window in the sequence $\mathcal{T}$. The products $X \otimes Y$ and $X \otimes Y'$ appear as $k \times n$ windows in the array $\mathcal{T} \otimes \mathcal{S}$. Both $k \times n$ windows differ only in the last column and it would be impossible to distinguish between the two windows if half of the bits in the last column are in error. If more than half of the bits in the last column are in error then an incorrect decoding of the sensor location will be made. The same arguments can be applied to any other column. □

We note that by Lemma 10.1 we cannot correct $\lceil \frac{k}{2} \rceil$ or more random errors in a $k \times n$ input array. The reason is that the array is highly redundant. This is quite weak from an error-correction point of view. However, by Theorem 10.2 we can correct about $\frac{kn}{4}$ errors in an $k \times n$ array if fewer than $\frac{n}{4}$ errors occur in a row and fewers than $\frac{k}{2}$ errors occur in a column. This result is quite strong from an error-correction point of view. Thus the weakness for one type of error becomes an advantage for another type of error.

Example 10.2. The following 7×9 input array of an 64×512 array $\mathcal{R}$

$$
\begin{array}{ccccccccc}
1 & 0 & 0 & 1 & 0 & 1 & 0 & 0 & 1 \\
0 & 0 & 0 & 0 & 0 & 1 & 1 & 1 & 0 \\
1 & 0 & 1 & 0 & 0 & 0 & 1 & 1 & 1 \\
0 & 0 & 1 & 0 & 0 & 0 & 1 & 0 & 0 \\
1 & 1 & 0 & 0 & 1 & 0 & 0 & 0 & 1 \\
0 & 0 & 1 & 0 & 1 & 0 & 1 & 1 & 0 \\
1 & 0 & 0 & 1 & 1 & 0 & 0 & 0 & 1 \\
\end{array}
$$

has no more than two errors in a row and no more than three errors in a column. The first row (from the top) has more than half bits in common only with the 5th and the 7th rows. Thus the vertical pattern (from the top) is $(b\bar{b}\bar{b}\bar{b}b\bar{b}b)^{tr}$. Suppose that $b = 0$, i.e., the vertical column is $(0111010)^{tr}$. We now compare all of the columns with $(0111010)^{tr}$. If more than half of the corresponding bits agree, the bit in the horizontal sequence is a *zero*; otherwise, it is a *one*. Thus the horizontal sequence is 110111001. The 7×9 sub-array with no errors is

$$
\begin{array}{ccccccccc}
1 & 1 & 0 & 1 & 1 & 1 & 0 & 0 & 1 \\
0 & 0 & 1 & 0 & 0 & 0 & 1 & 1 & 0 \\
0 & 0 & 1 & 0 & 0 & 0 & 1 & 1 & 0 \\
0 & 0 & 1 & 0 & 0 & 0 & 1 & 1 & 0 \\
1 & 1 & 0 & 1 & 1 & 1 & 0 & 0 & 1 \\
0 & 0 & 1 & 0 & 0 & 0 & 1 & 1 & 0 \\
1 & 1 & 0 & 1 & 1 & 1 & 0 & 0 & 1 \\
\end{array}
$$

■

10.2 Key predistribution for sensor networks

A *wireless sensor network* (WSN) is a large collection of small sensor nodes that are equipped with wireless-communication capability. Sensor nodes have limited communication range and thus data transmitted over the network is typically passed from node to node in a series of *hops* to reach its end destination. Such networks can be employed for a wide range of applications, whether scientific, commercial, humanitarian, or military. The data being transmitted over the wireless medium is frequently valuable or sensitive; hence, there is a need for cryptographic techniques to provide data integrity, confidentiality, and authentication.

On deployment, the sensor nodes aim to form a secure and connected network. In other words, we desire a significant proportion of nodes within the communication range to share cryptographic keys. The nodes' size limits their computational power and battery capacity, so it is assumed that the sensor nodes are unable to use public-key cryptography to establish shared keys. Hence, we should use symmetric cryptographic keys, where two parties can communicate only if they have the same key and are preloaded onto each node before deployment. A method for deciding which keys are assigned to a node is known as the *key-predistribution scheme* (KPS). The sensor nodes are assumed to be highly vulnerable to compromise, so a single key should not be given to too many nodes. Another constraint is that each node can only store a limited number of keys. The aim is to design an efficient and secure KPS so that a sensor node can establish secure wireless links with many of its neighbors: It is important to establish as many short secure links in the network as possible since the nodes' capacity to relay information is very limited.

KPSs for WSNs generally assume that the precise location of nodes is not known before deployment, hence such schemes aim to provide reasonable levels of "average" connectivity across the entire network. However, in many applications, the location of sensor nodes can be determined before deployment. In such cases, this knowledge can be used to improve the efficiency of the underlying KPS. One such scenario is that of networks consisting of a large number of sensor nodes arranged in a square grid. There are many potential applications for which such a pattern may be useful: monitoring vines in a vineyard or trees in a commercial plantation or reforestation project, studying traffic or pollution levels on city streets, measuring humidity and temperature at regular intervals in the library shelves, performing acoustic testing on each of the seats in a theater, monitoring goods in a warehouse, indeed any application where the objects being studied are naturally distributed in a grid. For purposes of commercial confidentiality or for protecting the integrity of scientific data it is necessary to secure communication between sensors, and thus it is important to have efficient methods of distributing keying material in such networks. The goal of this section is to provide a practical key-predistribution scheme designed specifically for square grids. We show that the highly structured topology of these networks can be exploited to develop schemes that perform significantly better for this application than more general techniques. The schemes are designed for homogeneous networks in which all sensors have the same capabilities. We assume that the nodes have no access to an external trusted authority (such as a base station) to establish keys once they have been deployed. We assume that the location of each node within the grid is known before deployment, and consider the problem of establishing pairwise keys between nodes within communication distance of one another.

We say that a WSN is *grid based* if it consists of a (potentially unbounded) number of identical sensors arranged in a square grid. If each sensor has a maximum transmission range r, then a sensor can communicate directly with all nodes within the circle of radius r that surrounds it. We say that two squares occur at distance r if the distance between the centers of the squares is r. W.l.o.g. we can scale our unit of distance so that adjacent nodes in the grid are at a distance of 1 from each other as it removes unnecessary complications from our discussions.

We refer to nodes within the circle of radius r centered at some node Ψ as *r-neighbors* of Ψ. For most applications, it is useful for any two neighboring nodes in a sensor network to be able to communicate securely. In designing a KPS, however, we are restricted by the limited storage capacity of the sensors: if a node has many neighbors, it may be unable to store enough keys to share a distinct key with each neighbor. We would like to design KPSs in which each node shares a key with as many of its r-neighbors as possible while taking storage constraints into account. (Note that we only require keys to be shared by nodes that are r-neighbors, in contrast to a randomly distributed sensor network that potentially requires all pairs of nodes to share keys.) One way of achieving this is for each key to be shared by several different nodes; however, it is necessary to restrict the extent to which each key is shared, to protect the network against key compromise through node capture.

Now, we provide basic definitions relating to KPS and examine certain properties that must be considered when designing such schemes, before proposing constructions of KPSs that are specifically adapted to grid-based networks.

Let $\mathcal{K}$ be a finite set whose elements are referred to as keys (whether they are either actual secret keys or quantities from which such keys may be derived). We consider a set $\mathcal{U}$ of wireless sensors, each of which has sufficient memory to store m keys; after deployment, the nodes $\mathcal{U}$ form a WSN $\mathcal{W}$.

Definition 10.1. A KPS for $\mathcal{W}$ is a mapping $\mathcal{U} \to \mathcal{K}^m$ that assigns up to m keys from $\mathcal{K}$ to each node of $\mathcal{U}$.

Although the number of sensor nodes is finite in practice, it is convenient to model the physical location of the nodes by the set of points of $\mathbb{Z}^2$. The scheme employs a DDC to create a key-predistribution scheme in the following way.

Definition 10.2. Let $\mathcal{D} = \{v_1, v_2, \ldots, v_\beta\} \subset \mathbb{Z}^2$ be a DDC. Allocate keys to nodes as follows:

- Label each node with its position in $\mathbb{Z}^2$.
- For every 'shift' $u \in \mathbb{Z}^2$, generate a new key k_u in $\mathcal{K}$ and assign k_u to the nodes labeled by $u + v_i$, for $i = 1, 2, \ldots, \beta$.

Each node stores the keys assigned to it in its memory before deployment. Once the nodes are deployed we have the following possible situations.

- Two nodes that share a common element of $\mathcal{K}$ can use it for secure communication.
- Two nodes that do not share a key may rely on an intermediate node with which they both share a key to communicate securely; this is referred to as a *two-hop path*.

If each $k \in \mathcal{K}$ is assigned to a set $S_k \subset \mathcal{U}$ of at most β nodes we refer to the KPS as an $[m, \beta]$-KPS. One of the goals in the design of an $[m, \beta]$-KPS is to enable each node to communicate directly with as many nodes as possible, hence we would like to maximize the expected number of neighboring nodes that share at least one key with a given node Ψ. We note that when evaluating properties of a grid-based network in which the network does not extend infinitely in all directions, complications may arise due to nodes on the edge of the network having a reduced number of neighbors. This can be avoided by restricting attention to properties of nodes on the interior of the network (node Ψ such that each grid position that is within range of Ψ contains a node of the network). This is a reasonable restriction to make as it greatly simplifies the analysis and comparison of KPSs, especially since for a grid-based network of any size the edge nodes will only represent a small proportion of the network.

Theorem 10.3. *When an $[m, \beta]$-KPS is used to distribute keys to nodes in a square-grid network, the expected number of r-neighbors of a node Ψ, in the interior of the network, that share at least one key with Ψ is at most $m(\beta - 1)$. The value $m(\beta - 1)$ is achieved precisely when the following conditions are met.*

1. *Each interior node stores exactly m keys, each one is shared by exactly β nodes.*
2. *No pair of nodes shares two or more keys.*
3. *The distance between any two nodes sharing a key is at most r.*

Proof. The maximum number of keys allocated to an interior node Ψ by an $[m, \beta]$-KPS is m; each of these keys is shared by at most β nodes (which may or may not be r-neighbors of Ψ). Hence, a given interior node shares keys with at most $m(\beta - 1)$ of its r-neighbors, and this maximum value is achieved if and only if no two nodes share more than one key with Ψ, and every node with which Ψ shares a key is in the r-neighbor of Ψ. Therefore the claims in the theorem follow directly. $\square$

Theorem 10.3 indicates that when distributing keys using an $[m, \beta]$-KPS, limiting the number of keys shared by each pair of nodes to at most one increases the number of pairs of neighboring nodes that share keys, hence this is desirable from the point of view of efficiency. Now, we describe a method of constructing $[m, \beta]$-KPSs with this property.

We propose a KPS for a grid-based network, in which the pattern of nodes that share a particular key is determined by a Costas array. The result is an $[n, n]$-KPS in which any two nodes have at most one key in common.

Construction 10.1. *Let $\mathcal{A}$ be a $n \times n$ Costas array. Use $\mathcal{A}$ to distribute keys from a key pool $\mathcal{K}$ to a set $\mathcal{U}$ of nodes arranged in a grid-based network.*

- *Arbitrarily choose one square of the grid to be the origin, and superimpose $\mathcal{A}$ on the grid, with its lower left-hand square over the origin. Select a key k_{00} from $\mathcal{K}$, and distribute it to nodes occurring in squares coinciding with a dot of $\mathcal{A}$ (so n nodes receive the key k_{00}).*
- *Similarly, for each square occurring at a position (i, j) in the grid, we place the lower left-hand square of $\mathcal{A}$ over that square, then assign a key $k_{ij} \in \mathcal{K}$ to the squares that are now covered by dots of $\mathcal{A}$.* ∎

If the dots of the Costas array occur in entries

$$(0, a_0), (1, a_1), \ldots, (n-1, a_{n-1})$$

of the array, then the above scheme associates a key k_{ij} with the nodes in squares $(i, a_0 + j), (i+1, a_1 + j), \ldots, (i+n-1, a_{n-1} + j)$ (where such nodes exist, i.e., the array is not periodic). We observe that the deterministic nature of this key allocation, together with the structured topology of a square grid means that nodes can simply store the coordinates in the grid of those nodes with which they share keys, thus obviating the need for a shared-key discovery process with ensuing communication overheads.

Example 10.3. Consider the 3×3 Costas array of Fig. 10.2.

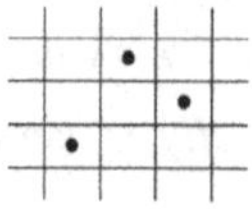

FIGURE 10.2   A 3×3 Costas array.

If we use this array for key distribution as described above, each node stores three keys. Fig. 10.3 illustrates this key distribution: each square in the grid represents a node, and each symbol contained in a square represents a key possessed by that node. The central square stores keys marked by the letters A, B, and C; two further nodes share each of these keys. These three keys are colored with underlined red, while the other keys are colored black. Note that only some of the keys are illustrated; the pattern of key sharing extends similarly throughout the entire network.

FIGURE 10.3   Key distribution using a 3×3 Costas array.

Theorem 10.4. *The key-predistribution scheme in Construction* 10.1 *has the following properties:*

1. *Each sensor is assigned n different keys.*
2. *Each key is assigned to n sensors.*

3. *Any two sensors have at most one key in common.*
4. *The distance between two sensors that have a common key is at most $\sqrt{2}(n-1)$.*

Proof.

1. There are n dots in $\mathcal{A}$. For each dot in turn, if we position $\mathcal{A}$ so that a dot lies over a given node Ψ, this determines the positioning of $\mathcal{A}$ for which the corresponding key is allocated to Ψ. Hence, Ψ stores n keys in total.
2. A key k_{ij} is assigned to n positions in the square grid, namely those that coincide with the n dots of a fixed shift of $\mathcal{A}$.
3. Suppose there exist two sensors A and B sharing (at least) two keys. These keys correspond to different translations of the array $\mathcal{A}$, hence, there exist two translations of $\mathcal{A}$ in which dots occur at the positions of both A and B. However, by the distinct differences property, two different translations of a Costas array $\mathcal{A}$ can coincide in at most one dot, thus contradicting the original assumption.
4. The two most distant sensors that have a key in common must correspond to two dots in the same translation of $\mathcal{A}$. The largest distance between two dots in $\mathcal{A}$ occurs if they are in two opposite corners of the array, i.e., at distance $\sqrt{2}(n-1)$. $\qquad\qquad\Box$

Corollary 10.3. *When the $[n, n]$-KPS of Construction 10.1 is applied to a grid-based network then a node on the interior of the network shares keys with $n(n-1)$ other nodes, the maximum possible for an $[n, n]$-KPS.*

Remark. Construction 10.1 can be generalized by using any DD(m, r).

Now, we describe a specific KPS for a grid-based network, in which the pattern of nodes that share a particular key is determined by a Costas array. The result is an $[n, n]$-KPS in which any two nodes have at most one key in common. For this construction, we will be able to prove a desirable two-hop coverage.

We now define a DD(m) by choosing a finite subset of the dots in the Welch periodic array $\mathcal{L}_p$ (see Definition 9.5). For this definition let α be a primitive root modulo a prime p and we will use a **transpose** of the array as follows:

$$\mathcal{W}_p \triangleq \{(i, j) \in \mathbb{Z}^2 \; : \; \alpha^j \equiv i \pmod{p}\}.$$

Construction 10.2. *Let p be an odd prime and let α be a primitive element modulo p. Let $(i, j) \in \mathbb{Z}^2$ be such that $\mathcal{W}_p$ has dots at (i, j) and $(i+1, j+1)$. Note that such a position (i, j) exists. To see this, let i and j be integers such that*

$$\alpha^j \equiv i \equiv \frac{1}{\alpha - 1} \pmod{p}.$$

The right-hand side of this equality is well defined and nonzero modulo p, and so there is a suitable choice for i and j. Clearly, $\mathcal{W}_p$ has a dot at the posi-

tion (i, j). However, there is also a dot at $(i + 1, j + 1)$ since

$$\alpha^{j+1} \equiv \frac{\alpha}{\alpha - 1} \equiv \frac{1}{\alpha - 1} + 1 \equiv i + 1 \pmod{p}.$$

Consider the $(p-1) \times p$ rectangle S bounded by the positions (i, j), $(i+p-1, j)$, $(i, j + p - 2)$, and $(i + p - 1, j + p - 2)$. By construction, $\mathcal{W}_p$ has $p - 1$ dots in S. Due to its periodic nature, $\mathcal{W}_p$ also has dots at positions $(i, j + p - 1)$, $(i + p, j)$, and $(i + p + 1, j + p)$. We construct a configuration $\mathcal{B}$ by adding these three dots to the set of dots in $\mathcal{W}_p \cap S$. ∎

Our configuration $\mathcal{B}$ is shown in Fig. 10.4. The configuration is contained in a $(p + 1) \times (p + 2)$ rectangle. The *border region* of width 2 contains exactly 5 dots: A, A', A'', B, and B'. The *central region* is a $(p - 3) \times (p - 2)$ rectangle. This region contains $p - 3$ dots: one column is empty, but every other column and every row contains exactly one dot. Note that $A \equiv A' \equiv A''$ and $B \equiv B'$, but there are no other equivalent pairs of dots in $\mathcal{B}$.

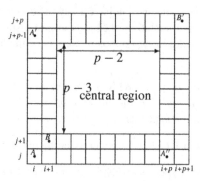

FIGURE 10.4 The configuration $\mathcal{B}$. The five dots shown are the dots that lie on the border of width 2 of the $(p + 1) \times (p + 2)$ rectangle containing the configuration.

Lemma 10.2. *The configuration $\mathcal{B}$ is a $\mathrm{DD}(p + 2)$, all of whose points lie in a $(p + 1) \times (p + 2)$ rectangle.*

Proof. We have already remarked that $\mathcal{B}$ contains $p + 2$ dots, all lying in a $(p+1) \times (p+2)$ rectangle. Hence, it remains to show that $\mathcal{B}$ satisfies the distinct differences property.

Suppose, for a contradiction, that $\{X, Y\}$, and $\{X', Y'\}$, are distinct pairs of dots in $\mathcal{B}$ with the same difference vector (d, e).

Suppose that $d \in \{0, -p, p\}$ or $e \in \{0, -(p - 1), (p - 1)\}$. A difference vector between a dot in the central region of our configuration and any other dot has x coordinate and y coordinate of absolute value at most $p - 1$ or $p - 2$, respectively. Moreover, a central dot is the only dot in its row and column. Hence, our assumption implies that none of X, X', Y, Y' can lie in the central region of our configuration. Moreover, the $5 \cdot 4 = 20$ ordered pairs of dots in the border region all have distinct difference vectors, and so we have a contradiction in this case.

Hence, we may assume that $d \notin \{0, -p, p\}$ and $e \notin \{0, -(p-1), (p-1)\}$. In particular, since all dots lie in a $(p+1) \times (p+2)$ rectangle, we see that $d \not\equiv 0 \pmod{p}$ and $e \not\equiv 0 \pmod{(p-1)}$. Lemma 9.17 now implies that $X \equiv X'$ and $Y \equiv Y$. If $X = X'$ then $Y = Y'$, which contradicts the fact that our pairs of dots are distinct. Hence, $X \neq X'$. The fact that $X \equiv X'$ now implies that X and X' must lie in the border of our configuration. A similar argument implies the same is true for Y and Y'. As in the previous case, we now have a contradiction. Thus the lemma follows. $\qquad \square$

It should be noted that the use of a $DD(m, r)$ maximizes the number of r-neighbors that share keys with a given node. Additionally, it is desirable to maximize the number of r-neighbors that can communicate securely with a given node Ψ via a one-hop or a two-hop path. We refer to this quantity as the **two-hop r-coverage** of a KPS. For our scheme, which is based on distinct-difference configurations, we refer to the two-hop r-coverage of a $DD(m, r)$ to indicate the two-hop r-coverage obtained by a KPS constructed from that configuration.

Our goal now is to show that $\mathcal{B}$ achieves complete two-hop coverage on a $(2p - 3) \times (2p - 1)$ rectangle relative to the central point of the rectangle. To demonstrate this, it is necessary to show that every vector (d, e) with $|d| \leq p - 1$ and $|e| \leq p - 2$ can be expressed as a difference vector or a two-hop path of difference vectors from $\mathcal{B}$. The following lemma proves this for the majority of such vectors of the form (d, e).

Lemma 10.3. *Any vector of the form (d, e), where d and e are nonzero integers satisfying $|d| \leq p - 1$ and $|e| \leq p - 2$, can be expressed as the sum of two difference vectors from $\mathcal{B}$.*

Proof. Consider the $(p - 1) \times p$ rectangle S defined in Construction 10.2, and let $\mathcal{A}$ be the restriction of $\mathcal{W}_p$ to the $(2p - 2) \times 2p$ sub-array whose lower leftmost corner coincides with that of S.

We partition $\mathcal{A}$ into four $(p - 1) \times p$ sub-arrays as follows:

$$\mathcal{A} = \begin{pmatrix} \mathcal{D}_3 & \mathcal{D}_4 \\ \hline \mathcal{D}_1 & \mathcal{D}_2 \end{pmatrix}.$$

The periodicity of $\mathcal{W}_p$ means that the set of dots of $\mathcal{W}_p$ contained in each sub-array is a translation of the set of dots of $\mathcal{W}_p$ contained in $\mathcal{D}_1$. Moreover, since $\mathcal{D}_1 = S$, it follows that all the dots in $\mathcal{D}_1$ are contained in $\mathcal{B}$.

We claim that each of the vectors (d, e) appears as the difference of two points in $\mathcal{A}$. Since the negative of a difference vector is always a difference vector, it follows that we may assume w.l.o.g. that $d > 0$. Suppose that $0 < e \leq p - 2$. There is a unique position $(i', j') \in \mathcal{D}_1$ such that

$$\alpha^{j'} \equiv i' \equiv \frac{d}{\alpha^e - 1} \pmod{p}.$$

It is easy to check, just as in Construction 10.2, that $\mathcal{W}_p$ has dots at (i', j') and $(i'+d, j'+e)$. Since d and e are both positive, it follows that $(i'+d, j'+e)$ lies in $\mathcal{A}$, and so our claim follows in this case. The argument for the case when $e < 0$ is the same, except that now we choose $(i', j') \in \mathcal{D}_3$. Hence, the claim follows.

To prove the lemma, we need to show that each difference vector (d, e) can be written as the sum of two difference vectors of $\mathcal{B}$. This follows from the paragraph above and the following observations:

- Any vector connecting two dots of $\mathcal{D}_1$ is a difference vector of $\mathcal{B}$ by construction.
- Due to the periodicity of $\mathcal{W}_p$, a vector connecting a dot in $\mathcal{D}_1$ with a dot in $\mathcal{D}_3$ (or, similarly, a dot in $\mathcal{D}_2$ with a dot in $\mathcal{D}_4$) can be expressed as the sum of the vector $(0, p-1)$ (which occurs as a difference between the dots A and A' in $\mathcal{B}$) and some other difference vector of $\mathcal{B}$.
- A vector connecting a dot in $\mathcal{D}_1$ with a dot in $\mathcal{D}_2$ (or, similarly, a dot in $\mathcal{D}_3$ with a dot in $\mathcal{D}_4$) can be expressed as the sum of the difference vector $(p, 0)$ (which occurs between A and A'') and some other difference vector of $\mathcal{B}$.
- A vector connecting a dot in $\mathcal{D}_1$ with a dot in $\mathcal{D}_4$ is the sum of the difference vector $(p, p-1)$ (which occurs between B and B') and some other difference vector of $\mathcal{B}$.
- A vector connecting a dot in $\mathcal{D}_3$ with a dot in $\mathcal{D}_2$ is the sum of the difference vector $(p, -(p-1))$ (which occurs between A' and A'') and some other difference vector of $\mathcal{B}$. $\qquad\square$

It remains to consider vectors that have a zero coordinate. We will use the following lemma in our proof that such vectors all occur as the sum of two difference vectors from $\mathcal{B}$.

Lemma 10.4. *Let t be a positive integer with $t \geq 3$. Let $\mathcal{F}$ be a set of integers satisfying the following properties:*

(p1) $|\mathcal{F}| = t + 1$;

(p2) $\mathcal{F} \subset \{-(t-1), -(t-2), \ldots, -1\} \cup \{1, 2, \ldots, t-1\} \cup \{t+1\}$;

(p3) $\{1, -(t-1), t+1\} \subset \mathcal{F}$;

(p4) $\exists i \in \mathcal{F} \setminus \{1, -(t-1), t+1\}$ with $i < 0$;

(p5) *if $i > 0$ and $i \in \mathcal{F} \setminus \{1, -(t-1), t+1\}$, then $i - t \notin \mathcal{F}$.*

Then, each positive integer γ with $1 \leq \gamma \leq t - 1$ has a representation of the form $\gamma = j - i$, where $i, j \in \mathcal{F}$.

Proof. Since by properties **(p1)** and **(p3)**, we have that $\mathcal{F} \setminus \{1, -(t-1), t+1\}$ contains $t - 2$ elements, it follows by property **(p5)** that $\mathcal{F}$ must contain precisely one element of each pair $\{i, i - t\}$ for each $i = 2, 3, \ldots, t - 1$. Suppose, for a contradiction, that there exists a positive integer $\gamma \leq t - 1$ that cannot be expressed as the difference between two elements of $\mathcal{F}$.

Suppose that $\gamma > 1$. Since $1, t + 1 \in \mathcal{F}$, our assumption implies that $1 - \gamma \notin \mathcal{F}$ and $t + 1 - \gamma \notin \mathcal{F}$. However, $1 - \gamma = (t + 1 - \gamma) - t$, and hence one of these integers must be contained in $\mathcal{F}$, which gives a contradiction in this case.

Suppose that $\gamma = 1$. The assumption implies that $\mathcal{F}$ does not contain a pair of integers that differ by 1. If t is odd, this implies that $\mathcal{F} \setminus \{t + 1\}$ contains at most $(t - 1)/2$ positive integers, and at most $(t - 1)/2$ negative integers. Hence, $\mathcal{F}$ contains at most $(t - 1) + 1 = t$ integers, which contradicts property **(p1)**. If t is even, then for the size of $\mathcal{F}$ to be $t + 1$, $\mathcal{F} \setminus \{t + 1\}$ must contain $t/2$ positive integers, all of which are odd, and $t/2$ negative integers that are also all odd. This implies that for each positive odd integer $1 < i < t$ we have that $i \in \mathcal{F}$ and $i - t \in \mathcal{F}$, which contradicts property **(p5)**. Hence, the lemma follows. $\square$

We can now combine these two lemmas to obtain our desired result:

Theorem 10.5. *Let p be a prime, $p \geq 5$. The distinct difference configuration $\mathcal{B}$ achieves complete two-hop coverage on a $(2p - 3) \times (2p - 1)$ rectangle relative to the central point of the rectangle.*

Proof. By Lemma 10.3, a vector (d, e) from the center of a $(2p - 3) \times (2p - 1)$ rectangle to another point of the rectangle can be expressed as the sum of two difference vectors of $\mathcal{B}$ if d and e are nonzero.

We now consider vectors of the form $(0, e)$ with $0 < e \leq p - 2$. Such a vector can be expressed as the sum of two difference vectors of $\mathcal{B}$ if $\mathcal{B}$ has difference vectors of the form $(1, y')$ and $(1, y)$ with $y' - y = e$. The second coordinates of the set of difference vectors of $\mathcal{B}$ of the form $(1, y)$ with $y \neq 0$ satisfy the conditions of Lemma 10.4 for $t = p - 1$, since:

(1) The leftmost column of the array contains two dots; all the other columns contain a single dot apart from a single central column that is empty. Hence, $\mathcal{B}$ has p difference vectors of the form $(1, y)$ with $y \neq 0$.

(2) Except for the vector $(1, p)$, all difference vectors of $\mathcal{B}$ of the form $(1, y)$ with $y \neq 0$ satisfy $|y| \leq p - 2$.

(3) The vectors $(1, 1)$, $(1, -(p - 2))$ and $(1, p + 1)$ are all difference vectors of $\mathcal{B}$ (as they occur as differences between dots in the border region of $\mathcal{B}$, as illustrated in Fig. 10.4).

(4) The difference vectors of $\mathcal{B}$ of the form $(1, y)$ cannot all satisfy $y > 0$. This is obvious if the rightmost central column contains a dot. If this column is empty and y is always positive, then the remaining $(p - 3) \times (p - 3)$ central region must contain dots along a lower-left to top-right diagonal. Since $p \geq 5$, it follows that the two central dots have the difference vector $(1, 1)$. Since dots A and B also have this difference vector, it follows that the distinct difference property is violated and so we have a contradiction, as required.

(5) If $(1, y)$ with $y \notin \{1, p\}$ is a difference vector of $\mathcal{B}$, then $(1, y - (p - 1))$ is not a difference vector of $\mathcal{B}$. Lemma 9.17 implies that the dots involved must be equivalent, and so must be in the border region of our construction.

Lemma 10.4 now implies that any vector $(0, e)$ with $0 < e \le p - 2$ has an expression in the form $(0, e) = (1, y') + (-1, -y)$, where $(1, y')$ and $(1, y)$ are difference vectors of $\mathcal{B}$. Vectors of the form $(0, e)$ with $-(p - 2) < e < 0$ can be written as $(1, y) + (-1, -y')$.

Similarly, we can show that the first coordinates of the difference vectors of $\mathcal{B}$ of the form $(x, 1)$ satisfy the conditions of Lemma 10.4 with $t = p$, and hence any vector of the form $(d, 0)$ with $0 < |d| \le p - 1$ can be written as the sum of two difference vectors of $\mathcal{B}$. Thus the result is proven. $\square$

We can thus apply the DD(m) specified in Construction 10.2 to the scheme presented in Definition 10.2 to establish a key-predistribution scheme that guarantees two-hop paths between a node and all of its neighbors within a surrounding rectangular region. This provides a powerful notion of local connectivity to facilitate connectivity across the wider network. The resulting scheme is also highly configurable since the value of p can be adjusted for a tradeoff of storage against the size of the fully connected local region.

10.3 Folding of one-dimensional sequences

In Section 9.3 we gave a construction of pseudo-random arrays based on the folding of M-sequences. Surprisingly, folding can be also used to form two-dimensional DDCs of various shapes from one-dimensional patterns with distinct differences. It can be used also to form pseudo-random two-dimensional arrays with a window property when the shape of the window is not necessarily a rectangle. The method of folding that will be discussed in this section generalizes the one in Section 9.3 so that the DDCs can have different shapes. This is especially useful for KPS when a sensor can communicate at radius r, i.e., to all sensors in a circle of radius r around it.

Let $\mathcal{S}$ be a shape (a set of positions) in the square grid. We are interested in finding DDCs that are contained in $\mathcal{S}$ and have many dots. We present a general technique for showing the existence of such DDCs, using doubly periodic constructions. The following lemma follows straightforwardly from our definitions:

Lemma 10.5. *Let $\mathcal{A}$ be a doubly periodic $\mathcal{S}$-DDC, and let $\mathcal{S}' \subseteq \mathcal{S}$. Then, $\mathcal{A}$ is a doubly periodic $\mathcal{S}'$-DDC.*

After we examine the density of various DDCs in Section 9.5 (a square and a circle) we intend to improve these results by using folding and also to consider DDCs whose shape is a regular polygon. Folding a rope, a ruler, or any other feasible object is a common action in everyday life. Folding a one-dimensional sequence into a multi-dimensional array is very similar, but there are a few variants. First, we will summarize three variants for folding a one-dimensional sequence into a two-dimensional array $\mathcal{A}$. The generalization for a D-dimensional array, where $D > 2$, is straightforward, while the description becomes more clumsy.

F1. $\mathcal{A}$ is considered as a cyclic array horizontally and vertically in such a way that a walk diagonally visits all the entries of the array. The elements of the sequence are written along the diagonal of the $r \times t$ array $\mathcal{A}$. This folding will work (i.e., all elements of the sequence are written into the array) if and only if r and t are co-prime.

F2. The elements of the sequence are written row after row (or column after column) in $\mathcal{A}$.

F3. The elements of the sequence are written diagonal after diagonal in $\mathcal{A}$.

The folding defined by **F1** was demonstrated in Example 9.5. The next two examples demonstrate the folding of **F2** and **F3**.

Example 10.4. The following sequence (ruler) of length 13 with five dots:

0	1	2	3	4	5	6	7	8	9	10	11	12
•	•			•						•		•

is folded, as described in **F2**, into a 3×5 array with positions ordered as follows:

10	11	12	13	14
5	6	7	8	9
0	1	2	3	4

and hence the dots in the array are as follows:

•		•		
	•	•		•

and form a DDC with five dots.   ∎

Example 10.5. The following sequence, which is the characteristic vector of a $(31, 5, 1)$ difference set, in $\mathbb{Z}_{31}$: $\{0, 1, 4, 10, 12, 17\}$ (can be viewed as a cyclic ruler) is folded, as defined by **F3**, into an infinite array (we demonstrate part of the array, where folding into a small rectangle is given in bold). Note that while the folding is carried out we should consider all the integers modulo 31 (see Fig. 10.5).

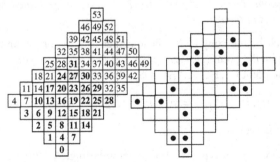

FIGURE 10.5   Folding by diagonals.

∎

Tiling is one of the most basic concepts in combinatorics. We say that a D-dimensional shape S tiles the D-dimensional space $\mathbb{Z}^D$ if disjoint copies of S cover $\mathbb{Z}^D$.

Remark. We assume that S is a discrete shape, i.e., it consists of discrete points of $\mathbb{Z}^D$ such that there is a path between any two points of S that consists only of points of S. The shape S in $\mathbb{Z}^D$ is usually not represented as a union of points in $\mathbb{Z}^D$, but rather as a union of unit cubes in $\mathbb{R}^D$ with 2^D vertices in $\mathbb{Z}^D$. Let A be the set of points in the first representation. The set of unit cubes by the second representation is

$$\{ \mathcal{U}_{(i_1, i_2, \ldots, i_D)} \, : \, (i_1, i_2, \ldots, i_D) \in A \} \, ,$$

where

$$\mathcal{U}_{(i_1, i_2, \ldots, i_D)} = \{ (i_1, i_2, \ldots, i_D) + \xi_1 \epsilon_1 + \xi_2 \epsilon_2 + \cdots + \xi_D \epsilon_D \, : \, 0 \le \xi_i < 1, \; 1 \le i \le D \}$$

and ϵ_i is a vector of length D and weight one with a *one* in the ith position. We omit the case of shapes in $\mathbb{R}^D$, which are not of interest to our discussion.

A cover for $\mathbb{Z}^D$ with disjoint copies of S is called a ***tiling*** of $\mathbb{Z}^D$ with S. For each shape S we distinguish one of the points of S to be the ***center*** of S. Each copy of S in a tiling has the center in the same related point. The set $\mathcal{T}$ of centers in a tiling defines the tiling, and hence the tiling is denoted by the pair $(\mathcal{T}, S)$. Given a tiling $(\mathcal{T}, S)$ and a grid point $(i_1, i_2, \ldots, i_D)$ we denote by $c(i_1, i_2, \ldots, i_D)$ the center of the copy of S for which $(i_1, i_2, \ldots, i_D) \in S$. We will also assume that the origin is a center of some copy of S.

Remark. It is easy to verify that any point of S can serve as the center of S. If $(\mathcal{T}, S)$ is a tiling then we can choose any point of S to serve as a center without affecting the fact that $(\mathcal{T}, S)$ is a tiling.

Lemma 10.6. *If $(\mathcal{T}, S)$ is a tiling, then for any given point $(i_1, i_2, \ldots, i_D)$ in $\mathbb{Z}^D$, the point $(i_1, i_2, \ldots, i_D) - c(i_1, i_2, \ldots, i_D)$ is contained in the shape S whose center is at the origin.*

Proof. Let S_1 be the copy of S whose center is at the origin and let S_2 be the copy of S with the point $(i_1, i_2, \ldots, i_D)$. Let $(x_1, x_2, \ldots, x_D)$ be the point in S_1 related to the point $(i_1, i_2, \ldots, i_D)$ in S_2. By definition,

$$(i_1, i_2, \ldots, i_D) = c(i_1, i_2, \ldots, i_D) + (x_1, x_2, \ldots, x_D)$$

and the lemma follows. $\qquad\qquad\qquad\qquad\qquad\qquad\qquad\qquad\qquad\qquad\square$

The most common type of tiling is lattice tiling. A ***lattice*** Λ is a discrete, additive subgroup of the real D-space $\mathbb{R}^D$. W.l.o.g., we can assume that

$$\Lambda = \{ u_1 v_1 + u_2 v_2 + \cdots + u_D v_D \, : \, u_1, \ldots, u_D \in \mathbb{Z} \} \, , \qquad (10.1)$$

where $\{v_1, v_2, \ldots, v_D\}$ is a set of linearly independent vectors in $\mathbb{R}^D$. A lattice Λ defined by Eq. (10.1) is a sublattice of $\mathbb{Z}^D$ if and only if $\{v_1, v_2, \ldots, v_D\} \subset \mathbb{Z}^D$. We will be interested solely in sublattices of $\mathbb{Z}^D$ since our shapes are defined in $\mathbb{Z}^D$. The vectors $v_1, v_2, \ldots, v_D$ are called a *base* for $\Lambda \subseteq \mathbb{Z}^D$, and the $D \times D$ matrix

$$
\mathbf{G} = \begin{bmatrix}
v_{11} & v_{12} & \cdots & v_{1D} \\
v_{21} & v_{22} & \cdots & v_{2D} \\
\vdots & \vdots & \ddots & \vdots \\
v_{D1} & v_{D2} & \cdots & v_{DD}
\end{bmatrix}
$$

having these vectors as its rows is said to be a ***generator matrix*** for Λ.

The ***volume*** of a lattice Λ, denoted $V(\Lambda)$, is inversely proportional to the number of lattice points per unit volume. More precisely, $V(\Lambda)$ may be defined as the volume of the ***fundamental parallelogram*** $\Pi(\Lambda)$ in $\mathbb{R}^D$, which is defined by

$$
\Pi(\Lambda) \triangleq \{\xi_1 v_1 + \xi_2 v_2 + \cdots + \xi_D v_D : 0 \le \xi_i < 1, 1 \le i \le D\}.
$$

There is a simple expression for the volume of Λ, namely, $V(\Lambda) = |\det \mathbf{G}|$.

We say that Λ is a ***lattice tiling*** for S if the lattice points can be taken as the set $\mathcal{T}$ to form a tiling $(\mathcal{T}, S)$. In this case, we have that $|S| = V(\Lambda) = |\det \mathbf{G}|$.

Remark. Note that different generator matrices for the same lattice will result in different fundamental parallelograms. This is related to the fact that the same lattice can induce a tiling for different shapes with the same volume. A fundamental parallelogram is always a shape in $\mathbb{R}^D$ that is tiled by Λ (usually this is not a shape in $\mathbb{Z}^D$ and as a consequence, most and usually all, of the shapes in $\mathbb{Z}^D$ are not fundamental parallelograms).

Now, we will generalize the definition of folding. All the previous three definitions (**F1, F2,** and **F3**) are special cases of the new definition. The new definition involves a lattice tiling Λ, for a shape S on which the folding is performed.

A ***direction*** of length D, $(d_1, d_2, \ldots, d_D)$, is a nonzero word of length D, where $d_i \in \mathbb{Z}$.

Let S be a D-dimensional shape and let $\delta = (d_1, d_2, \ldots, d_D)$ be a direction of length D. Let Λ be a lattice tiling for a shape S, and let S_1 be the copy of S, in the related tiling, which includes the origin. We define recursively a ***folded-row*** starting at the origin. If the point $(i_1, i_2, \ldots, i_D)$ is the current point of S_1 in the folded-row, then the next point on its folded-row is defined as follows:

- If the point $(i_1 + d_1, i_2 + d_2, \ldots, i_D + d_D)$ is in S_1 then it is the next point on the folded-row.
- If the point $(i_1 + d_1, i_2 + d_2, \ldots, i_D + d_D)$ is in $S_2 \ne S_1$ whose center is at the point $(c_1, c_2, \ldots, c_D)$ then $(i_1 + d_1 - c_1, i_2 + d_2 - c_2, \ldots, i_D + d_D - c_D)$ is the next point on the folded-row (by Lemma 10.6 this point is in S_1).

The new definition of folding is based on a lattice Λ, a shape $\mathcal{S}$, and a direction δ. The triple $(\Lambda, \mathcal{S}, \delta)$ defines a *folding* if the definition yields a folded-row that contains all the elements of $\mathcal{S}$. It will be proved that only Λ and δ determine whether the triple $(\Lambda, \mathcal{S}, \delta)$ defines a folding. The role of $\mathcal{S}$ is only in the order of the folded-row elements and Λ must define a lattice tiling for $\mathcal{S}$. Different lattice tilings for the same shape $\mathcal{S}$ can function completely differently in this respect. Also, not all directions for the same lattice tiling of the shape $\mathcal{S}$ should define (or not define) a folding.

Remark. It is not difficult to see that the three foldings defined earlier (**F1**, **F2**, and **F3**) are special cases of the new definition. The definition of the generator matrices for the three corresponding lattices is left as an exercise.

Lemma 10.7. *Let Λ be a lattice tiling for the shape $\mathcal{S}$. Let $(d_1, d_2, \ldots, d_D)$ be a direction, $(i_1, i_2, \ldots, i_D)$ be a lattice point, and the point $(d_1, d_2, \ldots, d_D)$ is in the shape $\mathcal{S}$ whose center is at the origin. Then, the folded-rows defined by the directions $(d_1, d_2, \ldots, d_D)$ and $(i_1 + d_1, i_2 + d_2, \ldots, i_D + d_D)$ are equivalent.*

Proof. This follows immediately from the observation that

$$c(i_1 + d_1, i_2 + d_2, \ldots, i_D + d_D) = (i_1, i_2, \ldots, i_D). \qquad \square$$

How many different folded-rows do we have? In other words, how many different folding operations are defined in this way? As a consequence of Lemma 10.7 we have that there are at most $|\mathcal{S}| - 1$ different folded-rows since each point, except for $(0, 0)$ can serve for the definition of a direction. Hence, in the following, each direction $\delta = (d_1, d_2, \ldots, d_D)$ will have the property that the point $(d_1, d_2, \ldots, d_D)$ will be contained in the copy of $\mathcal{S}$ whose center is at the origin. If Λ with the direction $(d_1, d_2, \ldots, d_D)$ define a folding then also Λ with the direction vector $(-d_1, -d_2, \ldots, -d_D)$ define a folding. The two folded-rows are in reverse order and they will be considered *equivalent*. If two folded-rows are not equal and not a reverse pair then they will be considered to be nonequivalent. The question whether for each D, there exists a D-dimensional shape $\mathcal{S}$ with $\left\lfloor \frac{|\mathcal{S}|-1}{2} \right\rfloor$ nonequivalent folded-rows, will be partially answered in the following.

How do we fold a sequence into a shape $\mathcal{S}$? Let Λ be a lattice tiling for the shape $\mathcal{S}$ for which $n = |\mathcal{S}|$. Let δ be a direction for which $(\Lambda, \mathcal{S}, \delta)$ defines a folding and let $\mathcal{B} = b_0 b_1 \ldots b_{n-1}$ be a sequence of length n. The folding of $\mathcal{B}$ induced by $(\Lambda, \mathcal{S}, \delta)$ is denoted by $(\Lambda, \mathcal{S}, \delta, \mathcal{B})$ and is defined as the shape $\mathcal{S}$ with the elements of $\mathcal{B}$, where b_i is in the ith entry of the folded-row of $\mathcal{S}$ defined by $(\Lambda, \mathcal{S}, \delta)$.

Next, we aim to find sufficient and necessary conditions that a triple $(\Lambda, \mathcal{S}, \delta)$ defines a folding. We start with a simple characterization of the order of the elements in a folded-row.

Lemma 10.8. *Let Λ be a lattice tiling for the shape S and let $\delta = (d_1, d_2, \ldots, d_D)$ be a direction. Let $g(i) = (i \cdot d_1, \ldots, i \cdot d_D) - c(i \cdot d_1, \ldots, i \cdot d_D)$ and let i, j be two integers. Then, $g(i) = g(j)$ if and only if $g(i + 1) = g(j + 1)$.*

Proof. The lemma follows immediately from the observation that $g(i) = g(j)$ if and only if $(i \cdot d_1, \ldots, i \cdot d_D)$ and $(j \cdot d_1, \ldots, j \cdot d_D)$ are associated with the same position in S, i.e., correspond to the same position of the folded-row.   □

The next two lemmas are immediate consequences of the definitions and provide a concise condition on whether the triple (Λ, S, δ) defines a folding.

Lemma 10.9. *Let Λ be a lattice tiling for the shape S and let $\delta = (d_1, d_2, \ldots, d_D)$ be a direction. (Λ, S, δ) defines a folding if and only if the set*

$$\{(i \cdot d_1, i \cdot d_2, \ldots, i \cdot d_D) - c(i \cdot d_1, i \cdot d_2, \ldots, i \cdot d_D) : 0 \leq i < |S|\}$$

contains $|S|$ distinct elements.

Proof. The lemma is an immediate consequence of Lemmas 10.6 and 10.8, and the definition of folding.   □

Lemma 10.10. *Let Λ be a lattice tiling for the shape S and $\delta = (d_1, d_2, \ldots, d_D)$ be a direction. (Λ, S, δ) defines a folding if and only if*

$$(|S| \cdot d_1, \ldots, |S| \cdot d_D) - c(|S| \cdot d_1, \ldots, |S| \cdot d_D) = (0, \ldots, 0)$$

and for each i, $0 < i < |S|$ we have

$$g(i) = (i \cdot d_1, \ldots, i \cdot d_D) - c(i \cdot d_1, \ldots, i \cdot d_D) \neq (0, \ldots, 0).$$

Proof. Assume first that (Λ, S, δ) defines a folding. If for some $0 < j < |S|$ we have $(j \cdot d_1, \ldots, j \cdot d_D) - c(j \cdot d_1, \ldots, j \cdot d_D) = (0, \ldots, 0)$, then $g(j) = g(0)$ and hence by Lemma 10.8 the folded-row will have at most j elements of S. Since $j < |S|$ we will have that (Λ, S, δ) does not define a folding, a contradiction. On the other hand, Lemma 10.8 also implies that if (Λ, S, δ) defines a folding, then $g(|S|) = (0, \ldots, 0)$.

Now, assume that $(|S| \cdot d_1, \ldots, |S| \cdot d_D) - c(|S| \cdot d_1, \ldots, |S| \cdot d_D) = (0, \ldots, 0)$ and for each i, $0 < i < |S|$ we have $(i \cdot d_1, \ldots, i \cdot d_D) - c(i \cdot d_1, \ldots, i \cdot d_D) \neq (0, \ldots, 0)$. Let $0 < i_1 < i_2 < |S|$; if $g(i_1) = g(i_2)$ then by Lemma 10.8 we have $g(i_2 - i_1) = g(0) = (0, \ldots, 0)$, a contradiction. Therefore the folded-row contains all the elements of S and hence by definition (Λ, S, δ) defines a folding.   □

Corollary 10.4. *If (Λ, S, δ), $\delta = (d_1, d_2, \ldots, d_D)$, defines a folding then the point $(|S| \cdot d_1, \ldots, |S| \cdot d_D)$ is a lattice point.*

Before considering the general D-dimensional case we want to give a simple condition to check whether the triple (Λ, S, δ) defines a folding in the two-dimensional case, which is the main theme of this section.

Lemma 10.11. *Let G be the generator matrix of a lattice* Λ *and let* $s = |\det G|$. *Then, the points* $(0, s)$, $(s, 0)$, (s, s), *and* $(s, -s)$ *are lattice points.*

Proof. It is sufficient to prove that the points $(0, s)$, $(s, 0)$ are lattice points. Let Λ be a lattice whose generator matrix is given by

$$G = \begin{bmatrix} v_{11} & v_{12} \\ v_{21} & v_{22} \end{bmatrix},$$

i.e., $s = v_{11}v_{22} - v_{12}v_{21}$. Since $v_{22}(v_{11}, v_{12}) - v_{12}(v_{21}, v_{22}) = (s, 0)$ and $v_{11}(v_{21}, v_{22}) - v_{21}(v_{11}, v_{12}) = (0, s)$, it follows that the points $(0, s)$, $(s, 0)$ are lattice points. $\qquad\square$

Theorem 10.6. *Let* Λ *be a lattice whose generator matrix is given by*

$$G = \begin{bmatrix} v_{11} & v_{12} \\ v_{21} & v_{22} \end{bmatrix}.$$

Let d_1 *and* d_2 *be two positive integers and* $\tau = \mathrm{g.c.d.}(d_1, d_2)$. *If* Λ *defines a lattice tiling for the shape* $\mathcal{S}$, *then the triple* $(\Lambda, \mathcal{S}, \delta)$ *defines a folding as follows:*

- *with* $\delta = (+d_1, +d_2)$ *if and only if* $\mathrm{g.c.d.}\left(\frac{d_1v_{22}-d_2v_{21}}{\tau}, \frac{d_2v_{11}-d_1v_{12}}{\tau}\right) = 1$ *and* $\mathrm{g.c.d.}(\tau, |\mathcal{S}|) = 1$;
- *with* $\delta = (+d_1, -d_2)$ *if and only if* $\mathrm{g.c.d.}\left(\frac{d_1v_{22}+d_2v_{21}}{\tau}, \frac{d_2v_{11}+d_1v_{12}}{\tau}\right) = 1$ *and* $\mathrm{g.c.d.}(\tau, |\mathcal{S}|) = 1$;
- *with* $\delta = (+d_1, 0)$ *if and only if* $\mathrm{g.c.d.}(v_{12}, v_{22}) = 1$ *and* $\mathrm{g.c.d.}(d_1, |\mathcal{S}|) = 1$;
- *with* $\delta = (0, +d_2)$ *if and only if* $\mathrm{g.c.d.}(v_{11}, v_{21}) = 1$ *and* $\mathrm{g.c.d.}(d_2, |\mathcal{S}|) = 1$.

Proof. We will prove the case where $\delta = (+d_1, +d_2)$; the other three cases are inferred or proved similarly.

Let Λ be a lattice tiling for the shape $\mathcal{S}$. By Lemma 10.11 we have that $(|\mathcal{S}| \cdot d_1, |\mathcal{S}| \cdot d_2)$ is a lattice point. Therefore there exist two integers α_1 and α_2 such that $\alpha_1(v_{11}, v_{12}) + \alpha_2(v_{21}, v_{22}) = (|\mathcal{S}| \cdot d_1, |\mathcal{S}| \cdot d_2)$, i.e., $\alpha_1 v_{11} + \alpha_2 v_{21} = d_1|\mathcal{S}|$, $\alpha_1 v_{12} + \alpha_2 v_{22} = d_2|\mathcal{S}|$, and $|\mathcal{S}| = v_{11}v_{22} - v_{12}v_{21}$. These equations have exactly one solution, $\alpha_1 = d_1v_{22} - d_2v_{21}$ and $\alpha_2 = d_2v_{11} - d_1v_{12}$. By Lemma 10.10, $(\Lambda, \mathcal{S}, \delta)$ defines a folding if and only if $(|\mathcal{S}| \cdot d_1, |\mathcal{S}| \cdot d_2) = c(|\mathcal{S}| \cdot d_1, |\mathcal{S}| \cdot d_2)$ and for each i, $0 < i < |\mathcal{S}|$ we have $(i \cdot d_1, i \cdot d_2) \neq c(i \cdot d_1, i \cdot d_2)$.

Assume first that $\mathrm{g.c.d.}\left(\frac{d_1v_{22}-d_2v_{21}}{\tau}, \frac{d_2v_{11}-d_1v_{12}}{\tau}\right) = 1$ and $\mathrm{g.c.d.}(\tau, |\mathcal{S}|) = 1$. Assume, on the contrary, that there exist three integers i, β_1, and β_2, such that $\beta_1(v_{11}, v_{12}) + \beta_2(v_{21}, v_{22}) = (i \cdot d_1, i \cdot d_2)$, $0 < i < |\mathcal{S}|$. Hence, we have $\frac{\beta_2}{\beta_1} = \frac{d_2v_{11}-d_1v_{12}}{d_1v_{22}-d_2v_{21}} = \frac{\alpha_2}{\alpha_1}$. Since $\mathrm{g.c.d.}\left(\frac{d_1v_{22}-d_2v_{21}}{\tau}, \frac{d_2v_{11}-d_1v_{12}}{\tau}\right) = 1$ it follows that $\beta_1 = \gamma\frac{d_1v_{22}-d_2v_{21}}{\tau}$ and $\beta_2 = \gamma\frac{d_2v_{11}-d_1v_{12}}{\tau}$, for some $0 < \gamma < \tau$. Therefore we have $i \cdot d_1 = \beta_1v_{11} + \beta_2v_{21} = \frac{\gamma d_1|\mathcal{S}|}{\tau}$, i.e., $i = \frac{\gamma|\mathcal{S}|}{\tau}$. However, since g.c.d.$(\tau, |\mathcal{S}|) = 1$ it follows that $\gamma = \rho\tau$, for some integer $\rho > 0$, a contradiction to the fact that $0 < \gamma < \tau$. Hence, our assumption on the existence of three integers i, β_1, and β_2 is false. Thus by Lemma 10.10 we have that if g.c.d. $\left(\frac{d_1v_{22}-d_2v_{21}}{\tau}, \frac{d_2v_{11}-d_1v_{12}}{\tau}\right) = 1$ and g.c.d.$(\tau, |\mathcal{S}|) = 1$, then $(\Lambda, \mathcal{S}, \delta)$ defines a folding with the direction $\delta = (+d_1, +d_2)$.

Assume that $(\Lambda, \mathcal{S}, \delta)$ defines a folding with the direction $\delta = (+d_1, +d_2)$. Assume now, on the contrary, that g.c.d. $\left(\frac{d_1v_{22}-d_2v_{21}}{\tau}, \frac{d_2v_{11}-d_1v_{12}}{\tau}\right) = \nu_1 > 1$ or g.c.d.$(\tau, |\mathcal{S}|) = \nu_2 > 1$. We distinguish now between two cases.

Case 1: If g.c.d. $\left(\frac{d_1v_{22}-d_2v_{21}}{\tau}, \frac{d_2v_{11}-d_1v_{12}}{\tau}\right) = \nu_1 > 1$ then $\beta_1 = \frac{d_1v_{22}-d_2v_{21}}{\tau\nu_1}$ and $\beta_2 = \frac{d_2v_{11}-d_1v_{12}}{\tau\nu_1}$ are integers. Therefore $\beta_1(v_{11}, v_{12}) + \beta_2(v_{21}, v_{22}) = \left(\frac{|\mathcal{S}|\cdot d_1}{\tau\nu_1}, \frac{|\mathcal{S}|\cdot d_2}{\tau\nu_1}\right)$. Hence, $\frac{|\mathcal{S}|}{\nu_1}$ is an integer and for the integers $\beta_1' = \frac{d_1v_{22}-d_2v_{21}}{\nu_1}$ and $\beta_2' = \frac{d_2v_{11}-d_1v_{12}}{\nu_1}$ we have $\beta_1'(v_{11}, v_{12}) + \beta_2'(v_{21}, v_{22}) = \left(\frac{|\mathcal{S}|}{\nu_1}d_1, \frac{|\mathcal{S}|}{\nu_1}d_2\right)$, i.e., $\left(\frac{|\mathcal{S}|}{\nu_1}d_1, \frac{|\mathcal{S}|}{\nu_1}d_2\right)$ is a lattice point, and as a consequence by Lemma 10.10 we have that $(\Lambda, \mathcal{S}, \delta)$ does not define a folding, a contradiction.

Case 2: If g.c.d.$(\tau, |\mathcal{S}|) = \nu_2 > 1$ then let $\beta_1 = \frac{d_1v_{22}-d_2v_{21}}{\nu_2}$ and $\beta_2 = \frac{d_2v_{11}-d_1v_{12}}{\nu_2}$. Hence, $\beta_1(v_{11}, v_{12}) + \beta_2(v_{21}, v_{22}) = \left(\frac{|\mathcal{S}|}{\nu_2}d_1, \frac{|\mathcal{S}|}{\nu_2}d_2\right)$. Clearly, β_1, β_2, and $\frac{|\mathcal{S}|}{\nu_2}$ are integers, and as a consequence by Lemma 10.10 we have that $(\Lambda, \mathcal{S}, \delta)$ does not define a folding, a contradiction.

Therefore if $(\Lambda, \mathcal{S}, \delta)$ defines a folding with the direction $\delta = (+d_1, +d_2)$ then g.c.d. $\left(\frac{d_1v_{22}-d_2v_{21}}{\tau}, \frac{d_2v_{11}-d_1v_{12}}{\tau}\right) = 1$ and g.c.d.$(\tau, |\mathcal{S}|) = 1$. □

The generalization of Theorem 10.6 for the D-dimensional case is presented in Theorem 10.12. The most important types of directions (used for **F1**, **F2**, and **F3**), are those in which the points P and $\delta + P$, where δ is the direction, are adjacent for any given point P, i.e., if $\delta = (d_1, d_2, \ldots, d_D)$, then $|d_i| \leq 1$ for each i, $1 \leq i \leq D$. For these types of directions, we have the following result.

Corollary 10.5. *Let* $\delta = (d_1, d_2)$ *be a direction and let* Λ *be a lattice whose generator matrix is given by*

$$G = \begin{bmatrix} v_{11} & v_{12} \\ v_{21} & v_{22} \end{bmatrix}.$$

If Λ *defines a lattice tiling for the shape* $\mathcal{S}$ *then the triple* $(\Lambda, \mathcal{S}, \delta)$ *defines a folding as follows:*

- *with* $\delta = (+1, +1)$ *if and only if* g.c.d.$(v_{22} - v_{21}, v_{11} - v_{12}) = 1$;
- *with* $\delta = (+1, -1)$ *if and only if* g.c.d.$(v_{22} + v_{21}, v_{11} + v_{12}) = 1$;
- *with* $\delta = (+1, 0)$ *if and only if* g.c.d.$(v_{12}, v_{22}) = 1$;
- *with* $\delta = (0, +1)$ *if and only if* g.c.d.$(v_{11}, v_{21}) = 1$.

There are cases when we can easily determine whether $(\Lambda, \mathcal{S}, \delta)$ defines a folding. It will be a consequence of the following lemmas.

Lemma 10.12.

- *The number of elements in a folded-row does not depend on the chosen point to be the center of $\mathcal{S}$.*
- *The number of elements in a folded-row is a divisor of $|\mathcal{S}| = V(\Lambda)$.*

Proof. By Lemmas 10.8 and 10.10 and the definition of the folded-row, if we start the folded-row at the origin then the number of elements in the folded-row is the smallest t such that $t \cdot \delta$ is a lattice point (since the folded-row starts at a lattice point and ends one step before it again reaches a lattice point). This implies that the number of elements in a folded-row does not depend on the point of $\mathcal{S}$ chosen to be the center of $\mathcal{S}$. We can make any point of $\mathcal{S}$ to be the center of $\mathcal{S}$ and hence any point can be at the origin. Therefore all folded-rows with the direction δ have t elements. For a given lattice Λ and a direction δ, any two folded-rows are either equal or disjoint and those that are disjoint have the same number of points. Hence, t must be a divisor $|\mathcal{S}|$ and t does not depend on which point of $\mathcal{S}$ is the center. $\qquad\square$

The next lemma is an immediate consequence of the definition of a folded-row.

Lemma 10.13. *The number of elements in a folded-row is one if and only if δ is a lattice point.*

Lemmas 10.12 and 10.13 lead to the following consequence.

Corollary 10.6. *Let Λ be a lattice tiling for a shape S. If the volume of Λ is a prime number then (Λ, S, δ) defines a folding with any direction δ, unless δ is a lattice point. If $|S|$ is a prime number, then there exists $\frac{|S|-1}{2}$ different directions that form $\frac{|S|-1}{2}$ nonequivalent folded-rows.*

Lemma 10.14. *Let Λ be a lattice tiling for the shape S, where $n = |S|$. Let $\delta = (d_1, d_2, \ldots, d_D)$ be a direction that defines a folding and let $f_0 f_1 \ldots f_{n-1}$ be its folded-row, where $f_0 = (0, 0, \ldots, 0)$ and $f_1 = (d_1, d_2, \ldots, d_D)$. Then, the direction $\delta' = f_i$ defines a folding if and only if g.c.d.$(i, n) = 1$. If the direction $\delta' = f_i$ defines a folding then its folded-row is $f_0 f_i f_{2i} \ldots f_{n-i}$, where indices are taken modulo n.*

Proof. By definition and by Lemma 10.8 we have that

$$\delta' = f_i = (i \cdot d_1, i \cdot d_2, \ldots, i \cdot d_D) - c(i \cdot d_1, i \cdot d_2, \ldots, i \cdot d_D)$$

and

$$f_{\ell \cdot i} = (\ell \cdot i \cdot d_1, \ell \cdot i \cdot d_2, \ldots, \ell \cdot i \cdot d_D) - c(\ell \cdot i \cdot d_1, \ell \cdot i \cdot d_2, \ldots, \ell \cdot i \cdot d_D).$$

Since the sequence $f_0 f_1 \ldots f_{n-1}$ contains n distinct points of $\mathbb{Z}^D$, it follows that the sequence $f_0 f_i f_{2i} \ldots f_{n-i}$ contains n distinct points of $\mathbb{Z}^D$ if and only if g.c.d.$(i, n) = 1$. Thus the lemma follows. $\qquad\square$

Corollary 10.7. *Let Λ be a lattice tiling for the shape $\mathcal{S}$. There exists at least one folding associated with Λ if and only if the number of nonequivalent folding operations associated with Λ is $\frac{\phi(|\mathcal{S}|)}{2}$.*

Corollary 10.7 implies that once we have one folding operation with its folded-row, then we can easily find and compute all the other folding operations with their folded-rows. It also implies that once the necessary and sufficient conditions for the existence of one folding in the related theorems are satisfied, then the necessary and sufficient conditions for the existence of other foldings are also satisfied. Nevertheless, there are cases in which no direction defines a folding.

Lemma 10.15. *Let $\gamma > 1$ be a positive integer, let $a_1, a_2, \ldots, a_D$, be nonzero integers, and let $b_1, b_2, \ldots, b_D$ be nonzero integers such that either $b_i = a_i$ or $b_i = a_i\gamma$, for each $1 \leq i \leq D$, and $|\{i : b_i = a_i\gamma, \ 1 \leq i \leq D\}| = r \geq 2$. Let $\mathcal{S}$ be a D-dimensional shape and Λ be a lattice tiling for $\mathcal{S}$ whose generator matrix is given by*

$$\begin{bmatrix} b_1 & 0 & \ldots & 0 \\ 0 & b_2 & \ldots & 0 \\ \vdots & \vdots & \ddots & \vdots \\ 0 & 0 & \ldots & b_D \end{bmatrix}.$$

Then, there is no direction δ for which the triple $(\Lambda, \mathcal{S}, \delta)$ defines a folding.

Proof. Let $\delta = (d_1, d_2, \ldots, d_D)$ be any direction and let $\sigma = \gamma \prod_{i=1}^{D} a_i$. Then, $\sigma < |\mathcal{S}| = \prod_{i=1}^{D} b_i = \gamma^r \prod_{i=1}^{D} a_i = \gamma^{r-1}\sigma$ and for any given shape $\mathcal{S}$ for which Λ is a lattice tiling we have

$$(\sigma \cdot d_1, \sigma \cdot d_2, \ldots, \sigma \cdot d_D) - c(\sigma \cdot d_1, \sigma \cdot d_2, \ldots, \sigma \cdot d_D) = (0, 0, \ldots, 0).$$

Hence, by Lemma 10.10, the triple $(\Lambda, \mathcal{S}, \delta)$ does not define a folding.   □

The motivation for the generalization of the folding operation came from the design of two-dimensional synchronization patterns. Given a grid and a shape $\mathcal{S}$ on the grid, we would like to find what is the largest set Δ of dots on grid points, where $|\Delta| = m$, located in $\mathcal{S}$, such that all the $\binom{m}{2}$ lines connecting dots in Δ are distinct in their length or in their slope. Such a shape $\mathcal{S}$ with dots is a DDC. In the application of these patterns to the design of KPS for WSNs various shapes might be required. This application requires in some cases to consider these shapes in the other grid, e.g., the hexagonal grid. **F3** can be used for this application to form a DDC whose shape is a rectangle rotated at 45 degrees on the square grid (see Fig. 10.5).

We will generalize some of the definitions given for DDCs in two-dimensional arrays for multi-dimensional arrays. The associated results (lemmas and theorems) are also generalized, but most of them will not be stated. Let $\mathcal{A}$ be a (generally infinite) D-dimensional array of dots in $\mathbb{Z}^D$, and let $\eta_1, \eta_2, \ldots, \eta_D$ be positive integers. We say that $\mathcal{A}$ is *multi-periodic* with period $(\eta_1, \eta_2, \ldots, \eta_D)$ if $\mathcal{A}(i_1, i_2, \ldots, i_D) = \mathcal{A}(i_1 + \eta_1, i_2, \ldots, i_D) = \mathcal{A}(i_1, i_2 + \eta_2, \ldots, i_D) = \cdots = \mathcal{A}(i_1, i_2, \ldots, i_D + \eta_D)$. We define the *density* of $\mathcal{A}$ to be $d/(\Pi_{j=1}^{D}\eta_j)$, where d is the number of dots in any $\eta_1 \times \eta_2 \times \cdots \times \eta_D$ sub-array of $\mathcal{A}$. Note that the period $(\eta_1, \eta_2, \ldots, \eta_D)$ might not be unique, but that the density of $\mathcal{A}$ does not depend on the period we choose. We say that a multi-periodic array $\mathcal{A}$ of dots is a *multi-periodic* $n_1 \times n_2 \times \cdots n_D$ **DDC** if every $n_1 \times n_2 \times \cdots n_D$ sub-array of $\mathcal{A}$ is a DDC.

We write $(i_1, i_2, \ldots, i_D) + \mathcal{S}$ for the shifted copy $\{(i_1 + i'_1, i_2 + i'_2, \ldots, i_D + i'_D) : (i'_1, i'_2, \ldots, i'_D) \in \mathcal{S}\}$ of $\mathcal{S}$. We say that a multi-periodic array $\mathcal{A}$ is a *multi-periodic* $\mathcal{S}$-**DDC** if the dots contained in every shift $(i_1, i_2, \ldots, i_D) + \mathcal{S}$ of $\mathcal{S}$ form a DDC.

Let $\mathcal{S}$ and $\mathcal{S}'$ be D-dimensional shapes in a grid. We will denote by $\Delta(\mathcal{S}, \mathcal{S}')$ the largest intersection between $\mathcal{S}$ and $\mathcal{S}'$ in all possible shifts. Bounds on the number of dots in a DDC with a given shape are based on the following result.

Theorem 10.7. *Assume we are given a multi-periodic $\mathcal{S}$-DDC array $\mathcal{A}$ with density μ. Let $\mathcal{Q}$ be another shape on $\mathbb{Z}^D$. Then, there exists a copy of $\mathcal{Q}$ on $\mathbb{Z}^D$ with at least $\lceil \mu \cdot \Delta(\mathcal{S}, \mathcal{Q}) \rceil$ dots.*

Proof. Let $\mathcal{Q}'$ be the shape such that $\mathcal{Q}' = \mathcal{S} \cap \mathcal{Q}$ and $|\mathcal{Q}'| = \Delta(\mathcal{S}, \mathcal{Q})$. By Lemma 10.5 we have that $\mathcal{A}$ is a multi-periodic $\mathcal{Q}'$-DDC. By Theorem 9.19, there exists a set of at least $\lceil \mu|\mathcal{Q}'| \rceil$ dots contained in $\mathcal{S}$ that form a DDC. Thus there exists a copy of $\mathcal{Q}$ on $\mathbb{Z}^D$ with at least $\lceil \mu \cdot \Delta(\mathcal{S}, \mathcal{Q}) \rceil$ dots. $\square$

To apply Theorem 10.7 we will use folding of the sequences defined as follows. Let A be an Abelian group, and let $\mathcal{B} = \{b_1, b_2, \ldots, b_m\} \subseteq A$ be a sequence of m distinct elements of A. We say that $\mathcal{B}$ is a B_2-*sequence* over A if all the sums $a_{i_1} + a_{i_2}$ with $1 \leq i_1 \leq i_2 \leq m$ are distinct.

Lemma 10.16. *A subset $\mathcal{B} = \{a_1, a_2, \ldots, a_m\} \subseteq A$ is a B_2-sequence over A if and only if all the differences $a_{i_1} - a_{i_2}$ with $1 \leq i_1 \neq i_2 \leq m$ are distinct in A.*

Proof. Assume first that $\mathcal{B}$ is a B_2-sequence. Assume, on the contrary, that there exists four elements $\alpha_1, \alpha_2, \alpha_3, \alpha_4$ in $\mathcal{B}$ such that

$$\alpha_4 - \alpha_3 = \alpha_2 - \alpha_1$$

in A, where $\alpha_4 \neq \alpha_2$ and also $\alpha_4 \neq \alpha_3$. This immediately implies that

$$\alpha_4 + \alpha_1 = \alpha_2 + \alpha_3,$$

a contradiction and hence all the differences are distinct in A.

Assume now that all the differences $a_{i_1} - a_{i_2}$ of distinct elements in $\mathcal{B}$ are distinct in A. Assume, on the contrary, that there exist four elements $\alpha_1, \alpha_2, \alpha_3, \alpha_4$ in $\mathcal{B}$ such that

$$\alpha_1 + \alpha_2 = \alpha_3 + \alpha_4$$

in A, where $\{\alpha_1, \alpha_2\} \neq \{\alpha_3, \alpha_4\}$. This immediately implies that

$$\alpha_4 - \alpha_1 = \alpha_2 - \alpha_3,$$

a contradiction and hence $\mathcal{B}$ is a B_2-sequences. □

Note that Lemma 10.16 implies that if A is a $(v, m, 1)$-difference set, then it is also a B_2-sequences, but the converse is not true.

Note that if $\mathcal{B}$ is a B_2-sequence over $\mathbb{Z}_n$ and $t, r \in \mathbb{Z}_n$, then so are the t-shift $t + B = \{t + b : b \in B\}$ and the r-decimation $r \cdot B = \{rb : b \in B\}$. The following theorem shows that large B_2-sequences over $\mathbb{Z}_n$ exist for many values of n.

Theorem 10.8. *Let q be a prime power. Then, there exists a B_2-sequence $a_1, a_2, \ldots, a_m$ over $\mathbb{Z}_n$, where $n = q^2 - 1$ and $m = q$.*

Proof. Let α be a primitive element in $\mathbb{F}_{q^2}$ and $\mathbb{F}_q$ its subfield. Define the set

$$\mathcal{B} \triangleq \{r : r \in \mathbb{Z}_n, \ \alpha^r - \alpha \in \mathbb{F}_q\}.$$

We claim that $\mathcal{B}$ is a B_2-sequence of size q in $\mathbb{Z}_n$, where $n = q^2 - 1$. Clearly, the set $\{\beta - \alpha : \beta \in \mathbb{F}_{q^2}\}$ contains exactly all the elements of $\mathbb{F}_{q^2}$ including the q elements of $\mathbb{F}_q$. Moreover, $0 - \alpha = -\alpha \notin \mathbb{F}_q$ and hence $\mathcal{B}$ contains q elements. Now, let $\mathcal{B} = \{r_1, r_2, \ldots, r_q\}$ and $\alpha^{r_i} - \alpha = \beta_i$, where $\beta_i \in \mathbb{F}_q$ for $1 \leq i \leq q$. Clearly, the set $\{\beta_i : 1 \leq i \leq q\}$ contains q elements.

Assume, on the contrary, that $\mathcal{B}$ is not a B_2 sequence over $\mathbb{Z}_{q^2-1}$. This implies that there exists in $\mathcal{B}$ two disjoint pairs, which can be taken w.l.o.g. as $\{r_1, r_2\}$ and $\{r_3, r_4\}$, such that

$$r_1 + r_2 \equiv r_3 + r_4 \pmod{q^2 - 1}.$$

Hence, we have that

$$\alpha^{r_1} \alpha^{r_2} = \alpha^{r_3} \alpha^{r_4}$$

and therefore

$$(\alpha + \beta_1)(\alpha + \beta_2) = (\alpha + \beta_3)(\alpha + \beta_4). \tag{10.2}$$

Eq. (10.2) is equivalent to

$$\alpha(\beta_1 + \beta_2) + \beta_1 \beta_2 = \alpha(\beta_3 + \beta_4) + \beta_3 \beta_4$$

and hence

$$\alpha(\beta_1 + \beta_2 - \beta_3 - \beta_4) = \beta_3 \beta_4 - \beta_1 \beta_2. \tag{10.3}$$

Assume now that the two sides of Eq. (10.3) are equal to *zero*. This implies that

$$\beta_1 + \beta_2 = \beta_3 + \beta_4 \quad \text{and} \quad \beta_1 \beta_2 = \beta_3 \beta_4$$

and as a consequence

$$\beta_3 \beta_4 = \beta_1 (\beta_3 + \beta_4 - \beta_1),$$

which is equivalent to

$$\beta_1 (\beta_1 - \beta_3) = \beta_4 (\beta_1 - \beta_3)$$

and since all the β_is are distinct this implies that $\beta_1 - \beta_3 \neq 0$ and hence $\beta_1 = \beta_4$, a contradiction. Hence, the two sides of (10.3) are not equal to *zero*.

Since $\beta_i \in \mathbb{F}_q$ for $1 \leq i \leq 4$, it follows that $\beta_3 \beta_4 - \beta_1 \beta_2 \in \mathbb{F}_q$ and $\beta_1 + \beta_2 - \beta_3 - \beta_4 \in \mathbb{F}_q$. Moreover, $\alpha \in \mathbb{F}_{q^2} \setminus \mathbb{F}_q$ and hence

$$\alpha(\beta_1 + \beta_2 - \beta_3 - \beta_4) \in \mathbb{F}_{q^2} \setminus \mathbb{F}_q.$$

Therefore the left side of (10.3) is an element in $\mathbb{F}_{q^2} \setminus \mathbb{F}_q$ while the right side of (10.3) is an element in $\mathbb{F}_q$, a contradiction.

Thus $\mathcal{B}$ is a B_2-sequence. $\qquad\qquad\qquad\qquad\qquad\qquad\qquad\qquad\qquad\square$

Now, we will describe how we apply folding to obtain a DDC with a shape $\mathcal{S}$ and a multi-periodic $\mathcal{S}$-DDC. Let Λ be a lattice tiling for $\mathcal{S}$ and let $\delta = (d_1, d_2, \ldots, d_D)$ be a direction such that $(\Lambda, \mathcal{S}, \delta)$ defines a folding. We assign an integer from $\mathbb{Z}_n$, $n = |\mathcal{S}|$, to each point of $\mathbb{Z}^D$. The **lattice coloring** $\mathcal{C}(\Lambda, \delta)$ is defined as follows. We assign 0 to the point $(0, 0, \ldots, 0)$ and 1 to the next element of the folded-row and so on until $|\mathcal{S}| - 1$ is assigned to the last element of the folded-row. This completes the coloring of the points in the shape $\mathcal{S}$ whose center is at the origin. To position $(i_1, i_2, \ldots, i_D)$ we assign the color of position $(i_1, i_2, \ldots, i_D) - c(i_1, i_2, \ldots, i_D)$. The color of position $(i_1, i_2, \ldots, i_D)$ will be denoted by $\mathcal{C}(i_1, i_2, \ldots, i_D)$.

The folding of the sequence $\mathcal{B} = b_0 b_1 \ldots b_{n-1}$ into an array colored by the elements of $\mathbb{Z}_n$ is defined by assigning the value b_i to all the points of the array colored with the color i. If the coloring was defined by the order of the folded-row as described in this section, we say that the array is defined by $(\Lambda, \mathcal{S}, \delta, \mathcal{B})$. Note that we use the same notation for folding the sequence $\mathcal{B}$ into the shape $\mathcal{S}$. The one to which we refer should be understood from the context.

Given a point $(i_1, i_2, \ldots, i_D) \in \mathbb{Z}^D$, we say that the set of points

$$\{(i_1 + \ell \cdot d_1, i_2 + \ell \cdot d_2, \ldots, i_D + \ell \cdot d_D) : \ell \in \mathbb{Z}\}$$

is a **row of** $\mathbb{Z}^D$ **defined by** $\delta = (d_1, d_2, \ldots, d_D)$. This is also the row of $(i_1, i_2, \ldots, i_D)$ defined by δ.

Lemma 10.17. *If the triple $(\Lambda, \mathcal{S}, \delta)$ defines a folding, then in any folded-row of $\mathbb{Z}^D$ defined by δ there are lattice points.*

Proof. Given a point $(i_1, i_2, \ldots, i_D)$ and its color $C(i_1, i_2, \ldots, i_D)$, then by the definitions of the folding and the coloring we have that

$$C(i_1 + d_1, i_2 + d_2, \ldots, i_D + d_D) \equiv C(i_1, i_2, \ldots, i_D) + 1 \ (\mathrm{mod} \ |\mathcal{S}|).$$

Hence, the folded-row defined by δ has all the values between 0 and $|\mathcal{S}| - 1$ in their natural order modulo $|\mathcal{S}|$. Therefore any folded-row defined by δ has lattice points (which are exactly the points of $\mathbb{Z}^D$, in this folded-row, which are colored with *zeros*). $\qquad\square$

Corollary 10.8. *If* $(i_1, i_2, \ldots, i_D)$, $(i_1 + e_1, i_2 + e_2, \ldots, i_D + e_D)$, $(j_1, j_2, \ldots, j_D)$, *and* $(j_1 + e_1, j_2 + e_2, \ldots, j_D + e_D)$ *are four points of* $\mathbb{Z}^D$, *then*

$$C(i_1 + e_1, i_2 + e_2, \ldots, i_D + e_D) - C(i_1, i_2, \ldots, i_D)$$
$$\equiv C(j_1 + e_1, j_2 + e_2, \ldots, j_D + e_D) - C(j_1, j_2, \ldots, j_D) \ (\mathrm{mod} \ |\mathcal{S}|).$$

Proof. By Lemma 10.17 to each one of these four points there exists a lattice point in its folded-row defined by δ. Let

- $P_1 = (i_1 + \alpha_1 \cdot d_1, i_2 + \alpha_1 \cdot d_2, \ldots, i_D + \alpha_1 \cdot d_D)$ be the lattice point in the folded-row of $(i_1, i_2, \ldots, i_D)$;
- $P_2 = (j_1 + \alpha_2 \cdot d_1, j_2 + \alpha_2 \cdot d_2, \ldots, j_D + \alpha_2 \cdot d_D)$ be the lattice point in the folded-row of $(j_1, j_2, \ldots, j_D)$;
- $P_3 = ((i_1 + e_1) + \alpha_3 \cdot d_1, (i_2 + e_2) + \alpha_3 \cdot d_2, \ldots, (i_D + e_D) + \alpha_3 \cdot d_D)$ be the lattice point in the folded-row of $(i_1 + e_1, i_2 + e_2, \ldots, i_D + e_D)$.

Therefore by the linearity of the lattice $P_4 = P_2 + P_3 - P_1 = ((j_1 + e_1) + (\alpha_2 + \alpha_3 - \alpha_1) \cdot d_1, (j_2 + e_2) + (\alpha_2 + \alpha_3 - \alpha_1) \cdot d_2, \ldots, (j_D + e_D) + (\alpha_2 + \alpha_3 - \alpha_1) \cdot d_D)$ is also a lattice point. P_4 is a lattice point in the folded-row, defined by δ, of $(j_1 + e_1, j_2 + e_2, \ldots, j_D + e_D)$. All these four points are colored with *zeroes*. Hence,

$$C(i_1, i_2, \ldots, i_D) \equiv -\alpha_1 \ (\mathrm{mod} \ |\mathcal{S}|),$$
$$C(i_1 + e_1, i_2 + e_2, \ldots, i_D + e_D) \equiv -\alpha_3 \ (\mathrm{mod} \ |\mathcal{S}|),$$
$$C(j_1, j_2, \ldots, j_D) \equiv -\alpha_2 \ (\mathrm{mod} \ |\mathcal{S}|),$$
$$C(j_1 + e_1, j_2 + e_2, \ldots, j_D + e_D) \equiv -(\alpha_2 + \alpha_3 - \alpha_1) \ (\mathrm{mod} \ |\mathcal{S}|).$$

Now, the claim of the corollary is readily verified. $\qquad\square$

Corollary 10.9. *If* δ' *is an integer vector of length* D, *then there exists an integer* $e(\delta')$ *such that for any given point* $P = (i_1, i_2, \ldots, i_D)$ *we have* $C(P + \delta') \equiv C(P) + e(\delta') \ (\mathrm{mod} \ |\mathcal{S}|)$.

Corollary 10.10. *If the triple* $(\Lambda, \mathcal{S}, \delta)$ *defines a folding and* $\mathcal{B}$ *is a* B_2-*sequence over* $\mathbb{Z}_n$, *where* $n = |\mathcal{S}|$, *then the array* $\mathcal{A}$ *defined by* $(\Lambda, \mathcal{S}, \delta, \mathcal{B})$ *is multiperiodic.*

Proof. Clearly, the array has period $(|\mathcal{S}|, |\mathcal{S}|, \ldots, |\mathcal{S}|)$ and the claim follows. □

Theorem 10.9. *If the triple* $(\Lambda, \mathcal{S}, \delta)$ *defines a folding and* $\mathcal{B}$ *is a* B_2-*sequence over* $\mathbb{Z}_n$, *where* $n = |\mathcal{S}|$, *then the pattern of dots defined by* $(\Lambda, \mathcal{S}, \delta, \mathcal{B})$ *is a multi-periodic* $\mathcal{S}$-*DDC.*

Proof. By Corollary 10.10 the constructed array is multi-periodic.

Since $(\Lambda, \mathcal{S}, \delta)$ defines a folding it follows that the $|\mathcal{S}|$ colors inside the shape $\mathcal{S}$ centered at the origin are all distinct. By Corollary 10.8, for the four positions $(i_1, i_2, \ldots, i_D)$, $(i_1 + e_1, i_2 + e_2, \ldots, i_D + e_D)$, $(j_1, j_2, \ldots, j_D)$, and $(j_1 + e_1, j_2 + e_2, \ldots, j_D + e_D)$, associated with two pairs of equal vectors, we have that

$$\mathcal{C}(i_1 + e_1, i_2 + e_2, \ldots, i_D + e_D) - \mathcal{C}(i_1, i_2, \ldots, i_D)$$
$$\equiv \mathcal{C}(j_1 + e_1, j_2 + e_2, \ldots, j_D + e_D) - \mathcal{C}(j_1, j_2, \ldots, j_D) \pmod{|\mathcal{S}|}.$$

Since the dots are distributed by the B_2-sequence $\mathcal{B}$, it follows that at most three of these integers (colors) are contained in $\mathcal{B}$. This implies that if these four points are contained in the same copy of $\mathcal{S}$ on the grid, then at most three of these points have dots. Thus any shape $\mathcal{S}$ on $\mathbb{Z}^D$ will define a DDC and the theorem follows. □

Corollary 10.11. *If the triple* $(\Lambda, \mathcal{S}, \delta)$ *defines a folding and* $\mathcal{B}$ *is a* B_2-*sequence over* $\mathbb{Z}_n$, *where* $n = |\mathcal{S}|$, *then the pattern of dots defined by* $(\Lambda, \mathcal{S}, \delta, \mathcal{B})$ *is a DDC.*

Note that the difference between Theorem 10.9 and Corollary 10.11 is related to the folding of $\mathcal{B}$ into $\mathbb{Z}^D$ and the folding of $\mathcal{B}$ into $\mathcal{S}$, respectively.

Lemma 10.18. *If* $(\Lambda, \mathcal{S}, \delta)$ *defines a folding then the* $|\mathcal{S}|$ *colors inside any copy of* $\mathcal{S}$ *on* $\mathbb{Z}^D$ *are all distinct.*

Proof. Let $\mathcal{S}_1$ and $\mathcal{S}_2$ be two distinct copies of $\mathcal{S}$ on $\mathbb{Z}^D$. Clearly, we have that $\mathcal{S}_2 = (e_1, \ldots, e_D) + \mathcal{S}_1$. By Corollary 10.8, we have the following equality for each two points $(i_1, \ldots, i_D)$, $(j_1, \ldots, j_D) \in \mathcal{S}_1$,

$$\mathcal{C}(i_1 + e_1, \ldots, i_D + e_D) - \mathcal{C}(i_1, \ldots, i_D)$$
$$\equiv \mathcal{C}(j_1 + e_1, \ldots, j_D + e_D) - \mathcal{C}(j_1, \ldots, j_D) \pmod{|\mathcal{S}|}.$$

Therefore if $\mathcal{S}_1$ contains $|\mathcal{S}|$ distinct colors then also $\mathcal{S}_2$ contains $|\mathcal{S}|$ distinct colors. The lemma follows now from the fact that $(\Lambda, \mathcal{S}, \delta)$ defines a folding and therefore all the colors in the shape $\mathcal{S}$ whose center is at the origin are distinct. □

Now, we will present some lower bounds on the number of dots in some two-dimensional DDCs with specific shapes. In the following, we will use Theorems 10.7 and 10.9, and Corollary 10.11 to form DDCs with various given shapes and a large number of dots. To examine how good the lower bounds on the number of dots, in a DDC whose shape is Q, we should know what is the upper bound on the number of dots in a DDC whose shape is Q. By Theorem 9.22, we have that for a DDC whose shape is a regular polygon or a circle, an upper bound on the number of dots is at most $\sqrt{s} + o(\sqrt{s})$, where the shape contains s points of $\mathbb{Z}^2$ and $s \to \infty$. One of the main keys of our constructions, and the usage of the given theory, is the ability to produce a multi-periodic S-DDC, where S is a rectangle, the ratio between its sides is close as much as we want to any given number γ, and if its area is s, then the number of dots in it is $\sqrt{s+1}$.

For the next theorem it will be required to use Theorems 1.5 and 1.25 (Euler Theorem and Dirichlet Theorem, respectively).

Theorem 10.10. *For each positive number γ and any $\epsilon > 0$, there exist two integers n_1 and n_2 such that $\gamma \leq \frac{n_1}{n_2} < \gamma + \epsilon$ and a multi-periodic S-DDC with $\sqrt{a \cdot b} R + o(R)$ dots, where S is an $n_1 \times n_2 = (aR + o(R)) \times (bR + o(R))$ rectangle, $n_1 n_2 = p^2 - 1$ for some prime p, and n_1 is an even integer.*

Proof. Given a positive number γ and an $\epsilon > 0$, it is easy to verify that there exist two integers α and β such that $\sqrt{\gamma} \leq \frac{\beta}{\alpha} < \sqrt{\gamma + \epsilon}$ and g.c.d.$(\alpha, \beta) = 2$. By Theorem 1.5, there exist two integers c_α, c_β such that either $c_\alpha \alpha + 2 = c_\beta \beta > 0$ or $c_\beta \beta + 2 = c_\alpha \alpha > 0$.

Assume $c_\alpha \alpha + 2 = c_\beta \beta > 0$ (the case where $c_\beta \beta + 2 = c_\alpha \alpha > 0$ is handled similarly). Any factor of α cannot divide $c_\alpha \alpha + 1$. Since β divides $c_\alpha \alpha + 2$, it follows that a factor of β cannot divide $c_\alpha \alpha + 1$. Hence, g.c.d.$(\alpha\beta, c_\alpha \alpha + 1) = 1$. Therefore by Theorem 1.25 there exist infinitely many primes in the sequence $\alpha\beta R + c_\alpha \alpha + 1$, $R = 1, 2, \ldots$.

Let p be a prime number of the form $\alpha\beta R + c_\alpha \alpha + 1$. Now,

$$p^2 - 1 = (p+1)(p-1)$$
$$= (\alpha\beta R + c_\alpha \alpha + 2)(\alpha\beta R + c_\alpha \alpha)$$
$$= (\alpha\beta R + c_\beta \beta)(\alpha\beta R + c_\alpha \alpha) = (\alpha^2 R + \alpha c_\beta)(\beta^2 R + \beta c_\alpha).$$

Thus a $(\beta^2 R + \beta c_\alpha) \times (\alpha^2 R + \alpha c_\beta)$ rectangle satisfies the size requirements for the $n_1 \times n_2$ rectangle of the theorem.

Let $a = \beta^2, b = \alpha^2, n_1 = \beta^2 R + \beta c_\alpha$ (n_1 is even as required), $n_2 = \alpha^2 R + \alpha c_\beta$, and let S be an $n_1 \times n_2$ rectangle. Let Λ be the lattice tiling for S with the generator matrix

$$G = \begin{bmatrix} n_2 & \frac{n_1}{2} + \theta \\ 0 & n_1 \end{bmatrix},$$

where $\theta = 1$ if $n_1 \equiv 0$ (mod 4) and $\theta = 2$ if $n_1 \equiv 2$ (mod 4). By Corollary 10.5, we have that (Λ, S, δ), where $\delta = (+1, 0)$, defines a folding.

The existence of a multi-periodic S-DDC with $\sqrt{a \cdot b} R + o(R)$ dots follows now from Theorems 10.8 and 10.9. $\square$

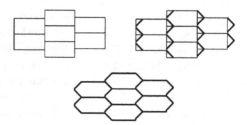

FIGURE 10.6   From a rectangle to a quasi-regular hexagon with the same lattice tiling.

The next key structure is a certain family of hexagons that are defined next. We consider hexagons with three disjoint pairs of parallel sides. If the four angles of two parallel sides (called the *bases* of the hexagon) are equal and the four other sides are equal, then the hexagon will be called a ***quasi-regular hexagon*** and will be denoted by $QRH(w, b, h)$, where b is the length of a base, h is the distance between the two bases, and $b + 2w$ is the length between the two vertices not on the bases. We will call the line that connects these two vertices, the ***diameter*** of the hexagon (even if it might not be the longest line between two points of the hexagon). A quasi-regular hexagon will be the shape that will have the role of $\mathcal{S}$ when we apply Theorem 10.7 to obtain a lower bound on the number of dots in a shape $\mathcal{Q}$ that will be either a regular polygon or a circle. In the following, we will say that $\frac{\beta}{\alpha} \approx \gamma$, when we mean that $\gamma \le \frac{\beta}{\alpha} < \gamma + \epsilon$.

The goal now is to show that there exists a $QRH(w, b, h)$ with approximately $\sqrt{(b + w)h} + o(\sqrt{(b + w)h})$ dots. By Theorem 10.10, there exists a doubly periodic $\mathcal{S}$-DCC, where $\mathcal{S}$ is an $n_1 \times n_2 = (\alpha R + o(R)) \times (\beta R + o(R))$ rectangle, such that $\frac{n_2}{n_1} \approx \frac{b+w}{h}$, $n_1 n_2 = p^2 - 1$ for some prime p, and n_1 is an even integer. The lattice Λ of Theorem 10.10 is also a lattice tiling for a shape $\mathcal{S}'$, where $\mathcal{S}'$ is a $QRH(w, b, h)$ (part of this lattice tiling is depicted in Fig. 10.6). By Corollary 10.5, $(\Lambda, \mathcal{S}, \delta)$, where $\delta = (+1, 0)$, defines a folding for this shape too. Hence, we obtain a doubly periodic $\mathcal{S}'$-DCC, where $\mathcal{S}'$ is a $QRH(w, b, h)$ with approximately $\sqrt{(b + w)h} + o(\sqrt{(b + w)h})$ dots. This construction implies the following theorem.

Theorem 10.11. *If $R \to \infty$ then there exists a regular hexagon with sides of length R and approximately $\frac{\sqrt{3\sqrt{3}}}{\sqrt{2}} R + o(R)$ dots.*

Now, we can give a few examples of other specific shapes, mostly, regular polygons. To have some comparison between the bounds for various shapes we will assume that the radius of the circle or the regular polygons is R (the ***radius*** is the distance from the center of the regular polygon to any one its vertices). We also define the ***packing ratio*** as the ratio between the lower and the upper bounds on the number of dots, which will be obtained. The shape $\mathcal{S}$ that we use will always be a multi-periodic $\mathcal{S}$-DCC on a multi-periodic array $\mathcal{A}$.

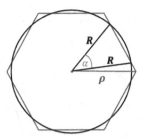

FIGURE 10.7   Intersection between a circle and a regular hexagon.

For a circle, we apply Theorem 10.7 with a multi-periodic $\mathcal{S}$-DDC $\mathcal{A}$, where $\mathcal{S}$ is a regular hexagon with radius ρ and $\mathcal{Q}$ is a circle with radius R, sharing the same center. By Theorem 9.22, the upper bound on the number of dots in $\mathcal{Q}$ is $\sqrt{\pi}R + o(R)$. By Theorem 10.10, a lower bound on the number of dots in $\mathcal{S}$ is approximately $\frac{\sqrt{3\sqrt{3}}}{\sqrt{2}}\rho + o(\rho)$ and hence the density of $\mathcal{A}$ is approximately $\frac{\sqrt{2}}{\sqrt{3\sqrt{3}}\rho}$. Let α be the angle between two radius lines to the two intersection points of the hexagon and the circle on one edge of the hexagon. The regular hexagon $\mathcal{S}$ and the circle $\mathcal{Q}$ are depicted in Fig. 10.7. We have that $\Delta(\mathcal{S}, \mathcal{Q}) = (\pi - 3\alpha + 3\sin\alpha)R^2$ and $\rho = \frac{\cos\frac{\alpha}{2}}{\cos\frac{\pi}{6}}R$. Thus a lower bound on the number of dots in $\mathcal{Q}$ is $\frac{\sqrt{3\sqrt{3}}\rho + o(\rho)}{\sqrt{2}|\mathcal{S}|}\Delta(\mathcal{S}, \mathcal{Q})$. The maximum is obtained when $\alpha = 0.536267$, yielding a lower bound of $1.70813R + o(R)$ on the number of dots in $\mathcal{Q}$ and a packing ratio of 0.9637.

We must note again that although this construction works for infinitely many values of R, the density for these values is quite low. This is a consequence of Theorem 10.10 that can be applied for an arbitrary ratio γ only when the corresponding integers obtained by Dirichlet's Theorem are primes. Of course, many possible ratios between the sides of the rectangle can be obtained for infinitely many values. A simple example is that for any factorization of $p^2 - 1 = n_1 n_2$ we can form an $n_1 \times n_2$ DDC and from it, we can form related quasi-regular hexagons. We are not going to describe in detail how to obtain all these bounds that hold asymptotically for any given R.

For regular polygons with a small number of sides, we have to use specific constructions. We consider for example a regular pentagon. Let $\mathcal{Q}$ be a pentagon with radius R. The area of $\mathcal{Q}$ is $\frac{5}{2}\sin\frac{2\pi}{5}R^2$ and, hence, by Theorem 9.22, an upper bound on the number of dots in $\mathcal{Q}$ is $1.54196R + o(R)$. Let $\mathcal{S}$ be a quasi-perfect hexagon having a joint base with $\mathcal{Q}$ and two short overlapping sides with $\mathcal{Q}$, where these sides are connected to this base (see Fig. 10.8). The distance between the base and the diameter of $\mathcal{S}$ is aR, $2\sin\frac{\pi}{10}\cos\frac{3\pi}{10} < a \leq \left(1 + \sin\frac{3\pi}{10}\right)/2$. The length of the base is $2R\sin\frac{\pi}{5}$ and the length of the diameter of $\mathcal{S}$ is $2R\sin\frac{\pi}{5} + 2aR\tan\frac{\pi}{10}$. Hence, the area of $\mathcal{S}$

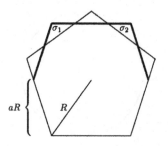

FIGURE 10.8 Quasi-regular hexagon intersecting a regular pentagon.

is $\left(4\sin\frac{\pi}{5} + 2a\tan\frac{\pi}{10}\right)aR^2$ and the density of the array is $\dfrac{1}{\sqrt{4a\sin\frac{\pi}{5}+2a^2\tan\frac{\pi}{10}}\,R}$.
The area of the intersection between $\mathcal{Q}$ and $\mathcal{S}$, $\Delta(\mathcal{S},\mathcal{Q})$, is computed by subtracting from the area of $\mathcal{S}$ the area of the two isosceles triangles σ_1 and σ_2. The lower bound on the number of dots is $\dfrac{1}{\sqrt{4a\sin\frac{\pi}{5}+2a^2\tan\frac{\pi}{10}}\,R}\Delta(\mathcal{S},\mathcal{Q})$. The maximum on this lower bound is obtained for $a = 0.814853$, i.e., the lower bound on the number of dots in a pentagon whose radius R is $1.45992R + o(R)$, yielding a packing ratio of 0.946795.

For some constructions, we need DDCs, which might need $\mathcal{S}$-DDCs of other shapes. If the number of sides in the polygon is large we will use Theorem 10.7, where $\mathcal{Q}$ will be the regular polygon and $\mathcal{S}$ is a regular hexagon (for a small number of sides quasi-regular hexagons will be used). A computer program was developed to compute the packing ratios, some of which can be also obtained by mathematical methods. Table 10.1 presents the results.

TABLE 10.1 The number of dots in an n-gon DDC.

n	upper bound	lower bound	packing ratio
3	1.13975R	1.02462R	0.899
4	1.41421R	1.41421R	1
5	1.54196R	1.45992R	0.946795
6	1.61185R	≈1.61185R	≈1
7	1.65421R	1.58844R	0.960241
8	1.68179R	1.62625R	0.966977
9	1.70075R	1.63672R	0.96235
10	1.71433R	1.65141R	0.963297
60	1.77083R	1.70658R	0.963718
96	1.77182R	1.70752R	0.96371
circle	1.77245R	1.70813R	0.963708

10.4 Notes

There are more applications for two-dimensional arrays besides those presented and discussed in this chapter. For example, perfect maps were used in pattern recognition for structured light systems, as described in Geng [19], Lin, Nie, and Song [26], Morano, Ozturk, Conn, Dubin, Zietz, and Nissanov [33], Salvi, Fernandez, Pribanic, and Llado [39], and Salvi, Pagès, and Batlle [40], in transferring planar surface into a sensitive touch-screen display, see Dai and Chung [14]. They are also used in camera localization, as described by Szentandrasi, Zachariáš, Havel, Herout, Dubska, and Kajan [45], in one-shot shape acquisition, see Pagès, Salvi, Collewet, and Forest [35], in surface measurements, see Kiyasu, Hoshino, Yano, and Fujimura [23] and Spoelder, Vos, Petriu, and Groen [43], and in coded aperture imaging, see Gottesman and Fenimore [21]. Patterns with distinct differences were used, for example, in radar, sonar, physical alignment, and time–position synchronization, see Golomb and Taylor [20].

Section 10.1. It is quite obvious why there are many applications for an instrument that provides your location in the area. However, sometimes the inherent accumulating error in relative self-location methods, or some other reasons, make them infeasible or unfit for certain applications, where we would want the capability to obtain instant and accurate absolute self-location. Given several visible landmarks of known locations, a mobile robot could calculate its position through a triangulation, as in Cohen and Koss [12]. Alternatively, cleverly designed space fiducials (e.g., see Bruckstein, Holt, Huang, and Netravali [9]), whose appearance changes with the angle of observation can also serve for self-location.

Much like street signs for people, there are absolute self-location methods that provide sufficient local information to the device sensors, such that absolute positioning can be attained. Specifically, planar patterns have been suggested, where a small local sample from anywhere in the pattern provides sufficient information for decoding the absolute position. A naive example could consist of a floor filled with densely packed miniature markings, in which the exact coordinates are inscribed inside each marking. Of course, that would require a high sensor resolution and character-recognition capabilities. Indeed, there are much more efficient methods, which do with considerably less geometric detail in the pattern. Some commercial products have been utilizing this approach, e.g., a pen with a small imaging device in its tip, writing on paper with a special pattern printed on it, which allows full tracking of the pen's position at any time.

A classic method for absolute self-location in one dimension is the use of span n de Bruijn sequences. Sampling n consecutive letters somewhere in the sequence is sufficient for perfect positioning of the sampled subsequence within the sequence. Perfect maps and shortened perfect maps can serve as the basis for absolute self-location on the plane. We note also that M-sequences can be used for robust one-dimensional location by using their error-correction properties, as analyzed by Kumar and Wei [24].

Product constructions such as the one used in this section were used to construct various types of two-dimensional arrays for error correction of two-dimensional shapes. For example, similar and more sophisticated product constructions to generate arrays with low redundancy and effective two-dimensional error-correction capabilities, were suggested in various papers, e.g., see Breitbach, Bossert, Zyablov, and Sidorenko [3], and Etzion and Yaakobi [17].

As was discussed in Chapter 4, the classic approach of creating one span n de Bruijn sequence, requires $O(n)$ space and $O(n \cdot 2^n)$ time to generate the whole sequence S. If the running time is an issue, one could create and store in advance a look-up table that lists the locations of all subsequences. This yields $O(n)$ time complexity, but requires $O(n \cdot 2^n)$ space for the table. For larger n, a more flexible tradeoff between time and space complexity was suggested by Petriu [36]. A partial look-up table of evenly spaced locations called *milestones* is created in advance. During runtime, the algorithm that generates the sequence is initialized with the query subsequence and then iterated until one of the milestones is encountered. For example, this can yield $O(n \cdot 2^{\frac{n}{2}})$ time complexity and will require $O(n \cdot 2^{\frac{n}{2}})$ space for the table.

In either case, implementation of the self-location process using modern computer systems is feasible, at least for reasonable and practical values of n, depending on the application. Take $n = 16$ for a concrete example. It allows a definition of 2^{16} locations, e.g., a resolution of 0.1mm over a range of about 6.5 meters. In the first approach, it would take, in the worst case, about 65k simple iterations (on a 16-bit register), which can be performed reasonably quickly on current modest embedded processors currently clocked at about tens or hundreds of Megahertz. In the second approach, the look-up table would consume about 128k bytes (each entry being a two-byte word), which is, again, a quite modest requirement given today's memory capabilities.

There are some more efficient methods to generate de Bruijn sequences, e.g., see Mitchell, Etzion, and Paterson [30] that can be used in the case of an application in which k and n are much larger. The problem of decoding perfect maps was considered, for example, in Mitchell and Paterson [31]. A comprehensive survey on this topic was given by Burns and Mitchell [10].

Self-location was considered in many papers. One-dimensional self-location was analyzed in Kumar and Wei [24], and Wang, Hu, and Shayevitz [49]. Finding a robot location using de Bruijn sequences and discrete optical sensors was considered by Scheinerman [41]. The material in this section on two-dimensional self-location was taken from Bruckstein, Etzion, Giryes, Gordon, Holt, and Shuldiner [8]. This work described in the section was followed by more papers, e.g., see Berkowitz and Kopparty [1], Chee, Dao, Kiah, Ling, and Wei [11], Horan and Stevens [22], Mitchell and Wild [32], and Wei [50].

Section 10.2. The material in this section is taken from the papers by Blackburn, Etzion, Martin, and Paterson [4–6].

Wireless sensors are small, battery-powered devices with the ability to take measurements of quantities such as temperature or pressure and to engage in wireless communication. When a collection of sensors is deployed the sensors can communicate with each other and thus form an ad hoc network, known as a WSN, to facilitate the transmission and manipulation of data by the sensors. Such networks have a wide range of potential applications, including wildlife monitoring or pollution detection. Examples of how they have been used in practice are described in Römer and Mattern [38].

Much of the literature on key predistribution in wireless sensor networks deals with the case where the physical topology of the network is completely unknown before deployment, e.g., see Delgosha and Fekri [18], Lee and Stinson [25], and Liu, Ning, and Li [27]. There are several examples of location-based schemes, but in many cases, the networks consist of randomly distributed nodes whose approximate location is known. Martin and Paterson [29] indicated types of networks that have been considered in the WSN key-predistribution literature and suggested that there is considerable scope for the development of schemes suited to specific network topologies, in situations where the topology is known before sensor deployment. The application in this section provides a solution for these scenarios.

Section 10.3. The idea of folding a one-dimensional code into a two-dimensional array is well known in coding theory, e.g., see Etzion and Yaakobi [17]. Foldings to construct pseudo-random arrays were carried out in MacWilliams and Sloane [28]. The material of this section is due to the work by Etzion [15].

F1 was used by MacWilliams and Sloane [28] to form the pseudo-random arrays presented in Section 9.3. Another construction of pseudo-random arrays based on **F2** was presented by Spann [44]. **F2** was also used by Robinson [37] to fold a one-dimensional ruler into a two-dimensional Golomb rectangle. The generalization to higher dimensions is straightforward. **F3** was used in Blackburn, Etzion, Martin, and Paterson [5] to obtain some synchronization patterns in $\mathbb{Z}^D$.

There is a large variety of literature about tiling and lattices. We will refer the reader to two of the most interesting and comprehensive books written by Conway and Sloane [13], and by Stein and Szabó [42].

Lattice is a very fundamental structure in various coding problems, e.g., see Tarokh, Vardy, and Zeger [46], Urbanke and Rimoldi [47], and Viterbo and Boutros [48]. This is a small sample that is not meant to be representative. Lattices are also applied in multi-dimensional coding, e.g., see Blaum, Bruck, and Vardy [2], and Etzion and Vardy [16]. These papers exhibit an application of lattices for multi-dimensional coding and discrete geometry problems.

Let Λ be a D-dimensional lattice tiling for the shape $\mathcal{S}$. Let G be the following generator matrix of Λ:

$$G = \begin{bmatrix} v_{11} & v_{12} & \cdots & v_{1D} \\ v_{21} & v_{22} & \cdots & v_{2D} \\ \vdots & \vdots & \ddots & \vdots \\ v_{D1} & v_{D2} & \cdots & v_{DD} \end{bmatrix}.$$

Given the direction $\delta = (d_1, d_2, \ldots, d_D)$, w.l.o.g. we assume that the first ℓ values of δ are nonzeros and the last $D - \ell$ values are zeros. By Lemma 10.10 and Corollary 10.4, if $(\Lambda, \mathcal{S}, \delta)$ defines a folding, then there exist D integer coefficients $\alpha_1, \alpha_2, \ldots, \alpha_D$ such that

$$\sum_{j=1}^{D} \alpha_j(v_{j1}, v_{j2}, \ldots, v_{jD}) = (|\mathcal{S}|d_1, \ldots, |\mathcal{S}|d_\ell, 0, \ldots, 0)$$

and there is no integer i, $0 < i < |\mathcal{S}|$, and D integer coefficients $\beta_1, \beta_2, \ldots, \beta_D$ such that

$$\sum_{j=1}^{D} \beta_j(v_{j1}, v_{j2}, \ldots, v_{jD}) = (i \cdot d_1, \ldots, i \cdot d_\ell, 0, \ldots, 0).$$

Hence, we have the following D equations:

$$\sum_{j=1}^{D} \alpha_j v_{jr} = |\mathcal{S}| \cdot d_r, \quad 1 \le r \le \ell, \tag{10.4}$$

$$\sum_{j=1}^{D} \alpha_j v_{jr} = 0, \quad \ell + 1 \le r \le D. \tag{10.5}$$

Let $\tau = d_1$ if $\ell = 1$ and let $\tau = \text{g.c.d.}(d_1, d_2, \ldots, d_\ell)$ if $\ell > 1$. The D equations defined in Eqs. (10.4) and (10.5) are equivalent to the following D equations:

$$\sum_{j=1}^{D} \alpha_j v_{j1} = |\mathcal{S}| \cdot d_1,$$
$$\sum_{j=1}^{D} \alpha_j \frac{d_1 v_{jr} - d_r v_{j1}}{\tau} = 0, \quad 2 \le r \le \ell,$$
$$\sum_{j=1}^{D} \alpha_j v_{jr} = 0, \quad \ell + 1 \le r \le D.$$

The analysis leads to the following theorem proved by Etzion [15].

Theorem 10.12. *If Λ is a lattice tiling for the shape $\mathcal{S}$, then the triple $(\Lambda, \mathcal{S}, \delta)$ defines a folding if and only if* g.c.d.$(\frac{\alpha_1}{\tau}, \frac{\alpha_2}{\tau}, \ldots, \frac{\alpha_D}{\tau}) = 1$ *and* g.c.d.$(\tau, |\mathcal{S}|) = 1$ *(for this purpose* g.c.d.$(a,0) = 0$*).*

The hexagonal grid considered in Section 9.6 can be more practical for many applications. For example, to simulate pixels in computers, the pixels that are circles, are compressed better in the hexagonal grid. Moreover, it was proved by Etzion [15] that the packing ratio for regular hexagons is asymptotically 1 as the packing ratio for squares (see Table 10.1). Therefore DDCs in this model are very interesting. DDCs in this model and their constructions by folding were considered by Etzion [15].

Finally, for a survey on B_2-sequences and their generalizations, the reader is referred to the work of O'Bryant [34]. The construction of B_2-sequences in Theorem 10.8 is due to Bose [7].

References

[1] R. Berkowitz, S. Kopparty, Robust positioning patterns, in: Proc. 27th Annu. ACM-SIAM Disc. Algorithms, 2016, pp. 1937–1951.

[2] M. Blaum, J. Bruck, A. Vardy, Interleaving schemes for multidimensional cluster errors, IEEE Trans. Inf. Theory 44 (1998) 730–743, 1998.

[3] M. Breitbach, M. Bossert, V. Zyablov, V. Sidorenko, Array codes correcting a two-dimensional cluster of errors, IEEE Trans. Inf. Theory 44 (1998) 2025–2031.

[4] S.R. Blackburn, T. Etzion, K.M. Martin, M.B. Paterson, Efficient key predistribution for grid-based wireless sensor networks, Lect. Notes Comput. Sci. 5155 (2008) 54–69.

[5] S.R. Blackburn, T. Etzion, K.M. Martin, M.B. Paterson, Two-dimensional patterns with distinct differences – constructions, bounds, and maximal anticodes, IEEE Trans. Inf. Theory 56 (2010) 1216–1229.

[6] S.R. Blackburn, T. Etzion, K.M. Martin, M.B. Paterson, Distinct difference configurations: multihop paths and key predistribution in sensor networks, IEEE Trans. Inf. Theory 56 (2010) 3961–3972.

[7] R.C. Bose, An affine analogue of Singer's theorem, J. Indian Math. Soc. 6 (1942) 1–15.

[8] A.M. Bruckstein, T. Etzion, R. Giryes, N. Gordon, R.J. Holt, D. Shuldiner, Simple and robust binary self-location patterns, IEEE Trans. Inf. Theory 58 (2012) 4884–4889.

[9] A.M. Bruckstein, R.J. Holt, T.S. Huang, A.N. Netravali, New devices for 3d pose estimation: Mantis eyes, Agam paintings, sundials, and other space fiducials, Int. J. Comput. Vis. 39 (2000) 131–140.

[10] J. Burns, C.J. Mitchell, Coding schemes for two-dimensional position sensing, in: M.J. Ganley (Ed.), Proc. 3rd IMA Cryptography and Coding Conf., Cirencester, U.K., Dec. 1991, 1991, pp. 31–66.

[11] Y.M. Chee, D.T. Dao, H.M. Kiah, S. Ling, H. Wei, Robust positioning with low redundancy, SIAM J. Comput. 49 (2020) 284–317.

[12] C. Cohen, F.V. Koss, A comprehensive study of three object triangulation, in: Proc. of the SPIE Conf. on Mobile Robots VII, 1992, pp. 95–106.

[13] J.H. Conway, N.J.A. Sloane, Sphere Packings, Lattices, and Groups, Springer-Verlag, New York, 1999.

[14] J. Dai, C.-K.R. Chung, Touchscreen everywhere: on transferring a normal planar surface to touch-sensitive display, IEEE Trans. Cybern. 44 (2014) 1383–1396.

[15] T. Etzion, Sequence folding, lattice tiling, and multidimensional coding, IEEE Trans. Inf. Theory 57 (2011) 4383–4400.

[16] T. Etzion, A. Vardy, Two-dimensional interleaving schemes with repetitions: constructions and bounds, IEEE Trans. Inf. Theory 48 (2002) 428–457.

[17] T. Etzion, E. Yaakobi, Error-correction of multidimensional bursts, IEEE Trans. Inf. Theory 55 (2009) 961–976.

364 Sequences and the de Bruijn Graph

[18] F. Delgosha, F. Fekri, Threshold Key-establishment in distributed sensor networks using a multivariate scheme, in: Proc. 25th Infocom, 2006, pp. 1–12.

[19] J. Geng, Structured-light 3D surface imaging: a tutorial, Adv. Opt. Photonics 3 (2011) 128–160.

[20] S.W. Golomb, H. Taylor, Two-dimensional synchronization patterns for minimum ambiguity, IEEE Trans. Inf. Theory 28 (1982) 600–604.

[21] S.R. Gottesman, E.E. Fenimore, New family of binary arrays for aperture imaging, Appl. Opt. 28 (1989) 4344–4352.

[22] V. Horan, B. Stevens, Locating patterns in the de Bruijn torus, Discrete Math. 339 (2016) 1274–1282.

[23] S. Kiyasu, H. Hoshino, K. Yano, S. Fujimura, Measurement of the 3-D shape of specular polyhedrons using an M-array coded light source, IEEE Trans. Instrum. Meas. 44 (1995) 775–778.

[24] P.V. Kumar, V.K. Wei, Minimum distance of logarithmic and fractional partial m-sequences, IEEE Trans. Inf. Theory 38 (1992) 1474–1482.

[25] J. Lee, D.R. Stinson, On the construction of practical Key predistribution schemes for distributed sensor networks using combinatorial designs, ACM Trans. Inf. Syst. Secur. 11 (2008) 1–35.

[26] H. Lin, L. Nie, Z. Song, A single-shot structured light means by encoding both color and geometrical features, Pattern Recognit. 54 (2016) 178–189.

[27] D. Liu, P. Ning, R. Li, Establishing pairwise keys in distributed sensor networks, ACM Trans. Inf. Syst. Secur. 8 (2005) 41–77.

[28] F.J. MacWilliams, N.J.A. Sloane, Pseudo-random sequences and arrays, Proc. IEEE 64 (1976) 1715–1729.

[29] K.M. Martin, M. Paterson, An application-oriented framework for wireless sensor network key establishment, Electron. Notes Theor. Comput. Sci. 192 (2008) 31–41.

[30] C.J. Mitchell, T. Etzion, K.G. Paterson, A method for constructing decodable de Bruijn sequences, IEEE Trans. Inf. Theory 42 (1996) 1472–1478.

[31] C.J. Mitchell, K.G. Paterson, Decoding perfect maps, Des. Codes Cryptogr. 4 (1994) 11–30.

[32] C.J. Mitchell, P.R. Wild, Constructing orientable sequences, IEEE Trans. Inf. Theory 68 (2022) 4782–4789.

[33] R.A. Morano, C. Ozturk, R. Conn, S. Dubin, S. Zietz, J. Nissanov, Structured light using pseudorandom codes, IEEE Trans. Pattern Anal. Mach. Intell. 20 (1998) 322–327.

[34] K. O'Bryant, A complete annotated bibliography of work related to Sidon sequences, Electron. J. Comb. DS11 (2004) 1–39.

[35] J. Pagès, J. Salvi, C. Collewet, J. Forest, Optimised de Bruijn patterns for one-shot shape acquisition, Image Vis. Comput. 23 (2005) 707–720.

[36] E.M. Petriu, New pseudorandom/natural code conversion methods, Electron. Lett. 24 (1988) 1358–1359.

[37] J.P. Robinson, Golomb rectangles as folded rulers, IEEE Trans. Inf. Theory 43 (1997) 290–293.

[38] K. Römer, F. Mattern, The design space of wireless sensor networks, Wirel. Commun. 11 (2004) 54–61.

[39] J. Salvi, S. Fernandez, T. Pribanic, X. Llado, A state of art in structured light patterns for surface profilometry, Pattern Recognit. 43 (2010) 2666–2680.

[40] J. Salvi, J. Pagès, J. Batlle, Pattern codification strategies in structured light systems, Pattern Recognit. 37 (2004) 827–849.

[41] E.R. Scheinerman, Determining planar location via complement-free de Bruijn sequences using discrete optical sensors, IEEE Trans. Robot. Autom. 17 (2001) 883–889.

[42] S.K. Stein, S. Szabó, Algebra and Tiling, The Mathematical Association of America, Washington, DC, 1994.

[43] H.J.W. Spoelder, F.M. Vos, E.M. Petriu, F.C.A. Groen, Some aspects of pseudo random binary array-based surface characterization, IEEE Trans. Instrum. Meas. 49 (2000) 1331–1336.

[44] R. Spann, A two-dimensional correlation property of pseudo-random maximal-length sequences, Proc. IEEE 53 (1963) 2137.

[45] I. Szentandrási, M. Zachariáš, J. Havel, A. Herout, M. Dubská, R. Kajan, Uniform marker fields: camera localization by orientable de Bruijn tori, in: Proc. IEEE Int. Symp. Mixed Augmented Reality, 2012, pp. 319–320.

[46] V. Tarokh, A. Vardy, K. Zeger, Universal bounds on the performance of lattice codes, IEEE Trans. Inf. Theory 45 (1999) 670–681.

[47] R. Urbanke, B. Rimoldi, Lattice codes can achieve capacity on AWGN channel, IEEE Trans. Inf. Theory 44 (1998) 273–278.

[48] E. Viterbo, J. Boutros, A universal lattice decoder for fading channels, IEEE Trans. Inf. Theory 45 (1999) 1639–1642.

[49] L. Wang, S. Hu, O. Shayevitz, Quickest sequence phase detection, IEEE Trans. Inf. Theory 63 (2017) 5834–5849.

[50] H. Wei, Nearly optimal robust positioning patterns, IEEE Trans. Inf. Theory 68 (2022) 193–203.

Chapter 11

Unique path property graphs

Properties, constructions, cycles, and factors

A graph with the *unique path property* is a digraph in which there is a unique directed path of a given length n from each vertex u to each vertex v. By Lemma 1.15 such unique paths exist in $G_{\sigma,n}$ for each $\sigma \geq 2$ and $n \geq 1$. In this chapter, we examine the existence of other graphs with this property, check their other properties, construct a large set of these graphs, and examine whether their factors have some resemblance to state diagrams of FSR_ns. In the next chapter, these graphs will be considered as interconnection networks for parallel computation.

In Section 11.1 we will show the representation of such a graph by its adjacency matrix and use this representation to prove that for some integer $\sigma > 1$, such a graph has σ^n vertices, σ self-loops, and the same in-degree and out-degree σ for all the vertices. This implies that problems on such graphs are equivalent to related problems on the associated matrices. We then consider only graphs for which $\sigma = 2$. We present an efficient algorithm that decides whether two such graphs are isomorphic and we consider properties that are common to the structure of all such graphs. We define the concept of an alternating cycle that plays an important role in these graphs.

In Section 11.2 we consider constructions of large sets of graphs with the unique path property. The constructions are based on removing some edges from G_n and adding other edges that are not contained in G_n. We enumerate the number of non-isomorphic graphs obtained by some of the proposed constructions.

In Section 11.3 we show that surprisingly the exact number of factors in a graph with the unique path property depends on the number of alternating cycles in the graph that for itself can be easily computed. Moreover, if all the alternating cycles in the graph are of length four, then the factors are similar to those of an FSR_n, associated functions can be defined, and similar properties to those of the feedback function of an FSR_n are demonstrated.

11.1 Basic properties of UPP graphs

A directed graph $G = (V, E)$ is called a *graph with the unique path property* (a **UPP** graph, in short) of order n, if for every two vertices (not necessarily distinct) $u, v \in V$ there exists a unique directed path of length n from u to v.

Sequences and the de Bruijn Graph. https://doi.org/10.1016/B978-0-44-313517-0.00017-2

Recall that if $G = (V, E)$ is a directed graph with an adjacency matrix A, then by Lemma 1.14 we have that $A^\ell(u, v)$ is the number of distinct directed paths of length ℓ from u to v. This implies that A is the adjacency matrix of a UPP graph of order n if and only if $A^n = J$, where J is the all-ones matrix. We want to prove now that the definition of a UPP graph of order n implies the number of vertices in the graph, the in-degree and the out-degree of each vertex, and also the number of self-loops in the graph.

Lemma 11.1. *If A is a $t \times t$ binary matrix for which $A^n = J$, then all the rows and all the columns of A have the same number of* ones.

Proof. Assume that the row with the most number of *ones* in A has σ *ones* in columns $p_1, p_2, \dots, p_\sigma$. Consider the product $A \cdot A^{n-1} = J$ and such a row k with σ *ones* in columns $p_1, p_2, \dots, p_\sigma$. Since $A \cdot A^{n-1} = J$ has exactly one *one* in each entry of the kth row, it follows that each column of A^{n-1} has exactly one *one* in exactly one of the rows $p_1, p_2, \dots, p_\sigma$, i.e., these rows in A^{n-1} have *ones* in distinct columns and a total of t *ones* in all these rows since A^{n-1} has exactly t columns.

Consider the product $A^{n-1} \cdot A = J$. Rows $p_1, p_2, \dots, p_\sigma$ in A^{n-1} are responsible for the *ones* in σ rows of J that in total have σt *ones*. Since rows $p_1, p_2, \dots, p_\sigma$ in A^{n-1} have *ones* in distinct columns and each column in these rows of A^{n-1} has exactly one *one*, it follows from $A^{n-1} \cdot A = J$ that A has exactly σt *ones* (each entry in A is multiplied exactly once by a *one* from one of these σ rows of A^{n-1}). Since no row has more than σ *ones*, it follows that each row of A has exactly σ *ones*.

By symmetric arguments, we have that each column of A has exactly σ *ones*. $\square$

Corollary 11.1. *If $G = (V, E)$ is a* UPP *graph of order n, then the in-degree and the out-degree of each vertex is the same integer σ.*

Lemma 11.2. *If $G = (V, E)$ is a* UPP *graph of order n with t vertices, where each vertex has in-degree σ and out-degree σ, then $t = \sigma^n$.*

Proof. Let $G = (V, E)$ be a UPP graph of order n with t vertices, where each vertex has in-degree σ and out-degree σ. Let v be a vertex in V. Since each vertex has the same out-degree σ, it follows that there are σ^n distinct directed paths of length n starting at v. Since there exists exactly one directed path of length n from v to each vertex of V, it follows that each such path ends in a distinct vertex. Hence, the total number of vertices in V is σ^n. $\square$

The following lemma is required for the proof that in a UPP graph with σ^n vertices, there are σ self-loops.

Lemma 11.3. *An edge e is a self-loop edge in a directed graph $G = (V, E)$ if and only if the vertex e is a self-loop vertex in the line graph $L(G) = (V' = E, E')$.*

Proof. If $e \in E$ is a self-loop edge in the directed graph $G = (V, E)$, then by the definition of the line graph the vertex $e \in V'$ is a self-loop vertex in the line graph $L(G)$.

Assume now that $e \in V' = E$ is a self-loop vertex in the line graph $L(G)$. Clearly, $e = (u_1, v)$ is an edge in G. Since e is a self-loop vertex in $L(G)$, it follows by definition that there exists in G a path of length two with the two consecutive edges $e = (u_1, v)$ and $e = (v, u_1)$. However, since e is a self-loop vertex in $L(G)$, it follows that $u_1 = v$ and hence $e = (v, v)$ is a self-loop edge in G. □

Finally, before the next lemma we generalize Theorem 1.16 for de Bruijn graphs to UPP graphs.

Theorem 11.1. *If $G = (V, E)$ is a* UPP *graph of order n, then $L(G)$ is a* UPP *graph of order $n + 1$.*

Proof. Let $G = (V, E)$ be a UPP graph of order n and $L(G) = (V' = E, E')$ its line graph. To prove the claim in the theorem it is sufficient to prove that for each two vertices $e_1, e_2 \in V'$ there exists a unique path of length $n + 1$ from e_1 to e_2 in $L(G)$. Let $e_1, e_2 \in V' = E$ be two vertices of $L(G)$, i.e., $e_1 = (u_1, u_2)$ and $e_2 = (u_3, u_4)$ are two edges in G, where $u_1, u_2, u_3, u_4 \in V$, are vertices of G. Since G is a UPP graph of order n, it follows that there exists a directed path of length n from u_2 to u_3. This path, represented by its $n + 1$ vertices, is $u_2, v_1, v_2, \ldots, v_{n-1}, u_3$ or, represented by its n edges, is $\varepsilon_1, \varepsilon_2, \ldots, \varepsilon_n$. These n edges of E are vertices in $L(G)$ and this path of length n with n edges in G is a path of length $n - 1$ with n vertices in $L(G)$. Therefore $e_1, \varepsilon_1, \varepsilon_2, \ldots, \varepsilon_n, e_2$ is a path of length $n + 1$ in $L(G)$ represented by its $n + 2$ vertices, i.e., $\varepsilon_j \in V' = E$, $1 \leq j \leq n$.

To complete the proof we have to show that this path is unique, i.e., there is no other path of length $n + 1$ from e_1 to e_2 in $L(G)$. Assume, on the contrary, that there exists another such path $e_1, e_3, e_4, \ldots, e_{n+1}, e_{n+2}, e_2$, where $e_j \in V' = E$, $1 \leq j \leq n + 2$, are $n + 2$ vertices in $L(G)$. This implies that $\varepsilon_1, \varepsilon_2, \ldots, \varepsilon_n$ and $e_3, e_4, \ldots, e_{n+1}, e_{n+2}$ are two distinct paths of length n from u_2 to u_3 (each path is represented by its edges) in G, contradicting the unique property of G. Therefore the path from e_1 to e_2 in $L(G)$ is unique.

Thus $L(G)$ is a UPP graph of order $n + 1$. □

It should be noted that given a graph with σ^n vertices, where each vertex has in-degree σ and out-degree σ if we prove that for each pair of vertices u and v, there exists a path of length n from u to v, then there is no other path of length n from u to v. This property will be used later without mentioning it.

Lemma 11.4. *If $G = (V, E)$ is a* UPP *graph of order n with σ^n vertices, then G has exactly σ self-loops.*

Proof. Assume first that n is a prime and $\sigma < n$. Consider the paths of length n from each vertex to itself. Each such path is a cycle. If a vertex u is on this cycle, then this cycle forms the path of length n from u to itself. We claim that such a cycle cannot use two distinct out-edges (or in-edges) for some vertex v. Assume, on the contrary, that the path of length n from v to v is

$$v, u_1, u_2, \ldots, u_\ell, v, \omega_1, \omega_2, \ldots, \omega_k, v,$$

where $u_1 \neq \omega_1$. This contradicts the unique path property since we also have that

$$v, \omega_1, \omega_2, \ldots, \omega_k, v, u_1, u_2, \ldots, u_\ell, v$$

is such a different path from v to v (it is the same cycle, but a different closed path). Hence, the length of such a cycle (from a vertex to itself) is a divisor of n (which can be only 1 or n since n is a prime). Since by Corollary 1.5, n divides $\sigma^n - \sigma$, $\sigma < n$, n is a prime, and n must divide the number of vertices that are not self-loops, it follows that the number of self-loops is $\sigma + i \cdot n$ for some nonnegative integer i (taking into account that all the other paths from a vertex to itself are simple cycles of length n).

By Lemma 11.3, an edge e is a self-loop in a directed graph G if an only if the vertex e is a self-loop vertex in the line graph $L(G)$. By Theorem 11.1, G is a UPP graph of order n if and only if $L(G)$ is a UPP graph of order $n + 1$. By Lemma 11.3, we have that the number of self-loops in G equals the number of self-loops in $L(G)$. Let p be a prime, such that $p > \sigma + i \cdot n$ whose existence is guaranteed by Theorem 1.3. Apply the line graph iteratively on G, $L(G)$, $L(L(G))$, and so on, until we have a UPP graph $L'(G)$ of order p. By the same arguments as at the start of the proof we have that the number of self-loops in $L'(G)$ is $\sigma + j \cdot p$. On the other hand, since the number of self-loops in each graph of this sequence of line graphs is the same as in G, it follows that $\sigma + i \cdot n = \sigma + j \cdot p$, which is possible only when $i = j = 0$, i.e., the number of self-loops is σ.

Assume now that n is any integer and the UPP graph G has δ self-loops. We construct again a sequence of line graphs until we obtain a UPP graph $L'(G)$ of order p, p prime, $p > \text{maximum}\{\delta, \sigma\}$ and reach the same contradiction as in the previous part of the proof. □

Corollary 11.2. *If $G = (V, E)$ is a UPP graph of order n with t vertices, then*

1. *The number of vertices in G is $t = \sigma^n$ for some integers $\sigma \geq 2$ and $n \geq 1$.*
2. *The in-degree and the out-degree of each vertex is σ.*
3. *There exist exactly σ self-loop vertices in G.*

Corollary 11.3. *If A is a binary $t \times t$ matrix such that $A^n = J$ (adjacency matrix of a UPP graph of order n) for some positive integer n, then*

1. *The number of rows in A is $t = \sigma^n$ for some integers $\sigma \geq 2$ and $n \geq 1$.*
2. *Each row and each column of A has exactly σ ones.*

3. *There exist exactly σ ones on the main diagonal of A.*

Throughout the rest of this chapter (in Sections 11.1, 11.2, and 11.3) we consider only UPP graphs with $\sigma = 2$. We will consider some generalizations for $\sigma > 2$ in Section 11.4, where such a graph is called a σ-UPP graph, and for $\sigma = 2$ the graph will be called only a UPP graph.

One of the main goals of this chapter is to construct non-isomorphic UPP graphs. Verifying that two UPP graphs are non-isomorphic is important for this enumeration. For this purpose, we will design an algorithm that for given two UPP graphs $G(1)$ and $G(2)$ determines whether $G(1)$ and $G(2)$ are isomorphic or not. In this algorithm, there will be an important role for a *breadth first search* (BFS) algorithm and BFS trees from the self-loops of the UPP graphs.

For a directed connected graph $G = (V, E)$, a BFS search starts at step 0 with a vertex v that is labeled *marked* and all the other vertices in the graph are labeled *unmarked*. At step i, $i > 0$ consider all the vertices that were found and marked at step $i - 1$. If u_1 is such a vertex, then we consider each edge $u_1 \rightarrow u_2$ for which u_2 is unmarked and replace the status of u_2 to marked. The algorithm terminates when all the vertices are marked. The algorithm also generates a BFS directed tree, whose root is v at layer 0. In layer i, $i > 0$ the tree contains all the vertices that were marked at step i. Between layer $i - 1$ and layer i we have all the edges that were used to mark the vertices in layer i.

For a UPP graph, we consider two BFS trees whose roots are the self-loops of the graph. If we remove the self-loop and its out-edge (which is not the self-loop) from the tree, then the graph that remains is a binary balanced tree. An example of the BFS trees of G_4 from the two self-loops is depicted in Fig. 11.1.

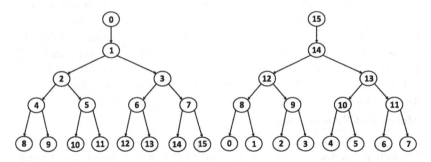

FIGURE 11.1 The BFS trees from the two self-loops of G_4.

Since there are exactly two self-loops in each UPP graph, it follows that there are two possible bijections for the self-loop vertices of $G(1)$ and $G(2)$. The algorithm decides if one of these two possible bijections can achieve isomorphism between the graphs. The algorithm uses the following property of UPP graphs.

Lemma 11.5. *Let v be a vertex in a* UPP *graph that has out-edges to two vertices u_1 and u_2. If r is a self-loop vertex then*

$$d(u_1, r) = n \quad \text{if and only if} \quad d(u_2, r) < n.$$

Proof. If $d(u_1, r) = n$ and $d(u_2, r) = n$, then there is no path of length n from v to r. If $d(u_1, r) < n$ and $d(u_2, r) < n$, then there are at least two paths of length n from v to r, by adding to the beginning of the paths, from u_1 to r and from u_2 to r, whose length is smaller than n, the edges $v \to u_1$ and $v \to u_2$, respectively, and possibly using the self-loop edge of r a few times at the end of the path. Hence, in both cases, we have a contradiction to the unique path property. Thus $d(u_1, r) = n$ if and only if $d(u_2, r) < n$. $\qquad \square$

Similarly to Lemma 11.5, we can prove the following lemma.

Lemma 11.6. *Let v be a vertex in a* UPP *graph that has in-edges from two vertices u_1 and u_2. If r is a self-loop vertex then*

$$d(r, u_1) = n \quad \text{if and only if} \quad d(r, u_2) < n.$$

As noted above, given two UPP graphs, there are two possible bijections for the self-loop vertices that can achieve isomorphism between the graphs. Algorithm UPP isomorphism, given below, accepts two self-loop vertices r_1 and r_2 of two UPP graphs $G(1)$ and $G(2)$, respectively, and tries to achieve isomorphism with the initial bijection $h(r_1) = r_2$. Then, it defines the bijection for all the other vertices of the graph by using Lemma 11.5 (or Lemma 11.6). Namely, if h is defined for a vertex $v \in G(1)$, which has outgoing edges to u_1 and u_2, then by Lemma 11.5, u_1 and u_2 can be uniquely mapped to the vertices in $G(2)$ that have incoming edges from the vertex $h(v) \in G(2)$. In algorithm UPP isomorphism, step (**I2**) assigns a necessary bijection h for all the vertices of the graph that was imposed by the initial bijection for the self-loop vertices. In step (**I3**) we check that h is a legal bijection by scanning the edge lists of $G(1)$ and $G(2)$. If the bijection is illegal, then the algorithm tries the other possible initial bijection for the self-loops. If, for the other possible bijection the algorithm finds that the bijection is illegal, then $G(1)$ and $G(2)$ are not isomorphic.

Algorithm UPP isomorphism:

Let r_1 and r_2 denote two self-loop vertices of $G(1)$ and $G(2)$, respectively. The initial bijection is $h(r_1) = r_2$; for each other vertex $v \in G(1)$, $h(v)$ is undefined. Initially, h is defined to be legal.

(**I1**) Compute the distance $d(v, r_1)$ and $d(u, r_2)$ for each vertex v and u in $G(1)$ and $G(2)$, respectively, by applying the BFS procedure on the graphs of $G(1)$ and $G(2)$ (for this purpose we have to reverse the direction of the edges in $G(1)$ and $G(2)$ since we need the distances to r_1 and r_2 and not the distances from r_1 and r_2).

(I2) Scan the vertices of $G(1)$ and $G(2)$ from r_1 and $r_2 = h(r_1)$ using the BFS procedure. Look at the stage when the BFS scans edges $v \to v_1$ and $v \to v_2$ in $G(1)$, where $d(v_1, r_1) < n$. Let $h(v) \to u_1$ and $h(v) \to u_2$ be the associated edges in $G(2)$, where $d(u_1, r_2) < n$.
If $d(v_1, r_1) = d(u_1, r_2) < n$, then set $h(v_1) := u_1$ and $h(v_2) := u_2$; else h is illegal.

(I3) Scan the edges list (only the edges from the vertices of the last layer of the two BFS trees have to be scanned as the other edges were assigned in the previous step) of $G(1)$ and $G(2)$ and check for every edge (u, v) in $G(1)$, whether $(h(u), h(v))$ is an edge in $G(2)$, until h is found to be illegal, i.e., $(h(u), h(v))$ is not an edge in $G(2)$; or all the edges were checked and h was not found to be illegal. If h is legal, then $G(1)$ and $G(2)$ are isomorphic; otherwise, try the other possible bijection for the self-loops with the same algorithm. ∎

Remark. Note that in **(I3)** it is not required to check all the distances (from r_1 and r_2) to verify if the bijection h is illegal or not, but for simplicity (of the claims and their proofs) we check all the distances.

Theorem 11.2. *Let $G(1)$ and $G(2)$ be two UPP graphs. By applying algorithm UPP isomorphism with the two possible bijections on the self-loop vertices, one decides in $O(|E|)$ time if $G(1)$ and $G(2)$ are isomorphic or not.*

Proof. The correctness of the algorithm follows immediately from the discussion before the description of the algorithm and from Lemma 11.5. The complexity of each BFS procedure in **(I1)** and **(I2)** is $O(|E|)$, and in **(I3)** each graph's edge is checked at most once. Thus the total complexity is $O(|E|)$. □

There are a few interesting properties associated with the set of vertices that are reachable with paths of length smaller than n from a given vertex v of a UPP graph $G = (V, E)$ of order n. The set of vertices that are **reachable** from a given vertex $v \in V$ with a path of length k will be denoted by $R_G^k(v)$ (or usually $R^k(v)$ if there is no ambiguity in G), i.e.,

$$R^k(v) \triangleq \{u \ : \ \text{there is a directed path of length } k \text{ in } G \text{ from } v \text{ to } u\}.$$

Lemma 11.7. *For each vertex v in a UPP graph $G = (V, E)$ of order n, and for each k, $0 \le k \le n$, we have that $\left|R^k(v)\right| = 2^k$.*

Proof. Since the out-degree of each vertex in G is two and $R^0(v) = \{v\}$, i.e., $\left|R^0(v)\right| = 1$, it follows that $\left|R^k(v)\right| \le 2^k$. It is easily verified that if there exist two distinct paths of length k from v to some vertex $u \in V$, then there exist two distinct paths of length n from v to each vertex of V reachable from u with a path of length $n - k$, violating the unique path property. Thus $\left|R^k(v)\right| = 2^k$. □

Lemma 11.8. *If v is a vertex in a UPP graph $G = (V, E)$ of order n, $v \to v_1$ and $v \to v_2$ are two distinct edges in G, then $R^{n-1}(v_1) \cup R^{n-1}(v_2) = V$ and $R^{n-1}(v_1) \cap R^{n-1}(v_2) = \varnothing$.*

Proof. By Lemma 11.7, $\left|R^{n-1}(v_1)\right| = \left|R^{n-1}(v_2)\right| = 2^{n-1}$ and $|R^n(v)| = 2^n$. Since we also have that $|V| = 2^n$, which implies that $R^n(v) = V$, it follows that $R^{n-1}(v_1) \cup R^{n-1}(v_2) = V$ and $R^{n-1}(v_1) \cap R^{n-1}(v_2) = \varnothing$. □

All the paths of length at most k from v to the vertices of $R^k(v)$, $0 \le k \le n$, are described by a balanced binary directed tree, $\mathcal{T}_G^k(v)$, or $\mathcal{T}^k(v)$ if G is understood from the context, called the **reachable tree** of v whose **depth (height)** is k. The vertices in $\mathcal{T}^k(v)$ are vertices from V and the edges in $\mathcal{T}^k(v)$ are edges in E. The tree $\mathcal{T}^k(v)$ has $k + 1$ layers (of distinct vertices in each layer), where v is the root of the tree in layer 0. In layer i, $i \le k$, of $\mathcal{T}^k(v)$ we have all the vertices that are reached by a path of length i from v. As argued in the proof of Lemma 11.7 all the vertices in layer i are distinct. For each vertex u, in layer i, $0 \le i \le k - 1$, of $\mathcal{T}^k(v)$, which has the two out-edges (u, u_1) and (u, u_2) in G, the same two edges also appear between layer i and layer $i + 1$ in $\mathcal{T}^k(v)$. Note that some vertices and some edges of G can appear more than once in $\mathcal{T}^k(v)$, but in different layers. Let $V(\mathcal{T}^k(v))$ denote the set of vertices in $\mathcal{T}^k(v)$, which are distinct vertices in V, i.e.,

$$V(\mathcal{T}^k(v)) = \bigcup_{i=0}^{k} R^i(v).$$

The reachable trees $\mathcal{T}^3(0)$ and $\mathcal{T}^3(1)$ of G_4 are depicted in Fig. 11.2.

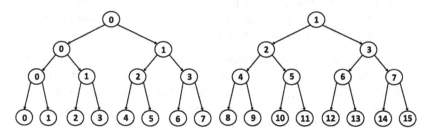

FIGURE 11.2 The reachable trees $\mathcal{T}^3(0)$ (left) and $\mathcal{T}^3(1)$ (right) of G_4.

An **alternating cycle** $\mathcal{C} = [e_1, e_2, \ldots, e_\ell]$, where $e_1, e_2, \ldots, e_\ell \in E$, in a digraph $G = (V, E)$ is an undirected edge-disjoint cycle (of even length) in the underline graph of G, such that every two consecutive edges in the cycle are in opposite directions in $\mathcal{C}$. In other words, either $e_1 = v_1 \to v_2$, $e_2 = v_3 \to v_2$, $e_3 = v_3 \to v_4$, and so on, or $e_1 = v_1 \to v_2$, $e_2 = v_1 \to v_3$, $e_3 = v_4 \to v_3$, and so on. W.l.o.g. we assume that the first two edges e_1, e_2 in the representation $\mathcal{C} = [e_1, e_2, \ldots, e_\ell]$, share their end-vertex, i.e., $e_1 = v_1 \to v_2$, $e_2 = v_3 \to v_2$, $e_3 = v_3 \to v_4$, and so on. Such an alternating cycle can be represented also by its set of consecutive vertices, i.e., $\mathcal{C} = [v_1 v_2 v_3 v_4 v_5 \cdots v_\ell]$, ℓ even, where $v_1 \to v_2$, $v_3 \to v_2$, $v_3 \to v_4$, $v_5 \to v_4$, and so on until $v_1 \to v_\ell$. This cycle $\mathcal{C}$ can be also represented by a bipartite digraph whose two sides have the same number of vertices. In such a bipartite digraph $G' = (V', E')$ there are two sides A and B, where each side has half of the vertices of C (note that in a UPP graph a vertex of V that is contained twice in C is contained in both A and B) and the edges of G' are exactly all the edges of the cycle C. Each edge $u \to v$ in C is also an edge in G', where $u \in A$, $v \in B$, and $(u, v) \in E'$. A digraph, with in-degree two and out-degree two for each vertex, in which each edge is on a unique alternating cycle of length four has the ***buddy property*** (also called the ***Heuchenne condition***). The family of UPP graphs that have the buddy property has many interesting and unique properties, some of which will be proven and used later in this chapter and also in the next chapter. Now, we present a simple result on the number of edges in an alternating cycle. By its definition, the number of edges in an alternating cycle is even. For any alternating cycle $C = [e_1, e_2, \ldots, e_\ell]$, where $e_i \in E$, ℓ is even, we define two sets of edges

$$E_o(C) \triangleq \{e_i \ : \ 1 \le i \le \ell, \ i \text{ is an odd integer}\}$$

and

$$E_e(C) \triangleq \{e_i \ : \ 1 \le i \le \ell, \ i \text{ is an even integer}\}.$$

Note that the alternating cycle C can be also written as $C = [e_\ell, \ldots, e_2, e_1]$. Hence, we can exchange between these two sets of edges (each one can be chosen as $E_o(C)$, which implies that the second set will be $E_e(C)$).

Lemma 11.9. *The number of edges in an alternating cycle* $C = [e_1, e_2, \ldots, e_\ell]$, *of a UPP graph is a multiple of 4, i.e., ℓ is divisible by 4.*

Proof. Let $C = [e_1, \ldots, e_\ell]$ be an alternating cycle in UPP graph $G = (V, E)$, where $e_1 = (v_1, v_2)$, $e_2 = (v_3, v_2)$, $e_3 = (v_3, v_4)$, $e_4 = (v_5, v_4)$, and so on. We distinguish between two cases depending on whether $v_5 = v_1$ or $v_5 \ne v_1$.
Case 1. If $v_5 = v_1$, then clearly the number of edges in this alternating cycle is four, i.e., $\ell = 4$.
Case 2. If $v_5 \ne v_1$, then by Lemma 11.8, we have that $R^{n-1}(v_2) \cup R^{n-1}(v_4) = V$ and $R^{n-1}(v_2) \cap R^{n-1}(v_4) = \varnothing$. Therefore $R^{n-1}(v_2) = V \setminus R^{n-1}(v_4)$. Consider now the next edge $e_5 = (v_5, v_6)$ of C. By Lemma 11.8, we have that

$$R^{n-1}(v_4) \cup R^{n-1}(v_6) = V \ \text{ and } \ R^{n-1}(v_4) \cap R^{n-1}(v_6) = \varnothing.$$

Therefore we have that $R^{n-1}(v_6) = V \setminus R^{n-1}(v_4)$. This immediately implies that $R^{n-1}(v_2) = R^{n-1}(v_6)$ and similarly we obtain that $R^{n-1}(v_2) = R^{n-1}(v_s)$, where $s \equiv 2 \pmod{4}$. Similarly, we have that $R^{n-1}(v_4) = R^{n-1}(v_s)$, where $s \equiv 0 \pmod{4}$. Moreover, similarly we have that $R^{n-1}(v_{\ell-2}) = R^{n-1}(v_2)$ and since $R^{n-1}(v_2) \cap R^{n-1}(v_4) = \varnothing$, it follows that ℓ is divisible by 4. $\square$

The next lemma provides an interesting property on the edge-disjoint alternating cycles in some digraphs.

Lemma 11.10. *If $G = (V, E)$ is a digraph (with no parallel edges), where each vertex has in-degree two and out-degree two, then E can be partitioned into edge-disjoint alternating cycles that contain all the edges in the graph.*

Proof. Given an edge $e_1 = (v_1, v_2)$ there is a unique edge $e_2 = (v_3, v_2)$ that can follow it in an alternating cycle since the in-degree of v_2 is two. The next edge that can follow e_2 is $e_3 = (v_3, v_4)$ since the out-degree of v_3 is two. This process of generating the alternating cycle continues in the same manner and can end only with an edge $e_\ell = (v_1, v_\ell)$. This process implies that each edge is contained in an alternating cycle. Since each edge is forced by the previous edge (and the following edge), it follows that each edge is contained in exactly one alternating cycle. Thus E can be partitioned into edge-disjoint alternating cycles that contain all the edges in the graph. $\square$

11.2 Constructions for UPP graphs

How many non-isomorphic UPP graphs of order n exist? There are three such graphs of order 3 depicted in Fig. 11.3.

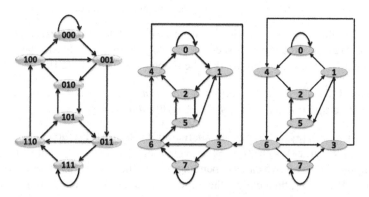

FIGURE 11.3 The three non-isomorphic UPP graphs of order 3.

The *line graph*, will be called now the *integral graph*, and will be denoted by $I(G)$, i.e., $I(G) = L(G)$. By Theorem 11.1, the integral graph of a UPP graph of order n is a UPP graph of order $n + 1$. Moreover, by the definition of the line graph, the following lemma is an immediate consequence.

Lemma 11.11. *If G is a UPP graph of order n, then $I(G)$ is a UPP graph of order $n + 1$ with the buddy property.*

Given a graph $G = (V, E)$ with the buddy property, we define the *derivative graph* $D(G) = (\hat{V}, \hat{E})$, where

$$\hat{V} \triangleq \{c : c \text{ is an alternating cycle in } G\},$$

$$\hat{E} \triangleq \{(c_1, c_2) : \exists (v_1, v_2), (v_2, v_3) \in E, (v_1, v_2) \in c_1, (v_2, v_3) \in c_2, c_1, c_2 \in \hat{V}\}.$$

By Theorem 11.1, Lemma 11.11, the definitions of the integral graph and the derivative graph, and using arguments similar to those used in the proof of Theorem 11.1, we have the following theorem.

Theorem 11.3.

- *If G is a* UPP *graph of order n, then I(G) is a* UPP *graph of order n + 1 with the buddy property and G = D(I(G)).*
- *If G is a* UPP *graph of order n, with the buddy property, then D(G) is a* UPP *graph of order n − 1 and G = I(D(G)).*

Corollary 11.4. *The number of* UPP *graphs of order n equals the number of* UPP *graphs of order n + 1 with the buddy property. There is a one-to-one correspondence between these two sets of graphs.*

Fig. 11.4 depicts a UPP graph G of order 3 and its integral graph $I(G)$, where if (u, v) is an edge in G, then uv denotes the associated vertex in $I(G)$.

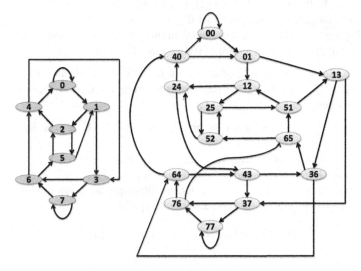

FIGURE 11.4 A UPP graph of order 3 and its integral graph of order 4.

We can define the kth integral of a graph G, $I^{(k)}(G)$, recursively as follows. The first integral of G, $I^{(1)}(G)$, is just $I(G)$. Given the kth integral of G, $I^{(k)}(G)$, $k \geq 1$, the $(k + 1)$th integral of G, $I^{(k+1)}(G)$, is defined as the integral of $I^{(k)}(G)$, i.e., $I^{(k+1)}(G) \triangleq I(I^{(k)}(G))$. Similarly, the first derivative $D^{(1)}(G)$ of a graph G with the buddy property is just $D(G)$. If $D^{(k)}(G)$, $k \geq 1$, is defined and it has the buddy property, then $D^{(k+1)}(G)$ is defined by $D^{(k+1)}(G) \triangleq D(D^{(k)}(G))$.

The integral can be defined as a mapping that can be applied to any subgraph of a digraph $G = (V, E)$. If $G' = (V', E')$, where $V' \subseteq V$ and $E' \subseteq E$ is a subgraph of G, then clearly $I(G')$ is a subgraph of $I(G)$. In particular, the integral can be applied on simple paths and simple cycles of a digraph. If P is a simple directed path of length ℓ in G, then $I(P)$ is a directed path of length $\ell - 1$ in $I(G)$. It can be also applied to cycles (not necessarily simple) of a digraph $G = (V, E)$ to obtain cycles of the same length in the integral graph $I(G)$. Given a directed cycle $C = [e_0, e_1, \ldots, e_{\ell-1}]$, $e_i \in E$, $0 \le i \le \ell - 1$, its integral is the cycle $I(C) = [(e_0, e_1), (e_1, e_2), \ldots, (e_{\ell-1}, e_0)]$. The kth integral of a cycle C, $I^{(k)}(C)$ is defined similarly.

The definition of the integral graph implies the following simple lemma.

Lemma 11.12. $C = [e_0, e_1, \ldots, e_{\ell-1}]$, where $e_i \in E$, $0 \le i \le \ell - 1$, is a cycle in a digraph $G = (V, E)$, if and only if

$$I(C) = [(e_0, e_1), (e_1, e_2), \ldots, (e_{\ell-2}, e_{\ell-1}), (e_{\ell-1}, e_0)]$$

is a cycle in $I(G) = (V' = E, E')$.

Let $C = [v_0, v_1, \ldots, v_{t-1}]$ be a cycle of length t in a graph $G = (V, E)$, where $v_i \in V$, $0 \le i \le t - 1$. The cycle C has a ***repeated path*** of length ℓ if there exist two integers m_1, m_2 such that $0 \le m_1 < m_2 \le t$ and $v_{m_1+i} = v_{m_2+i}$ for each $0 \le i \le \ell$, where subscripts are taken modulo t. The repeated path can be represented also by the sequence of its associated ℓ edges.

Lemma 11.13. *The cycle C, of a digraph $G = (V, E)$, does not contain a repeated path of length ℓ, $\ell \ge 1$, if and only if the cycle $I(C)$, of the digraph $I(G)$, does not contain a repeated path of length $\ell - 1$.*

Proof. Let $G = (V, E)$ be a digraph and let $I(G) = (E, E')$ be its integral graph.

Assume first that $C = [e_0, e_1, \ldots, e_{t-1}]$, $e_i \in E$, $0 \le i \le t - 1$, is a cycle of length t in G with a repeated path of length ℓ, $e_i, e_{i+1}, \ldots, e_{i+\ell-1}$ that equals the path $e_j, e_{j+1}, \ldots, e_{j+\ell-1}$, $i \ne j$, where subscripts are taken modulo t. The cycle $I(C)$ in $I(G) = (E, E')$ contains the two paths

$$(e_i, e_{i+1}), (e_{i+1}, e_{i+2}), \ldots, (e_{i+\ell-2}, e_{i+\ell-1})$$

and

$$(e_j, e_{j+1}), (e_{j+1}, e_{j+2}), \ldots, (e_{j+\ell-2}, e_{j+\ell-1}),$$

of length $\ell - 1$, where the e_ms are edges in G and vertices in $I(G)$. These two paths are equal and hence they represent a repeated path in $I(G)$.

Assume now that $I(C) = [e'_0, e'_1, \ldots, e'_{t-1}]$ is a cycle of length t, in $I(G)$, where $e'_m = (e_m, e_{m+1}) \in E'$, $e_m \in E$, $0 \le m \le t - 1$, and subscripts are taken modulo t. Let $e'_i, e'_{i+1}, \ldots, e'_{i+\ell-2}$ and $e'_j, e'_{j+1}, \ldots, e'_{j+\ell-2}$, $i \ne j$, where subscripts are taken modulo t, be a repeated path of length $\ell - 1$ in $I(C)$. Since $e'_m = (e_m, e_{m+1})$, $0 \le m \le t - 1$, where subscripts are taken modulo t, it follows that $C = [e_0, e_1, \ldots, e_{t-1}]$ is a cycle of length t in G, which contains the repeated path $e_i, e_{i+1}, \ldots, e_{i+\ell-1}$ that is equal to the path $e_j, e_{j+1}, \cdots, e_{j+\ell-1}$ of length ℓ. $\square$

Corollary 11.5. *The cycle C, in a digraph $G = (V, E)$, does not contain a repeated path of length two if and only if the cycle $I(C)$, in the graph $I(G)$, does not contain a repeated edge.*

Corollary 11.6. *The cycle C, in a digraph $G = (V, E)$, does not contain a repeated edge if and only if the cycle $I(C)$, in the graph $I(G)$, does not contain a repeated vertex, i.e., $I(C)$ is a simple cycle.*

Corollary 11.7. *If C is a cycle in a digraph $G = (V, E)$, for which the longest repeated path contained in C has length ℓ, then the smallest k, for which $I^{(k)}(C)$ is a simple cycle in the graph $I^{(k)}(G)$, is $k = \ell + 1$.*

The relation between cycles in the UPP graph G and the cycles in its integral graph $I(G)$ can be used to enumerate the number of some cycles in such graphs. In this context, we point to an important observation from the definition of the integral graph $I(G)$ and the cycles of G (which was already stated for G_n) that is given in the following theorem.

Theorem 11.4. *The cycle C is an Eulerian cycle in a digraph G if and only if $I(C)$ is a Hamiltonian cycle in $I(G)$.*

Corollary 11.8. *The number of Eulerian cycles in a digraph G is equal to the number of Hamiltonian cycles in $I(G)$.*

The integral graph yields just one UPP graph of order $n + 1$ from a UPP graph of order n. However, our goal is to construct a large set of UPP graphs. The main constructions to obtain many non-isomorphic UPP graphs are based on modifications of the de Bruijn graph (by exchanging the location of some edges, i.e., by removing some edges and adding others instead). The idea of exchanging the location of edges in UPP graphs is defined as follows.

Definition 11.1. Let $G = (V, E)$ be a UPP graph. Two vertices u and v in G are *input-compatible* if there exist another two vertices z_1 and z_2 such that all the following four vertices are distinct and (z_1, u), (z_1, v), (z_2, u), (z_2, v) are edges in G, i.e., the vertices z_1, z_2, u, v define an alternating cycle of length 4. In G_n, the vertices u and v are companion states, i.e., they differ exactly in the last bit of their binary representation.

Definition 11.2. Let $S_{uv} = \{u, v\}$ and $S_{xy} = \{x, y\}$ be two pairs of input-compatible vertices with outgoing edges to the set $S_{ouv} = \{u_1, u_2, v_1, v_2\}$ and $S_{oxy} = \{x_1, x_2, y_1, y_2\}$, respectively. These pairs are called *independent* if

$$S_{uv} \cap S_{oxy} = \varnothing \quad \text{and} \quad S_{xy} \cap S_{ouv} = \varnothing.$$

Otherwise, the pair is called *dependent*. A set S of input-compatible vertex pairs is called *independent* if every two pairs of S are independent.

Definition 11.3. Two input-compatible vertices u and v satisfy the **reach property** when the following condition holds. If (u,u_1), (u,u_2), (v,v_1), (v,v_2) are the out-edges of u and v, then there exists $i, j \in \{1, 2\}$ such that $R^{n-1}(u_i) = R^{n-1}(v_j)$. By Lemma 11.8, this also implies that $R^{n-1}(u_{3-i}) = R^{n-1}(v_{3-j})$.

We are now going to use the unique structure of input-compatible vertices and the reach property to form a large set of non-isomorphic UPP graphs of the same order. The idea will be to have a large number of independent input-compatible pairs of vertices and on each such set we replace two edges with two other edges. This should be done in a way that the distances between all the pairs of vertices that are at distance n apart will remain the same.

Let u, v be input-compatible vertices in G that satisfy the reach property, and suppose that (u, u_2) and (v, v_1) are edges in E such that $R^{n-1}(u_2) = R^{n-1}(v_1)$. The **exchange operation** for these edges is defined by removing the edges (u, u_2) and (v, v_1) from G and by adding the edges (u, v_1) and (v, u_2) to the obtained graph. Equivalently, we define the graph $G' = G(ex(u, v)) = (V, E')$, where

$$E' \triangleq E \setminus \{(u, u_2), (v, v_1)\} \cup \{(u, v_1), (v, u_2)\}.$$

This scenario of the exchange operation is depicted in Fig. 11.5. An isomorphic graph is obtained if the following set of edges E'' is defined instead of E', where

$$E'' \triangleq E \setminus \{(u, u_1), (v, v_2)\} \cup \{(u, v_2), (v, u_1)\}$$

and the graph $G'' = G(ex(u, v)) = (V, E'')$. It can be immediately observed by drawing this scenario, as is done in Fig. 11.5, that G' and G'' are isomorphic graphs. Hence, there is no need to distinguish between the exchange operation that yields G', and the exchange operation that yields G''.

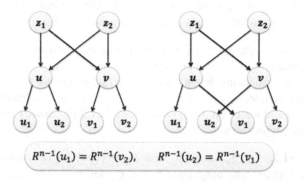

$$R^{n-1}(u_1) = R^{n-1}(v_2), \qquad R^{n-1}(u_2) = R^{n-1}(v_1)$$

FIGURE 11.5 The exchange operation: before on the left and after on the right.

In the rest of this section the structures that contain $z_1, z_2, u, v, u_1, u_2, v_1, v_2$ will be used with these symbols without reminders for the edges of these two structures of Fig. 11.5. The next lemma is an immediate observation from the exchange operation and it can be readily observed from Fig. 11.5.

Lemma 11.14. *For each path of length two in a* UPP *graph* $G = (V, E)$ *that starts in a vertex* α, *which is either* z_1 *or* z_2 *and ends in a vertex* β, *there exists a path of length two in* $G(ex(u, v))$ *which starts with* α *and ends in* β.

Lemma 11.14 considers specific paths of length two. The lemma will help to prove the existence of alternative paths of length n in $G' = G(ex(u, v))$ for each path of length n in the UPP graph $G = (V, E)$ of order n. The existence of such alternative paths leads to the next lemma.

Lemma 11.15. *If* $G = (V, E)$ *is a* UPP *graph of order n, then* $G' = G(ex(u, v))$ *is also a* UPP *graph of order n.*

Proof. Given the unique path P of length n from a vertex α to a vertex β is G we have to show such a unique path of length n from α to β in $G(ex(u, v))$. Note that the existence of such a path implies its uniqueness. If the path P in G does not contain the edges (u, u_2) and (v, v_1), then clearly the same path P exists also in $G(ex(u, v))$. Assume now that the edge (u, u_2) or the edge (v, v_1) (or both) is an edge in the path P. By Lemma 11.14, any part of the path P whose length is two and starts with either z_1 or z_2 can be ignored. This implies that we have to consider the edges (u, u_2) and (v, v_1) only in paths that start with either (u, u_2) or (v, v_1).

Assume first that the first edge in P is (u, u_2), i.e., $\alpha = u$ and

$$P = u \to u_2, u_2 \to \beta_1, \beta_1 \to \beta_2, \dots, \beta_{n-2} \to \beta_{n-1} = \beta.$$

Since $R^{n-1}(u_2) = R^{n-1}(v_1)$, it follows that there exists a path P' of length $n-1$ in G that starts with v_1 and ends in β. By Lemma 11.14 any part of the path P' whose length is two and starts with either z_1 or z_2 can be ignored as it can be replaced by another path of length two. Hence, in G' there exists a path P'' of length $n - 1$ that starts with v_1 and ends in β. Therefore there exists a path of length n in G', which starts with the edge $u \to v_1$ and continues with the path P''. This path in G' starts with α and ends in β.

If the first edge in P is (v, v_1), i.e., $\alpha = v$, then we continue in a similar way to the case when the first edge in P is (u, u_2). $\square$

The idea of Lemma 11.15 can be further extended to make several exchanges of edges in parallel and thus obtain a large set of different UPP graphs.

Lemma 11.16. *Let* G *be a* UPP *graph and* G' *be the graph derived from* G *by an exchange operation with an input-compatible pair of vertices* u *and* v. *Then, for each* $i \geq 2$, *every vertex* $\alpha \notin \{u, v\}$, *and every vertex* β, *we have*

$$\beta \in R^i_{G'}(\alpha) \quad \text{if and only if} \quad \beta \in R^i_G(\alpha).$$

Lemma 11.17. *Let* G *be a* UPP *graph and let* S *be an independent set of input-compatible vertex pairs. If* $\{u, v\} \in S$, *then in* $G' = G(ex(u, v))$ *the set* $S' = S \setminus \{\{u, v\}\}$ *is an independent set of input-compatible vertex pairs.*

Proof. Since the edges $\{u, u_2\}$ and $\{v, v_1\}$ were removed from G and the edges $\{u, v_1\}$ and $\{v, u_2\}$ were added to obtain G', it follows by Lemma 11.16 that each pair $\{x, y\} \in S'$ satisfies the reach property. Hence, all the pairs of vertices that were input-compatible in G except for the pairs with vertices from $S_{ouv} = \{u_1, u_2, v_1, v_2\}$ remain input-compatible. $\qquad\square$

Let $G(ex(S))$ be the UPP graph obtained from G by performing all the exchanges defined by a set S of input-compatible vertex pairs in an arbitrary order.

Let S be a set of input-compatible vertex pairs in G_n. The **complement** $\bar{S}$ of S is defined as

$$\bar{S} \triangleq \{\{u, v\} \ : \ \{N - 1 - v, N - 1 - u\} \in S\} = \{\{u, v\} \ : \ \{\bar{u}, \bar{v}\} \in S\},$$

whereby abuse of notation u and v are considered as integers and as binary words. The following lemma can be easily verified from Lemma 1.16 and its proof.

Lemma 11.18. *Let S be an independent set of input-compatible vertex pairs. Then, $G_n(ex(S))$ is isomorphic to $G_n(ex(\bar{S}))$ under the bijection $h : V_n \longrightarrow V_n$, where $h(v) = N - 1 - v$ for all $v \in V_n$. In other words,*

$$\forall \{u, v\} \in G_n(ex(\bar{S})), \quad h(\{u, v\}) = \{N - 1 - u, N - 1 - v\}.$$

The goal of the following lemmas is to show that if we are given two different independent sets of input-compatible vertex pairs, S_1 and S_2 such that $S_1 \neq \bar{S}_2$, then the graphs $G(1) = G_n(ex(S_1))$ and $G(2) = G_n(ex(S_2))$ are non-isomorphic.

Lemma 11.19. *In G_n, we have that*

$$d(2i, 0) < n, \quad d(2i, N - 1) = n, \quad d(2i + 1, 0) = n, \quad \text{and} \quad d(2i + 1, N - 1) < n.$$

Proof. This follows immediately from the observation that the binary representation of $2i$ ends with a *zero* and that the binary representation of $2i + 1$ ends with a *one*. $\qquad\square$

Lemma 11.20. *Let S be an independent set of input-compatible vertex pairs. In $G_n(ex(S))$, for each $\{2i, 2i + 1\} \in S$ we have that $d(2i, 0) = n$, $d(2i + 1, 0) < n$, and all the other distances to the vertices 0 and $N - 1$ remain as in G_n.*

Proof. Consider a vertex whose integer value is j, where $d(j, 0) = \ell_1$ and $d(j, N - 1) = \ell_2$. This vertex will be labeled by $j_{(\ell_1, \ell_2)}$. By Lemma 11.19, we can deduce that for each pair of vertices $\{2i, 2i + 1\}$ the labeling is as illustrated on the left in Fig. 11.6. By Lemma 11.16, all the distances from vertices beside vertices $2i$ and $2i + 1$, such that $\{2i, 2i + 1\} \in S$ remain the same as in G_n, when the exchange operation is applied on $\{2i, 2i + 1\}$. Considering this labeling and the exchange that was performed, one can easily verify that the new labeling is as in the right side of Fig. 11.6. Performing all the exchanges and applying Lemma 11.16 iteratively implies the required result. $\qquad\square$

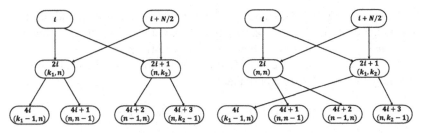

FIGURE 11.6 The exchange operation in terms of integer representation.

Note that after the exchange operation, we have that

$$d(2i, 0) = d(2i, N - 1) = n, \quad d(2i + 1, 0) < n, \quad \text{and} \quad d(2i + 1, N - 1) < n.$$

These distances will be used in the proof of the non-isomorphism, for the obtained graphs with different sets of independent input-compatible pairs.

Assume now that the two self-loops of two isomorphic UPP graphs, $G(1)$ and $G(2)$, are 0 and $N - 1$. There are two possible bijections between $G(1)$ and $G(2)$. Let h_0 be a bijection in which $h_0(0) = 0$ and $h_0(N - 1) = N - 1$, and let h_1 be the bijection in which $h_1(0) = N - 1$ and $h_1(N - 1) = 0$.

Lemma 11.21. *Let S_1 and S_2 be two different sets of input-compatible vertex pairs. Then, there is no legal bijection under h_0 for which $G(1) = G_n(ex(S_1))$ is isomorphic to $G(2) = G_n(ex(S_2))$*

Proof. Assume that Algorithm UPP isomorphism finds a legal bijection h between $G(1)$ and $G(2)$. If h is the identity mapping, namely for each vertex $v \in V_n$ we have $h(v) = v$, then let $\{u = 2i, v = 2i + 1\}$ be an input-compatible vertex pair such that $\{u, v\} \in S_1$ and $\{u, v\} \notin S_2$. The out-edges of u in $G(1)$ are to vertices $4i + 1$ and $4i + 2$, while in $G(2)$ the out-edges of u are to vertices $4i$ and $4i + 1$, and hence h (the identity bijection) is not a legal bijection.

Assume h is not the identity mapping. Let v be the first vertex in $G(1)$ that is assigned by h an assignment for which $h(v) \neq v$. We claim that this assignment is due to the exchange operation that is only in one of the graphs $G(1)$ or $G(2)$. Let $f(v) \to v$ be the related edge in the BFS tree of $G(1)$. $h(v)$ is defined by traversing the edge $f(v) \to v$ in the BFS tree of $G(1)$ and its associated edge in $G(2)$. By our assumption $h(f(v)) = f(v)$.

If $f(v)$ and its companion (namely $f(v) + 1$ if $f(v)$ is even and $f(v) - 1$ if $f(v)$ is odd) are not in S_1 and not in S_2, then both out-edges of $f(v)$ in $G(1)$ and $G(2)$ are the same and going to the same vertices. By Lemmas 11.19 and 11.20 their distances to the self-loop vertices 0 and $N - 1$ are the same as in G_n. If $f(v)$ and its companion are contained in S_1 and S_2, then again in both graphs the out-edges from $f(v)$ are the same and going to the same vertices. Their distance to the self-loop vertices 0 and $N - 1$ are changed as described

in Lemma 11.20. In both cases, algorithm UPP isomorphism would result in
the assignment $h(v) = v$, a contradiction. Therefore we have that $f(v)$ and its
companion are contained in either S_1 or S_2 since $h(v) \neq v$.

W.l.o.g. assume that the pair $f(v)$ and its companion are contained in S_1.
Then, by Lemma 11.20, we have that

$$d_{G(1)}(f(v), 0) = d_{G(1)}(f(v), N - 1) = n,$$

or

$$d_{G(1)}(f(v), 0) < n \text{ and } d_{G(1)}(f(v), N - 1) < n.$$

Similarly, by Lemma 11.20, we have that

$$d_{G(2)}(f(v), 0) = n \text{ and } d_{G(2)}(f(v), N - 1) < n,$$

or

$$d_{G(2)}(f(v), 0) < n \text{ and } d_{G(2)}(f(v), N - 1) = n.$$

This implies that h is not a legal bijection, contradicting our assumption, and
hence under h_0, $G_n(ex(S_1))$ and $G_n(ex(S_2))$ are not isomorphic. $\quad\square$

Lemma 11.22. *Let S_1 and S_2 be two independent sets of input-compatible
vertex pairs. If $S_1 \neq S_2$ and $S_1 \neq \bar{S}_2$, then $G_n(ex(S_1))$ is not isomorphic to
$G_n(ex(S_2))$.*

Proof. The lemma is a consequence of the following three results.

(1) By Lemma 11.21, there is no legal bijection between $G_n(ex(S_1))$ and
$G_n(ex(S_2))$ under h_0.
(2) By Lemma 11.18, we have that $G_n(ex(S_2))$ is not isomorphic to $G_n(ex(\bar{S}_2))$
under h_1.
(3) By Lemma 11.21, we have that $G_n(ex(S_1))$ is not isomorphic to $G_n(ex(\bar{S}_2))$
under h_0.

Hence, by **(1)**, **(2)**, and **(3)**, $G_n(ex(S_1))$ is not isomorphic to $G_n(ex(S_2))$ $\quad\square$

Now, given two sets of independent-compatible vertex pairs S_1 and S_2, it
is clear that $G_n(ex(S_1))^R$ is not isomorphic to $G_n(ex(S_2))^R$, unless $S_1 = S_2$
or $S_1 = \bar{S}_2$. We also want to show that the intersection between the set of UPP
graphs, obtained from all the independent sets and the set of their reverse graphs,
consists only of one UPP graph, G_n.

Lemma 11.23. *Let S_1 and S_2 be two sets of independent input-compatible
vertex pairs. Then, $G' = G_n(ex(S_1))$ is not isomorphic to $G_n(ex(S_2))^R$, unless $S_1 = S_2 = \varnothing$.*

Proof. If $S_2 \neq \varnothing$ then by Lemmas 11.19 and 11.20, there exists a vertex u in $G_n(ex(S_2))$ such that $d(u, 0) = d(u, N - 1) = n$. Therefore in $G_n(ex(S_2))^R$ we have that $d(0, u) = d(N - 1, u) = n$. Similarly to Lemma 11.19 we have that for each vertex v in G_n, $d(0, v) = n$ if and only if $d(N - 1, v) < n$. By Lemma 11.16, we have that $R^i_{G'}(0) = R^i_{G_n}(0)$ and $R^i_{G'}(N - 1) = R^i_{G_n}(N - 1)$ for $2 \leq i \leq n$. Also, it is clear by the definition of the exchange operation that $R^1_{G'}(0) = R^1_{G_n}(0)$ and $R^1_{G'}(N - 1) = R^1_{G_n}(N - 1)$. Hence, in G' for each v, $d(0, v) = n$ if and only if $d(N - 1, v) < n$. Since in $G_n(ex(S_2))^R$ we have that $d(0, u) = d(N - 1, u) = n$, it follows that $G_n(ex(S_1))$ is not isomorphic to $G_n(ex(S_2))^R$, unless $S_1 = S_2 = \varnothing$. □

From Lemmas 11.22 and 11.23 we infer a concluding theorem.

Theorem 11.5. *Let S_1 and S_2 be two independent sets of input-compatible pairs.*

- *The graph $G_n(ex(S_1))$ is not isomorphic to the graph $G_n(ex(S_2))$ unless $S_1 = S_2$ or $S_1 = \bar{S}_2$;*
- *The graph $G_n(ex(S_1))$ is not isomorphic to the graph $G_n(ex(S_2))^R$ unless $S_1 = S_2 = \varnothing$.*

Using Theorem 11.5 we want to derive a lower bound on the number of non-isomorphic UPP graphs. We construct a ***dependency graph***, in which each vertex represents an input-compatible vertex pair $\{u, v\}$ in G_n. Between the vertex pairs $\{u, v\}$ and $\{u', v'\}$ there is an undirected edge if and only if $\{u, v\}$ and $\{u', v'\}$ are dependent, i.e., each vertex $\{i, i + 1\}$, where i is even, $i \neq 0$, $i \neq N - 2$, is connected to vertices $\{2i, 2i + 1\}$ and $\{2i + 2, 2i + 3\}$. One can easily verify, by using the substitution $f(i, i + 1) = i/2$, i even, $i \neq 0$, and $i \neq N - 2$, that this dependency graph is the underline graph of G_{n-1}, without the vertices 0 and $N/2 - 1$. This graph for G_4 is depicted in Fig. 11.7.

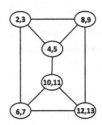

FIGURE 11.7   The dependency graph for G_4.

By Theorem 11.5 every two independent sets of vertices in the dependency graph are associated with two non-isomorphic UPP graphs, unless their corresponding sets of input-compatible vertices are complements. Those non-isomorphic graphs and their reverses (except for G_n^R) induce a set of non-isomorphic UPP graphs. Let GIS (Graphs from Independent Sets) denote the set of non-isomorphic UPP graphs obtained by Theorem 11.5. Let $G_n \setminus \{0, N-1\}$ be the subgraph of G_n without the vertices 0 and $N-1$ and their adjacent vertices. Therefore we have the following theorem.

Theorem 11.6. *The number of non-isomorphic* UPP *graphs of order $n+1$ is at least $2^{\theta(n)}$, where $\theta(n)$ is the maximum size of an independent set in $G_n \setminus \{0, N-1\}$.*

Theorem 11.7. *The size $\theta(n)$ of the largest independent set in G_n, where n is even, is at least*

$$\frac{2^n - \binom{n}{n/2}}{2}.$$

Proof. Partition the set of vertices in G_n, i.e., V_n into three sets, S_1, S_2, and S_3, as follows:

$$S_1 \triangleq \left\{ (x_1, x_2, \ldots, x_n) \ : \ x_i \in \{0, 1\}, \ \sum_{i=1}^{n/2} x_{2i} > \sum_{i=1}^{n/2} x_{2i-1} \right\},$$

$$S_2 \triangleq \left\{ (x_1, x_2, \ldots, x_n) \ : \ x_i \in \{0, 1\}, \ \sum_{i=1}^{n/2} x_{2i} < \sum_{i=1}^{n/2} x_{2i-1} \right\},$$

$$S_3 \triangleq \left\{ (x_1, x_2, \ldots, x_n) \ : \ x_i \in \{0, 1\}, \ \sum_{i=1}^{n/2} x_{2i} = \sum_{i=1}^{n/2} x_{2i-1} \right\}.$$

We claim that S_1 is an independent set of vertices in $G_n \setminus \{0, N-1\}$. Assume that $(x_1, x_2, \ldots, x_n) \in S_1$, i.e., $\sum_{i=1}^{n/2} x_{2i} > \sum_{i=1}^{n/2} x_{2i-1}$. Consider the four adjacent vertices of $(x_1, x_2, \ldots, x_{n-1}, x_n)$,

$$Y_1 = (x_2, \ldots, x_{n-1}, x_n, 0), \quad Y_2 = (x_2, \ldots, x_{n-1}, x_n, 1),$$
$$Y_3 = (0, x_1, x_2, \ldots, x_{n-1}), \quad Y_4 = (1, x_1, x_2, \ldots, x_{n-1}).$$

It is easily observed that $Y_1 \in S_2$ and $Y_4 \in S_2$, while $Y_2 \in S_2 \cup S_3$ and $Y_3 \in S_2 \cup S_3$, and hence S_1 is an independent set (and also S_2 is an independent set). Clearly,

$$S_2 = \{(x_1, x_2, \ldots, x_n) \ : \ (x_n, \ldots, x_2, x_1) \in S_1\}$$

and hence $|S_1| = |S_2|$. We also have by the definition of S_3 that

$$|S_3| = \sum_{i=0}^{n/2} \binom{n/2}{i}^2 = \binom{n}{n/2}.$$

Therefore

$$\theta(n) \geq |S_1| = \frac{2^n - \binom{n}{n/2}}{2}. \qquad \square$$

Theorem 11.8. *Asymptotically, we have*

$$\lim_{n \to \infty} \frac{\theta(n)}{2^n} = \frac{1}{2}.$$

Proof. First, note that on each necklace no more than half of the vertices can be contained in an independent set of $G_n \setminus \{0, N - 1\}$ and hence $\theta(n) \leq 2^{n-1}$. Assume now that S is an independent set in $G_n \setminus \{0, N - 1\}$. We claim that

$$S' \triangleq \{(x_1, x_2, \ldots, x_{n-1}, x_n, b) : (x_1, x_2, \ldots, x_{n-1}, x_n) \in S, \ b \in \{0, 1\}\}$$

is an independent set in $G_{n+1} \setminus \{0, 2N - 1\}$. We first observe that since $0, N - 1 \notin G_n \setminus \{0, N - 1\}$, it follows that there is no edge between the vertices $(x_1, x_2, \ldots, x_n, 0)$ and $(x_1, x_2, \ldots, x_n, 1)$. Assume, on the contrary, that S' is not an independent set. This implies that there exists a vertex $(x_1, x_2, \ldots, x_{n-1}, x_n) \in S$ and a vertex $(x_1, x_2, \ldots, x_{n-1}, x_n, b) \in S'$, where $b \in \{0, 1\}$, for which one of its adjacent vertices is also in S'. We distinguish between two cases:

Case 1: If $(0, x_1, x_2, \ldots, x_{n-1}, x_n) \in S'$ or $(1, x_1, x_2, \ldots, x_{n-1}, x_n) \in S'$, then this implies that $(0, x_1, x_2, \ldots, x_{n-1}) \in S$ or $(1, x_1, x_2, \ldots, x_{n-1}) \in S$, respectively. However, $(x_1, x_2, \ldots, x_{n-1}, x_n) \in S$, a contradiction since S is an independent set in $G_n \setminus \{0, N - 1\}$.

Case 2: If $(x_2, \ldots, x_{n-1}, x_n, b, 0) \in S'$ or $(x_2, \ldots, x_{n-1}, x_n, b, 1) \in S'$, then this implies that $(x_2, \ldots, x_{n-1}, x_n, b) \in S$. This is a contradiction since we also have that $(x_1, x_2, \ldots, x_{n-1}, x_n) \in S$ and S is an independent set $G_n \setminus \{0, N - 1\}$.

Therefore S' is an independent set in $G_{n+1} \setminus \{0, 2N - 1\}$ that implies that $\theta(n + 1) \geq 2 \cdot \theta(n)$. Therefore the limit $\frac{\theta(n)}{2^n}$, where $n \to \infty$, exists. Since $\theta(n) \leq 2^{n-1}$, $\theta(n + 1) \geq 2 \cdot \theta(n)$, and also by Theorem 11.7 we have that $\theta(n) \geq \frac{2^n - \binom{n}{n/2}}{2}$, it follows that

$$\lim_{n \to \infty} \frac{\theta(n)}{2^n} = \frac{1}{2}. \qquad \square$$

Lemma 11.24. *All the alternating cycles in a* UPP *graph of order n contained in the* GIS *have lengths 4 or 8.*

Proof. Let S be an independent set of input-compatible vertex pairs. Consider the two input-compatible vertex pairs $\{i, i + 1\}$ and $\{i + N/2, i + 1 + N/2\}$, where i is even. Before any exchange operation, the alternating cycles associated with the out-edges of the four vertices $i, i + 1, i + N/2, i + 1 + N/2$ have length 4, as depicted in Fig. 11.8(a). By an exchange operation on G_n, which includes the input-compatible vertex pair $\{i, i + 1\}$ and does not include any vertex pair $\{i + N/2, i + 1 + N/2\}$, we obtain an alternating cycle of length 8, as depicted in Fig. 11.8(b). Since the edges that are contained in this alternating cycle of length 8 are out-edges of these four vertices, it follows that some of these 8 out-edges can be removed only by an exchange on these two pairs of

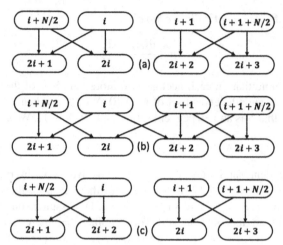

FIGURE 11.8   Two exchange operations to form an alternating cycle of length 8 and splitting it into two alternating cycles of length 4.

input-compatible vertices. Hence, we have to use the input-compatible vertex pair $\{i + N/2, i + 1 + N/2\}$ to make another exchange. An exchange operation on this input-compatible changes this alternating cycle by splitting it into the two alternating cycles of length 4, as depicted in Fig. 11.8(c).   □

By Lemma 11.24 we have that all the alternating cycles in graphs that are contained in the GIS have length 4 or 8. We will describe now another relatively simple construction for UPP graphs based on another exchange of edges in G_n. One of the advantages of this construction is that it can also yield alternating cycles of various lengths, especially alternating cycles of long lengths. Before introducing the general construction we will show an instance of this construction that produces alternating cycles of length 12.

Let **x** be a binary sequence of length $n - 3 \geq 1$. Consider the structure $\mathcal{E}_{\mathbf{x}}$ in G_n that is depicted in Fig. 11.9, and the exchange of the location of edges in this structure, as depicted in Fig. 11.10. Assume further that the edges that were removed from Fig. 11.9 do not appear between the 12 vertices in the top three layers of vertices of these two structures. We claim that the exchange of edges carried out on G_n, to obtain the structure in Fig. 11.10 from that in Fig. 11.9, yields a UPP graph of order n. We omit the proof as we are going to give a more general construction with a detailed proof. This exchange of edges yields an alternating cycle of length 12, as depicted in Fig. 11.11 (the vertices colored with orange and their out-edges colored with dashed red) do not appear in the structure of Fig. 11.10). The first n on which the construction can work is $n = 4$, where the binary sequence of length one is $\mathbf{x} = 1$.

We continue first with a simple generalization associated only with the structure of G_n depicted in Fig. 11.9. The 12 vertices in the top three layers with the

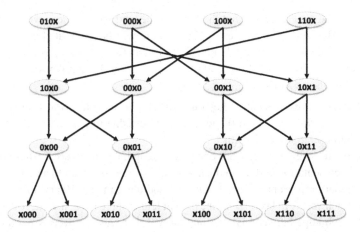

FIGURE 11.9 The 20 vertices of the structure $\mathcal{E}_x$.

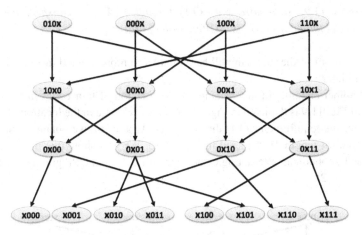

FIGURE 11.10 Removing four edges and adding another four in the structure $\mathcal{E}_x$ of Fig. 11.9.

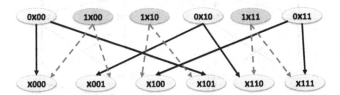

FIGURE 11.11 The alternating cycle of length 12 after the exchange of four edges.

edges between them should remain unchanged in the construction. Let the top layer of the first four vertices be called layer 0, the layer of the next four vertices will be called layer 1 and the layer of the next four vertices will be called layer 2. The last eight vertices of the structure will be associated with layer 3. The exchange of edges, to obtain a UPP graph G of order n, will be carried out between vertices of layer 2 and vertices of layer 3, such that the following requirements are satisfied.

(T1) Each vertex in layer 2 has one out-edge to a vertex whose binary representation ends with 0 and a second out-edge to a vertex whose binary representation ends with 1.

(T2) Each vertex in layer 1 has one path of length two to a vertex whose binary representation ends with 00, and similarly one such path to a vertex whose binary representation ends with 01 (and the same for 10 and 11).

(T3) Each vertex in layer 0 has one path of length three to each vertex of layer 3.

(T4) Each edge between layer 2 and layer 3 that was removed is not an edge between vertices in the first three layers (0, 1, and 2).

Theorem 11.9. *If requirements* **(T1)** *through* **(T4)** *are satisfied, then the graph G obtained from G_n is a UPP graph of order n.*

The proof of the theorem will be omitted as a more general theorem will be proved later.

Another example of an exchange in the location of four edges in the structure of Fig. 11.9 is depicted in Fig. 11.12. This exchange in the location of edges satisfies the requirements **(T1)** through **(T4)**. As a result, we obtain an alternating cycle of length 16 depicted in Fig. 11.13. We will show that this idea can be used to obtain an alternating cycle of length 2^{k+1} in a UPP graph of order n, where $n \geq 2k - 1$.

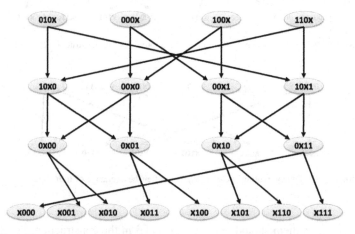

FIGURE 11.12 The exchange is done by removing four edges and adding another four.

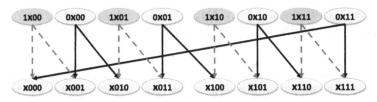

FIGURE 11.13 The alternating cycle of length 16 after the exchange of four edges.

The same idea of this construction can be generalized to a larger structure, which will be also called $\mathcal{E}_{\mathbf{x}}$, and is obtained from G_n, where $\mathbf{x}$ is a binary string of length $n - k$. The structure $\mathcal{E}_{\mathbf{x}}$ has $k + 1$ layers, denoted by layer 0 through layer k, where each layer from layer 0 through layer $k - 1$ has 2^{k-1} vertices and layer k has 2^k vertices (the vertices in each layer are distinct, but the same vertex can appear in a few layers). All the 2^{k-1} vertices in layer 0 share the last $n - k + 1$ bits in their binary representation. These $n - k + 1$ bits start with a *zero* followed by the string $\mathbf{x}$. All the 2^{k-1} vertices in layer $k - 1$ share their first $n - k + 1$ bits in their binary representation. These $n - k + 1$ bits start with a *zero* followed by the string $\mathbf{x}$. All the vertices on each path P of length $k - 1$ in G_n between vertices in layer 0 and vertices in layer $k - 1$ are contained in the structure $\mathcal{E}_{\mathbf{x}}$. Each vertex on each such path P is contained in the layer associated with its position on the path P. These edges of the path P will not be removed in the construction. In other words, all the in-edges and out-edges between vertices of the first k layers are as in G_n and they will remain unchanged in the construction. All the out-edges of vertices in layer $k - 1$ are those that are out-edges of vertices in G_n associated with the vertices of layer $k - 1$. Hence, layer k has 2^k vertices. Now, we exchange the location of edges between vertices of layer $k - 1$ and vertices of layer k, such that the following two requirements are satisfied:

(S1) Each vertex in layer m, $0 \leq m \leq k - 1$, has one path of length $k - m$ to a vertex whose binary representation ends with $b_1 b_2 \cdots b_{k-m}$, for each one of the 2^{k-m} such possible binary $(k - m)$-tuples.

(S2) Each edge between layer $k - 1$ and layer k that was removed is not an edge between vertices of the first top k layers.

One way to satisfy **(S2)** is to require that no vertex of layer $k - 1$ will be a vertex of layers 0 through layer $k - 2$. However, note that this is a stronger requirement. It is not satisfied in Fig. 11.10, but **(S2)** is satisfied (consider the out-edges of vertex 0x01, where $n = 4$, and $\mathbf{x} = 1$). For this reason, we have required that $n \geq 2k - 1$, which guarantees that **(S2)** is satisfied for all choices of $\mathbf{x}$, except for the all-zeros word $\mathbf{x}$.

Note also that requirements **(S1)** and **(S2)** are equivalent to requirements **(T1)** through **(T4)** that are more detailed. The following theorem is a generalization of Theorem 11.9.

Theorem 11.10. *If requirements* (**S1**) *and* (**S2**) *are satisfied (and the other edges in G_n remained unchanged), then the obtained graph G is a* UPP *graph of order n.*

Proof. We have to show that after the exchange of the location of edges, there is a directed path of length n between any two vertices of G. Let u, v be two vertices of G (and G_n). There is such a directed path from u to v in G_n and if none of its edges are among the edges that were exchanged, then this path of length n exists also in G. Now, assume u and v are not vertices in the first k layers of the structure $\mathcal{E}_\mathbf{x}$. If the path passes through all the layers of $\mathcal{E}_\mathbf{x}$, then this part of the path P starts at layer 0, vertex i and ends at layer k, vertex j. By requirement (**S1**) (using $m = 0$) the path P can be replaced by another path of length k that starts at vertex i and ends at vertex j.

We continue and assume that u or v (and maybe both) is a vertex in the first k layers of the structure. We distinguish between two cases depending on whether u or v is a vertex in the structure.

Case 1: v is a vertex in the structure and u is not a vertex in the structure. This implies that the path P from u to v must pass through the structure. It can go several times from layer 0 to layer k and it must go exactly once from layer 0 and stop at v. If the path P passed from layer 0 to layer k, then by (**S1**) this part of P can be replaced by an alternative path of length k. Since the path ends at v it should pass through layer 0 before it reaches v, but by (**S2**) the edges of this part of the path were not changed and hence this part of the path will remain unchanged.

Case 2: The path starts at vertex u that is a vertex in one of the first k layers of the structure. The path is of length $n > k$ and hence it reaches layer k. When the path reaches layer k for the first time (it might reach layer k more than once) it has length ℓ, for some $1 \le \ell \le k$, and it reaches a vertex ω, in layer k, whose binary representation ends with $b_1 b_2 \cdots b_\ell$. This implies that the binary representation of v starts with $b_1 b_2 \cdots b_\ell$ since the path is from u to v and its section from ω to v has length $n - \ell$. By (**S1**) there is an alternative path from u that reaches a vertex ω' in layer k whose binary representation ends with $b_1 b_2 \cdots b_\ell$. Clearly, in G_n there exists a path of length $n - \ell$ from ω' to v (since ω' ends with $b_1 b_2 \cdots b_\ell$ and v starts with $b_1 b_2 \cdots b_\ell$). This path might still have some edges that were exchanged, but alternate sections to replace these edges can be found based on the arguments used in the previous parts of the proof.

Thus G is a UPP graph of order n. $\qquad\qquad\qquad\qquad\qquad\qquad$ $\square$

We will describe now how to form a graph with an alternating cycle of length 2^{k+1} in a UPP graph of order $n \ge 2k - 1$, where $k \ge 3$. This construction will be called the *Long Cycle Construction*. We choose $\mathbf{x}$ to be the all-ones string of length $n - k$. We have to describe the exchange of edges between layer $k - 1$ and layer k. A vertex in layer $k - 1$ has the binary representation $0\mathbf{x}b_1 b_2 \cdots b_{k-1}$ and it will have two out-edges, one whose end-vertex is $\mathbf{x}b_1 b_2 \cdots b_{k-1}1$ and a second whose end-vertex is $\mathbf{x}c_1c_2\cdots c_{k-1}0$, where

$$c_1c_2\cdots c_{k-1}=b_1b_2\cdots b_{k-1}+0\ldots01.$$

The addition in this computation is an integer addition (and not a vector addition), where $1\cdots11+0\cdots01=00\cdots0$.

Theorem 11.11. *The alternating cycle obtained via the long cycle construction has length* 2^{k+1}.

Proof. Consider an ordering of the following 2^{k+1} vertices in two lists, list $\mathcal{A}$ and list $\mathcal{B}$. List $\mathcal{A}$ contains the 2^{k-1} vertices of layer $k-1$ and their conjugates. List $\mathcal{B}$ contains the 2^k vertices of layer k.

In list $\mathcal{A}$, the binary representation of the $(2i)$th vertex, $0\le i\le 2^{k-1}-1$, starts with $1\mathbf{x}$ followed by the binary representation of the integer i. These vertices are not contained in the $(k-1)$th layer of $\mathcal{E}_\mathbf{x}$. In list $\mathcal{A}$, the binary representation of the $(2i+1)$th vertex, $0\le i\le 2^{k-1}-1$, starts with $0\mathbf{x}$ followed by the binary representation of the integer i. These vertices whose binary representation starts with $0\mathbf{x}$ are exactly the vertices in the $(k-1)$th layer of $\mathcal{E}_\mathbf{x}$.

In list $\mathcal{B}$, the binary representation of the ith vertex, $0\le i\le 2^k-1$, starts with $\mathbf{x}$ followed by the binary representation of the integer i. These vertices are exactly the vertices in the kth layer of $\mathcal{E}_\mathbf{x}$.

In the long cycle construction, the edges from vertices of list $\mathcal{A}$, which contains the vertices of the $(k-1)$th layer and their conjugates, and the vertices of list $\mathcal{B}$, which contains the vertices of the kth layer are defined as follows. There is an edge from the ith vertex of list $\mathcal{A}$ to the ith vertex of list $\mathcal{B}$ for each i, $0\le i\le 2^k-1$. There is an edge from the ith vertex of list $\mathcal{A}$ to the $(i+1)$th vertex of list $\mathcal{B}$ for each i, $0\le i<2^k-1$. There is an edge from the last vertex of list $\mathcal{A}$ to the first vertex of list $\mathcal{B}$. It is now easily verified that the edges defined by the two lists form an alternating cycle of length 2^{k+1}. $\square$

An example of the alternating cycle of length 2^{k+1} obtained via the construction is depicted in Fig. 11.13 for $k=3$. One can use the exchange of edges in this new construction on a few structures using a set S of strings of length $n-k$ and for each string $\mathbf{x}\in S$ to form the structure $\mathcal{E}_\mathbf{x}$. Requirements **(S1)** and **(S2)** must be satisfied for each exchange of edges done on each structure $\mathcal{E}_\mathbf{x}$. However, it is also required that the edges that were removed in structure $\mathcal{E}_{\mathbf{x}_1}$ for some $\mathbf{x}_1\in S$ are not edges between the vertices of the first k layers of any structure $\mathcal{E}_{\mathbf{x}_2}$ for some $\mathbf{x}_2\in S$. We will not do the exact computation of the exact number of non-isomorphic UPP graphs of order n obtained by this construction. This number is similar to that obtained for the first construction exchange of edges. This construction generalizes the construction with an exchange of out-edges of input-compatible vertex pair. Moreover, we can combine the two constructions and also use generalized constructions with structures of different sizes. The following problems are left as research problems.

Problem 11.1. Find weak conditions to combine all the constructions that involve the exchanging of edges. What is the lower bound on the number of UPP graphs of order n obtained by such a combined construction?

Problem 11.2. Do there exist UPP graphs of order $n \le 2k - 3$ with alternating cycles of length larger than 2^k?

11.3   Cycles and factors in UPP graphs

We turn now to two basic concepts in digraphs; cycles and factors. Recall that a factor in a digraph is a set of vertex-disjoint directed cycles that contain all the vertices in the graph. A factor $\mathcal{F}$ in a graph G can be represented by its set of cycles. The in-degree and the out-degree of each vertex in a factor $\mathcal{F}$ (considered as a subgraph of G) is one. Since the in-degree and the out-degree of each vertex in $\mathcal{F}$ is one, it follows that $\mathcal{F}$ can be represented with no ambiguity only by the set of edges that are contained in the cycles of $\mathcal{F}$. Hence, another equivalent definition for a factor in a graph $G = (V, E)$ is as follows: a *factor* in a graph $G = (V, E)$ is a subgraph $G' = (V, E')$, where $E' \subseteq E$ and each vertex in G' has in-degree one and out-degree one. The concept of a factor is extremely important in UPP graphs.

The first interesting factor in a UPP graph $G = (V, E)$ of order n, is the equivalent of the state diagram of the PCR_n. The cycles in this factor are based on the paths of length n from each vertex $v \in V$ to itself. Each vertex $v \in V$ is on exactly one such path and all these paths are vertex-disjoint cycles. Therefore these cycles form a factor. If $G = G_n$, then this factor is the state diagram of the PCR_n. Hence, in a UPP graph, this factor will be called the *pure cycling factor*. With similar arguments as in the proof of Lemma 3.3, it can be proved that all these cycles in the factor are of a length that is a divisor of n.

There are interesting results on factors in UPP graphs that connect them to the alternating cycles in the graph. Moreover, we can easily find the number of factors in a UPP graph.

Lemma 11.25. *Let $G = (V, E)$ be a UPP graph and let $\mathcal{C} = [e_1, e_2, e_3, e_4, \cdots, e_\ell]$ be an alternating cycle in G. If $\mathcal{F}$ is a factor in G, then, exactly one set of edges from the two sets $E_e(\mathcal{C})$, $E_o(\mathcal{C})$ is contained in $\mathcal{F}$ and no edge from the other set is contained in $\mathcal{F}$.*

Proof. Since the in-degree and the out-degree of a vertex in a UPP graph is two and in-degree and the out-degree of a vertex in any factor is one, it follows that half of the edges from E are contained in the cycles of $\mathcal{F}$. Moreover, since the out-degree of each vertex in $\mathcal{F}$ is one, it follows that exactly half of the edges of an alternating cycle $\mathcal{C}$ are contained in $\mathcal{F}$, where from each two consecutive edges of $\mathcal{C}$ one edge must appear in $\mathcal{F}$ and the second one cannot appear in $\mathcal{F}$. Since the in-degree of each vertex in $\mathcal{F}$ is also one and the consecutive edges in $\mathcal{C}$ are in opposite directions, it follows that exactly one set of edges from the two sets $E_e(\mathcal{C})$, $E_o(\mathcal{C})$ is contained in $\mathcal{F}$ and no edge from the other set is contained in $\mathcal{F}$. □

The fact that all the edges of a UPP graph can be partitioned into edge-disjoint alternating cycles (see Lemma 11.10) guarantees the existence of factors in the graph. By Corollary 1.14, the number of distinct factors in G_n is $2^{2^{n-1}}$ and this result is generalized for any UPP graph as follows.

Lemma 11.26. *The number of distinct factors in a* UPP *graph $G = (V, E)$ is a power of two. A* UPP *graph G has k alternating cycles if and only if G has 2^k distinct factors.*

Proof. Note again that half of the edges from E are contained in any factor $\mathcal{F}$.

By Lemma 11.25, for each alternating cycle $\mathcal{C}$ exactly one set of edges of either $E_e(\mathcal{C})$ or $E_o(\mathcal{C})$ is contained in $\mathcal{F}$. If we choose the edges of either $E_e(\mathcal{C})$ or $E_o(\mathcal{C})$ from each alternating cycle $\mathcal{C}$, to form a subgraph G' of G, then each vertex of G' will have an in-degree one and an out-degree one and a factor $\mathcal{F} \triangleq G'$ will be obtained. It follows that if G has k alternating cycles, then there are 2^k such choices and hence there exist 2^k distinct factors in G.

From each alternating cycle, no two consecutive edges (which have different directions) can be in a factor $\mathcal{F}$. Hence, all the factors in G are constructed in the same way of taking either $E_e(\mathcal{C})$ or $E_o(\mathcal{C})$ from each alternating cycle $\mathcal{C}$. This implies also that if there exist 2^k distinct factors in G, then there are k alternating cycles in G. Thus the claims of the theorem follow. □

Corollary 11.9. *The number of distinct factors in a* UPP *graph of order n that has the buddy property is $2^{2^{n-1}}$.*

Proof. A UPP graph G of order n has 2^n vertices and 2^{n+1} edges. Each edge is contained in exactly one alternating cycle and since G has the buddy property, it follows that this alternating cycle has length 4. Hence, the number of alternating cycles in the graph is 2^{n-1}. Therefore Lemma 11.26 implies that the number of distinct factors in G is $2^{2^{n-1}}$. □

There are tight connections between some of the factors in a UPP graph. Let $\mathcal{F}$ be a factor in a UPP graph $G = (V, E)$. Let $\mathcal{F}^c \triangleq \{e : e \in E, e \notin \mathcal{F}\}$, i.e., $\mathcal{F}^c$ contains all the edges of G that are not contained in $\mathcal{F}$.

Theorem 11.12. *If $\mathcal{F}$ is a factor in a* UPP *graph G, then $\mathcal{F}^c$ is also a factor in G.*

Proof. Since the in-degree and the out-degree of each vertex of G is two and the in-degree and the out-degree of each vertex of G in $\mathcal{F}$ is one, it follows that the in-degree and the out-degree of each vertex of G in $\mathcal{F}^c$ is also one. Thus $\mathcal{F}^c$ is a factor in G. □

In view of Theorem 11.12 the factor $\mathcal{F}^c$ will be called the **complement factor** of $\mathcal{F}$. In G_n, the complement factor of the pure cycling register is the factor of the complemented cycling register whose cycles have lengths that divide $2n$. Unfortunately, this is not the case in general UPP graphs (see the UPP graphs in Fig. 11.4).

Problem 11.3. Can factors and cycles of a UPP graph be analyzed in the same way as those in G_n?

By Theorem 11.3, the integral graph $I(G)$ of a UPP graph G of order n is a UPP graph of order $n + 1$. Moreover, by Lemma 11.12, $\mathcal{C}$ is a cycle in G if and only if $I(\mathcal{C})$ is a cycle in $I(G)$. This is the motivation to define the integral of a factor $\mathcal{F}$, $I(\mathcal{F})$, by $I(\mathcal{F}) \triangleq \{I(\mathcal{C}) \ : \ \mathcal{C} \in \mathcal{F}\}$.

Lemma 11.27. *If $\mathcal{F}$ is a factor in a UPP graph $G = (V, E)$ of order n, then $I(\mathcal{F}) \cup I(\mathcal{F}^c)$ is a factor in the UPP graph $I(G)$ of order $n + 1$.*

Proof. By Theorem 11.3, if G is a UPP graph of order n, then $I(G)$ is a UPP graph of order $n + 1$. By the definition of the integral graph, we have that $I(G) = (V', E')$, where $V' = E$. By Corollary 11.6, if $\mathcal{C}$ is a simple cycle in G, then $I(\mathcal{C})$ is a simple cycle in $I(G)$, where the edges of G are the vertices of $I(G)$. This implies that the cycles of a factor in G form a set of vertex-disjoint cycles in $I(G)$. The factors $\mathcal{F}$ and $\mathcal{F}^c$ in G contain disjoint sets of edges and hence in $I(G)$ their two associated sets of vertices are disjoint. Moreover, $E = \mathcal{F} \cup \mathcal{F}^c$ and $\mathcal{F} \cap \mathcal{F}^c = \varnothing$ and hence in $I(G)$ the two related sets of cycles $I(\mathcal{F})$ and $I(\mathcal{F}^c)$ contain all the vertices of $I(G)$ and no repeated vertices. Thus $I(\mathcal{F}) \cup I(\mathcal{F}^c)$ is a factor in $I(G)$. $\qquad\Box$

There is another possible definition for complement factors that is the usual one for complements in the de Bruijn graphs. If there exists an isomorphism h from G to itself under which $r_2 = h(r_1)$, where r_1 and r_2 are the two self-loops of G, then the digraph G will be called a **self-complement** UPP graph. Let G be a self-complement UPP graph with the isomorphism h from G to itself, which is not the identity isomorphism. If $\mathcal{F}$ is a factor in G, then the **binary complement factor** $\bar{\mathcal{F}}$ of $\mathcal{F}$ is defined by

$$\bar{\mathcal{F}} \triangleq \{h(\mathcal{C}) \ : \ \mathcal{C} \text{ is a cycle in } \mathcal{F}\},$$

where

$$h(\mathcal{C}) = [(h(v_0), h(v_1)), (h(v_1), h(v_2)), \ldots, (h(v_{r-1}), h(v_0))]$$

for each cycle

$$\mathcal{C} = [(v_0, v_1), (v_1, v_2), \ldots, (v_{r-1}, v_0)]$$

of $\mathcal{F}$.

11.4 Notes

The first comprehensive work on UPP graphs was done by Mendelsohn [26] who also pointed to the connection to the adjacency matrices. His work started with properties of the de Bruijn graph $G_{\sigma,n}$ and followed by generalizations (if possible) of these properties for σ-UPP graphs. UPP graphs were considered later in the connection of multistage interconnection networks by Sridhar and Raghavendra [29,30] and further discussed in Goldfeld [11] and Goldfeld and Etzion [12]. Another application of these graphs for cellular automata was found by Boykett [3]. A comprehensive study on these graphs was given later in Etzion [7]. All these studies considered the properties of the graph mainly from a graph point of view, although some results were proved using algebraic techniques, mainly in Mendelsohn [26]. In parallel, there has been much work on $n \times n$ matrices A for which $A^n = J$, which represent adjacency matrices of UPP graphs of order n. Such a direction of study was carried out, for example, in Boykett [4], Curtis, Drew, Li, and Pragel [6], King and Wang [14], Knuth [15], Kündgen, Leander, and Carsten [16], Ma and Waterhouse [22], Ryser [27], Shader [28], Wang [32,33], and Wu, Jia, and Li [34]. Similar matrices were also considered, for example, by Lam [19] and Lam and van Lint [20].

Section 11.1. Reachable trees have some interesting properties and it is questionable if the structure of these trees characterizes the associated UPP graphs. The following related questions were discussed in Etzion [7]. How many distinct vertices of V can be reached from a vertex $v \in V$ with paths of length at most k, where $k < n$, i.e., what is that size of $V(\mathcal{T}^k(v))$? What is the minimum number of such vertices, and what is the maximum number? These questions were considered for $k = n - 1$, which is the most interesting case since $|V(\mathcal{T}^n(v))| = 2^n$ by the unique path property. These two questions are associated with the set of trees $\{\mathcal{T}^{n-1}(v) : v \in V\}$ of a UPP graph $G = (V, E)$.

Theorem 11.13. Let $G = (V, E)$ be a UPP graph of order n,

1. For each vertex $v \in V$, $|V(\mathcal{T}^{n-1}(v))| \le 2^n - 1$. Moreover, we have that $|V(\mathcal{T}^{n-1}(v))| = 2^n - 1$ if and only if $r \to v$, where r is a self-loop in G (and clearly v is not a self-loop in G).
2. For each vertex $v \in V$, $|V(\mathcal{T}^{n-1}(v))| \ge 2^{n-1}$. Moreover, we have that $|V(\mathcal{T}^{n-1}(v))| = 2^{n-1}$ if and only if v is a self-loop in G.

Theorem 11.13 can be examined in the context of de Bruijn graph G_n. Clearly, from the self-loop vertex $(\alpha, \alpha, \dots, \alpha) = (\alpha^n)$, $\alpha \in \{0, 1\}$, of G_n, paths of length $n - 1$ or less cannot reach vertices that start with an $\bar{\alpha}$, but reach all the vertices that start with an α with such a path. Hence, $|V(\mathcal{T}^{n-1}(\alpha^n))| = 2^{n-1}$. For any other vertex $v = (v_1, \dots, v_{n-1}, \alpha)$, where one of the v_is is not an α, all the vertices whose representation starts with an α are reached by paths of length $n - 1$ and at least one vertex that starts with an $\bar{\alpha}$ is reached from v via a path whose length is smaller than $n - 1$. Hence, $|V(\mathcal{T}^{n-1}(v))| > 2^{n-1}$. From the vertex $(\alpha, \alpha, \ldots, \alpha, \bar{\alpha}) = (\alpha^{n-1}\bar{\alpha})$ all vertices except for (α^n) are reached with paths whose length is at most $n - 1$ and, hence, $\left| V(\mathcal{T}^{n-1}(\alpha^{n-1}\bar{\alpha}))) \right| = 2^n - 1$. Similarly, for each other vertex v in G_n we have that $\left| V(\mathcal{T}^{n-1}(v)) \right| \leq 2^n - 1$.

We would like also to apply the results from the previous sections for σ-UPP graphs of order n for which $\sigma > 2$. The line graph of a σ-UPP graph of order n is a σ-UPP graph of order $n + 1$ with a similar proof to that of Theorem 11.1. The definition of an alternating cycle for a larger alphabet is the same as for the binary alphabet, but for $\sigma > 2$ a more generalized definition will be required and it will be given when factors in these graphs will be considered.

The proof for the length of an alternating cycle cannot be generalized based on the same arguments given in the proof of Lemma 11.9. By definition, the length of an alternating cycle is even, but the proof of Lemma 11.9 might not be generalized at all. The proof of Lemma 11.9 is based on the definition of a reachable tree and on Lemma 11.8 that are generalized straightforwardly for a σ-UPP graph, where the only difference is that the out-degree of each vertex in the tree is σ compared to two in the binary case. Lemma 11.8 has a straightforward generalization with a similar proof to that of Lemma 11.8.

Lemma 11.28. *If v is a vertex of a σ-UPP graph $G = (V, E)$ and $v \to v_1, v \to v_2, \ldots, v \to v_\sigma$ are σ distinct edges in G, then $\bigcup_{i=1}^{\sigma} R^{n-1}(v_i) = V$ and $R^{n-1}(v_i) \cap R^{n-1}(v_j) = \varnothing$, for each $1 \leq i < j \leq \sigma$.*

All these generalizations were considered in Etzion [7]. As noted before, the first paper, written by Mendelsohn [26], examined UPP graphs and their properties from both algebraic and combinatoric points of view. More algebraic properties of such graphs were analyzed in Malyshev [23] and Malyshev and Tarakanov [24]. The algorithm to check whether two UPP graphs are isomorphic was given in Goldfeld and Etzion [12]. Corollaries 11.2 and 11.3 were stated in Mendelsohn [26] who claimed that they can be proved by elementary matrix theory (for example, the connection between the trace of a matrix and its eigenvalues are required). The proofs that we gave are combinatorial, as was done throughout the book, and they do not require to define more elements of matrix theory and linear algebra. The buddy property was defined by Agrawal [2] in the context of interconnection networks, which is the topic of the next chapter. It was called the Heuchenne condition since it was also defined by Heuchenne [13]. This was independently considered before in graph theory and especially in connections to line graphs. Alternating cycles were considered first in Sridhar and Raghavendra [29] who also defined the set of reachable vertices $R^i(v)$. Reachable trees and their properties are considered by Etzion [7].

The adjacency matrix A of a UPP graph of order n satisfies $A^n = J$. This type of matrix and its associated matrices for which $A^n = \lambda J$ were extensively studied, especially those with some circulant structure, e.g., see King and Wang [14], Ma and Waterhouse [22], Trefois, van Dooren, and Delvenne [31], Wang [32,33], and Wu, Jia, and Li [34]. Similar graphs for which the adjacency matrix A satisfies $A^n = J - I$ were studied, for example, in Lam and van Lint [20]. These papers are more concerned with the algebraic structure of the adjacency matrix A and not with the graphical properties or constructions of non-isomorphic UPP graphs. Other papers, like the one by Gimbert [10], considered other types of adjacency matrices, like an adjacency matrix A for which $A^k + A^{k+1} + \cdots + A^n = J$. The solution for the matrix equation $A^1 = J$ implies that $A = J$ and the associated UPP graph is an n-UPP graph of order 1, which is a complete graph with n vertices, in which each vertex is a self-loop vertex. It was proved by Knuth [15] that solutions for the matrix equation $A^2 = J$, where A is a binary matrix, is equivalent to the existence of a central groupoid that is an algebraic system with one binary operation satisfying the identity $(x \cdot y) \cdot (y \cdot z) = y$ for all x, y, and z in the system. These matrices, groupoids, and other generalizations were also considered in a few papers, e.g., Curtis, Drew, Li, and Pragel [6] Fletcher [8,9], Lam [19], and Ryser [27]. They also have some surprising connections to cyclic difference sets, as was presented in Lam [17,18]. The idea of a groupoid was generalized for a matrix A for which $A^n = J$ and $n > 2$ in Mendelsohn [25]. It was further analyzed in Fletcher [8].

Section 11.2. Non-isomorphic σ-UPP graphs of order n can be constructed similarly to those in Section 11.2. In the paper of Mendelsohn [26], it was already mentioned that the number of solutions for $A^n = J$ is very large. For example, he mentioned that for $\sigma = 3$ and $n = 2$, there are 6 non-isomorphic solutions given by the following six 9×9 adjacency matrices over $\mathbb{F}_3$:

$$A_1 = \begin{pmatrix} 1 & 1 & 1 & 0 & 0 & 0 & 0 & 0 & 0 \\ 0 & 0 & 0 & 1 & 1 & 1 & 0 & 0 & 0 \\ 0 & 0 & 0 & 0 & 0 & 0 & 1 & 1 & 1 \\ 1 & 1 & 1 & 0 & 0 & 0 & 0 & 0 & 0 \\ 0 & 0 & 0 & 1 & 1 & 1 & 0 & 0 & 0 \\ 0 & 0 & 0 & 0 & 0 & 0 & 1 & 1 & 1 \\ 1 & 1 & 1 & 0 & 0 & 0 & 0 & 0 & 0 \\ 0 & 0 & 0 & 1 & 1 & 1 & 0 & 0 & 0 \\ 0 & 0 & 0 & 0 & 0 & 0 & 1 & 1 & 1 \end{pmatrix} \quad A_2 = \begin{pmatrix} 1 & 1 & 1 & 0 & 0 & 0 & 0 & 0 & 0 \\ 0 & 0 & 0 & 1 & 1 & 1 & 0 & 0 & 0 \\ 0 & 0 & 0 & 0 & 0 & 0 & 1 & 1 & 1 \\ 1 & 1 & 1 & 0 & 0 & 0 & 0 & 0 & 0 \\ 0 & 0 & 0 & 0 & 1 & 1 & 1 & 0 & 0 \\ 0 & 0 & 0 & 1 & 0 & 0 & 0 & 1 & 1 \\ 1 & 1 & 1 & 0 & 0 & 0 & 0 & 0 & 0 \\ 0 & 0 & 0 & 0 & 1 & 1 & 1 & 0 & 0 \\ 0 & 0 & 0 & 1 & 0 & 0 & 0 & 1 & 1 \end{pmatrix}$$

$$A_3 = \begin{pmatrix} 1 & 1 & 1 & 0 & 0 & 0 & 0 & 0 & 0 \\ 0 & 0 & 0 & 1 & 1 & 1 & 0 & 0 & 0 \\ 0 & 0 & 0 & 0 & 0 & 0 & 1 & 1 & 1 \\ 1 & 1 & 1 & 0 & 0 & 0 & 0 & 0 & 0 \\ 0 & 0 & 0 & 1 & 1 & 1 & 0 & 0 & 0 \\ 0 & 0 & 0 & 0 & 0 & 0 & 1 & 1 & 1 \\ 1 & 1 & 1 & 0 & 0 & 0 & 0 & 0 & 0 \\ 0 & 0 & 0 & 0 & 1 & 1 & 1 & 0 & 0 \\ 0 & 0 & 0 & 1 & 0 & 0 & 0 & 1 & 1 \end{pmatrix} \quad A_4 = \begin{pmatrix} 1 & 1 & 1 & 0 & 0 & 0 & 0 & 0 & 0 \\ 0 & 0 & 0 & 1 & 1 & 1 & 0 & 0 & 0 \\ 0 & 0 & 0 & 0 & 0 & 0 & 1 & 1 & 1 \\ 1 & 1 & 1 & 0 & 0 & 0 & 0 & 0 & 0 \\ 0 & 0 & 0 & 0 & 1 & 1 & 1 & 0 & 0 \\ 0 & 0 & 0 & 1 & 0 & 0 & 0 & 1 & 1 \\ 1 & 0 & 1 & 0 & 1 & 0 & 0 & 0 & 0 \\ 0 & 1 & 0 & 1 & 0 & 1 & 1 & 0 & 0 \\ 0 & 0 & 0 & 0 & 0 & 0 & 1 & 1 & 1 \end{pmatrix}$$

$$A_5 = \begin{pmatrix} 1 & 1 & 1 & 0 & 0 & 0 & 0 & 0 & 0 \\ 0 & 0 & 0 & 1 & 1 & 1 & 0 & 0 & 0 \\ 0 & 0 & 0 & 0 & 0 & 0 & 1 & 1 & 1 \\ 1 & 1 & 1 & 0 & 0 & 0 & 0 & 0 & 0 \\ 0 & 0 & 0 & 1 & 1 & 0 & 0 & 0 & 1 \\ 0 & 0 & 0 & 0 & 0 & 1 & 1 & 1 & 0 \\ 0 & 1 & 1 & 1 & 0 & 0 & 0 & 0 & 0 \\ 1 & 0 & 0 & 0 & 1 & 0 & 0 & 0 & 1 \\ 0 & 0 & 0 & 0 & 0 & 1 & 1 & 1 & 0 \end{pmatrix} \qquad A_6 = \begin{pmatrix} 1 & 1 & 1 & 0 & 0 & 0 & 0 & 0 & 0 \\ 0 & 0 & 0 & 1 & 1 & 1 & 0 & 0 & 0 \\ 0 & 0 & 0 & 0 & 0 & 0 & 1 & 1 & 1 \\ 1 & 1 & 0 & 0 & 0 & 1 & 0 & 0 & 0 \\ 0 & 0 & 1 & 1 & 1 & 0 & 0 & 0 & 0 \\ 0 & 0 & 0 & 0 & 0 & 0 & 1 & 1 & 1 \\ 1 & 0 & 1 & 0 & 0 & 0 & 0 & 1 & 0 \\ 0 & 0 & 0 & 1 & 1 & 1 & 0 & 0 & 0 \\ 0 & 1 & 0 & 0 & 0 & 0 & 1 & 0 & 1 \end{pmatrix}.$$

Construction for a large set of UPP graphs of order n, using graph theory, i.e., using the exchange operation, was given first by Sridhar and Raghavendra [29], who defined the concept of input-compatible vertex pairs. An improvement based on independent sets of input-compatible vertex pairs was carried out in Goldfeld and Etzion [12]. They found the connection of their asymptotic number to the size of the largest independent set in the de Bruijn graph. In parallel, Fletcher [8,9] considered the same method of exchange, but using the adjacency matrix. He has considered switching *zeros* and *ones* in a 2×2 sub-matrix of A to obtain non-isomorphic adjacency matrices of UPP graphs, but the number of adjacency graphs obtained was not computed. The size of the largest independent set in the de Bruijn graph was considered first in Bryant and Fredricksen [5] who looked at a slightly different problem that also yields an independent set. The bound given in Theorem 11.8 was proved by Ahlswede, Balkenhol, and Khachatrian [1]. The same bound was found later in Lichiardopol [21] with a more general result on the size of independent sets of iterated line graphs. Further construction with alternating cycles of length 12 was given by Goldfeld [11]. The more generalized construction that yields UPP graphs with alternating cycles of different sizes was presented in Etzion [7].

Section 11.3. The first question to be answered is the number of factors in a σ-UPP graph of order n. For $\sigma = 2$ the answer is related to the number of alternating cycles in the graph. Hence, we would like to generalize the definition of alternating cycles for σ-UPP graphs.

The definition of an alternating cycle for a larger alphabet is the same as for the binary alphabet. The definition of the integral graph (line graph) is the same for all σ-UPP graphs. However, for the derivative of a UPP graph, we have to generalize the definition of the buddy property for σ-UPP graphs, where $\sigma > 2$, since the definition referring to alternating cycles in this case, is not good enough.

Definition 11.4. A $\sigma \times \sigma$ *alternating subgraph*, in a digraph G, is a complete bipartite digraph with two sides A and B, where each one of the two sides has σ vertices. The edges are directed from vertices of A to vertices of B.

In the binary case such a 2×2 alternating subgraph is exactly an alternating cycle of length four. A σ-UPP graph G has the **buddy property** if each edge of G

is contained in exactly one $\sigma \times \sigma$ alternating subgraph (note, that if $\sigma > 2$, then each edge is contained in a few alternating cycles.). Now, the definition of the derivative graph for a UPP graph with the buddy property is generalized trivially to σ-UPP graphs from that for the binary case, where the $\sigma \times \sigma$ alternating subgraphs take the role of the alternating cycles of length four. Note that a vertex of the σ-UPP graph G can appear on the two sides of some $\sigma \times \sigma$ alternating subgraph associated with G.

In the binary case, the set of edges in each alternating cycle is partitioned into two subsets, where each factor contains all the edges from one of these subsets and no edge from the other subset. For σ-UPP graphs with the buddy property each perfect matching of a $\sigma \times \sigma$ alternating subgraph can be used to generate a factor. If we choose a perfect matching from each $\sigma \times \sigma$ alternating subgraph, then the union of the chosen perfect matchings forms a factor. Moreover, all the factors in a σ-UPP graph with the buddy property are formed in this way. There are $\sigma!$ distinct perfect matchings in any $\sigma \times \sigma$ alternating subgraph. Therefore while the generalization of Lemma 11.26 is not straightforward (since not all the edges lie in a $\sigma \times \sigma$ alternating subgraph), Corollary 11.9 is generalized as follows.

Theorem 11.14. *The number of distinct factors in a σ-UPP graph of order n that has the buddy property is $(\sigma!)^{\sigma^{n-1}}$.*

Proof. A σ-UPP graph of order n contains σ^{n+1} edges and each $\sigma \times \sigma$ alternating subgraph contains σ^2 edges. Furthermore, since the graph has the buddy property, it follows that each edge is contained in exactly one $\sigma \times \sigma$ alternating subgraph. Hence, a σ-UPP graph of order n with the buddy property has σ^{n-1} pairwise edge-disjoint $\sigma \times \sigma$ alternating subgraphs. Each perfect matching in an alternating subgraph can be used to form a factor. There are $\sigma!$ distinct perfect matchings, from the σ inputs to the σ outputs, in each such alternating subgraph. Therefore the number of distinct factors in a σ-UPP graph of order n that has the buddy property is $(\sigma!)^{\sigma^{n-1}}$. $\Box$

For each factor $\mathcal{F}$ of a binary UPP graph, there exists a complement factor $\mathcal{F}^c$. There is no obvious generalization for the concept of the complement factor for σ-UPP graphs (although some generalizations can be given). The complement factor is used in Lemma 11.27. Fortunately, the lemma can be generalized for σ-UPP graphs without the definition of complement factors. This generalization is based on a partition of the edges in each $\sigma \times \sigma$ alternating subgraph into σ pairwise disjoint perfect matchings.

Lemma 11.29. *Assume G is a σ-UPP graph of order n with the buddy property. Assume further that each $\sigma \times \sigma$ alternating subgraph of G is partitioned into σ pairwise disjoint perfect matchings. For each i, $1 \leq i \leq \sigma$, let $\mathcal{F}_i$ be the union of the edges in the ith perfect matching (of this partition) from each alternating subgraph of G. Then,*

1. $\mathcal{F}_i$ is a factor of G.
2. $\bigcup_{i=1}^{\sigma} I(\mathcal{F}_i)$ is a factor in $I(G)$.

Proof.

1. Each vertex of G is contained exactly once in the left side of one $\sigma \times \sigma$ alternating subgraph and exactly once in the right side of one $\sigma \times \sigma$ alternating subgraph (when each alternating subgraph is considered as a complete bipartite digraph). Hence, each vertex has in-degree one and out-degree one in $\mathcal{F}_i$, and therefore $\mathcal{F}_i$ is a factor of G.
2. For the given partition, each edge of G is contained in exactly one factor $\mathcal{F}_i$, $1 \le i \le \sigma$, of G. This implies that each vertex of $I(G)$ is contained in exactly one set $I(\mathcal{F}_i)$. Moreover, for each cycle $\mathcal{C}$ of G, $I(\mathcal{C})$ is a cycle of $I(G)$. Thus $\bigcup_{i=1}^{\sigma} I(\mathcal{F}_i)$ is a factor in $I(G)$. $\square$

In Chapter 3 we considered the number of simple cycles of a given length in G_n. It is natural to ask whether the results in Chapter 3 can be generalized to any UPP graph of order n. If $1 \le \ell \le n$, then the number of simple cycles (either all the cycles or just the simple ones) of length ℓ in a UPP graph of order n does not depend on the graph. The analysis of cycles and factors in UPP graphs was done in Etzion [7]. In Mendelsohn [26] it was proved that the number of simple cycles of length $\ell \le n$ in a UPP graph of order n is the same for all UPP graphs. This proof is presented in the following results.

Lemma 11.30. *If $G = (V, E)$ is a σ-UPP graph of order n, A is its adjacency matrix, and ℓ is an integer, $1 \le \ell \le n$, then A^{ℓ} is a binary matrix, where in the main diagonal of A^{ℓ} there are exactly σ^{ℓ} ones.*

Proof. Let A be the adjacency matrix of G. Since $A^n = J$, it follows that

$$A^{n+1} = AJ = \sigma J = \sigma A^n.$$

Hence, A satisfies the polynomial equation $x^{n+1} - \sigma x^n = 0$ and therefore A has two characteristic roots, 0 and σ, where the root σ has multiplicity 1 and the root 0 has multiplicity $\sigma^n - 1$. For $\ell \le n$ we have that the characteristic roots of A^{ℓ} are σ^{ℓ} and 0. Since A^{ℓ} is a binary matrix, it follows that A^{ℓ} has exactly σ^{ℓ} ones on the main diagonal. $\square$

Corollary 11.10. *The number of distinct closed paths of length $\ell \le n$ in a σ-UPP graph of order n is σ^{ℓ}.*

Now, Theorem 3.8 and its proof can be generalized, as was pointed out in Mendelsohn [26].

Theorem 11.15. *The number of simple cycles of length ℓ in a σ-UPP graph $G = (V, E)$ of order n, where $\ell \le n$, is*

$$\frac{1}{\ell} \sum_{d \mid \ell} \mu \left(\frac{\ell}{d} \right) \cdot \sigma^d.$$

Corollary 11.11. *The number of cycles in the pure cycling factor of a σ-UPP graph of order n is*

$$\frac{1}{n}\sum_{d|n}\phi(d)\cdot\sigma^{n/d}.$$

For $\ell > n$, Theorem 11.15 cannot be extended. It was shown in Etzion [7] that the number of cycles of length ℓ is not the same for all the UPP graphs. This claim is already true for $\ell = n + 1$. On the other hand, the number of closed paths of length $\ell > n$ that starts with a given vertex v of the UPP graph does not depend on the graph.

Lemma 11.31. *If $G = (V, E)$ be a σ-UPP graph of order n, then the number of closed paths of length $\ell \geq n$ that start with a given vertex $v \in V$ is $\sigma^{\ell-n}$.*

Proof. By definition, the claim is true for $\ell = n$ and hence we assume that $\ell > n$. By Lemma 1.14, if A is the adjacency matrix of G, then $A^\ell(v, u) = k$, $v, u \in V$, if there exist exactly k closed paths that start with the vertex v and end with the vertex u. By definition, $A^n = J$ and hence

$$A^\ell = A^{\ell-n} \cdot A^n = A^{\ell-n-1} \cdot A \cdot J = A^{\ell-n-1} \cdot \sigma \cdot J = \sigma^{\ell-n}J$$

and the claim of the lemma follows. □

It appears that usually the number of cycles of a given length ℓ, where $\ell > n$ in a UPP graph of order n, is not the same as in G_n. Another distinction between UPP graphs and the de Bruijn graph is the factor with the largest number of cycles. Factors with more cycles than in the PCR_n of G_n (see Section 3.3) exist in other UPP graphs of order n. The first result is demonstrated by using the exchange operation on G_3 and constructing the line graph from the obtained UPP graph of order 3 (see Fig. 11.4), which can be also obtained by two exchange operations on G_4. By taking the line graph of the obtained UPP graph of order 4 the second result is obtained.

References

[1] R. Ahlswede, B. Balkenhol, L. Khachatrian, Some properties of fixed-free codes, in: Proc. 1st Int. Seminar on Coding Theory and Combinatorics, Thahkadzor, Armenia, 1996, pp. 20–33.
[2] D.P. Agrawal, Graph theoretical analysis and design of multistage interconnection networks, IEEE Trans. Comput. 32 (1983) 637–648.
[3] T. Boykett, Efficient exhaustive listings of reversible one dimensional cellular automata, Theor. Comput. Sci. 325 (2004) 215–247.
[4] T. Boykett, Orderly algorithm to enumerate central groupoids and their graphs, Acta Math. Sin. 23 (2007) 249–264.
[5] R.D. Bryant, H. Fredricksen, Covering the de Bruijn graph, Discrete Math. 89 (1991) 133–148.
[6] F. Curtis, J. Drew, C.-K. Li, D. Pragel, Central groupoids, central digraphs, and zero-one matrices A satisfying $A^2 = J$, J. Comb. Theory, Ser. A 105 (2004) 35–50.
[7] T. Etzion, Graphs with the unique path property: structure, cycles, factors, and constructions, J. Graph Theory (2023), https://doi.org/10.1002/jgt.23007.

[8] R.R. Fletcher, Unique path property digraphs, Ph.D. thesis, Emory University, Michigan, USA, 1991.

[9] R.R. Fletcher, Using the theory of groups to construct unique path property digraphs, in: Proc. of the 32nd Southeastern Int. Conf. on Combinatorics, Graph Theory and Computing, Baton Rouge, LA, in: Congr. Numer., vol. 153, 2001, pp. 193–209.

[10] J. Gimbert, On digraphs with unique walks of closed lengths between vertices, Aust. J. Comb. 20 (1999) 77–90.

[11] D. Goldfeld, Equivalence of Interconnection Networks, M.Sc. thesis, Technion, Haifa, Israel, 1989.

[12] D. Goldfeld, T. Etzion, UPP graphs and UMFA networks – architecture for parallel systems, IEEE Trans. Comput. 41 (1992) 1479–1483.

[13] C. Heuchenne, Sur une certaine correspondance entre graphes, Bull. Soc. R. Sci. Liège 33 (1964) 743–753.

[14] F. King, K. Wang, On the g-circulant solutions to the matrix equation $A^m = \lambda J$, II, J. Comb. Theory, Ser. A 38 (1985) 182–186.

[15] D.E. Knuth, Notes on central groupoids, J. Comb. Theory 8 (1970) 376–390.

[16] A. Kündgen, G. Leander, C. Carsten, Switchings, extensions, and reductions in central digraphs, J. Comb. Theory, Ser. A 118 (2011) 2025–2034.

[17] C.W.H. Lam, A generalization of cyclic difference sets I, J. Comb. Theory, Ser. A 19 (1975) 51–65.

[18] C.W.H. Lam, A generalization of cyclic difference sets II, J. Comb. Theory, Ser. A 19 (1975) 177–191.

[19] C.W.H. Lam, On some solutions of $A^k = dI + \lambda J$, J. Comb. Theory, Ser. A 23 (1977) 140–147.

[20] C.W.H. Lam, J.H. van Lint, Directed graphs with unique paths of fixed length, J. Comb. Theory, Ser. B 24 (1978) 331–337.

[21] N. Lichiardopol, Independence number of iterated line digraphs, Discrete Math. 293 (2005) 185–193.

[22] S.L. Ma, W.C. Waterhouse, The g-circulant solutions of $A^m = \lambda J$, Linear Algebra Appl. 85 (1987) 211–220.

[23] F.M. Malyshev, Generalized de Bruijn graphs, Discrete Math. Appl. 32 (2022) 11–38.

[24] F.M. Malyshev, V.E. Tarakanov, Generalized de Bruijn graphs, Math. Notes 62 (1997) 449–456.

[25] N.S. Mendelsohn, An application of matrix theory to a problem in universal algebra, Linear Algebra Appl. 1 (1968) 471–478.

[26] N.S. Mendelsohn, Directed graphs with unique path property, in: P. Erdos, A. Renyi, V.T. Sos (Eds.), Combinatorial Theory and Its Appl., North-Holland, Amsterdam, 1970, pp. 783–799.

[27] H.J. Ryser, A generalization of the matrix $A^2 = J$, Linear Algebra Appl. 3 (1970) 451–460.

[28] L.E. Shader, On the existence of finite central groupoids of all possible ranks. I, J. Comb. Theory, Ser. A 16 (1974) 221–229.

[29] M.A. Sridhar, C.S. Raghavendra, Uniform minimal full-access networks, J. Parallel Distrib. Comput. 5 (1988) 383–403.

[30] M.A. Sridhar, C.S. Raghavendra, Fault-tolerant networks based on the de Bruijn graph, IEEE Trans. Comput. 40 (1991) 1167–1174.

[31] M. Trefois, P. van Dooren, J.-C. Delvenne, Binary factorizations of the matrix of all ones, Linear Algebra Appl. 468 (2015) 63–79.

[32] K. Wang, On the matrix equation $A^m = \lambda J$, J. Comb. Theory, Ser. A 29 (1980) 134–141.

[33] K. Wang, On the g-circulant solutions to the matrix equation $A^m = \lambda J$, J. Comb. Theory, Ser. A 33 (1982) 287–296.

[34] Y.-K. Wu, R.-Z. Jia, Q. Li, g-circulant solutions to the (0, 1) matrix equation $A^m = J_n$, Linear Algebra Appl. 345 (2002) 195–224.

Chapter 12

Interconnection networks

Shuffle-exchange and permutation networks, layouts

Starting towards the end of the 1970s, parallel computations were becoming more and more attractive for the computer-science community. To apply parallel computations we have to design a model for the implementation of such computations. Such a model can be only of theoretical value or can be implemented with associated computer architecture. The models that were proposed are completely different from those that are being implemented in the 21st century. Nevertheless, there was an increasing interest in a model that is closely related to or derived from the de Bruijn graph. An *interconnection network* is just a network (graph) for parallel computation in which each vertex is a processing unit. Such a processing unit could be either a processor with some limited computational power or just a switching element (box). As a switching element, the vertex receives a few inputs and transfers a few outputs. The inputs come either from the system or outputs of other switching elements. The outputs are transferred either to the system or to the inputs of other switching elements. Usually, each vertex has a small in-degree and a small out-degree. When the vertices are processors that have limited computational power, they transfer information (or packets) on their edges to their adjacent vertices according to some specified rules.

Section 12.1 is devoted to the shuffle-exchange network, which is essentially a de Bruijn network (graph) from a different point of view. The network will be defined and the operations performed with the network will be discussed. The main part of this section will be devoted to the realization of permutations with the shuffle-exchange network. In particular, permutations that are associated with linear transformations will be implemented on the network.

The implementation of permutations on the shuffle-exchange network is best understood when the network is presented as a multistage interconnection network. This representation is the topic of Section 12.2, but it is important to make it clear that the shuffle-exchange network and its multistage version are different networks. The unit of computation in a multistage interconnection network is a switching box whose computation power is much weaker than the computation power of each processor in the shuffle-exchange network. It will be discussed how switching boxes are used in these networks. A few more interesting multistage interconnection networks will be defined in Section 12.2. Routing of the information from the inputs of the whole network to the outputs of the whole

Sequences and the de Bruijn Graph. https://doi.org/10.1016/B978-0-44-313517-0.00018-4

network will be discussed for each of these networks. It will be shown that although the definitions of these networks are different, all these networks are isomorphic.

In Section 12.3, it will be discussed how to implement permutations on concatenations of the multistage interconnection networks that were defined. A formal definition for the concept of a permutation network will be given. We will consider the minimum number of switching boxes that are required to realize all the permutations on such a network.

Networks are implemented on chips and for this, a layout of the network is required. Various types of layouts are considered in Section 12.4. The layouts are demonstrated on the shuffle-exchange network and the de Bruijn graph.

12.1 The shuffle-exchange network

The *shuffle-exchange* (SE in short) network is a graph that is very similar to the de Bruijn graph. The SE network $G = (V, E)$ is a directed graph with 2^n vertices represented by the 2^n binary words of length n. There are two types of edges in E. The first type of edges are the *shuffle* edges. A shuffle edge is from a vertex $(x_1, x_2, \ldots, x_n)$ to the vertex $(x_2, \ldots, x_n, x_1)$. These edges are also edges in de Bruijn graph G_n, where they form the edges in the state diagram of the PCR_n, and the associated factor is the necklaces factor. The second type of edges are *exchange* edges. An exchange edge is between any two vertices represented by the binary n-tuples $(x_1, \ldots, x_{n-1}, 0)$ and $(x_1, \ldots, x_{n-1}, 1)$, where $x_i \in \{0, 1\}$, $1 \le i \le n - 1$. This edge can be viewed as an undirected edge or as two antiparallel edges (an edge in each direction between the associated vertices). This edge connects between companion vertices and hence for a vertex X for which $X \to Y$ and $X \to Y'$ in the de Bruijn graph, the SE network contains one of these edges as a shuffle edge, and the second edge $Y \leftrightarrow Y'$ as an exchange edge. The network is designed to implement parallel computations. Each vertex is regarded as a processor that can perform some basic operations like comparison, addition, multiplication, etc. The network operates in passes, where each **pass** is composed of two operations associated with the two types of edges, shuffle and exchange. At each step, each processor holds some information (a packet). In the shuffle phase of a pass, each processor $(x_1, x_2, \ldots, x_n)$ transfers its information to the vertex represented by $(x_2, \ldots, x_n, x_1)$ via the shuffle edge. In the exchange phase, the processors $(x_1, \ldots, x_{n-1}, 0)$ and $(x_1, \ldots, x_{n-1}, 1)$ may exchange their information, independent of other pairs of this form. One can use the de Bruijn graph G_n as a network of processors for the same purpose. In one pass, each vertex (processor) delivers its information (the packet) via one of its outgoing edges in a way that each vertex receives one packet. This is equivalent to one pass in the SE network. It is also easy to verify that each pass can be associated with a factor of G_n whose edges are those on which the processors of G_n deliver their packets. Each processor delivers a packet associated with an out-degree of one and each processor receives a packet associated with an in-degree of one. The number of possible distinct passes is $2^{2^{n-1}}$ since the shuffle operation is deterministic, while on each one of the 2^{n-1} exchange edges an exchange can be either performed or not performed resulting in $2^{2^{n-1}}$ possible distinct passes that by Corollary 1.14 is the number of distinct factors in G_n.

Between the shuffle phase and the exchange phase of a pass, there is a computational phase during which the active pairs of the upcoming exchange are determined. Before the first pass, there is a preprocessing stage. The overall procedure consisting of the preprocessing stage and the passes is often referred to as the *routing algorithm*.

An important problem in this context is the design of an efficient routing algorithm that implements permutations in the SE network with a minimal number of passes. In general, a transformation in the SE network associates with each processor a destination processor for information transfer. Each destination processor will receive information from exactly one source processor and hence the transformation defines a permutation. We will start first to describe how to realize nonsingular linear transformation, i.e., permutations for which each bit of the destination processor is a predetermined linear combination of the bits of the source processor. This linear combination is the same one for each processor. Using the ideas of such a routing algorithm for bit permutation, will enable us to design a simple routing algorithm that realizes any given permutation in $3n - 3$ passes.

A routing of information from a source processor $(x_1, x_2, \ldots, x_n)$ to a destination processor $(y_1, y_2, \ldots, y_n)$ can be done with exactly n passes. We just have to look at the path $x_1 x_2 \cdots x_n y_1 y_2 \cdots y_n$, where each $n + 1$ consecutive digits represent a pass. If the $n + 1$ consecutive bits are $z_1 z_2 \cdots z_n z_{n+1}$, then a shuffle operation will transfer the information from vertex $(z_1, z_2, \ldots, z_n)$ to the vertex $(z_2, \ldots, z_n, z_{n+1})$ if $z_{n+1} = z_1$. A shuffle operation followed by an exchange will transfer the information from the vertex $(z_1, z_2, \ldots, z_n)$ to the vertex $(z_2, \ldots, z_n, z_{n+1})$ if $z_{n+1} \neq z_1$. This routing path is equivalent to a path of length n in G_n.

Routing information from a source processor to a destination processor is one possible task. A more complicated task is to route information in parallel, where each processor has a destination, and every two distinct processors have two distinct destinations, and during this routing at each step, no two packets are stored in the same processors. This routing is, in simple words, a *realization of a permutation*. The realization of permutations in the SE network will be described in terms of some matrices as follows.

Definition 12.1. A binary matrix A, of size $N \times k$, $N = 2^n$, $k \geq n$, is *balanced* if all the rows in any projection of n consecutive columns of A are distinct.

Definition 12.2. The *standard* matrix is an $N \times n$ matrix M whose ith row is the binary representation of i, $0 \leq i \leq N - 1$.

The general problem of realizing a permutation in the SE network can be written as follows. Given a balanced $N \times n$ matrix A, find a matrix X (possibly empty) such that the matrix

$$[M \; X \; A]$$

is a balanced matrix. Each $n + 1$ consecutive columns in such a matrix represents a pass in the SE network, where in each row of such $n + 1$ columns, the first n bits represent the source processor and the last $n + 1$ bits represent the destination processor in one pass.

Theorem 12.1. *An* $N \times k$, $k > n$, *balanced matrix is equivalent to* $k - n$ *passes in the* SE *network.*

Lemma 12.1. *There are permutations that for their realization with an* $N \times k$ *matrix* X, *we must have* $k \geq n - 1$,

Proof. Consider the matrix M^R, in which each row is reversed compared to M. Since the last column of M is equal to the first column of M^R, it follows that if $[M \; X \; M^R]$ is balanced, where X is an $N \times k$ matrix, then $k \geq n - 1$. □

Lemma 12.1 implies that even a simple permutation associated with a bit reversal of the representation of each processor cannot be obtained with less than $2n - 1$ passes.

The $n \times n$ identity matrix $\mathbf{I}$ will be identified with its column vectors, i.e., $\mathbf{I} = [I(1) \; I(2) \cdots I(n)]$, where in $I(j)$ the unique *one* is in the jth row.

Lemma 12.2. *If* A *is an* $N \times n$ *matrix and* $\mathbf{T}$ *is an* $n \times n$ *nonsingular matrix, then* $A \cdot \mathbf{T}$ *is balanced if an only if* A *is balanced.*

Proof. Assume first that A is balanced and assume, on the contrary, that $A \cdot \mathbf{T}$ is not balanced, i.e., there exist two vector rows u and v, $u \neq v$, in A such that $u \cdot \mathbf{T} = v \cdot \mathbf{T}$, i.e., the multiplication of two different rows from A by $\mathbf{T}$ yields the same row. Since $\mathbf{T}$ is nonsingular, it follows that there exists an $n \times n$ invertible matrix $\mathbf{T}^{-1}$. Since $u \cdot \mathbf{T} = v \cdot \mathbf{T}$, it follows that $u \cdot \mathbf{T} \cdot \mathbf{T}^{-1} = v \cdot \mathbf{T} \cdot \mathbf{T}^{-1}$. However, since $\mathbf{T} \cdot \mathbf{T}^{-1} = \mathbf{I}$, it follows that $u = v$, a contradiction. Hence, $A \cdot \mathbf{T}$ is balanced.

Assume now that $A \cdot \mathbf{T}$ is balanced. Since $\mathbf{T}^{-1}$ is also a nonsingular matrix and $A \cdot \mathbf{T}$ is a balanced matrix, it follows by the first part of the proof that $A \cdot \mathbf{T} \cdot \mathbf{T}^{-1} = A$ is a balanced matrix. □

Definition 12.3. Let $\mathbf{T}$ be an $n \times n$ nonsingular binary matrix and let M be the $N \times n$ standard matrix. The permutation defined by the matrix $M \cdot \mathbf{T}$ is called a *nonsingular linear transformation*

Definition 12.4. An $n \times m$, $n \leq m$, matrix $\mathbf{R}$ is *n-regular* if every n consecutive columns of $\mathbf{R}$ are linearly independent.

Our main task in this section is to present an algorithm for the realization of linear transformations in the SE network and in particular to realize bit permutations.

Given Lemma 12.2 and Theorem 12.1, our task to find a balanced matrix $[M \ X \ A]$, given the two $N \times n$ balanced matrices M and A such that $A = M \cdot T$, where T is a nonsingular matrix. We will try to solve the following equivalent problem. Given a nonsingular $n \times n$ matrix T, find a matrix Y (possibly empty) such that $[I \ Y \ T]$ is an n-regular matrix.

Theorem 12.2. *The $N \times k$ matrix $[M \ X \ A]$ is a balanced matrix if and only if the $n \times k$ matrix $[I \ Y \ T]$ is n-regular, where $[M \ X \ A] = M \cdot [I \ Y \ T]$.*

For the rest of this section, consider an $n \times n$ matrix $\mathbf{B} = [B(1) \ B(2) \cdots B(n)]$, where each column $B(i)$ has either one or two nonzero entries. The matrix $\mathbf{B}$ can be viewed as the incidence matrix of the undirected graph $G(\mathbf{B})$ defined as follows:

Definition 12.5. The graph $G(\mathbf{B})$ has $n + 1$ vertices $0, 1, 2, \ldots, n$ and n edges $e(1), e(2), \ldots, e(n)$, where $e(k)$ joins vertices $i > 0$ and $j > 0$ if $B(k)$ has nonzero entries in rows i and j, and $e(k)$ joins vertices $i > 0$ and 0 if $B(k)$ has a nonzero entry only in row i.

Lemma 12.3. *The vectors $B(1), B(2), \ldots, B(n)$ are linearly independent if and only if $G(\mathbf{B})$ is a tree.*

Proof. Assume first, on the contrary, that $B(1), B(2), \ldots, B(n)$ are linearly independent and $G(\mathbf{B})$ is not a tree. Hence, $G(\mathbf{B})$ contains a simple cycle C. Each vertex $v \neq 0$ on the cycle C is contained in exactly an even number (0 or 2) of edges on the cycle. Hence, the entry v is *one* in an even number of column vectors associated with the edges of C. Therefore the columns associated with the edges of C are linearly dependent, a contradiction. Thus $G(\mathbf{B})$ is a tree.

Assume now that the graph $G(\mathbf{B})$ is a tree and consider the n column vectors $B(1), B(2), \ldots, B(n)$. Let v be a leaf in $G(\mathbf{B})$, let $\{u, v\}$ be the only edge that contains v in $G(\mathbf{B})$, and let $B(v)$ be its associated column vector. Clearly, $B(v)$ is linearly independent of the other columns since the entry of v is *one* only in $B(v)$. Hence, we can remove $B(v)$ and the edge $\{u, v\}$ from $G(\mathbf{B})$ to obtain a subtree of $G(\mathbf{B})$ and continue with this subtree and a smaller set of column vectors. Thus by induction, we will obtain that the vectors $B(1), B(2), \ldots, B(n)$ are linearly independent. $\square$

Lemma 12.4. *If $B(1), B(2), \ldots, B(n)$ are linearly independent vectors, then there exist an integer k, $1 \le k \le n$, and binary coefficients b_j, $1 \le j \le n - 1$, such that*

$$I(k) = B(n) + \sum_{j=1}^{n-1} b_j B(j). \tag{12.1}$$

Proof. The matrix $\mathbf{B} = [B(1) \ B(2) \cdots B(n)]$ is nonsingular. Hence, there exists a matrix $\mathbf{Q} = [Q(1) \ Q(2) \cdots Q(n)]$ such that $\mathbf{B} \cdot \mathbf{Q} = \mathbf{I}$, i.e., $\mathbf{Q} = \mathbf{B}^{-1}$. Since $\mathbf{Q}$ is nonsingular, it follows that there exists at least one k such that the last entry of $Q(k)$ is 1. Since $\mathbf{B} \cdot \mathbf{Q} = \mathbf{I}$, it follows that $I(k) = \mathbf{B} \cdot Q(k)$. If the kth column of Q is $(q_{1,k}, q_{2,k}, \ldots, q_{n,k} = 1)^{\text{tr}}$, then we have

$$I(k) = B(n) + \sum_{j=1}^{n-1} q_{j,k} B(j)$$

and hence the claim of the lemma follows. $\qquad \square$

Lemma 12.5. *If* $B(1), B(2), \ldots, B(n)$ *are linearly independent vectors, and* k, $1 \leq k \leq n$, *is an integer satisfying Lemma 12.4, then,* $B(0), B(1), \ldots, B(n-1)$ *are linearly independent, where*

$$B(0) = I(k) + \sum_{j=1}^{n-1} c_j B(j) \qquad (12.2)$$

and $c_j \in \{0, 1\}$, $1 \leq j \leq n-1$, *are any* $n-1$ *coefficients.*

Proof. Assume, on the contrary, that $B(0), B(1), B(2), \ldots, B(n-1)$ are linearly dependent. Then, since the last $n-1$ vectors are linearly independent, it follows that there exist $n-1$ coefficients d_j, $1 \leq j \leq n-1$, $d_j \in \{0, 1\}$, such that

$$B(0) = \sum_{j=1}^{n-1} d_j B(j). \qquad (12.3)$$

From Eqs. (12.1), (12.2), and (12.3), we obtain that

$$B(n) = \sum_{j=1}^{n-1} (b_j + c_j + d_j) B(j),$$

which contradicts the linear independence of the $B(j)$s, $1 \leq j \leq n$. $\qquad \square$

Based on Lemmas 12.3, 12.4, and 12.5, we have the following construction of a matrix $\mathbf{Y} = [Y(1) \ Y(2) \cdots Y(n-1)]$ such that $[\mathbf{I} \ \mathbf{Y} \ \mathbf{T}]$ is an n-regular matrix for a given $n \times n$ nonsingular matrix $\mathbf{T} = [T(1) \ T(2) \cdots T(n)]$.

Construction 12.1. *Let* $\mathbf{B}_0 = \mathbf{T}$ *and let*

$$\mathbf{B}_m = [Y(n-m) \ \cdots \ Y(n-1) \ T(1) \ \cdots \ T(n-m)], \quad 1 \leq m \leq n-1.$$

Given $\mathbf{B}_m$, $0 \leq m < n-1$, *construct* $Y(n-m-1)$ *as follows:*

1. *If* $k = n-m-1$ *satisfies Lemma 12.4, then set* $Y(n-m-1) := I(n-m-1)$.
2. *If* $k = n-m-1$ *does not satisfy Lemma 12.4, then find an integer* ℓ *that satisfies Lemma 12.4 and set* $Y(n-m-1) := I(n-m-1) + I(\ell)$.

Lemma 12.6. *The matrix*

$$[I(1) \cdots I(n) \, Y(1) \cdots Y(n-1) \, T(1) \cdots T(n)]$$

obtained via Construction 12.1 is n-regular.

Proof. The n-regularity of $[Y(1) \cdots Y(n-1) \, T(1) \cdots T(n)]$ follows directly from Lemma 12.5. To complete the proof, it suffices to show that the matrix $[I(1) \cdots I(n) \, Y(1) \cdots Y(n-1)]$ is n-regular. Let $\mathbf{C}_1 = \mathbf{I}$ and let

$$\mathbf{C}_m = [I(m) \cdots I(n) \, Y(1) \cdots Y(m-1)], \quad 1 \le m \le n.$$

We will show that linear independence among the columns of $\mathbf{C}_m$, $1 \le m < n$ implies the same for $\mathbf{C}_{m+1}$. Clearly, the columns of $\mathbf{C}_1 = \mathbf{I}$ are linearly independent. Suppose $\mathbf{C}_m$, $m \ge 1$, is nonsingular and consider

$$\mathbf{C}_{m+1} = [I(m+1) \cdots I(n) \, Y(1) \cdots Y(m)].$$

By Construction 12.1, either $Y(m) = I(m)$ or $Y(m) = I(m) + I(\ell)$ for some $\ell \neq m$. In the first case, it is clear that $\mathbf{C}_{m+1}$ is nonsingular. In the latter case note that $\mathbf{C}_r$, $1 \le r < n$, has at most two nonzero entries in every column, and, hence, we can view $\mathbf{C}_r$ as the incidence matrix of the graph $G(\mathbf{C}_r)$ according to Definition 12.5. By Lemma 12.3, since $\mathbf{C}_m$ is nonsingular, it follows that $G(\mathbf{C}_m)$ is a tree. $G(\mathbf{C}_{m+1})$ is obtained from $G(\mathbf{C}_m)$ by deleting the edge $\{0, m\}$ (associated with the column $I(m)$) and inserting the edge $\{m, \ell\}$ (corresponding to the column $I(m) + I(\ell)$). If $G(\mathbf{C}_{m+1})$ contains a cycle, then, since $Y(1), \ldots, Y(n-1)$ are linearly independent, it follows that the cycle must include the vertex 0. Since $G(\mathbf{C}_m)$ is a tree, it follows that deleting the edge $\{0, m\}$ from $G(\mathbf{C}_m)$ leaves a graph with no path between the vertices 0 and m. Hence, inserting the edge $\{m, \ell\}$ cannot generate a cycle that contains the vertex 0. Thus $G(\mathbf{C}_{m+1})$ is a tree, and $\mathbf{C}_{m+1}$ is a nonsingular matrix. $\square$

To find an integer k that satisfies Lemma 12.4, we need an efficient algorithm to invert a matrix. There are such algorithms, but the processors of the SE network are too weak in their computational power to perform such a task. Hence, it is assumed that a slightly stronger machine is connected to all the processors of the SE network. This machine is capable of implementing efficiently an algorithm to invert a matrix and send the relevant information to all the processors of the SE network. Now, we can propose a procedure to realize linear transformations. In this procedure each processor of the SE network has a packet and the following information:

1. An $(n-1)$-tuple $U = (u(1), u(2), \ldots, u(n-1))$, where $u(j) = 0$ if $Y(j) = I(j)$ and $u(j) = k$ if $Y(j) = I(j) + I(k)$.
2. Two n-tuples S and $F = S \cdot \mathbf{T}$, $S = (s(1), \ldots, s(n))$, $F = (f(1), \ldots, f(n))$, whose initial values represent, respectively, the ID of the said processor and

that of the destination processor, as defined by the given linear transformation. In the shuffle and the exchange operations that follow, each processor transfers its current S and F and receives new values for S and F. In other words, the packet of the processor contains also the source of the packet S and its destination F.

Procedure 1 (linear transformations):

Given the linear transformation that is defined by a nonsingular matrix

$$\mathbf{T} = [T(1) \; T(2) \; \cdots \; T(n)],$$

let $\mathbf{B}_0 = \mathbf{T}$ and let

$$\mathbf{B}_m = [Y(n - m) \; \cdots \; Y(n - 1) \; T(1) \; \cdots \; T(n - m)].$$

Having computed

$$\mathbf{B}_r = [Y(n - r) \; \cdots \; Y(n - 1) \; T(1) \; \cdots \; T(n - r)], \quad r \geq 0,$$

apply the algorithm to generate the inverse $\mathbf{Q} = [Q(1) \; Q(2) \; \cdots \; Q(n)]$ of $\mathbf{B}_r$. If the last entry of $Q(n - r - 1)$ equals 1, then set $Y(n - r - 1) := I(n - r - 1)$ and $u(n - r - 1) := 0$; otherwise, find an integer k such that the last entry of $Q(k)$ equals 1; set $Y(n - r - 1) := I(n - r - 1) + I(k)$ and $u(n - r - 1) := k$. After $u(1), u(2), \ldots, u(n - 1)$ are generated, they are transferred to each of the N processors of the SE network. Regarding the n-tuples $S = (s(1), \ldots, s(n))$ and $F = (f(1), \ldots, f(n))$, stored with each processor, perform the following:
$s(0) := 0$;
for $i := 1$ to $n - 1$ do
    {shuffle;
    if $s(u(i)) \neq 0$ then exchange}
shuffle;
if $s(n) \neq f(1)$ then exchange;
for $i := 1$ to $n - 1$ do
    {shuffle;
    if $f(i + 1) \neq s(i) + s(u(i))$ then exchange}   ■

We recall again that the shuffle and exchange are executed in parallel by all the processors.

Theorem 12.3. *Procedure 1 realizes a linear transformation in $2n - 1$ passes using a routing algorithm whose number of steps depends on inverting an $n \times n$ matrix. The number of shuffle operations is $2n - 1$ and there are $2n - 1$ comparisons per processor to decide whether to perform an exchange.*

Proof. To show that Procedure 1 realizes the linear transformation associated with the matrix $\mathbf{T}$, it suffices to show that it implements the moves implied by the balanced matrix

$$M \cdot \hat{\mathbf{B}} = M[\mathbf{I}\ Y(1) \cdots Y(n-1)\ \mathbf{T}] = [M\ (M \cdot Y(1)) \cdots (M \cdot Y(n-1))\ (M \cdot \mathbf{T})].$$

That is, for a given processor $S = (s(1), \ldots, s(n))$ and its destination processor $F = (f(1), \ldots, f(n))$, the path in the SE network via which the transformation $F = S \cdot \mathbf{T}$ is implemented by Procedure 1 is given by the sequence of processors corresponding to successive n-tuples from the row

$$S\hat{\mathbf{B}} = s(1), \cdots, s(n), s(1)+s(u(1)), \ldots, s(n-1)+s(u(n-1)), f(1), \ldots, f(n).$$

To this end, note that for each row $S \cdot \hat{\mathbf{B}}$, Procedure 1 performs an exchange if and only if the leading bit of the current processor differs from the last bit of the succeeding processor.

The claimed complexity of Procedure 1 is obtained as follows. The n-regular matrix $\hat{\mathbf{B}} = [\mathbf{I}\ Y(1) \cdots Y(n-1)\ \mathbf{T}]$ is generated by $n-1$ applications of the inversion of an $n \times n$ matrix. Therefore this algorithm dictates the complexity of this part. The $(n-1)$-tuple $U = (u(1), \ldots, u(n-1))$ is transferred to each of the N processors of the SE network on a bus in $O(n)$ steps. The $2n-1$ passes correspond to the last $2n-1$ columns of $M \cdot \hat{\mathbf{B}}$ and each pass is executed in constant time. Thus the overall complexity of the procedure is n times the complexity of the inversion algorithm. $\qquad\square$

The permutations on the N processors defined by a linear transformation are important, but for some practical reasons, the linear transformations defined by permutation matrices are the most important ones. In this case, where the matrix $\mathbf{T}$ is a permutation matrix, we can speed the process of realizing the linear transformation. We will show that we can realize the associated linear transformation defined as a bit permutation transformation in $O(n)$ steps. The main reason that the realization will be more efficient is that there will be no need to invert matrices.

Definition 12.6. The $n \times n$ matrix $\mathbf{T} = [T(1)\ T(2) \cdots T(n)]$ is called a permutation matrix if $T(j) = I(p(j))$, $1 \leq j \leq n$, where $p(1), p(2), \ldots, p(n)$ is an arbitrary permutation on the integers $1, 2, \ldots, n$. In other words, the matrix $\mathbf{T}$ is obtained from a permutation on the columns of the $n \times n$ identity matrix.

Based on Lemma 12.3, we use the following construction of the matrix $\mathbf{Y} = [Y(1)\ Y(2) \cdots Y(n-1)]$ such that $[\mathbf{I}\ \mathbf{Y}\ \mathbf{T}]$ is an n-regular matrix for a given permutation matrix $\mathbf{T} = [I(p(1))\ I(p(2)) \cdots I(p(n))]$.

Construction 12.2. *Let* $\mathbf{B}_0 = \mathbf{T}$ *and let*

$$\mathbf{B}_m = [Y(n-m) \cdots Y(n-1)I(p(1)) \cdots I(p(n-m))], \quad 1 \leq m \leq n-1.$$

Along with the columns of $\mathbf{Y}$ *we construct a sequence of graphs* $\mathcal{G}_i$, $0 \leq i \leq n-1$. $\mathcal{G}_0$ *is the edgeless graph of* n *isolated vertices* $1, 2, \ldots, n$. *Given* $\mathbf{B}_m$ *and* $\mathcal{G}_m$, $0 \leq m \leq n-1$, *construct* $Y(n-m-1)$ *and* $\mathcal{G}_{m+1}$ *as follows.*

If the addition of edge $\{n - m - 1, p(n - m)\}$ to $\mathcal{G}_m$ creates a cycle, then set $Y(n - m - 1) := I(n - m - 1)$ and $\mathcal{G}_{m+1} := \mathcal{G}_m$; if it does not create a cycle in $\mathcal{G}_m$, then set $Y(n - m - 1) := I(n - m - 1) + I(p(n - m))$ and obtain $\mathcal{G}_{m+1}$ by adding the edge $\{n - m - 1, p(n - m)\}$ to $\mathcal{G}_m$.

Lemma 12.7. *The matrix*

$$[I(1) \cdots I(n) \ Y(1) \cdots Y(n - 1) \ I(p(1)) \cdots I(p(n))]$$

obtained via Construction 12.2 is an n-regular matrix.

Proof. First, observe that every column of the matrix $\mathbf{B}_r$, $0 \le r \le n - 1$, has at most two nonzero entries, and thus it can be viewed as the incidence matrix of the graph $G(\mathbf{B}_r)$ defined in Definition 12.5. Note that $\mathcal{G}_r$, as defined in Construction 12.2, can be obtained from $G(\mathbf{B}_r)$ by deleting from the latter the vertex 0 and all the edges incident with this vertex. Note further that $G(\mathbf{B}_{m+1})$ is obtained from $G(\mathbf{B}_m)$ by the following two operations:

(i) Deletion of the edge $\{0, p(n - m)\}$.

(ii) Addition of either the edge $\{0, n - m - 1\}$ or the edge $\{p(n-m), n - m - 1\}$.

Assume that $G(\mathbf{B}_m)$, $m \ge 0$, is a tree. Then, operation **(i)** results in two pieces of $G(\mathbf{B}_m)$, with no path between vertices 0 and $p(n - m)$. Hence, if at this stage connecting vertex $p(n - m)$ to vertex $n - m - 1$ creates a cycle, it follows that operation **(i)** leaves vertex $n - m - 1$ in the same place with vertex $p(n - m)$, namely, with no path between vertex 0 and vertex $n - m - 1$. Therefore in this case, the graph $G(\mathbf{B}_{m+1})$ obtained in operation **(ii)** by adding the edge $\{0, n - m - 1\}$ is a tree.

If, on the other hand, connecting the vertex $p(n - m)$ to vertex $n - m - 1$, after operation **(i)**, does not create a cycle in the subgraph containing vertex $p(n - m)$, it certainly does not create a cycle with vertex 0 and the resulting graph is again a tree.

Since $G(\mathbf{B}_0)$ is a tree, it follows that $G(\mathbf{B}_m)$ is a tree for $0 \le m \le n - 1$, which implies by Lemma 12.3 that the matrix $[Y(1) \cdots Y(n - 1) \ I(p(1)) \cdots I(p(n))]$ is an n-regular matrix.

The n-regularity of the matrix $[I(1) \cdots I(n) \ Y(1) \cdots Y(n - 1)]$ follows in the same manner as in the proof of Lemma 12.6. $\qquad\square$

Construction 12.2 leads to Procedure 2, given below for realizing bit permutations. In this procedure, which is simpler than Procedure 1, each processor has at each stage a packet that contains the following information:

1. An $(n - 1)$-tuple $U = (u(1), \ldots, u(n - 1))$ as in Procedure 1.
2. An n-tuple S (the source of the packet) as in Procedure 1.
3. The permutation $P = (p(1), \ldots, p(n))$.

Procedure 2 (bit permutations):

Part 1
for $i := 1$ to $n - 1$ do
$\quad \{u(i) := p(i + 1);$
$\quad check(i) := false\}$
$check(n) := false;$
for $i := 1$ to $n - 1$ do
$\quad \{cycle := false;$
$\quad current := i;$
$\quad$ while $((cycle = false) \wedge (current < n) \wedge (check(current) = false))$ do
$\quad\quad \{check(current) := true;$
$\quad\quad$ if $u(current) \neq i$ then $current := u(current)$
$\quad\quad\quad$ else $cycle := true\}$
$\quad$ if $cycle = true$ then $u(i) := 0\}$

Part 2
$s(0) := 0;$
for $i := 1$ to $n - 1$ do
$\quad \{$shuffle;
$\quad$ if $s(u(i)) \neq 0$ then exchange$\}$
shuffle;
if $s(n) \neq s(p(1))$ then exchange;
for $i := 1$ to $n - 1$ do
$\quad \{$shuffle;
$\quad$ if $s(p(i + 1)) \neq s(i) + s(u(i))$ then exchange$\}$ ∎

Theorem 12.4. *Procedure 2 realizes bit permutations in $2n - 1$ passes and $O(n)$ steps.*

Proof. In Part 1 of Procedure 2 each processor computes the $(n - 1)$-tuple $U = (u(1), \ldots, u(n - 1))$. Initially, $u(n - m - 1)$ is set to be $p(n - m)$, which corresponds to the setting of $Y(n - m - 1)$ to $I(n - m - 1) + I(p(n - m))$. Then, $u(n - m)$ is set to be 0 if the insertion of the edge $\{n - m - 1, p(n - m)\}$ creates a cycle in the corresponding graph $\mathcal{G}_m$. Part 2 of Procedure 2 is identical to Procedure 1, with $s(p(i))$ substituting for $f(i)$.

The claimed complexity of Procedure 2 is obtained as follows. Part 1 consists of $O(n)$ steps since the variables $check(i)$, $1 \leq i \leq n$, ensure that for each i, the variable $current$ takes the value i at most once in the while loop. As in Procedure 1, each of the $2n - 1$ passes is executed in constant time. Thus the overall complexity of the procedure is $O(n)$. □

Thus far, we have provided efficient algorithms to realize linear transformations and bit permutations in the SE network. We return now to the more general problem. Given a balanced $N \times n$ matrix A, find a matrix X (possibly empty) such that

$$[M \; X \; A]$$

is a balanced matrix.

We will provide now a simple solution that can be efficiently implemented, but it will not produce a matrix X with the smallest possible number of columns. Given two balanced $N \times n$ matrices P and Q we form an undirected graph $G(P, Q)$. The vertices of $G(P, Q)$ are the N distinct integers $0, 1, \ldots, 2^n - 1$. The set of edges in $G(P, Q)$ contains 2^n edges formed by the following two rules.

1. If the words in rows i and j of P share their last $n - 1$ entries, then vertices i and j of $G(P, Q)$ are connected by an edge.
2. If the words in rows i and j of Q share their first $n - 1$ entries, then vertices i and j of $G(P, Q)$ are connected by an edge.

Lemma 12.8. *The graph $G(P, Q)$ is a union of vertex disjoint cycles that cover all the vertices of the graph.*

Proof. Each vertex has one incident edge associated with a pair of words in the matrix P and one incident edge associated with the matrix Q. Therefore each vertex has degree two and this implies that the graph $G(P, Q)$ is a union of vertex disjoint cycles that cover all the vertices in the graph. $\square$

Lemma 12.9. *Each cycle in $G(P, Q)$ has an even length.*

Proof. Each vertex in a cycle of $G(P, Q)$ has one edge of the cycle associated with a row of P and one edge of the same cycle associated with a row of Q. These two edges are consecutive in the cycle for each vertex of $G(P, Q)$. Hence, the number of edges in the cycle associated with rows of P equals the number of edges in the cycle associated with rows of Q. Therefore each cycle of $G(P, Q)$ has an even length. $\square$

Since the cycles in $G(P, Q)$ are vertex disjoint and each cycle of $G(P, Q)$ has even length, it follows that vertices of $G(P, Q)$ can be colored in two colors, 0 and 1. Consider now the graph $G(M, A)$ and form a vector column B_1, of length 2^n, whose ith entry is the color of vertex i in $G(M, A)$.

Lemma 12.10. *The matrices $[M \; B_1]$ and $[B_1 \; A]$ are balanced matrices.*

Proof. Two vertices connected by an edge in $G(M, A)$ have different colors. If the edge was obtained from M it implies that the related two rows in $[M \; B_1]$ are $(bx_1x_2 \cdots x_n0)$ and $(\bar{b}x_1x_2 \cdots x_n1)$ and, hence, $[M \; B_1]$ is a balanced matrix. For a similar argument, we have also that $[B_1 \; A]$ is a balanced matrix. $\square$

Let M_1 be the matrix formed from the last n columns of $[M \; B_1]$ and A_1 be the matrix formed from the first n columns of $[B_1 \; A]$. Form the graph $G(M_1, A_1)$ and color its vertices with two colors 0 and 1. Form a column B_2 whose ith entry is the color of vertex i in $G(M_1, A_1)$. Continue with the same process to obtain n columns $B_1, B_2, \ldots, B_{n-1}$. As a direct consequence from the construction and Lemma 12.10 we have:

Lemma 12.11. *The matrices* $[M \ B_1 B_2 \cdots B_{n-1}]$ *and* $[B_{n-1} \cdots B_2 B_1 \ A]$ *are balanced matrices.*

Remark. To generate the graph $G(P, Q)$ it was sufficient to use $N \times (n - 1)$ matrices and observe the last $n - 1$ columns of P and the first $n - 1$ columns of Q.

We continue with a simple method for the realization of permutations in the SE network. The method is described in the following theorem.

Theorem 12.5. *There exists an* $N \times (n - 1)$ *matrix X such that the matrix*

$$[M \ B_1 B_2 \cdots B_{n-1} \ X \ B_{n-1} \cdots B_2 B_1 A]$$

is a balanced matrix.

Proof. Let B_0 be a column vector for which the matrix $[B_0 B_1 B_2 \cdots B_{n-1}]$ is balanced. This implies that also $[B_{n-1} \cdots B_2 B_1 B_0]$ is a balanced matrix. Now, the claim of the theorem follows directly from Lemma 12.11, the fact that B_0 has no effect on the correctness of all the associated proofs, and the matrix $[B_{n-1} \cdots B_2 B_1 B_0]$ is obtained from the matrix $[B_0 B_1 B_2 \cdots B_{n-1}]$ using a bit permutation that was solved with $2n - 1$ passes (see Theorem 12.4). $\square$

Theorem 12.5 can be considerably improved using the same technique and Lemma 12.2.

Theorem 12.6. *Let* $M = M'X$, *where X is the rightmost column of M. Then, the* $N \times (4n - 3)$ *matrix*

$$[M'X \ D_1 \cdots D_{n-3} D_{n-2} B_{n-1} B_{n-2} \cdots B_2 B_1 A],$$

where $D_i = B_i$, $1 \leq i \leq \left\lfloor \frac{n-2}{2} \right\rfloor$, $D_{n-i} = B_{n-i} + B_{i-2}$, $3 \leq i \leq \left\lfloor \frac{n+1}{2} \right\rfloor$, *and* $D_{n-2} = B_{n-2} + X$, *is a balanced matrix.*

Proof. We just have to note that every n consecutive columns are either n consecutive columns from the matrices defined in Lemma 12.11 or linear combinations of n consecutive columns of these matrices.

Let $\ell = \left\lfloor \frac{n-2}{2} \right\rfloor$ and distinguish between odd and even n.

Case 1: If n is odd then $2\ell = n - 3$ and consider the $N \times (4n - 3)$ matrix

$$[M'X \ B_1 \ B_2 \cdots B_\ell \ B_{\ell+1} + B_\ell \ B_{\ell+2} + B_{\ell-1} \cdots$$
$$B_{n-4} + B_2 \ B_{n-3} + B_1 \ B_{n-2} + X \ B_{n-1} \ B_{n-2} \cdots B_2 \ B_1 \ A].$$

Case 2: If n is even then $2\ell = n - 2$ and consider the $N \times (4n - 3)$ matrix

$$[M'X \ B_1 \ B_2 \cdots B_\ell \ B_{\ell+1} + B_{\ell-1} \ B_{\ell+2} + B_{\ell-2} \cdots$$
$$B_{n-4} + B_2 \ B_{n-3} + B_1 \ B_{n-2} + X \ B_{n-1} \ B_{n-2} \cdots B_2 \ B_1 \ A].$$

It is easily verified that by the definitions of the B_is, we have that every n consecutive columns, in the matrices defined in both cases, form a balanced matrix. □

Corollary 12.1. *The* SE *network of order n can realize every permutation using at most* $3n - 3$ *passes.*

12.2 Multistage interconnection networks

Definition 12.7. A $t \times t$ *switching element (box)* is a unit with t inputs and t outputs. The unit performs a permutation on its t inputs and sends them to its t outputs. In other words, the unit can perform all the $t!$ permutations on its t inputs and sends one of these permutations to its t outputs.

A 3×3 switching box with two of the six permutations that it can perform is depicted in Fig. 12.1. There is some similarity between a $t \times t$ switching box and a $t \times t$ alternating subgraph. In this section, only 2×2 switching boxes will be considered. However, the definitions and the results can be generalized for any $t \times t$ switching box, where $t > 2$. The 2×2 switching box receives two inputs and delivers them to two outputs either as received or switched, as depicted in Fig. 12.2. Such a switching element can be implemented by a flip-flop.

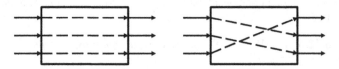

FIGURE 12.1 Two possible permutations in a 3×3 switching box.

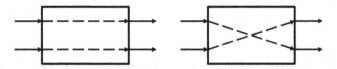

FIGURE 12.2 The two possible permutations in a 2×2 switching box.

A *multistage interconnection network* is a graph with N inputs and N outputs that consists of k stages of vertices, where each vertex is a switching box. Each stage has a certain number of switching boxes. Each switching box receives two pieces of data information as inputs, each one either from an input of the whole network or from an output of a switching box from one of the previous stages. Each switching box delivers two pieces of data information on its outputs, each one either to an output of the whole network or to an input of a switching box of one of the next stages. If there are k stages in the network, then the network will be called a *k-stage interconnection network*.

We will continue with two general definitions related to multistage interconnection networks.

Definition 12.8. A *banyan network* is a network with a unique path from each input to each output.

Definition 12.9. The process of sending the information of an input of the network to an output of the network along a path (a unique path for a banyan network) is called *routing*. The path of the routing can be described either by the sequence of consecutive edges that it traverses along the path or by the sequence of consecutive switching boxes that it passes along the path.

A banyan network by its definition is a network in which there are no redundant switching boxes for routing information from inputs to outputs. The next property makes the network simple to understand. If in the banyan network, each stage has the same number of switching boxes and all the edges are between consecutive stages, then the network is called a *Minimal Full-Access* network (MFA network in short)

Definition 12.10. A *Uniform Minimal Full-Access* network (UMFA network in short) is an MFA network with a uniform structure (the same structure) between any two consecutive stages.

All the multistage interconnection networks that will be considered in this chapter have $2N = 2^{n+1}$ inputs and the same number of outputs. Each such network will be said to be of *order* $n + 1$ (or multistage networks of order n with $N = 2^n$ inputs and 2^n outputs). Each network will have $k + 1$ stages, numbered from the first stage labeled by 0 up to the last stage labeled by k, and hence the network is a $(k + 1)$-*stage interconnection network* of order $n + 1$. In this section, each network will have $n + 1$ stages, i.e., $k = n$. Each stage in these networks has $N = 2^n$ switching boxes that are represented by the 2^n binary n-tuples from $00 \cdots 0$ to $11 \cdots 1$. The links between the switching boxes are always from stage i to stage $i + 1$, $0 \leq i \leq n - 1$. Each switching box at stage 0 receives two inputs and each switching box in stage n delivers information to two outputs. Each input at a switching box at stage 0 can be routed to 2^n switching boxes at stage n since the out-degree of each switching box is 2. Hence, when $k = n$, if each input can be routed to each output, then the network is an MFA network.

The first network that we define is the *omega network* of order $n + 1$. From switching box $x_1 x_2 \cdots x_n$ in stage i, $0 \leq i \leq n - 1$ there are two links to switching boxes in stage $i + 1$, one to switching box $x_2 \cdots x_n 0$ and one to switching box $x_2 \cdots x_n 1$. The link to switching box $x_2 \cdots x_n x_1$ will be called a *shuffle edge*, while the other link to switching box $x_2 \cdots x_n \bar{x}_1$ will be called an *exchange edge*. Clearly, these two edges are defined exactly as the two out-edges of a vertex in the de Bruijn graph G_n with the difference that in the omega network, all the vertices of G_n are defined in each stage. The omega network has the same edges between any two consecutive stages and hence it is a *uniform*

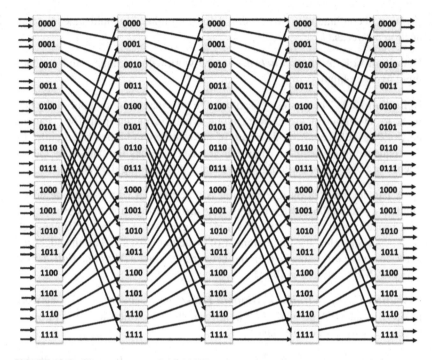

FIGURE 12.3 The omega network of order 5.

network. It is also called the $(n + 1)$-stage SE network since a switching box $x_1 x_2 \cdots x_n$ in stage i can deliver its information to the two vertices from which vertex $x_1 x_2 \cdots x_n$ in the SE network can deliver its packet in one pass. An example of the omega network of order 5 is presented in Fig. 12.3.

Now, we will elaborate more on the routing process. In each one of the networks that will be defined, there are $2N = 2^{n+1}$ inputs before stage 0, where each one of the N switching boxes in stage 0 receives two distinct inputs. Similarly, there are $2N = 2^{n+1}$ outputs, where each one of the N switching boxes of the last stage delivers two distinct outputs. Therefore the routing of an input to an output can be described as the sequence of switching boxes along the path in which the information of the input should be transferred to its associated output. The input starts at switching box $x_1 x_2 \cdots x_n$ at stage 0 and travels along the path to switching box $y_1 y_2 \cdots y_n$ at the last stage and our target is to describe this path of switching boxes.

Routing for the omega network:

The input starts from switching box $x_1 x_2 \cdots x_n$ at stage 0 to switching box $y_1 y_2 \cdots y_n$ at stage n. Consider the sequence $x_1 x_2 \cdots x_n y_1 y_2 \cdots y_n$ that describes the consecutive switching boxes in the path. At stage i, $0 \le i \le n - 1$, the information along this routing path will be at switching box $x_{i+1} \cdots x_n y_1 \cdots y_i$ and it will continue to switching box $x_{i+2} \cdots x_n y_1 \cdots y_{i+1}$ at stage $i + 1$ (with natural adjustment when $i = 0$ or $i = n - 1$; this adjustment will be required for the other networks too).

This path through $n + 1$ switching boxes is also unique since each input of switching box $x_1 x_2 \cdots x_n$ at stage 0 will pass through n stages before it reaches switching box $y_1 y_2 \cdots y_n$ at stage n. At each one of the n stages, it has two choices and hence it can reach at most 2^n switching boxes at stage n. Since the described routing path can reach each one of the 2^n switching boxes, it follows that each such a routing path is unique. This argument will hold for all the six multistage networks that will be defined in this section. However, the uniqueness of the path in the omega network is also a direct consequence of Lemma 1.15 on the uniqueness of a path of length n between any two vertices of G_n. The routing path that was described is the same as the associated path in G_n between the two vertices labeled by $x_1 x_2 \cdots x_n$ and $y_1 y_2 \cdots y_n$. Therefore the omega network is an MFA network. Moreover, since the patterns of links between any two consecutive stages are the same, the omega network is a UMFA network. We will elaborate more on this in Section 12.3 when the routing will be done in parallel on all the inputs, i.e., the routing will define a permutation. We will also explain how the routing of permutations that was done in Section 12.1 is implemented on its associated multistage network.

The simple analysis of the omega network can lead to a simple construction of UMFA networks whose structure between consecutive stages is the same as in a UPP graph. In other words, each UPP graph yields an associated UMFA network. Moreover, it is easily verified that each UMFA network can be used to construct a UPP graph. Thus there is a one-to-one correspondence between the set of UPP graphs of order n and the set of UMFA networks of order $n + 1$.

The second network we would like to consider is the *flip network* of order $n + 1$. In this network, from switching box $x_1 x_2 \cdots x_{n-1} x_n$ in stage i, $0 \le i \le n - 1$ there are two links to switching boxes in stage $i + 1$, one to switching box $0 x_1 x_2 \cdots x_{n-1}$ and one to switching box $1 x_1 x_2 \cdots x_{n-1}$. It is easily verified that routing on these edges is like performing un-shuffle (traversing the shuffle edge backward, i.e., an edge from vertex $(x_1, \ldots, x_{n-1}, x_n)$ to vertex $(x_n, x_1, \ldots, x_{n-1})$) and after that making exchange between vertices that are conjugates (differ exactly in their first position). It can also be described as performing an un-shuffle to reach one switching box and performing an exchange followed by an un-shuffle to reach the second switching box. In other words, if the links in the omega network are obtained from G_n, then the links of the flip network are obtained from G_n^R. The link to switching box $x_n x_1 \cdots x_{n-1}$ will be called an *un-shuffle edge*, while the other link to switching box $\bar{x}_n x_1 \cdots x_{n-1}$ will be called an *exchange edge*. The flip network is uniform and it is the reverse of the omega network. It is straightforward now to see that the flip network is a UMFA network and routing between an input to an output is done by taking the reverse path of length n in G_n between the associated vertices for the two switching boxes to which the input and the output are attached. The routing can

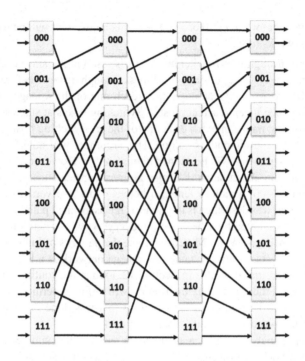

FIGURE 12.4   The flip network of order 4.

be also defined directly from the associated path in G_n^R. The flip network of order 4 is presented in Fig. 12.4.

The next network we would like to consider is the ***modified data manipulator network*** of order $n + 1$. In this network, from switching box $x_1 x_2 \cdots x_{n-1} x_n$ in stage i, $0 \leq i \leq n - 1$, there are two links to switching boxes in stage $i + 1$. One link is to switching box $x_1 x_2 \cdots x_{n-1} x_n$ and it will be called a ***line edge***. The second link is to switching box $y_1 y_2 \cdots y_{n-1} y_n$, where $y_j = x_j$ for $1 \leq j \leq n$, $j \neq i + 1$, and $y_{i+1} = \bar{x}_{i+1}$. This link will be called an ***exchange edge***. The modified data manipulator network of order 4 is presented in Fig. 12.5.

Routing for the modified data manipulator network:

The routing from an input to an output is rather simple. Assume we want to route an input from switching box $x_1 x_2 \cdots x_{n-1} x_n$ at stage 0 to switching box $y_1 y_2 \cdots y_{n-1} y_n$ at stage n. Assume further that during this routing from switching box $x_1 x_2 \cdots x_{n-1} x_n$ we reached switching box $y_1 y_2 \cdots y_i x_{i+1} \cdots x_n$ at stage i, $0 \leq i \leq n - 1$. If $y_{i+1} = x_{i+1}$, then we simply rout from switching box $y_1 y_2 \cdots y_i x_{i+1} \cdots x_n$ at stage i to switching box $y_1 y_2 \cdots y_i y_{i+1} x_{i+2} \cdots x_n$ at stage $i + 1$ using a line edge. If $y_{i+1} \neq x_{i+1}$, then similarly we rout from switching box $y_1 y_2 \cdots y_i x_{i+1} \cdots x_n$ at stage i to switching box $y_1 y_2 \cdots y_i y_{i+1} x_{i+2} \cdots x_n$ at stage $i + 1$ using an exchange edge. Since both associated edges exist between the switching boxes of stage i and stage $i + 1$, it follows that this routing is pos-

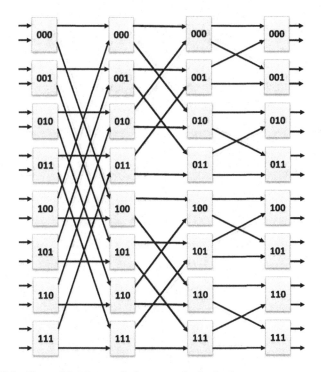

FIGURE 12.5 The modified data manipulator network of order 4.

sible. Now, by using induction, it is clear that at stage n the routing ends at switching box $y_1 y_2 \cdots y_{n-1} y_n$.

The next network is the ***indirect binary cube network*** of order $n + 1$. This network is just the reverse of the modified data manipulator network of order n. Hence, the definitions of the links are reversed. From switching box $x_1 x_2 \cdots x_{n-1} x_n$ in stage i, $0 \le i \le n - 1$, there are two links to switching boxes in stage $i + 1$. One link is to switching box $x_1 x_2 \cdots x_{n-1} x_n$ and it will be called a ***line edge***. The second link is to switching box $y_1 y_2 \cdots y_{n-1} y_n$, where $y_j = x_j$ for $1 \le j \le n$, $j \ne n - i$, and $y_{n-i} = \bar{x}_{n-i}$. This link will be called an ***exchange edge***. The indirect binary cube network of order 4 is presented in Fig. 12.6. The simple routing in the network is carried out in reverse order to that carried out for the modified data manipulator network and it will be described as follows.

Routing for the indirect binary cube network:

Assume we want to route an input from switching box $x_1 x_2 \cdots x_{n-1} x_n$ at stage 0 to switching box $y_1 y_2 \cdots y_{n-1} y_n$ at stage n. Assume further that during this routing from switching box $x_1 x_2 \cdots x_{n-1} x_n$ we reached switching box $x_1 x_2 \cdots x_{n-i} y_{n-i+1} \cdots y_n$ at stage i, $0 \le i \le n - 1$. If $y_{n-i} = x_{n-i}$, then we simply rout from switching box $x_1 x_2 \cdots x_{n-i} y_{n-i+1} \cdots y_n$ at stage i to switching box $x_1 x_2 \cdots x_{n-i-1} y_{n-i} y_{n-i+1} \cdots y_n$ at stage $i + 1$ using a line edge. If $y_{n-i} \ne$

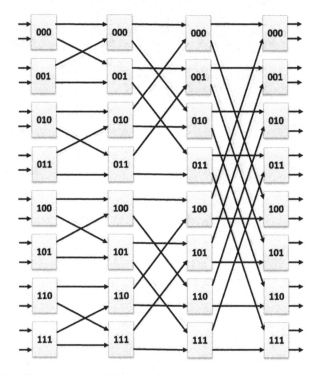

FIGURE 12.6 The indirect binary cube network.

x_{n-i}, then similarly we rout from switching box $x_1x_2 \cdots x_{n-i}y_{n-i+1} \cdots y_n$ at stage i to switching box $x_1x_2 \cdots x_{n-i-1}y_{n-i}y_{n-i+1} \cdots x_n$ at stage $i+1$ using an exchange edge. Since both associated edges exist between the switching boxes of stage i and stage $i+1$, it follows that this routing is possible. Now, by using induction, it is clear that at stage n the routing ends at switching box $y_1y_2 \cdots y_{n-1}y_n$.

The next network that will be considered is the ***baseline network*** of order $n+1$. From switching box $x_1x_2 \cdots x_{n-1}x_n$ in stage i, $0 \leq i \leq n-1$, there are two links to switching boxes in stage $i+1$, one link to switching box $y_1y_2 \cdots y_{n-1}y_n$, where $y_j = x_j$ for $1 \leq j \leq i$, $y_{i+1} = 0$, and $y_j = x_{j-1}$ for $i+2 \leq j \leq n$. The second link is to switching box $y_1y_2 \cdots y_{n-1}y_n$, where $y_j = x_j$ for $1 \leq j \leq i$, $y_{i+1} = 1$, and $y_j = x_{j-1}$ for $i+2 \leq j \leq n$. This is the same as fixing the first i bits and after that un-shuffle on the other $n-i$ bits to reach one switching box at stage $i+1$ and exchange followed by un-shuffle on these $n-i$ bits to reach the second switching box on stage $i+1$. The baseline network of order 4 is presented in Fig. 12.7.

Routing for the baseline network:

The routing from an input to an output is rather simple. Assume we want to route an input from switching box $x_1x_2 \cdots x_{n-1}x_n$ at stage 0 to switching

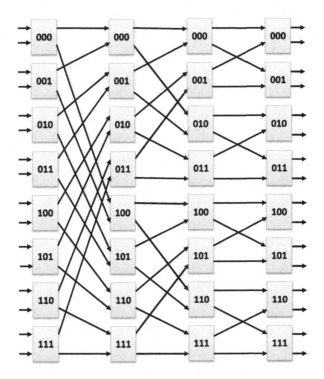

FIGURE 12.7 The baseline network of order 4.

box $y_1 y_2 \cdots y_{n-1} y_n$ at stage n. Assume that during the routing from switching box $x_1 x_2 \cdots x_{n-1} x_n$ we reached switching box $y_1 y_2 \cdots y_i z_1 \cdots z_{n-i}$ for some $z_j \in \{0, 1\}$, $1 \le j \le n - i$, at stage i, $0 \le i \le n - 1$. We rout from $y_1 y_2 \cdots y_i z_1 \cdots z_{n-i}$ at stage i to $y_1 y_2 \cdots y_i y_{i+1} z_1 \cdots z_{n-i-1}$ at stage $i + 1$. This routing is possible by the definition of the network since the first i bits are unchanged and bit $i + 1$ can be chosen as *zero* or *one* by the definition of the baseline network. Clearly, by induction at stage n the routing ends at switching box $y_1 y_2 \cdots y_{n-1} y_n$.

The last network to be considered is the ***reverse baseline network*** of order $n + 1$. This network is just the reverse of the baseline network of order n. Hence, the definitions of the links are reversed and the routing is done in reverse order to the one done for the baseline network. Instead of the un-shuffle used for the last part (of length $n - i$, $0 \le i \le n - 1$) of the binary representation, a shuffle is used for the first part (of length $i + 1$). The reverse baseline network of order 4 is depicted in Fig. 12.8.

The definitions and the analysis that we have done for each one of the six defined networks imply the following theorem.

Theorem 12.7. *The omega network of order n, the flip network of order n, the modified data manipulator of order n, the indirect binary cube of order n, the*

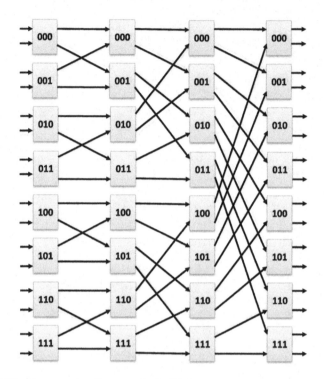

FIGURE 12.8   The reverse baseline network of order 4.

baseline network of order n, and the reverse baseline network of order n, are all MFA *networks. Moreover, the omega network and the flip network are* UMFA *networks.*

A natural question is to characterize the differences between the six defined networks. They all can route each input to each output, and by Theorem 12.7 they are all MFA networks. The omega network and the flip network are uniform and the other networks are not uniform. In each network, the routing is performed in a slightly different way. In the rest of this section, it will be proved that all six networks are isomorphic as graphs. In other words, by an assignment of different labels to the switching boxes in the various stages, all the networks will have the same labeling and links between the same switching boxes.

Theorem 12.8. *The six networks, i.e., the omega network, the flip network, the modified data manipulator network, the indirect binary cube network, the baseline network, and the reverse baseline network, are all isomorphic.*

Proof. We start by considering the omega network of order $n + 1$ and the flip network of order $n + 1$. The links between the switching boxes of stage i and stage $i + 1$, $0 \leq i \leq n - 1$, in the omega network are defined exactly as for the edges between the associated vertices in G_n. Similarly, the links between the switching boxes of stage i and stage $i + 1$, $0 \le i \le n - 1$, in the flip network are defined exactly as for the edges between the associated vertices in G_n^R. By Lemma 1.17, G_n and G_n^R are isomorphic graphs. Using the mapping $g_R : V_n \to V_n$, defined in Lemma 1.17, to prove the isomorphism between G_n^R and G_n, on the switching boxes of the flip network of order $n + 1$ will achieve its isomorphism to the omega network of order $n + 1$. Hence, the omega network of order $n + 1$ and the flip network of order $n + 1$ are isomorphic networks.

We continue and prove that the modified data manipulator network of order $n + 1$ and the baseline network of order $n + 1$ are isomorphic. The proof will be by induction. The basis is $n = 1$, where the two networks of order 2 with four inputs and three stages, each one with two switching boxes, and between two stages there is an alternating cycle of length 4. Assume now that the claim is true for order n, i.e., the modified data manipulator network of order n and the baseline network of order n are isomorphic. The networks have in each stage 2^{n-1} switching boxes. The induction step is for networks of order $n + 1$. Note first that the subgraph of the baseline network of order $n + 1$ induced by the switching boxes and edges of stages 1 through n consists of two identical baseline networks of order n, each one is isomorphic to the baseline network of order n, which by the induction hypothesis is also isomorphic to the modified data manipulator of order n. Similarly, the subgraph of the modified data manipulator network of order $n + 1$ induced by the switching boxes and edges of stages 1 through n consists of two identical modified data manipulator networks of order n. The edges between stage 0 to stage 1 in both networks of order $n + 1$ form 2^{n-1} alternating cycles of length 4. These edges and their associated vertices in both networks of order $n + 1$ at stage 0 can be rearranged in a way that switching boxes i and $2^{n-1} + i$ in stage 0 have edges to switching boxes i and $2^{n-1} + i$ in stage 1 (switching boxes labeled by i in the four networks of order n, two baseline networks and two modified data manipulators induced from stage 1 through stage n). This completes the proof of the induction step.

Next, we will show isomorphism between the flip network of order $n + 1$ and the baseline network of order $n + 1$. The proof will be again by induction, where the basis for $n = 1$ is trivial as in the previous case. Now, note that also the subgraph, of the flip network, induced from the switching boxes of stages 1 through n with the edges between them is combined of two flip networks of order n, where one network contains all the switching boxes of stage 1 that start with a *zero* in the flip network of order $n + 1$. In stage i, $1 \le i \le n$, it contains all the switching boxes whose labeling has a *zero* in the ith digits (this is the same bit in all stages since an un-shuffle is performed from stage i to stage $i + 1$). The second network contains all the other switching boxes, e.g., the switching boxes of stage 1 that start with a *one* in the flip network of order $n + 1$. By the induction hypothesis, each of these two networks is the flip network of order n and it is isomorphic to the baseline network of order n. Now, it is easy to verify that the induction step is the same as that between the modified data manipulator networks and the baseline network.

Since the omega network and its reverse, the flip network, are isomorphic networks, and also the flip network is isomorphic to the baseline network, it follows that the baseline network is isomorphic to the reverse baseline network. For similar arguments the modified data manipulator network is isomorphic to its reverse, the indirect binary cube network. This sequence of isomorphism between these networks implies that all six networks are isomorphic. □

We have proved that the six MFA networks that were defined in this section are isomorphic. Are there MFA networks that are not isomorphic to these six networks? The answer is positive and there are such UMFA networks. For example, all the UMFA networks of order $n + 1$ that can be constructed from non-isomorphic UPP graphs of order n are non-isomorphic UMFA networks. An example of such a network of order 4 is depicted in Fig. 12.9.

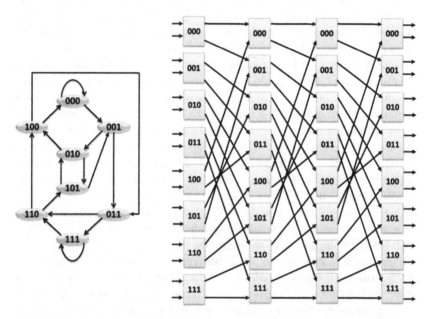

FIGURE 12.9   UMFA network of order 4 based on a UPP graph of order 3.

12.3   Multistage permutation networks

In Section 12.1 it was described how to realize permutations on the SE network with 2^n processors (vertices). For an arbitrary permutation, it was explained how to realize the permutation in $3n - 3$ passes. In this section, the task will be to realize permutations using a multistage network with switching boxes. We start by translating the model with the 2^n processors into the model of the multistage network. First, we will define the k-multistage SE network. Since we want to simulate a pass in the SE network by the edges of the multistage network, we consider the de Bruijn graph G_n rather than the SE network with 2^n vertices. In general, we can consider any UPP graph for this purpose as depicted in Fig. 12.9. The network has k stages, numbered from 0 to $k - 1$, where each stage has 2^n switching boxes that represent the 2^n vertices in the graph. The switching boxes are labeled (numbered) from 0 to $2^n - 1$ and in binary representation, from the all-zeros n-tuple to the all-ones n-tuple. The links between the switching boxes of two consecutive stages are exactly as the edges between the vertices of G_n, i.e., if there exists an edge from vertex u to vertex v in the graph, then there exists a link between switching box u in stage i to switching box v in stage $i + 1$, $0 \le i \le k - 2$. For the multistage SE network, assume there exists an edge from switching box $(x_1, x_2, \ldots, x_n)$ in one stage to switching box $(x_2, \ldots, x_n, x_{n+1})$ in the next stage. This edge will be labeled by $(x_1, x_2, \ldots, x_n, x_{n+1})$. Now, we will consider realizing permutations in the multistage network. For this purpose, we will distinguish between two possible translation models from the SE network to its multistage network.

The first one-to-one translation ignores the inputs to the network. The information (packets) will be stored in the switching boxes in the same way that they were stored in the processors of the network. The routing of a permutation in the multistage network will be precisely the same as was done in the SE network, where the two switching boxes $(0, x_2, \ldots, x_n)$ and $(1, x_2, \ldots, x_n)$ in stage i send their information to switching boxes in stage $i + 1$. One sends its information to switching box $(x_2, \ldots, x_n, 0)$ and the second to switching box $(x_2, \ldots, x_n, 1)$. Hence, a pass in the multistage network is implemented by sending the information from stage i to stage $i + 1$, $0 \le i \le k - 2$. Therefore $3n - 3$ passes can be implemented by a $(3n - 2)$-multistage SE network. It should be clear that this translation uses the switching box as a hardware element to transfer one input to one output and not to transfer two elements from two inputs to two outputs.

The second one-to-one translation considers 2^{n+1} inputs to the network and 2^{n+1} outputs from the network. The packets considered now are the inputs to the network. The switching boxes perform the shuffle and the possible exchange. Consider switching box $(x_1, x_2, \ldots, x_{n-1}, x_n)$ on stage i that receives its information from two edges, $(0, x_1, x_2, \ldots, x_{n-1}, x_n)$ and $(1, x_1, x_2, \ldots, x_{n-1}, x_n)$, coming from stage $i - 1$ or from the inputs if $i = 0$, and delivers the two pieces of information on edges, $(x_1, x_2, \ldots, x_{n-1}, x_n, 0)$ and $(x_1, x_2, \ldots, x_{n-1}, x_n, 1)$, going to stage $i + 1$ or to the outputs if stage i is the last stage. If the information coming from the edge $(0, x_1, x_2, \ldots, x_{n-1}, x_n)$ is delivered on the edge $(x_1, x_2, \ldots, x_{n-1}, x_n, 0)$ (which implies that the information coming from the edge $(1, x_1, x_2, \ldots, x_{n-1}, x_n)$ is delivered on the edge $(x_1, x_2, \ldots, x_{n-1}, x_n, 1)$), then only shuffle is performed. If the information coming from the edge $(0, x_1, x_2, \ldots, x_{n-1}, x_n)$ is delivered on the edge $(x_1, x_2, \ldots, x_{n-1}, x_n, 1)$ (which implies that the information coming from the edge $(1, x_1, x_2, \ldots, x_{n-1}, x_n)$ is delivered on the edge $(x_1, x_2, \ldots, x_{n-1}, x_n, 0)$)

then shuffle followed by exchange are performed. Therefore this translation model is for the SE network with 2^{n+1} packets (associated with 2^{n+1} processors). Each stage with 2^n switching boxes will be associated with one pass and therefore to realize a permutation on the 2^{n+1} inputs, by the simple algorithm, which by Corollary 12.1 requires $3(n + 1) - 3 = 3n$ passes, it will be required to have a $(3n)$-multistage network. Hence, to realize all the permutations on 2^n inputs with $3n - 3$ passes, it will require to have a $(3n - 3)$-multistage network.

We continue and consider the realization of permutations using the six networks that were defined, where the number of inputs is 2^{n+1}. We start with the following simple lemma regarding an $(n + 1)$-stage MFA network of order $n + 1$.

Lemma 12.12. *Each permutation that is realized by an $(n + 1)$-stage MFA network can be realized uniquely.*

Proof. This follows immediately from the fact that in an $(n + 1)$-stage MFA network of order $n + 1$ there is a unique path from each input of the network (before stage 0) to each output of the network (after stage n). This unique path is induced by the unique path from any switching box in stage 0 to any switching box in stage n. $\square$

Lemma 12.13. *The number of distinct permutations on the 2^{n+1} inputs that can be realized on an $(n + 1)$-stage MFA network is $2^{2^n(n+1)}$.*

Proof. In an $(n + 1)$-stage MFA network each stage has 2^n switching boxes. Each switching box can perform the two possible permutations on its two inputs to its two outputs independent of the other switching boxes. Hence, in each stage 2^{2^n} permutations can be performed from the inputs to the outputs of the switching boxes. There are $n + 1$ stages and hence $2^{2^n(n+1)}$ permutations can be realized with the $n + 1$ stages. By Lemma 12.12 all these permutations are distinct. $\square$

Theorem 12.9. *The number of distinct permutations that can be realized by a $(2n - 1)$-stage interconnection network of order $n + 1$ is smaller than $2^{n+1}!$.*

Proof. The number of permutations that can be realized in each stage is 2^{2^n} and hence with $2n - 1$ stages we can realize at most $2^{2^n(2n-1)}$ permutations. Since $\log_2 2^{2^n(2n-1)} = 2^n(2n - 1)$ and by the Stirling formula (see Theorem 1.21) we have that

$$\log_2(2^{n+1})! > 2^{n+1}(n + 1) - 2^{n+1}\log_2 e = 2^n(2n - 1) + 2^n(3 - 2\log_2 e) > 2^n(2n - 1)$$

and it follows that the number of permutations that can be realized by a $(2n - 1)$-stage interconnection network of order $n + 1$ is smaller than $2^{n+1}!$. $\square$

Corollary 12.2. *The minimum number of stages, required for a multistage interconnection network of order $n + 1$ to realize all the $2^{n+1}!$ permutations of the 2^{n+1} inputs, is $2n$.*

Since $2^{2^n 2n} > 2^{n+1}!$, it follows that theoretically a $(2n)$-stage interconnection network of order $n + 1$ might be able to realize all the $2^{n+1}!$ permutations on its inputs.

Problem 12.1. Does there exist a $(2n)$-stage interconnection network of order $n + 1$ that realizes all the $2^{n+1}!$ possible permutations in 2^{n+1} inputs?

The *concatenation* of two $(n + 1)$-multistage interconnection networks is done by joining together stage n of the first network with stage 0 of the second network. Switching box $x_1 x_2 \cdots x_n$ in stage n of the first network will coincide in this concatenation with switching box $x_1 x_2 \cdots x_n$ in stage 0 of the second network. The network that is obtained by this construction has $2n + 1$ stages.

Theorem 12.10. *By concatenating the baseline of order $n + 1$ with the reverse baseline of order $n + 1$ all the $2^{n+1}!$ permutations of the inputs can be realized.*

Proof. The proof is done by induction on n. The basis for the induction is $n = 1$, i.e., a network with 4 inputs, for which the proof is depicted with a picture for the routing of 6 permutations out of the $4! = 24$ permutations. Each permutation realized by such a network can be used to realize four permutations (each trivial permutation of the two outputs out of a switching box in the last stage yields 4 permutations for each routing). Hence, it is easily verified from Fig. 12.10 that the basis of the induction is solved.

Assume now that we can realize the $2^n!$ permutations for the concatenated network, of the baseline network of order n with the reverse baseline network of order n. The concatenated network has $2n - 1$ stages.

Finally, we will show the step of the induction. In the induction step, we have to construct the concatenated network of order $n + 1$ and to prove that all permutations can be realized on this concatenated network. The concatenation of the baseline network of order $n + 1$ with its reverse, the reverse baseline network of order $n + 1$ can be described as follows. Consider two networks NET0 and NET1, each one is a concatenation of the networks of order n. The inputs and outputs of these two networks are omitted. The whole network has 2^{n+1} inputs and 2^{n+1} outputs, numbered as integers from 0, 1, up to $2^{n+1} - 1$ and also as binary $(n + 1)$-tuples, where each binary $(n + 1)$-tuple represents its integer value. Each stage in the network has 2^n switching boxes numbered by the integers 0, 1, up to $2^n - 1$. The $2n - 1$ stages of the two networks of order n will be numbered from stage 1 to stage $2n - 1$. Inputs $2i$ and $2i + 1$ will be the inputs to switching box i of stage 0 and switching box i of stage $2n$ will deliver the information to outputs $2i$ and $2i + 1$. From switching boxes $(x_1, x_2, \ldots, x_{n-1}, 0)$ and $(x_1, x_2, \ldots, x_{n-1}, 1)$ of stage 0 there is an edge to switching box $(x_1, x_2, \ldots, x_{n-1})$ at the first stage of NET0 and an edge to switching box $(x_1, x_2, \ldots, x_{n-1})$ at the first stage of NET1 that will be stage 1 of the concatenated network. In any stage, the switching box $(x_1, x_2, \ldots, x_{n-1})$ on NET0 will be labeled by $(0, x_1, x_2, \ldots, x_{n-1})$ and the switching box $(x_1, x_2, \ldots, x_{n-1})$ on NET1 will be labeled by $(1, x_1, x_2, \ldots, x_{n-1})$. Similarly,

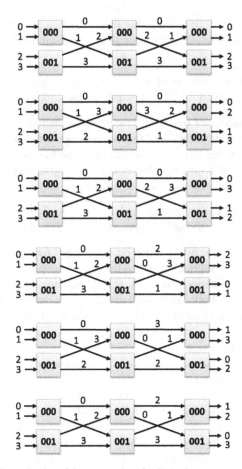

FIGURE 12.10 The realization of the permutations for the basis.

the switching boxes of the last stage of NET0 and NET1 deliver their information to the new last stage of the concatenated network of order $n + 1$. From switching box $(0, x_1, x_2, \ldots, x_{n-1})$ of the last stage in NET0 and switching box $(1, x_1, x_2, \ldots, x_{n-1})$ of the last stage in NET1, there are edges to switching boxes $(x_1, x_2, \ldots, x_{n-1}, 0)$ and $(x_1, x_2, \ldots, x_{n-1}, 1)$ of the last stage in the concatenated network. The construction of the network in the induction step is depicted in Fig. 12.11. It is easy to verify that this network is the concatenation of the baseline network of order $n + 1$ with the reverse baseline network of order $n + 1$.

We continue with the induction to show that all the $(2N)! = 2^{n+1}!$ permutations of the 2^{n+1} inputs can be implemented on the concatenated network, we construct an undirected graph $G = (V, E)$. The graph G has 2^{n+1} vertices numbered by $0, 1, \ldots, 2^{n+1} - 1$ associated with the 2^{n+1} inputs, and each integer

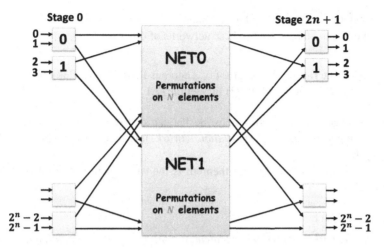

FIGURE 12.11   A network for realization of all $(2N)!$ permutations.

is also associated with its binary representation as a binary $(n + 1)$-tuple. The edges of the graph, are defined as follows. There is an edge between vertex $2i$ (represented by $(x_1 x_2 \cdots x_n 0)$) and vertex $2i + 1$ (represented by $(x_1 x_2 \cdots x_n 1)$) for each $0 \le i \le 2^n - 1$. There is an edge between vertex α and vertex β if α and β are inputs that by the permutation are set to two outputs from the same switching box. The degree of each vertex in the graph is 2. This implies that the graph consists of vertex-disjoint cycles. Since for any cycle, vertex $2i$ is in a cycle if and only if vertex $2i + 1$ is in the cycle, it follows that each cycle is of even length. Therefore the vertices of the graph can be colored with two colors, "0" and "1", which indicates to which part of the network the information will be delivered.

Based on this coloring, the routing is now simple. An input colored by "0" is routed to a switching box of NET0 whose binary representation starts with a *zero*, while an input that is colored by "1" is routed to a switching box of NET1 whose binary representation starts with a *one*. By the induction hypothesis, the two parts of the networks can realize $N! = 2^n!$ permutations (associated with the inputs of stage 1 and the outputs of stage $2n - 1$ is our concatenated network with $2n + 1$ stages numbered from stage 0 to stage $2n$). The first part of the network contains all those switching boxes whose binary representation starts with a *zero*. The second part of the network contains all those switching boxes whose binary representation starts with a *one*. Two inputs whose destination is to the same switching box of stage $2n$, where they will be delivered as outputs, are colored with different colors. Hence, NET0 has 2^n inputs that should be delivered to distinct switching boxes of stage $2n$ and the same is true for NET1. Therefore in each network, we should order its inputs in a permutation by the order in which they should be delivered from the last stage of NET0 and the last stage of NET1 to stage $2n$ of the concatenated network. This order defines a permutation for NET0 and a permutation for NET1, which by the induction hypothesis can be realized by these networks of order n.   □

Remark. The routing in the proof of Theorem 12.10 is essentially the same as that based on the graph $G(P, Q)$ in Section 12.1.

A k-stage interconnection network that can realize all the $N!$ permutations of its N inputs is called a *permutation network*. It is also called a *rearrangeable network*.

The proof of the following theorem is the same as in the proof of Theorem 12.10.

Theorem 12.11.

1. *The network obtained by concatenating the modified data manipulator network of order $n + 1$ with the indirect binary cube network of order $n + 1$ is a permutation network.*
2. *The network obtained by concatenating the modified data manipulator network of order $n + 1$ with the reverse baseline network of order $n + 1$ is a permutation network.*
3. *The network obtained by concatenating the baseline network of order $n + 1$ with the indirect binary cube network of order $n + 1$ is a permutation network.*

We have concatenated 4 pairs of the possible 36 pairs and proved in Theorems 12.10 and 12.11 that the concatenation of the networks in each pair, where stage n of one network coincides with stage 0 of the second network, yields a network that realizes all $2^{n+1}!$ permutations. The number of stages in each such concatenated network is $2n + 1$. What about the concatenation of the other 32 pairs? It is not difficult to prove that we can concatenate these 32 pairs to form $(2n + 1)$-stage interconnection networks that can realize all the $2^{n+1}!$ permutations. However, for this task sometimes it will be required to permute the switching boxes in stage 0 of the second network (or the nth one of the first network) to achieve this goal. By Theorem 12.8 the six networks (omega, flip, modified data manipulator, indirect binary cube, baseline, and reverse baseline) are isomorphic and hence we can rearrange the switching boxes in the last stage of the first network (or in the first stage of the second network) to concatenate them and obtain a permutation network. However, such a permutation of the switching boxes in the last stage is not required for all 36 pairs, as proved in Theorems 12.10 and 12.11. For example, we have the following results.

Theorem 12.12. *The network obtained by a trivial concatenation of each of the six networks (omega, flip, modified data manipulator, indirect binary cube, baseline, and reverse baseline) with its reverse is a permutation network.*

Corollary 12.3.

1. *The network obtained by concatenating the omega network of order $n + 1$ with the flip network of order $n + 1$ is a permutation network.*
2. *The network obtained by concatenating the flip network of order $n + 1$ with the omega network of order $n + 1$ is a permutation network.*

The networks in Corollary 12.3 were highlighted from the other pairs since the structure between consecutive stages in these networks is that of the de Bruijn graph (for the omega network) or its reverse (for the flip network). Other concatenations can be also analyzed in the same way, but this is left as an exercise. One of the most celebrated open problems is the next question.

Problem 12.2. Is the concatenation of the omega network of order n with the omega network of order n a rearrangeable network?

Problem 12.2 is equivalent to the following problem.

Problem 12.3. Can a SE network with 2^n processor realize any given permutation in $2n - 1$ passes?

A more general question than the one in Problem 12.2 is the following.

Problem 12.4. Which concatenation from the possible 36 of the defined six network is a rearrangeable network and which one is not, where switching box i of the last stage in the first network coincides with switching box i of the first stage in the second network, for each i, $0 \le i \le 2^{n-1} - 1$?

Note that by Corollary 12.1 concatenation of three omega networks yields a rearrangeable network.

We have seen that there are many $(2n + 1)$-stage interconnection networks of order $n + 1$ that form permutation networks. By Corollary 12.2 we have that theoretically at least $2n$ stages are required in a multistage permutation network of order $n + 1$. It is believed that such a network with $2n$ stages does not exist (see Problem 12.1). The number of switching boxes in such networks that were constructed in this section is $(2n + 1)2^n$. Can we construct a permutation network for 2^{n+1} inputs, with a smaller number of switching boxes?

A permutation network that realizes $N!$ permutations must make at least $\log_2 N!$ binary decisions to realize the $N!$ permutation and hence at least $\log_2 N!$ switching boxes are required to realize $N!$ permutations. Hence, we have the following theorem.

Theorem 12.13. *A permutation network with N inputs must have at least $N \cdot \log_2 N - N + \Theta(\log_2 N)$ switching boxes. If $N = 2^n$, then a permutation network with N inputs must have at least $2^n n - 2^n + \Theta(n)$ switching boxes.*

Proof. At least $\log_2 N!$ switching boxes are required. The proof follows now from Stirling's approximation (see Theorem 1.21). □

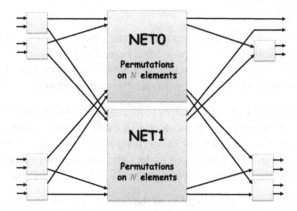

FIGURE 12.12 A network for realization of all $(2N)!$ permutations omitting switching boxes.

We will discuss now the general problem of realization of $N!$ permutations of N inputs into N outputs where the information is routed through a network that consists of 2×2 switching elements (boxes). If we want to keep the structure of Fig. 12.11 to realize $(2N)!$ permutations, then the routing will be done as was explained in the proof of Theorem 12.10. The routing was done by assigning two colors "0" and "1" to a graph $G = (V, E)$. There are at least two possible colorings as we can switch between the two colors. This implies that we can either choose one pair of inputs $2i$ and $2i + 1$ and decide which one will be routed to NET0 and which one to NET1. By doing so, we omit switching box i in stage 0 of the network and save one switching box. Similarly, we can consider the outputs instead of the inputs. If inputs α and β are the output of a chosen switching box i in stage $2n$, then we can decide which one will be delivered from NET0 and which one will be delivered from NET1, omitting switching box i in stage $2n$. We can omit one switching box either from the first stage or from the last stage. We will choose to omit the first switching box (number 0) in stage $2n$, as depicted in Fig. 12.12.

The network with $2n + 1$ stages that have 2^{n+1} inputs is constructed recursively as follows. For the basis of the recursion, $n + 1 = 2$, we have $2^{n+1} = 4$ inputs, a 3-stage network, and 5 switching boxes. The required network is depicted in Fig. 12.13 where the realization of 4 permutations is illustrated. For larger n, the recursion illustrated in Fig. 12.12 is applied. Let Ψ_{n+1} denote the network constructed in this way.

Theorem 12.14. *The network Ψ_{n+1} is a permutation network that has 2^{n+1} inputs, $2n + 1$ stages, and $2^{n+1} \cdot n + 1$ switching boxes.*

Proof. Clearly, by the recursive construction, the network Ψ_{n+1} has 2^{n+1} inputs and $2n + 1$ stages. The number of switching boxes is proved by induction.

The basis is $n = 1$, where the number of switching boxes is $2^2 + 1 = 5$ (see Fig. 12.13).

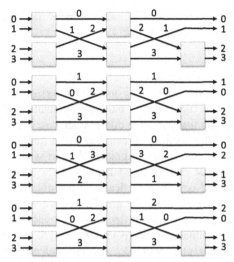

FIGURE 12.13   A network with 5 switching boxes realizing all permutations with 4 inputs (4 permutations are demonstrated).

For the induction hypothesis, assume that for $n - 1$, the number of switching boxes is $2^n \cdot (n - 1) + 1$.

For the induction step n we have the networks NET0 and NET1 on which we can apply the induction hypothesis. To these two networks we can add 2^n switching boxes in stage 0 and $2^n - 1$ switching boxes in stage $2n$. The total number of switching boxes in Ψ_{n+1} is

$$2(2^n(n-1)+1) + 2^n + 2^n - 1 = 2^{n+1} \cdot (n-1) + 2^{n+1} + 1 = 2^{n+1} \cdot n + 1.$$

Now, to complete the proof we have to show that each permutation of its 2^{n+1} inputs can be realized by the network. For this, the same recursive technique as in the proof of Theorem 12.10 is used. The colors "0" and "1" are decided to ensure that the first output, which was supposed to be delivered from the omitted switching box, will be delivered from NET0, i.e., its color is "0". As a consequence, the second output, which was supposed to be delivered from the omitted switching box, will be delivered from NET1, i.e., its color is "1".   □

By Theorem 12.13 we have that a permutation network with 2^{n+1} inputs must have at least $2^{n+1} \cdot n + \Theta(n)$ switching boxes. This implies the following observation.

Corollary 12.4. *Asymptotically, the network Ψ_n is an optimal permutation network.*

Corollary 12.5. *If the network Ψ_n has the smallest number of switching boxes for 2^n inputs, then a multistage network of order n must have at least $2n - 1$ stages to be rearrangeable.*

Problem 12.5. Does there exist a permutation network with 2^{n+1} inputs and less than $2^{n+1} \cdot n + 1$ switching boxes?

12.4 Layouts

In this section, we design embedding for the SE network and the de Bruijn graph. To implement the networks for parallel computation, the network has to be embedded on a chip that can have different shapes, but usually, the chip is a rectangular grid. In this section, we examine two different methods to embed the SE network and/or the de Bruijn graph. We start with a layout for the SE network on a rectangular grid. In each such layout, the vertices will be located on the grid points, and each two vertices that are connected by an edge are connected by wires that go along the grid lines. Two wires can intersect, but cannot go in parallel along the same line. The efficiency of the layout can be measured by a few parameters, such as the total length of the wires, the width of the grid or its height, etc. Of course, also the tradeoff between the various parameters can be taken into account. The parameter that will be considered to be essential in this section is the area of the grid that is just the number of vertical lines multiplied by the number of horizontal lines that are used by the wires in the grid.

It is quite natural to distinguish between the shuffle edges and the exchange edges when a layout for the SE network is designed. Furthermore, it is also quite natural to have the shuffle edges based on the necklaces placed on a cycle for each necklace, where a cycle will occupy either two vertical lines or two horizontal lines. We will distinguish now between two methods of designing such a layout. The two methods differ in the way in which the necklaces are placed in the layout.

Layout with all vertices on the same line:

All the vertices will be located on the first horizontal track of the rectangular grid. They are ordered from left to right by ascending weights in a way that the vertices of each necklace are consecutive in this order. Each set of n vertices (or less) in a necklace is connected by a line (a wire) on the first track. Given an exchange edge, the associated vertices will be connected by taking two vertical lines on the track, one from each vertex, and connecting these two lines with the first empty horizontal track between these two vertical tracks. An example of this layout for the SE network with 8 vertices is depicted in Fig. 12.14(a). The area of this layout is 48. Note that the layout can be improved to a layout with area 24 if horizontal lines are used for exchange edges in adjacent necklaces, as depicted in Fig. 12.14(b).

Theorem 12.15. *When N is large enough, the area required for a layout with all the vertices on the first track is at most $\frac{N^2}{\sqrt{\log N}}$.*

Proof. The vertices in the first track occupy $N = 2^n$ vertical tracks. Since the exchange edges are between vertices whose weight differs by one, it follows

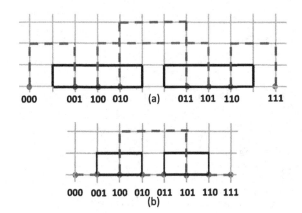

000   001 100 010   (a)   011 101 110   111

000 001 100 010 011 101 110 111
(b)

FIGURE 12.14   Layouts for the SE network with 8 vertices placed on the same horizontal line.

that an upper bound on the number of horizontal tracks that are occupied by exchange edges is the number of vertices in the weight with the most vertices. This weight is clearly $\lfloor \frac{n}{2} \rfloor$ (or $\lceil \frac{n}{2} \rceil$) and by Theorem 1.21 we have that $n! \sim \sqrt{2\pi n} \left(\frac{n}{e}\right)^n$ and hence the number of horizontal tracks that are occupied by exchange edges is at most

$$\binom{n}{\lfloor n/2 \rfloor} = \frac{n!}{\lfloor n/2 \rfloor! \lceil n/2 \rceil!} \sim \frac{\sqrt{2\pi n} \left(\frac{n}{e}\right)^n}{\sqrt{\frac{2\pi n}{2}} \left(\frac{n}{2e}\right)^{n/2} \sqrt{\frac{2\pi n}{2}} \left(\frac{n}{2e}\right)^{n/2}} = \sqrt{\frac{2}{\pi n}} 2^n$$

and since

$$N \cdot \sqrt{\frac{2}{\pi n}} 2^n < \frac{N^2}{\sqrt{\log N}},$$

the claim of the theorem follows.   □

Layout based on the complex plane:

A much more efficient layout for the SE network will be obtained if we consider the mappings of the necklaces to the complex plane. A necklace will be represented as $\langle \alpha \rangle$, where α is the integer value of one of its vertices. A necklace that will be represented by $\langle \alpha \rangle$ contains the vertices $\alpha, 2\alpha, 4\alpha$, and so on, where all these integers are taken modulo $2^n - 1$ and α is the smallest integer that represents one of these vertices.

Let $\delta_n = e^{2\pi i/n}$ be the nth primitive root of unity and denote a vertex by $x = (x_1, x_2, \ldots x_{n-1} x_n)$. Define the mapping

$$\xi(x) = \delta_n^{n-1} x_1 + \delta_n^{n-2} x_2 + \cdots + \delta_n x_{n-1} + x_n.$$

This mapping maps each n-tuple $x = (x_1, x_2, \ldots x_{n-1} x_n)$ to the complex plane.

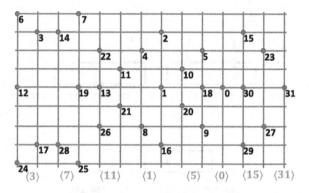

FIGURE 12.15 The level-necklace grid of the SE network with 32 vertices.

Consider now the following equalities.

$$\delta_n \cdot \xi(x) = \delta_n \cdot \xi(x_1 x_2 \cdots x_{n-1} x_n)$$
$$= \delta_n^n x_1 + \delta_n^{n-1} x_2 + \cdots + \delta_n^2 x_{n-1} + \delta_n x_n$$
$$= x_1 + \delta_n^{n-1} x_2 + \cdots + \delta_n^2 x_{n-1} + \delta_n x_n$$
$$= \delta_n^{n-1} x_2 + \cdots + \delta_n^2 x_{n-1} + \delta_n x_n + x_1$$
$$= \xi(x_2 \cdots x_{n-1} x_n x_1).$$

These equalities imply that the traversal of the shuffle edge corresponds to a $\frac{2\pi}{n}$ rotation in the complex plane. Hence, the vertices and the edges associated with the same necklace are symmetrically placed about the origin of the complex plane.

Since the edges associated with the same necklace have symmetry about the origin, it follows that degenerated necklaces are mapped to the origin of the complex plane. As for the exchange edge, we consider the following equation

$$\xi(x_1 x_2 \cdots x_{n-1} 0) + 1 = \delta_n^{n-1} x_1 + \cdots + \delta_n x_{n-1} + 0 + 1 = \xi(x_1 x_2 \cdots x_{n-1} 1).$$

Therefore exchange edges are contained in the same horizontal line in the complex plane. Such a line is called a *level*.

The *level-necklace grid* is a grid whose rows are associated with the levels of the exchange edges in the complex plane. The order of the exchange edges on this grid is the same as their order in the complex plane. Each full-order necklace is represented by a pair of consecutive columns, the left column for the vertices that are in the left half of the complex plane, and the right column for vertices in the right half of the complex plane. In Fig. 12.15 the level-necklace grid of the SE network with 32 vertices is depicted.

Theorem 12.16. *The area required for a layout with the level-necklace grid based on the complex plane is smaller than $N^2/\log N + N\sqrt{N}$.*

Proof. By Lemma 3.20 we have that there are at most $\sqrt{N}$ degenerated necklaces and $\frac{N}{\log N} - O\left(\frac{\sqrt{N}}{\log N}\right)$ full-order necklaces of order n. Hence, in total there are less than $\frac{N}{\log N} + \sqrt{N}$ necklaces that require fewer than $\frac{2N}{\log N} + 2\sqrt{N}$ vertical columns in the rectangular grid. There are $\frac{N}{2}$ exchange edges and hence they require fewer than $\frac{N}{2}$ horizontal lines. Thus the theorem is proved. $\square$

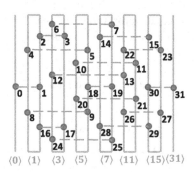

FIGURE 12.16 A layout based on the complex plane of the SE network with 32 vertices.

The layout of the SE network with 32 vertices based on the complex plane is presented in Fig. 12.16. The area of this layout is 140 with 14 vertical lines and 10 horizontal lines. However, this layout can be improved, e.g., there exists a layout of the SE network with 32 vertices, whose area is 84.

Another type of layout for graphs is called *embedding in books*. For this last topic covered in the book, we leave some of the proofs as exercises for the reader. A book has a *spine* on which we order all the vertices of the graph. The book also has some *pages* that share the spine as a common boundary. On the pages, the edges of the graph are drawn in a way that there is no crossing of edges on a page. The first target is to embed the graph in a book with the smallest number of pages.

A *book embedding* of a graph G consists of a linear ordering of the vertices of G on the spine and an assignment for the edges of G to pages such that there is no crossing of edges on each page. The *pagenumber* of G is the minimum number of pages of books in which G can be embedded. The *pagenumber* of a set S of graphs is the minimum number of pages of books in which each graph in S can be embedded.

The *width* of a page is the maximum number of edges that cross any line perpendicular to the spine of the book. The *cumulative pagewidth* is the sum of the widths of all the pages of the book.

Now, we will describe a book embedding for de Bruijn graph G_n using only three pages. Let $T_\alpha(n)$ be the subgraph (spanning tree) induced by all the edges of G_n that start with the symbol α, $\alpha \in \{0, 1\}$, except for the self-loops edges. These trees are exactly the BFS trees from the self-loops. For G_4 these trees are depicted in Fig. 11.1.

Lemma 12.14. $T_\alpha(n)$ *is a directed tree.*

Proof. We prove that claim for $T_0(n)$ and the exact arguments will hold for $T_1(n)$. $T_0(n)$ contains all the vertices of G_n and only nonzero edges that start with a *zero*. Hence, the number of vertices in $T_0(n)$ is 2^n and the number of edges is $2^n - 1$. Each vertex $(x_1, x_2, \ldots, x_{n-1}, x_n)$ has a unique in-going edge from a vertex whose representation starts with a *zero*, namely $(0, x_1, x_2, \ldots, x_{n-1})$. Hence, if we remove the all-zeros self-loop edge, then the in-degree of each vertex in $T_0(n)$, except the all-zeros vertex, is one. To complete the proof we have to show that in $T_0(n)$ there is a path from the all-zeros vertex to each other vertex in $T_0(n)$.

Given a vertex $(x_1, x_2, \ldots, x_{n-1}, x_n) \neq (0^n)$, for which i is the first index for which $x_i \neq 0$, there exists a unique directed path in $T_0(n)$ from $(0, 0, \ldots, 0)$ to $(x_1, x_2, \ldots, x_{n-1}, x_n)$,

$$(0, 0, \ldots, 0) \longrightarrow (0, \ldots, 0, x_i) \longrightarrow (0, \ldots, 0, x_i, x_{i+1}) \longrightarrow \cdots$$
$$\cdots \longrightarrow (0, \ldots, 0, x_i, x_{i+1}, \ldots, x_{n-1}, x_n) = (x_1, x_2, \ldots, x_{n-1}, x_n).$$

Therefore $T_0(n)$ is a directed tree. Similarly $T_1(n)$ is a directed tree, where the vertices and the edges are exactly the complements of the vertices and edges of $T_0(n)$. $\square$

The edges of $T_0(n)$ start with a *zero*, while the edges of $T_1(n)$ start with a *one* and each edge of G_n starts either with a *zero* or a *one*. Hence, the following lemma is implied.

Lemma 12.15. *For any given $n \geq 2$, we have that for $G_n = (V_n, E_n)$*

$$E(T_0(n)) \cap E(T_1(n)) = \varnothing$$

and

$$E(T_0(n)) \cup E(T_1(n)) = E_n \setminus \{(0^{n+1}), (1^{n+1})\},$$

where $E(G)$ is the set of edges of the graph G.

The *layers of the edges* in $T_\alpha(n)$ will be numbered from layer 1 (the bottom layer of the tree that contains all the edges to its leaves, which are the vertices whose binary representation starts with $\bar{\alpha}$) to layer n (contains the edge from $(\alpha, \alpha, \ldots, \alpha, \alpha)$ to $(\alpha, \alpha, \ldots, \alpha, \bar{\alpha})$. The *levels of the vertices* (levels are used for vertices to distinguish from the layers of edges) will be numbered from 0 to n, where layer i, $1 \leq i \leq n$, of edges, contains edges from vertices of level i to vertices of level $i - 1$. In level i, $0 \leq i \leq n$, we have all the vertices that start with exactly i *zeros* followed by a *one*.

Before the description of the embedding of the edges of G_n in the pages of the book, an ordering of the vertices of G_n for the spine of the book should be given. Let $S = s_0, s_1, \ldots, s_{2^n-1}$ be the sequence that forms the order of these vertices on the spine. The order of the vertices is based on their order in the BFS tree $T_0(n)$ (from the root to the leaves, a level by level of vertices, from level n to level 0). The order should satisfy a few properties as follows:

(**P1**) The sequence of vertices is a complement reverse sequence of integers, i.e., it is a CR sequence. In other words,

$$S = s_0, s_1, \ldots, s_{2^n-1} = \bar{s}_{2^n-1}, \ldots, \bar{s}_1, \bar{s}_0 = \bar{S}^R.$$

(**P2**) Each vertex of level i, $0 \le i \le n - 1$, is some s_j, $2^{n-i-1} \le j \le 2^{n-i} - 1$. The vertex 0^n of level n is s_0.

(**P3**) If $s_{2i} = 2j$ for some $0 \le i, j \le 2^{n-1} - 1$, then $s_{2i+1} = 2j + 1$.

(**P4**) If $s_{2i} = 2j + 1$ for some $0 \le i, j \le 2^{n-1} - 1$, then $s_{2i+1} = 2j$.

(**P5**) If $s_{i_1} = j$ for $2^\ell \le i_1, j \le 2^{\ell+1} - 1$, where $2 \le \ell \le n - 1$ and $s_{r_1} = 2j$, $s_{r_1+1} = 2j+1$, then for $i_1 < i_2 \le 2^{\ell+1} - 1$ we have that $s_{i_2} = t$ for some t, and $s_{r_2} = 2t$, $s_{r_2+1} = 2t + 1$, where $r_2 < r_1$.
If $s_{i_1} = j$ for $2^\ell \le i_1, j \le 2^{\ell+1} - 1$, where $2 \le \ell \le n - 1$ and $s_{r_1} = 2j+1$, $s_{r_1+1} = 2j$, then for $i_1 < i_2 \le 2^{\ell+1} - 1$ we have that $s_{i_2} = t$ for some t, and $s_{r_2} = 2t + 1$, $s_{r_2+1} = 2t$, where $r_2 < r_1$.

The properties (**P1**) through (**P5**) that the sequence S must satisfy will guarantee the following associated properties that enable embedding G_n in three pages. Property (**P1**) enables us to apply the same embedding for the edges of $T_0(n)$ and also for the edges of $T_1(n)$ in reverse order. By property (**P2**) we have that on the spine the vertices are ordered by their levels. In other words, we have the following lemma.

Lemma 12.16.

- *For each i, $0 \le i \le n$, the vertices of level i in $T_\alpha(n)$ are in consecutive entries of S.*
- *For each i, $1 \le i \le n$, the vertices of level i in $T_0(n)$ are in entries after the vertices of level $i - 1$.*
- *For each i, $0 \le i \le n - 1$, the vertices of level i in $T_1(n)$ are in entries before the vertices of level $i + 1$.*

By properties (**P3**) and (**P4**) we have that the two edges $j \to 2j$ and $j \to 2j + 1$ will be consecutive edges in the embedding (for the page in which they will be contained). Property (**P5**) guarantees that the edges of any given layer in $T_\alpha(n)$ will not intersect any other edges, of the same layer, in the page to which they will be contained. The full arguments for this assertion are left as an exercise to the reader.

The next step is to define a sequence S for each G_n, where $n \ge 4$, which satisfies properties (**P1**) through (**P5**). The sequence will be defined recursively as follows. The initial conditions for the recursion are the sequences for $n = 4$ and $n = 5$ presented in Example 12.1.

Assume now that the sequence S' of length 2^n, $n \ge 4$, is given by

$$S' = s_0', s_1', \ldots, s_{2^n-1}'.$$

For length 2^{n+2}, the sequence

$$S = s_0, s_1, \ldots, s_{2^{n+2}-1}$$

is constructed as follows.

(S1) For each $0 \leq i \leq 2^n - 1$, assign $s_i := s'_i$.

(S2) For each $0 \leq i \leq 2^n - 1$, assign $s_{2^{n+2}-1-i} := \bar{s}_i$.

(S3) Given the assignment of s_{2i} and s_{2i+1}, where $\{s_{2i}, s_{2i+1}\} = \{2r, 2r + 1\}$ and $2^n \leq i \leq 2^{n+1} - 1$, let $j = 2^{n+1} - 1 - i$ and assign $s_{2^n+j} := r$.

(S4) Given an assignment $s_{2i} = 2r$, assign $s_{2i+1} := 2r + 1$; given an assignment $s_{2i} = 2r + 1$, assign $s_{2i+1} := 2r$.

(S5) Given an assignment of some s_i, where $0 \leq i \leq 2^{n+1} - 1$, assign $s_{2^{n+2}-1-i} := \bar{s}_i$.

(S6) Continue to apply steps **(S3)**, **(S4)**, and **(S5)**, in arbitrary order, as long as new entries are assigned by one of these steps.

The following theorem makes the necessary relations between the sequence S and properties **(P1)** through **(P5)** and it is also left as an exercise.

Theorem 12.17. *The sequences S obtained from steps* **(S1)** *through* **(S6)** *satisfies properties* **(P1)** *through* **(P5)**.

Example 12.1. For each $n = 4$, 5, and 6, a sequence $S = s_0, s_1, s_2, \ldots$ that satisfies properties **(P1)** through **(P5)** is given as follows:
For $n = 4$, the sequence S is

$$0, 1, 2, 3, 7, 6, 4, 5, 10, 11, 9, 8, 12, 13, 14, 15.$$

For $n = 5$, the sequence S is

$$0, 1, 2, 3, 5, 4, 6, 7, 15, 14, 13, 12, 8, 9, 11, 10,$$
$$21, 20, 22, 23, 19, 18, 17, 16, 24, 25, 27, 26, 28, 29, 30, 31.$$

For $n = 6$, the sequence S will be constructed according to steps **(S1)** through **(S6)**. Step **(S1)** yields the first 16 elements of S according to the sequence for $n = 4$ and step **(S2)** continues to generate the last 16 elements induced by the first 16 elements as follows:

$$0, 1, 2, 3, 7, 6, 4, 5, 10, 11, 9, 8, 12, 13, 14, 15,$$
$$s_{16}, s_{17}, s_{18}, s_{19}, s_{20}, s_{21}, s_{22}, s_{23}, s_{24}, s_{25}, s_{26}, s_{27}, s_{28}, s_{29}, s_{30}, s_{31},$$
$$s_{32}, s_{33}, s_{34}, s_{35}, s_{36}, s_{37}, s_{38}, s_{39}, s_{40}, s_{41}, s_{42}, s_{43}, s_{44}, s_{45}, s_{46}, s_{47},$$
$$48, 49, 50, 51, 55, 54, 52, 53, 58, 59, 57, 56, 60, 61, 62, 63.$$

Now, we apply step **(S3)** and step **(S4)** 8 times each as follows. $s_{48} = 48$ and $s_{49} = 49$ yield $s_{23} = 24$ and $s_{22} = 25$. $s_{52} = 55$ and $s_{53} = 54$ yield $s_{21} = 27$ and $s_{20} = 26$. $s_{56} = 58$ and $s_{57} = 59$ yield $s_{19} = 29$ and $s_{18} = 28$. $s_{60} = 60$ and

$s_{61} = 61$ yield $s_{17} = 30$ and $s_{16} = 31$. On these 8 new values we apply step (S5) to obtain more 8 values and the sequence is now

$$0, 1, 2, 3, 7, 6, 4, 5, 10, 11, 9, 8, 12, 13, 14, 15,$$
$$31, 30, 28, 29, 26, 27, 25, 24, s_{24}, s_{25}, s_{26}, s_{27}, s_{28}, s_{29}, s_{30}, s_{31},$$
$$s_{32}, s_{33}, s_{34}, s_{35}, s_{36}, s_{37}, s_{38}, s_{39}, 39, 38, 36, 37, 34, 35, 33, 32,$$
$$48, 49, 50, 51, 55, 54, 52, 53, 58, 59, 57, 56, 60, 61, 62, 63.$$

We continue with this sequence of steps and the new values that were obtained. $s_{40} = 39$ and $s_{41} = 38$ yield $s_{27} = 19$, $s_{26} = 18$, $s_{36} = 44$, and $s_{37} = 45$. $s_{44} = 34$ and $s_{45} = 35$ yield $s_{25} = 17$, $s_{24} = 16$, $s_{38} = 46$, and $s_{39} = 47$. On the new 8 values, we can continue with the same steps as follows. $s_{36} = 44$ and $s_{37} = 45$ yield $s_{29} = 22$, $s_{28} = 23$, $s_{34} = 41$, and $s_{35} = 40$. Finally, we have one more sequence of these steps. $s_{34} = 41$ and $s_{35} = 40$ yield $s_{30} = 20$, $s_{31} = 21$, $s_{33} = 43$, and $s_{32} = 42$. The final sequence for $n = 6$ is

$$0, 1, 2, 3, 7, 6, 4, 5, 10, 11, 9, 8, 12, 13, 14, 15,$$
$$31, 30, 28, 29, 26, 27, 25, 24, 16, 17, 18, 19, 23, 22, 20, 21,$$
$$42, 43, 41, 40, 44, 45, 46, 47, 39, 38, 36, 37, 34, 35, 33, 32,$$
$$48, 49, 50, 51, 55, 54, 52, 53, 58, 59, 57, 56, 60, 61, 62, 63.$$

∎

Lemma 12.17. *When the recursion step from a sequence of length 2^n to a sequence of length 2^{n+2} is finished all the 2^{n+2} entries of S are assigned with some values.*

Proof. After steps (S1) and (S2) of the recursion all the entries s_i, where $0 \le i \le 2^n - 1$ or $2^{n+1} + 2^n \le i \le 2^{n+2} - 1$ are assigned (the first 2^n entries and the last 2^n entries). We refer to the first 2^{n+1} entries of S as the first half and to the last 2^{n+1} entries of S as the second half.

In either step (S2) or step (S5), 2^ℓ new assignments are carried out for the second half of S for some ℓ, where $1 \le \ell \le n$. These 2^ℓ new assignments immediately implies $2^{\ell-1}$ new assignments to the first half of S in step (S3) that follows by $2^{\ell-1}$ new assignments to the second half of S in step (S5). This process continues and there is no need to apply step (S4) until $2i = 2^{n+1} - 2$. When step (S4) is applied it follows by new assignment to $s_{2^{n+1}}$ and $s_{2^{n+1}+1}$. When these assignments are carried out the number of new assignments for the first half in steps (S3) and (S4) is

$$2^{n-1} + 2^{n-2} + \cdots + 2 + 1 + 1 = 2^n$$

and to the second half in steps (S2) and (S5)

$$2^n + 2^{n-1} + 2^{n-2} + \cdots + 4 + 2 + 2 = 2^{n+1}.$$

Combining these numbers with the 2^n assignments in step **(S1)** we have that all the 2^{n+2} entries of S are assigned. ☐

Lemma 12.18. *For each ℓ, $2 \leq \ell \leq n+1$, the entries of S, s_i, $2^\ell \leq i \leq 2^{\ell+1} - 1$ are assigned with distinct values in the range between 2^ℓ and $2^{\ell+1} - 1$.*

Proof. The proof is again by induction, where the basis is again the sequence S of length n for $n = 4$ and $n = 5$, as presented in Example 12.1.

Assume now that the claim is true for a sequence S' of length 2^n, $n \geq 4$ and we will provide a proof for a sequence S of length 2^{n+2}. By the induction hypothesis, the claim is true for the first 2^n elements of S as they are the same as the elements of S' in the same order. It remains to prove the claim for the next 2^n elements of S and the last 2^n elements of S. By steps **(S2)** and **(S5)**, as long as the first 2^{n+1} elements of S are distinct and smaller than 2^{n+1}, it follows that the last 2^{n+1} elements are distinct and in the range between 2^{n+1} and $2^{n+2} - 1$.

Step **(S3)** will obtain distinct elements in the required range as long as the elements obtained in step **(S5)** are distinct and in the required range. Therefore the induction step follows immediately. ☐

Corollary 12.6. *The edges between two levels of the same parity (their number modulo 2) in $T_0(n)$ cannot intersect in a page whose spine is the sequence S of length 2^n.*

To complete the proof for the embedding of G_n with only three pages, it is required to show that the edges of G_n can be partitioned into three subsets and each one can use the order of the vertices on the spine, defined by the sequence S, to embed the edges with no intersection.

Lemma 12.19. *The tree $T_\alpha(n)$ is a subtree of $T_\alpha(n+2)$ induced by the vertices of levels 2 through $n + 2$.*

Proof. By the definition of $T_\alpha(n)$, we have that the edges in $T_0(n)$ are $i \to 2i$ and $i \to 2i + 1$ for each $0 \leq i \leq 2^{n-1} - 1$. The edges in $T_0(n+2)$ are $i \to 2i$ and $i \to 2i + 1$ for each $0 \leq i \leq 2^{n+1} - 1$. This immediately implies the claim of the lemma for $\alpha = 0$.

The tree $T_1(n)$ is the complement of the tree $T_0(n)$. Similarly, the tree $T_1(n+2)$ is the complement of the tree $T_0(n+2)$ and hence the correctness of the claim for $\alpha = 0$ implies its correctness for $\alpha = 1$. ☐

Corollary 12.7. *The parity of the layers of edges in $T_\alpha(n)$ is the same as the parity of the layers of edges, from layer 3 through layer $n + 2$ in $T_\alpha(n+2)$.*

Theorem 12.17 implies the following lemma.

Lemma 12.20. *The sequence S satisfies property* **(P5)**.

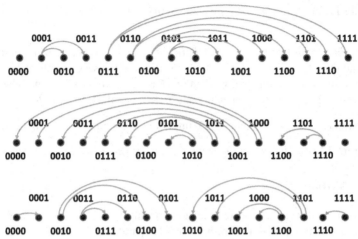

FIGURE 12.17   Book embedding for G_4 with three pages, the embedding of $A_0(4)$ on the top, the embedding of $A_1(4)$ on the middle, and the embedding of $B_0(4) \cup B_1(4)$ on the middle.

Corollary 12.8. *For any given layer i, $1 \leq i \leq n$, of edges of $T_0(n)$, there is no intersection between the different edges of layer i of $T_0(n)$, in a page of the book, where the order of the vertices on the spine is given by the sequence S of length 2^n.*

Now, we partition the edges of $T_\alpha(n)$ into two subsets $A_\alpha(n)$ and $B_\alpha(n)$. The subset $A_\alpha(n)$ will contain all the edges in the odd layers of $T_\alpha(n)$ and the subset $B_\alpha(n)$ will contain all the edges in the even layers of $T_\alpha(n)$. The following lemmas will form the proof that G_n can be embedded in a book with three pages, which is demonstrated in Fig. 12.17 on G_4.

Lemma 12.21. *The edges of $A_0(n)$ can be embedded in one page if the spine is ordered by the sequence S.*

Proof. By Corollary 12.6, there is no intersection between the edges of two different layers and the same parity. By Corollary 12.8 there is no intersection between the edges of the same layer. Since all the edges of $A_0(n)$ are from layers of the same parity the claim of the lemma follows.   □

Lemma 12.22. *The edges of $A_1(n)$ can be embedded in one page.*

Proof. Recall that if there is an edge $x \rightarrow y$ in G_n, then there is also the edge $\bar{x} \rightarrow \bar{y}$ in G_n. Moreover, the edge $x \rightarrow y$ is in layer i of $T_\alpha(n)$ if and only if the edge $\bar{x} \rightarrow \bar{y}$ is in layer i of $T_{\bar{\alpha}}(n)$. By Lemma 12.21, this implies that the same embedding for the edges of $A_0(n)$ will work the same for the complement vertices and the edges in $A_1(n)$.   □

To complete the book embedding of G_n we have the following lemma.

Lemma 12.23. *The edges of $B_0(n)$ and $B_1(n)$ can be embedded together in one page.*

Proof. First, note that the edges of $B_0(n)$ are between vertices of the first half of S and the edges of $B_1(n)$ are between vertices of the second half of S.

Now, the same claims that worked on $A_0(n)$ in Lemma 12.21 and on $A_1(n)$ in Lemma 12.22 will work on $B_0(n)$ and $B_1(n)$, and the lemma follows. ☐

Lemmas 12.21, 12.22, and 12.23, imply the following concluding result.

Theorem 12.18. *The de Bruijn graph G_n can be embedded in three pages.*

12.5 Notes

Interconnection networks were considered from the 1970s and the interest in them did not stop. Definitions of networks for parallel processing, their use for routing, and realizing permutations are only a few of the problems that were considered in the literature. For example, another important problem that was extensively studied is the fault tolerance of the defined network as it is quite natural that a few vertices (processors) or edges (communication links between the processors) will not work either for technical reasons or by an attack from a third party. Hence, it was asked whether when a processor or a link is not functioning, there exists an alternative communication path between processors that does not use this processor or link in their communication. Such consideration of the fault tolerance in de Bruijn networks and related networks was discussed, for example, in Bermond, Homobono, and Peyrat [10], Bruck, Cypher, and Ho [14], Du, Lyuu, and Hsu [24], Homobono and Peyrat [33], Mao and Yang [57], Rowley and Bose [68,69], and Sridhar and Raghavendra [73]. From the de Bruijn graph, many related networks were defined, e.g., the UPP graphs discussed in Chapter 11, the SE network discussed in this chapter, and other networks. One such example is the Kautz network defined as follows. The vertices of the network are n-tuples over an alphabet Σ with σ letters, where there are no identical two consecutive symbols in an n-tuple. There is an edge from the vertex labeled by $(x_1, x_2, \ldots, x_n)$ to each vertex labeled by $(x_2, \ldots, x_n, x_{n+1})$, where $x_{n+1} \in \Sigma \setminus \{x_n\}$. Therefore the Kautz graph has $\sigma \cdot (\sigma - 1)^{n-1}$ vertices and $\sigma \cdot (\sigma - 1)^n$ edges. Each vertex has in-degree $\sigma - 1$ and out-degree $\sigma - 1$. The Kautz network is a subgraph of the de Bruijn graph and it was also extensively considered by, e.g., Bermond, Homobono, and Peyrat [11], Du, Lyuu, and Hsu [24], Hasunuma and Shibata [30], Heydemann, Opatrny, and Sotteau [31], Homobono and Peyrat [33].

Section 12.1. The SE network was defined by Stone [76] and later discussed in many papers, e.g., see Abedini and Ravanmehr [1], Ansari, Sharma, and Mishra [2], Awerbuch and Shiloach [3], Chen, Lawrie, Yew, and Podera [16], Chen, Liu, and Qiu [17], Khosravi, Khosravi, and Khosravi [37], Kumar, Dias, and Jump [42], Lang [43], Lang and Stone [44], Liew [53], Liew and Lee [54],

Linial and Tarsi [55], Parker [63], Raghavendra and Varma [66], Schwartz [72], Steinberg [74], and Wu and Feng [80]. The name *shuffle* came from the concept of shuffling a deck of cards. Assume we are given a deck of $2n$ cards labeled $(0, 1, \ldots, n - 1, n, n + 1, \ldots, 2n - 1)$. There are two ways to have a perfect shuffle of the cards. First, the deck is cut into two halves, one with the cards labeled by $(0, 1, \ldots, n - 1)$ and a second one with the cards labeled by $(n, n + 1, \ldots, 2n - 1)$. There are two types of shuffle; in-shuffle and out-shuffle. For the in-shuffle of the two halves, the new order of the cards is

$$(n, 0, n + 1, 1, n + 2, 2, \ldots, 2n - 1, n - 1)$$

and the order after the out-shuffle of the two halves is

$$(0, n, 1, n + 1, 2, n + 2, \ldots, n - 1, 2n - 1).$$

The out-shuffle is associated with the permutation implied by shuffle only, while the in-shuffle is associated with the permutation implied by first applying a shuffle that is followed by an exchange performed by all pairs of companion processors. The in-shuffle and the out-shuffle have interesting mathematical and group structures that were studied in Diaconis, Graham, and Kantor [22]. However, the interest was high before as these two permutations are associated with distributing cards in many card games. A comprehensive description of the history, mostly connected with cards, is also described by Diaconis, Graham, and Kantor [22]. The literature is very rich and interesting, e.g., permutations obtained by shuffles and cuts were extensively studied by Golomb [29].

The first study that offered these permutations as a tool for parallel processing, i.e., the design of the SE network was presented by Stone [76]. The algorithms to generate linear transformations and bit permutations in the SE network were taken from Etzion and Lempel [25]. It was shown by Stone [76] that n^2 passes are enough to generate any permutation on the SE network of order n. Parker [63] improved the bound to $3n$ passes, Wu and Feng [80] improved it to $3n - 1$ passes, and Huang and Tripathi [34] improved the bound to $3n - 3$. The representation with balanced matrices is due to Linial and Tarsi [55] and using these matrices they proved that $3n - 4$ passes are enough to obtain any permutation in the SE network. Realizing linear transformation in a network with $p^t \times p^t$ switching boxes was carried out in Huang, Tripathi, Chen, and Tseng [35].

Section 12.2. Switching networks were defined first by Clos [19] and comprehensive work for their application in telephone networks was carried out in Beneš [7]. The omega network was defined by Lawrie [45], the flip network by Batcher [5], the modified data manipulator network by Feng [26], the indirect binary cube network by Pease [64], the baseline network and the reverse baseline network by Wu and Feng [79].

The equivalence between the omega network, the flip network, the modified data manipulator network, the indirect binary cube network, the baseline network, and the reverse baseline network was first proved by Wu and Feng [79] and later considered by other methods in Bermond, Fourneau, and Jean-Marie [8] and Kruskal and Snir [41]. The equivalence can be proved in various ways different from the one presented in this section. For example, Bermond, Fourneau, and Jean-Marie [8,9] proved this equivalence as follows.

Definition 12.11. A multistage interconnection network of order $n + 1$ has property $P(i, j)$ for $0 \leq i \leq j \leq n$ if the subgraph induced from the switching boxes from i to j has exactly $2^{n-(j-i)}$ connected components.

Property $P(i, i)$ implies that stage i has 2^n vertices. Property $P(i, i + 1)$ implies that the subgraph induced by stages i and $i + 1$ has exactly 2^n connected components that are alternating cycles of length 4.

Definition 12.12. A multistage interconnection network of order $n + 1$ has property $P(*, *)$ if and only if it satisfies Property $P(i, j)$ for every ordered pair (i, j), such that $0 \leq i \leq j \leq n$.

Theorem 12.19. *All the* MFA *networks of order n, which satisfy the banyan property and property $P(*, *)$, are isomorphic.*

It can be verified that the six networks (the omega network, the flip network, the modified data manipulator network, the indirect binary cube network, the baseline network, and the reverse baseline network) are all banyan networks (see Theorem 12.7) and also satisfy property $P(*, *)$ and hence the conditions of Theorem 12.19 are satisfied. Therefore these six networks are isomorphic.

Section 12.3. Permutation networks were considered throughout the years. The type of permutation networks defined in this section was suggested by Beneš [6] and these types of networks are contained in the family of networks defined by Clos [19]. Based on the work of Beneš [6], the idea of the permutation network of order n with $n \cdot 2^n - 2^n + 1$ switching boxes was presented first by Waksman [78]. Other algorithms to realize all permutations in the Beneš network are presented in Kannan [36], Lee [46], Lenfant [50], Nassimi and Sahni [58,59], Nikolaidis, Groumas, Kouloumentas, and Avramopoulos [61], and Raghavendra and Boppana [65]. The hierarchy between different types of permutations that are important to realize in multistage interconnection networks was given in Das, Bhattacharya, and Dattagupta [21]. Other interesting algorithms to realize permutations in related networks can be found, for example, in Freund Lev, Pippenger, and Valiant [51] and Li and Tan [52].

Waksman [78] claimed that it can be argued that $2^n n - 2^n + 1 = 2^n(n-1)+1$ switching boxes are required for a network to realize all the $2^n!$ permutation on 2^n inputs. This can be observed by noting that binary decision trees in the network can resolve individual terminal assignments only and not the partitioning of the permutation set itself that requires only $\log 2^n! = \sum_{k=1}^{2^n} \log_2 k$ binary decisions. This implies that the network Ψ_n is optimal. Moreover, it implies that the minimum number of stages, required for a multistage interconnection network of order $n + 1$ to realize all the $2^{n+1}!$ permutations of the 2^{n+1} inputs, is $2n + 1$. This result was used in the literature, but unfortunately, it seems that the arguments of Waksman [78] are not solid and many attempts to reproduce his proof were unsuccessful and the proof is considered to be incorrect. Therefore the open problems we raised ignore this result and seek a proof of this result and/or its consequences. Li and Tan [52] characterized such networks that can realize the $2^{n+1}!$ permutations on 2^{n+1} inputs with $2n + 1$ stages.

There were many attempts to solve Problem 12.2 (which is equivalent to Problem 12.3). A proof for the rearrangeability of such a network was given in Çam [15], but the analysis of the proof that was carried out in Bao, Hwang, and Li [4] showed that the proof is incomplete. Bao, Hwang, and Li [4] also gave some ideas on how to solve the problem as well as to rearrange other networks. The work has extended the ideas of balanced matrices. Other contributions that thought to solve the problem unsuccessfully are given, for example, in Feng and Seo [27], and in Kim, Yoon, and Maeng [38].

The basic questions that we asked can be asked for any UPP graph as follows.

Problem 12.6. Is there some a UPP graph of order n that cannot realize $2^n!$ permutations? There should be a distinction between graphs with the buddy property and those without the buddy property since a graph without the buddy property can implement fewer permutations in one pass.

Problem 12.7. Is there some $(2n - 1)$-stage UPP network of order n that is not rearrangeable? There should be a distinction between networks with the buddy property and those without the buddy property since a network without the buddy property can implement fewer permutations in one stage.

Problem 12.8. Is there some $(2n - 1)$-stage UPP network of order n that is rearrangeable?

There is an important difference between realizing permutations in a UPP graph G of order n and realizing permutations on its multistage version of order $n + 1$. Realizing permutations on the multistage network is equivalent to realizing permutations on the line graph $L(G)$. Hence, it is very important to distinguish between the two networks.

Realizing permutations in a UPP network (graph) $\mathcal{N}$ with 2^n vertices can be described in terms of a graph $G(\mathcal{N}) = (V, E)$ whose set of vertices V has size $2^n!$ vertices. Each vertex represents a different permutation on the set $\{0, 1, \ldots, 2^n - 1\}$. There is a directed edge $(u, v) \in E$, if u and v represent two permutations for which v can be obtained from u with the network $\mathcal{N}$ using one pass. Let $G(\mathcal{N})$ be called the **permutation graph** of the network $\mathcal{N}$. The first obvious question for a network $\mathcal{N}$ is whether the graph $G(\mathcal{N})$ is strongly connected.

Problem 12.9. Characterize which permutation graph $G(\mathcal{N})$ is strongly connected and which permutation graph $G(\mathcal{N})$ is not strongly connected, where $\mathcal{N}$ is a UPP network?

If $G(\mathcal{N})$ is a strongly connected graph, then each permutation can be realized by $\mathcal{N}$ and the next question for such a network is related to the minimum number of passes required to realize any given permutation. In other words, we ask the following question.

Problem 12.10. Given a UPP network $\mathcal{N}$, what is the diameter of a strongly connected graph $G(\mathcal{N})$, i.e., what is maximum$\{d(u, v) : u, v \in V\}$?

If the graph $G(\mathcal{N})$ is strongly connected, then there is another interesting property that is derived from the graph $G(\mathcal{N})$. There exists a factor $\mathcal{F}$ of $G(\mathcal{N})$ in which each cycle has n vertices. This factor is obtained by using the permutations obtained where each vertex of $\mathcal{N}$ is performing only "shuffle", i.e., taking the route on the cycle in $\mathcal{N}$ associated with the unique path of length n from the vertex to itself. The factor obtained can be called a necklaces factor of the permutations. For simplicity, we assume now that the UPP graph $\mathcal{N}$ has the buddy property. In this case, the in-degree of a vertex in $G(\mathcal{N})$ is $2^{2^{n-1}}$ and this is also the out-degree of a vertex in the graph. Therefore there exists an Eulerian cycle in the graph. However, if the permutation graph has enough alternating cycles of length 4, then it can be easily proved that there is also a Hamiltonian cycle in the graph.

In a very similar way to merging cycles in factors of G_n as was done in Section 4.3 using the merge-or split method, we can join all the cycles in the necklaces factor of $G(\mathcal{N})$ to form a Hamiltonian cycle. This Hamiltonian cycle can be realized by the network $\mathcal{N}$ to form all the $N!$ permutations using $N!$ passes. This process can be applied, for example, on the SE network with N vertices.

Theorem 12.20. *All the $2^n!$ permutations of $\{0, 1, \cdots, 2^n - 1\}$ can be realized in $2^n!$ passes using the SE network of order n.*

Problem 12.11. Consider UPP networks for realizing all the $2^n!$ permutations using $2^n!$ passes. Distinguish between networks with the buddy property and those without the buddy property.

Section 12.4. Various computational problems regarding interconnection networks in VLSI implementation, including bounds on the layout area, sorting with the SE network, etc. were discussed in Thompson [77] as well as others, e.g., Brent and Kung [13]. The work of Thompson [77] also contains a layout whose area is $O\left(N^2/\sqrt{\log N}\right)$, as proved in Theorem 12.15. Some considerations should be taken into account when designing a layout, as well as a layout for the SE network, were given in Hoey and Leiserson [32]. Layouts for the SE networks were extensively discussed by Leighton [47,48]. Layouts of the

SE network based on the complex plane were discussed in Leighton, Lepley, and Miller [49]. The idea of the level-necklace grid presented in Fig. 12.15 and its associated layout presented in Fig. 12.16 are taken from Leighton, Lepley, and Miller [49]. The layout based on the complex plane that requires area $O\left(N^2/log N\right)$ and was given in Theorem 12.16 is also from their paper. They also described in their paper layouts based on the complex plane whose area is $O\left(N^2/\log^{3/2} N\right)$. Another layout with the same area was presented by Steinberg and Rodeh [75]. A different approach was taken by Kleitman, Leighton, Lepley, and Miller [39] to obtain a layout that asymptotically requires area $O\left(N^2/\log^2 N\right)$. This area meets the asymptotic lower bound for the area of the SE network obtained by Thompson [77].

Book embedding was mentioned only briefly in graph theory, e.g., Bernhart and Kainen [12], Malitz [56], Muder, Weaver, and West [60], and Yannakakis [81]. The motivation to apply book embedding on interconnection networks was from an approach for fault-tolerant processor arrays. The motivation was presented by Rosenberg [67], where the processing units are laid out in a logical line, and some number of bundles of wires run in parallel with this logical line. The faulty units are bypassed, and those that are functioning well are interconnected through the bundles. If the bundles work as stacks, then the realization of an interconnection network requires book embedding. A comprehensive work on book embedding for interconnection networks was carried out by Chung, Leighton, and Rosenberg [18]. Optimal book embedding for the Beneš network as well as other networks was presented before by Games [28]. A book embedding of G_n and the SE network in five pages was generated in Obrenić [62]. The idea of the book embedding for the de Bruijn graph $G_{\sigma,n}$ was discussed in Hasunuma and Shibata [30] who showed that $G_{\sigma,n}$ can be embedded in $\sigma + 1$ pages, where the cumulative pagewidth is $\frac{1}{4}\sigma^{n-2}(3\sigma^3 - 2\sigma^2 + 4\sigma - \sigma(\sigma \pmod 2) - 4)$. Associated book embedding with $\sigma + 1$ pages, for the Kautz network over an alphabet with σ symbols, was also presented as well as a book embedding with three pages for the SE network.

In another type of important layout, called VLSI decomposition, the graph is decomposed into isomorphic subgraphs that are connected with other edges of the graph. The motivation for the VLSI decomposition of the de Bruijn graph is a consequence of the interest in building a large Viterbi decoder for use in deep-space communication. The decoder that was used in NASA's Galileo mission was based on VLSI decomposition of G_{13}. Fig. 12.18 illustrates how two copies of a subgraph H of G_3 are wired together to form G_4.

The ideas of the VLSI decomposition for the de Bruijn graph were presented first by Collins, Dolinar, McEliece, and Pollara [20] and Dolinar, Ko, and McEliece [23]. A comprehensive work on VLSI decompositions for various graphs was presented in the Ph.D. work of Ko [40]. The optimality of the VLSI decomposition for the de Bruijn graph was discussed by Schwabe [70,71].

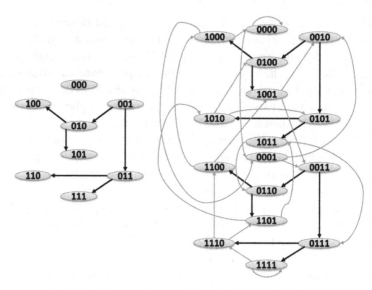

FIGURE 12.18   Two copies of a subgraph of G_3 relabeled and wired together to form G_4.

References

[1] R. Abedini, R. Ravanmehr, Parallel SEN: a new approach to improve the reliability of shuffle-exchange network, J. Supercomput. 76 (2020) 10319–10325.

[2] A.Q. Ansari, V. Sharma, R. Mishra, A 3-disjoint path design of non-blocking shuffle exchange network by extra port alignment, J. Supercomput. 78 (2022) 14381–14401.

[3] B. Awerbuch, Y. Shiloach, New connectivity and MSF algorithms for shuffle-exchange network and PRAM, IEEE Trans. Comput. 36 (1987) 1258–1263.

[4] X. Bao, F.K. Hwang, Q. Li, Rearrangeability of bit permutation networks, Theor. Comput. Sci. 352 (2006) 197–214.

[5] K.E. Batcher, The flip network in STARAN, in: Proc. 1976 Internat. Conf. Parallel Processing, Detroit, MI, U.S.A., 1976, pp. 65–71.

[6] V.E. Beneš, Permutation group, complexes, and rearrangeable connecting networks, Bell Syst. Tech. J. 43 (1964) 1619–1640.

[7] V.E. Beneš, Mathematical Theory of Connecting Networks and Telephone Traffic, Academic Press, New York, 1965.

[8] J.C. Bermond, J.M. Fourneau, A. Jean-Marie, Equivalence of multistage interconnection networks, Inf. Process. Lett. 26 (1987/88) 45–50.

[9] J.C. Bermond, J.M. Fourneau, A. Jean-Marie, A graph theoretical approach to equivalence of multistage interconnection networks, Discrete Appl. Math. 22 (1988/89) 201–214.

[10] J.-C. Bermond, N. Homobono, C. Peyrat, Large fault-tolerance interconnection networks, Graphs Comb. 5 (1989) 107–123.

[11] J.-C. Bermond, N. Homobono, C. Peyrat, Connectivity of Kautz networks, Discrete Math. 114 (1993) 51–62.

[12] F. Bernhart, P.C. Kainen, The book thickness of a graph, J. Comb. Theory, Ser. B 27 (1979) 320–331.

[13] R.P. Brent, H.T. Kung, A regular layout for parallel adders, IEEE Trans. Comput. 31 (1982) 260–264.

[14] J. Bruck, R. Cypher, C.-T. Ho, Fault-tolerant de Bruijn and shuffle-exchange networks, IEEE Trans. Parallel Distrib. Syst. 5 (1994) 548–553.

[15] H. Çam, Rearrangeability of $(2n - 1)$-stage shuffle-exchange networks, SIAM J. Comput. 32 (2003) 557–585.

[16] P.Y. Chen, D.H. Lawrie, P.C. Yew, D.A. Podera, Interconnection networks using shuffles, Computer (1981) 55–64.

[17] Z. Chen, Z.-J. Liu, Z.-L. Qiu, Bidirectional shuffle-exchange network and tag-based routing algorithm, IEEE Commun. Lett. 7 (2003) 121–123.

[18] F.K.R. Chung, F.T. Leighton, A.L. Rosenberg, Embedding graphs in books: a layout problem with applications to VLSI design, SIAM J. Algebraic Discrete Methods 8 (1987) 33–58.

[19] C. Clos, A study of non-blocking switching networks, Bell Syst. Tech. J. 32 (1953) 406–424.

[20] O. Collins, S. Dolinar, R. McEliece, F. Pollara, A VLSI decompositions of the de Bruijn graph, J. ACM 39 (1992) 931–948.

[21] N. Das, B.B. Bhattacharya, J. Dattagupta, Hierarchical classification of permutation classes in multistage interconnection networks, IEEE Trans. Comput. 43 (1994) 1439–1444.

[22] P. Diaconis, R.L. Graham, W.M. Kantor, The mathematics of perfect shuffles, Adv. Appl. Math. 4 (1983) 175–196.

[23] S. Dolinar, T.-M. Ko, R. McEliece, Some VLSI decompositions of the de Bruijn graph, Discrete Math. 106/107 (1992) 189–198.

[24] D.-Z. Du, Y.-D. Lyuu, D.F. Hsu, Line digraph iterations and connectivity analysis of de Bruijn and Kautz graphs, IEEE Trans. Comput. 42 (1993) 612–616.

[25] T. Etzion, A. Lempel, An efficient algorithm for generating linear transformation in a shuffle-exchange network, SIAM J. Comput. 15 (1986) 216–221.

[26] T.-Y. Feng, Data manipulating functions in parallel processors and their implementations, IEEE Trans. Comput. 23 (1974) 309–318.

[27] T.-Y. Feng, S.-W. Seo, A new routing algorithm for class of rearrangeable networks, IEEE Trans. Comput. 43 (1994) 1270–1280.

[28] R.A. Games, Optimal book embeddings of the FFT, Beneš, and barrel shifter networks, Algorithmica 1 (1986) 233–250.

[29] S.W. Golomb, Permutations by cutting and shuffling, SIAM Rev. 3 (1961) 293–297.

[30] T. Hasunuma, Y. Shibata, Embedding de Bruijn, Kautz and shuffle-exchange networks in books, Discrete Appl. Math. 78 (1997) 103–116.

[31] M.C. Heydemann, J. Opatrny, D. Sotteau, Broadcasing and spanning trees in de Bruijn and Kautz networks, Discrete Appl. Math. 37/38 (1992) 297–317.

[32] D. Hoey, C.E. Leiserson, A layout for the shuffle-exchange network, in: Proc. 1980 Internat. Conf. Parallel Processing, Boyne, MI, U.S.A., 1980, pp. 329–336.

[33] N. Homobono, C. Peyrat, Fault-tolerant routings in Kautz and de Bruijn networks, Discrete Appl. Math. 24 (1989) 179–186.

[34] S.T. Huang, S.K. Tripathi, Finite state model and compatibility theory: new analysis tools for permutation networks, IEEE Trans. Comput. 35 (1986) 591–601.

[35] S.-T. Huang, S.K. Tripathi, N.-S. Chen, Y.-C. Tseng, An efficient routing algorithm for realizing linear permutations on p^t-shuffle-exchange networks, IEEE Trans. Comput. 40 (1991) 1292–1298.

[36] R. Kannan, The KR-Beneš network: a control-optimal rearrangeable permutation network, IEEE Trans. Comput. 54 (2005) 534–544.

[37] B. Khosravi, B. Khosravi, B. Khosravi, Routing algorithms for the shuffle-exchange permutation network, J. Supercomput. 77 (2021) 11556–11574.

[38] M.K. Kim, H. Yoon, S.R. Maeng, On the correctness of inside-out routing algorithm, IEEE Trans. Comput. 46 (1997) 820–823.

[39] D. Kleitman, F.T. Leighton, M. Lepley, G.L. Miller, An asymptotically optimal layouts for the shuffle-exchange graph, J. Comput. Syst. Sci. 26 (1983) 339–361.

[40] T.-M. Ko, On the VLSI decompositions for complete graphs, de Bruijn graphs, hypercubes, hyperplanes, meshes, and shuffle-exchange graphs, Ph.D. thesis, Californian Institute of Technology, Los Angeles, CA, USA, 1993.

[41] C.P. Kruskal, M. Snir, A unified theory of interconnection network structure, Theor. Comput. Sci. 48 (1986) 75–94.

[42] M. Kumar, D.M. Dias, J.R. Jump, Switching strategies in shuffle-exchange packet-switched networks, IEEE Trans. Comput. 34 (1985) 180–186.

[43] T. Lang, Interconnections between processors and memory modules using shuffle-exchange network, IEEE Trans. Comput. 25 (1976) 496–503.

[44] T. Lang, H.S. Stone, A shuffle-exchange network with simplified control, IEEE Trans. Comput. 25 (1976) 55–65.

[45] D.H. Lawrie, Access and alignment of data in an array processor, IEEE Trans. Comput. 24 (1975) 1145–1155.

[46] K.Y. Lee, A new Beneš network control algorithm, IEEE Trans. Comput. 36 (1987) 768–772.

[47] F.T. Leighton, Layout for the shuffle-exchange graph and lower bounds techniques for VLSI, Ph.D. thesis, MIT, Cambridge, MA, USA, 1982.

[48] F.T. Leighton, Complexity Issues in VLSI: Optimal Layouts for the Shuffle-Exchange Graph and Other Networks, MIT Press, Cambridge, MA, USA, 1983.

[49] F.T. Leighton, M. Lepley, G.L. Miller, Layouts for the shuffle-exchange graph based on the complex plane diagram, SIAM J. Algebraic Discrete Methods 5 (1984) 202–215.

[50] J. Lenfant, Parallel permutation of data: a Beneš network control algorithm for frequently used permutations, IEEE Trans. Comput. 27 (1978) 637–647.

[51] G. Freund Lev, N. Pippenger, L.G. Valiant, A fast parallel algorithm for routing in permutation networks, IEEE Trans. Comput. 30 (1981) 93–100.

[52] S.-Y.R. Li, X.J. Tan, On rearrangeability of tandem connection of banyan-type networks, IEEE Trans. Commun. 57 (2009) 164–170.

[53] S.C. Liew, A general packet replication scheme for multicasting with application to shuffle-exchange networks, IEEE Trans. Commun. 44 (1996) 1021–1033.

[54] S.C. Liew, T.T. Lee, $N \log N$ dual shuffle-exchange network with error-correcting routing, IEEE Trans. Commun. 42 (1994) 754–766.

[55] N. Linial, M. Tarsi, Interpolation between bases and the shuffle exchange network, Eur. J. Comb. 10 (1989) 29–39.

[56] S.M. Malitz, Genus g graphs have pagenumber $O(\sqrt{g})$, J. Algorithms 17 (1994) 85–109.

[57] J.-W. Mao, C.-B. Yang, Shortest path routing and fault-tolerant routing on de Bruijn networks, Networks 35 (2000) 207–215.

[58] D. Nassimi, S. Sahni, A self-routing Beneš network and parallel permutation algorithms, IEEE Trans. Comput. 30 (1981) 332–340.

[59] D. Nassimi, S. Sahni, Parallel algorithms to set up the Beneš permutation network, IEEE Trans. Comput. 31 (1982) 148–154.

[60] D.J. Muder, M.L. Weaver, D.B. West, Pagenumber of complete bipartite graphs, J. Graph Theory 12 (1988) 469–489.

[61] D. Nikolaidis, P. Groumas, C. Kouloumentas, H. Avramopoulos, Novel Beneš network routing algorithm and hardware implementation, Technologies 10 (2002) 16.

[62] B. Obrenić, Embedding de Bruijn and shuffle-exchange graphs in five pages, SIAM J. Discrete Math. 6 (1993) 642–654.

[63] D.S. Parker Jr., Notes on shuffle/exchange-type switching networks, IEEE Trans. Comput. 29 (1980) 213–222.

[64] M.C. Pease, The indirect binary n-cube microprocessor array, IEEE Trans. Comput. 26 (1977) 458–473.

[65] C.S. Raghavendra, R.V. Boppana, On self-routing in Beneš and shuffle-exchange networks, IEEE Trans. Comput. 40 (1991) 1057–1064.

[66] C. Raghavendra, A. Varma, Rearrangeability of the five-stage shuffle/exchange network for $N = 8$, IEEE Trans. Commun. 35 (1987) 808–812.

[67] A.L. Rosenberg, The diogenes approach to testable fault-tolerant arrays of processors, IEEE Trans. Comput. 32 (1983) 902–910.

[68] R.A. Rowley, B. Bose, Fault-tolerant ring embedding in de Bruijn networks, IEEE Trans. Comput. 42 (1993) 1480–1486.

[69] R.A. Rowley, B. Bose, Distributed ring embedding and faulty de Bruijn networks, IEEE Trans. Comput. 46 (1997) 187–190.

[70] E.J. Schwabe, Efficient embeddings and simulations for hypercubic networks, Ph.D. thesis, MIT, Cambridge, MA, USA, 1992.

[71] E.J. Schwabe, Optimality of VLSI decomposition scheme for the de Bruijn graph, Parallel Process. Lett. 3 (1993) 261–265.

[72] J.T. Schawrtz Ultracompluers, ACM Trans. Program. Lang. Syst. 2 (1980) 484–521.

[73] M.A. Sridhar, C.S. Raghavendra, Fault-tolerant networks based on the de Bruijn graph, IEEE Trans. Comput. 40 (1991) 1167–1174.

[74] D. Steinberg, Invariant properties of the shuffle-exchange and a simplified cost-effective version of the omega network, IEEE Trans. Comput. 32 (1983) 444–450.

[75] D. Steinberg, M. Rodeh, A layout for the shuffle-exchange network with $O(N^2/\log^{3/2} N)$ area, IEEE Trans. Comput. 30 (1981) 977–982.

[76] H.S. Stone, Parallel processing with the perfect shuffle, IEEE Trans. Comput. 20 (1971) 153–161.

[77] C.D. Thompson, A complexity theory for VLSI, Ph.D. thesis, Carnegie-Mellon University, Pittsburgh, PA, USA, 1980.

[78] A. Waksman, A permutation network, J. ACM 15 (1968) 159–163.

[79] C.-L. Wu, T.-Y. Feng, On a class of multistage interconnection networks, IEEE Trans. Comput. 29 (1980) 694–702.

[80] C.-L. Wu, T.-Y. Feng, The universality of the shuffle-exchange network, IEEE Trans. Comput. 30 (1981) 324–332.

[81] M. Yannakakis, Embedding planar graphs in four pages, J. Comput. Syst. Sci. 38 (1989) 36–67.

Index